Neurobiology of TRP Channels

FRONTIERS IN NEUROSCIENCE

Series Editor
Sidney A. Simon, PhD

Published Titles

Apoptosis in Neurobiology
Yusuf A. Hannun, MD, Professor of Biomedical Research and Chairman, Department of
Biochemistry and Molecular Biology, Medical University of South Carolina, Charleston,
South Carolina
Rose-Mary Boustany, MD, tenured Associate Professor of Pediatrics and Neurobiology, Duke
University Medical Center, Durham, North Carolina

Neural Prostheses for Restoration of Sensory and Motor Function
John K. Chapin, PhD, Professor of Physiology and Pharmacology, State University of New York
Health Science Center, Brooklyn, New York
Karen A. Moxon, PhD, Assistant Professor, School of Biomedical Engineering, Science, and
Health Systems, Drexel University, Philadelphia, Pennsylvania

Computational Neuroscience: Realistic Modeling for Experimentalists
Eric DeSchutter, MD, PhD, Professor, Department of Medicine, University of Antwerp, Antwerp,
Belgium

Methods in Pain Research
Lawrence Kruger, PhD, Professor of Neurobiology (Emeritus), UCLA School of Medicine and
Brain Research Institute, Los Angeles, California

Motor Neurobiology of the Spinal Cord
Timothy C. Cope, PhD, Professor of Physiology, Wright State University, Dayton, Ohio

Nicotinic Receptors in the Nervous System
Edward D. Levin, PhD, Associate Professor, Department of Psychiatry and Pharmacology and
Molecular Cancer Biology and Department of Psychiatry and Behavioral Sciences, Duke
University School of Medicine, Durham, North Carolina

Methods in Genomic Neuroscience
Helmin R. Chin, PhD, Genetics Research Branch, NIMH, NIH, Bethesda, Maryland
Steven O. Moldin, PhD, University of Southern California, Washington, D.C.

Methods in Chemosensory Research
Sidney A. Simon, PhD, Professor of Neurobiology, Biomedical Engineering, and Anesthesiology,
Duke University, Durham, North Carolina
Miguel A.L. Nicolelis, MD, PhD, Professor of Neurobiology and Biomedical Engineering,
Duke University, Durham, North Carolina

The Somatosensory System: Deciphering the Brain's Own Body Image
Randall J. Nelson, PhD, Professor of Anatomy and Neurobiology,
University of Tennessee Health Sciences Center, Memphis, Tennessee

The Superior Colliculus: New Approaches for Studying Sensorimotor Integration
William C. Hall, PhD, Department of Neuroscience, Duke University, Durham, North Carolina
Adonis Moschovakis, PhD, Department of Basic Sciences, University of Crete, Heraklion, Greece

New Concepts in Cerebral Ischemia

Rick C.S. Lin, PhD, Professor of Anatomy, University of Mississippi Medical Center, Jackson, Mississippi

DNA Arrays: Technologies and Experimental Strategies

Elena Grigorenko, PhD, Technology Development Group, Millennium Pharmaceuticals, Cambridge, Massachusetts

Methods for Alcohol-Related Neuroscience Research

Yuan Liu, PhD, National Institute of Neurological Disorders and Stroke, National Institutes of Health, Bethesda, Maryland

David M. Lovinger, PhD, Laboratory of Integrative Neuroscience, NIAAA, Nashville, Tennessee

Primate Audition: Behavior and Neurobiology

Asif A. Ghazanfar, PhD, Princeton University, Princeton, New Jersey

Methods in Drug Abuse Research: Cellular and Circuit Level Analyses

Barry D. Waterhouse, PhD, MCP-Hahnemann University, Philadelphia, Pennsylvania

Functional and Neural Mechanisms of Interval Timing

Warren H. Meck, PhD, Professor of Psychology, Duke University, Durham, North Carolina

Biomedical Imaging in Experimental Neuroscience

Nick Van Bruggen, PhD, Department of Neuroscience Genentech, Inc.

Timothy P.L. Roberts, PhD, Associate Professor, University of Toronto, Canada

The Primate Visual System

John H. Kaas, Department of Psychology, Vanderbilt University, Nashville, Tennessee

Christine Collins, Department of Psychology, Vanderbilt University, Nashville, Tennessee

Neurosteroid Effects in the Central Nervous System

Sheryl S. Smith, PhD, Department of Physiology, SUNY Health Science Center, Brooklyn, New York

Modern Neurosurgery: Clinical Translation of Neuroscience Advances

Dennis A. Turner, Department of Surgery, Division of Neurosurgery, Duke University Medical Center, Durham, North Carolina

Sleep: Circuits and Functions

Pierre-Hervé Luppi, Université Claude Bernard, Lyon, France

Methods in Insect Sensory Neuroscience

Thomas A. Christensen, Arizona Research Laboratories, Division of Neurobiology, University of Arizona, Tuscon, Arizona

Motor Cortex in Voluntary Movements

Alexa Riehle, INCM-CNRS, Marseille, France

Eilon Vaadia, The Hebrew University, Jerusalem, Israel

Neural Plasticity in Adult Somatic Sensory-Motor Systems

Ford F. Ebner, Vanderbilt University, Nashville, Tennessee

Advances in Vagal Afferent Neurobiology

Bradley J. Undem, Johns Hopkins Asthma Center, Baltimore, Maryland

Daniel Weinreich, University of Maryland, Baltimore, Maryland

The Dynamic Synapse: Molecular Methods in Ionotropic Receptor Biology

Josef T. Kittler, University College, London, England

Stephen J. Moss, University College, London, England

Animal Models of Cognitive Impairment
Edward D. Levin, Duke University Medical Center, Durham, North Carolina
Jerry J. Buccafusco, Medical College of Georgia, Augusta, Georgia

The Role of the Nucleus of the Solitary Tract in Gustatory Processing
Robert M. Bradley, University of Michigan, Ann Arbor, Michigan

Brain Aging: Models, Methods, and Mechanisms
David R. Riddle, Wake Forest University, Winston-Salem, North Carolina

Neural Plasticity and Memory: From Genes to Brain Imaging
Frederico Bermudez-Rattoni, National University of Mexico, Mexico City, Mexico

Serotonin Receptors in Neurobiology
Amitabha Chattopadhyay, Center for Cellular and Molecular Biology, Hyderabad, India

TRP Ion Channel Function in Sensory Transduction and Cellular Signaling Cascades
Wolfgang B. Liedtke, MD, PhD, Duke University Medical Center, Durham, North Carolina
Stefan Heller, PhD, Stanford University School of Medicine, Stanford, California

Methods for Neural Ensemble Recordings, Second Edition
Miguel A.L. Nicolelis, MD, PhD, Professor of Neurobiology and Biomedical Engineering,
    Duke University Medical Center, Durham, North Carolina

Biology of the NMDA Receptor
Antonius M. VanDongen, Duke University Medical Center, Durham, North Carolina

Methods of Behavioral Analysis in Neuroscience
Jerry J. Buccafusco, PhD, Alzheimer's Research Center, Professor of Pharmacology and Toxicology,
    Professor of Psychiatry and Health Behavior, Medical College of Georgia, Augusta, Georgia

In Vivo Optical Imaging of Brain Function, Second Edition
Ron Frostig, PhD, Professor, Department of Neurobiology, University of California,
Irvine, California

Fat Detection: Taste, Texture, and Post Ingestive Effects
Jean-Pierre Montmayeur, PhD, Centre National de la Recherche Scientifique, Dijon, France
Johannes le Coutre, PhD, Nestlé Research Center, Lausanne, Switzerland

The Neurobiology of Olfaction
Anna Menini, PhD, Neurobiology Sector International School for Advanced Studies, (S.I.S.S.A.),
    Trieste, Italy

Neuroproteomics
Oscar Alzate, PhD, Department of Cell and Developmental Biology, University of North Carolina,
    Chapel Hill, North Carolina

Translational Pain Research: From Mouse to Man
Lawrence Kruger, PhD, Department of Neurobiology, UCLA School of Medicine, Los Angeles,
    California
Alan R. Light, PhD, Department of Anesthesiology, University of Utah, Salt Lake City, Utah

Advances in the Neuroscience of Addiction
Cynthia M. Kuhn, Duke University Medical Center, Durham, North Carolina
George F. Koob, The Scripps Research Institute, La Jolla, California

Neurobiology of Huntington's Disease: Applications to Drug Discovery
Donald C. Lo, Duke University Medical Center, Durham, North Carolina
Robert E. Hughes, Buck Institute for Age Research, Novato, California

Neurobiology of TRP Channels

Edited by
Tamara Luti Rosenbaum Emir

CRC Press
Taylor & Francis Group
Boca Raton London New York

CRC Press is an imprint of the
Taylor & Francis Group, an **informa** business

Cover figures:

Structures for the TRPA1 (PDB: 3J9P; Paulsen et al., *Nature* 2015:520); TRPV1 (PDB: 3J5P; Liao et al., *Nature* 2013:504); TRPV2 (PDB: 5HI9; Huynh et al., *Nature Communications* 2016:7); and the TRPP2 Ion Channels (PDB: 5MKE; Wilkes et al., *Nature Structural and Molecular Biology* 2017:24).

CRC Press
Taylor & Francis Group
6000 Broken Sound Parkway NW, Suite 300
Boca Raton, FL 33487-2742

First issued in paperback 2020

© 2010 by Taylor & Francis Group, LLC
CRC Press is an imprint of Taylor & Francis Group, an Informa business

No claim to original U.S. Government works

ISBN-13: 978-1-4987-5524-5 (hbk)
ISBN-13: 978-0-367-73580-7 (pbk)

Library of Congress Cataloging-in-Publication Data

Names: Rosenbaum Emir, Tamara Luti.
Title: Neurobiology of TRP channels / [edited by] Tamara Luti Rosenbaum Emir.
Other titles: Neurobiology of transient receptor potential channels
Description: Boca Raton : CRC Press, 2017. | Includes bibliographical references.
Identifiers: LCCN 2017008322| ISBN 9781498755245 (hardback) | ISBN 9781315152837 (e-book)
Subjects: LCSH: TRP channels. | Membrane proteins. | Ion channels. | Neurobiology. | Biochemistry.
Classification: LCC QP552.T77 N48 2017 | DDC 572/.696—dc23
LC record available at https://lccn.loc.gov/2017008322

Visit the Taylor & Francis Web site at
http://www.taylorandfrancis.com

and the CRC Press Web site at
http://www.crcpress.com

Dedication

To Maia, Meryem, Marcos, León, and Mustafa

Table of Contents

Preface

The field pertaining to the study of transient receptor potential (TRP) channels has grown vastly and each chapter in this volume embodies the most relevant and recent findings about their structures and their function. These beautiful proteins continue to astonish us with their complexity and their roles in several physiological and pathophysiological processes. Detailed understanding of their properties will not only shed light on how they work but also allow us to understand how other complex proteins function.

With 16 chapters, this book intends to summarize our current understanding of the neurobiology of these TRP channels. From their structure to functional features to the roles they play in a variety of processes that include vision; mechanical and thermal transduction; skin homeostasis; taste, renal, endocrine, airway, immune, cardiovascular, central nervous systems function; and fertility and aging, the contents of this book highlight the importance of TRP channels in organismal homeostasis.

This volume is made possible thanks to all of the authors, who are internationally renowned researchers in their fields and who contributed carefully written and interesting chapters. I appreciate that they have taken the time to objectively analyze and summarize relevant information on each topic and in the process they have successfully depicted the true spirit of a scientist: love for knowledge.

Science should, by all means, be supported as an act of faith for the future development of humankind. In the light of globally extended cuts to the funding of research activities and the lack of vision that these measures exemplify, we scientists would like to urge funding agencies and governments to carefully evaluate these actions since only the continued effort to engage and attain in scientific discovery, based on uncompromising ethics, will allow humanity to address and overcome all of the problems we face.

Tamara Luti Rosenbaum Emir
Departamento de Neurociencia Cognitiva
Instituto de Fisiología Celular
Universidad Nacional Autónoma de México
Mexico

Editor

Tamara Luti Rosenbaum Emir is a professor in the Department of Cognitive Neurosciences at the Institute for Cellular Physiology (IFC) of the National Autonomous University of Mexico (UNAM). She earned her B.Sc. in biology and her doctorate in biomedical sciences at UNAM and was a post-doctoral researcher at the University of Washington in the laboratory of Dr. Sharona Gordon. She joined the IFC as a researcher in 2004, where she works on several aspects of TRP channel function including structure–function relationships and modulation of these proteins.

Tamara Luti Rosenbaum Emir is a member of the Biophysical Society and the Mexican Society for Neurosciences and has received several awards for her research.

List of Contributors

Yeranddy A. Alpizar
Department of Cellular and Molecular
  Medicine
University of Leuven
Leuven, Belgium

Lucas Bacmeister
Pharmakologisches Institut
Ruprecht-Karls-Universität Heidelberg
Heidelberg, Germany
and
DZHK (German Centre for Cardiovascular
  Research)
Heidelberg/Mannheim, Germany

Michael Berlin
Pharmakologisches Institut
Ruprecht-Karls-Universität Heidelberg
Heidelberg, Germany
and
DZHK (German Centre for Cardiovascular
  Research)
Heidelberg/Mannheim, Germany

Ida Björkgren
Department of Molecular and Cell Biology
University of California
Berkeley, California

Brett Boonen
Department of Cellular and Molecular
  Medicine
University of Leuven
Leuven, Belgium

Juan E. Camacho Londoño
Pharmakologisches Institut
Ruprecht-Karls-Universität Heidelberg
Heidelberg, Germany
and
DZHK (German Centre for Cardiovascular
  Research)
Heidelberg/Mannheim, Germany

Yong Chen
Department of Neurology
Duke University Medical Center
Durham, North Carolina

Vladimir Chubanov
Ludwig-Maximilian University of Munich
DZHK (German Center for Cardiovascular
  Research) and DZL (German Center for
  Lung Research)
Munich, Germany

Alexander Dietrich
Ludwig-Maximilian University of Munich
DZHK (German Center for Cardiovascular
  Research) and DZL (German Center for
  Lung Research)
Munich, Germany

Jing Feng
Department of Anesthesiology
Center for the Study of Itch
Washington University School of Medicine
St. Louis, Missouri

Susanne Fiedler
Ludwig-Maximilian University of Munich
DZHK (German Center for Cardiovascular
  Research) and DZL (German Center for
  Lung Research)
Munich, Germany

Wiebke Frede
Pharmakologisches Institut
Ruprecht-Karls-Universität Heidelberg
Heidelberg, Germany
and
DZHK (German Centre for Cardiovascular
  Research)
Heidelberg/Mannheim, Germany

Marc Freichel
Pharmakologisches Institut
Ruprecht-Karls-Universität Heidelberg
Heidelberg, Germany
and
DZHK (German Centre for Cardiovascular
  Research)
Heidelberg/Mannheim, Germany

Shana Geffeney
Utah State University Uintah Basin
Roosevelt, Utah

Ricardo González-Ramírez
Departamento de Biología Molecular e
 Histocompatibilidad
Hospital General Dr. Manuel Gea González,
 Secretaría de Salud
Tlalpan, Mexico

Sharona E. Gordon
Department of Physiology and Biophysics
University of Washington
Seattle, Washington

Thomas Gudermann
Ludwig-Maximilian University of Munich
DZHK (German Center for Cardiovascular
 Research) and DZL (German Center for
 Lung Research)
Munich, Germany

Ranier Gutierrez
Department of Pharmacology
Laboratory of Neurobiology of Appetite
CINVESTAV
Mexico City, Mexico

Peter W. Hellings
Department of Microbiology and Immunology
University of Leuven
Leuven, Belgium
and
Clinical Department of Otorhinolaryngology
University Hospitals Leuven
Leuven, Belgium

Hongzhen Hu
Department of Anesthesiology
Center for the Study of Itch
Washington University School of Medicine
St. Louis, Missouri

León D. Islas
Departmento de Fisiología, Facultad de
 Medicina
Universidad Nacional Autónoma de México
 (UNAM)
Mexico City, Mexico

Shuji Kaneko
Department of Molecular Pharmacology
Graduate School of Pharmaceutical Sciences
Kyoto University
Kyoto, Japan

Ben Katz
Department of Medical Neurobiology
The Institute of Medical Research Israel-
 Canada (IMRIC)
The Edmond and Lily Safra Center for Brain
 Sciences (ELSC)
Faculty of Medicine of the Hebrew University
Jerusalem, Israel

Sebastian Kubanek
Ludwig-Maximilian University of Munich
DZHK (German Center for Cardiovascular
 Research) and DZL (German Center for
 Lung Research)
Munich, Germany

Sara L. Morales-Lázaro
Departamento de Neurociencia Cognitiva.
 Instituto de Fisiología Celular
Universidad Nacional Autónoma de México
Mexico City, Mexico

Wolfgang B. Liedtke
Departments of Neurology/ Neurobiology and
 Anesthesiology
Duke University Medical Center
Durham, North Carolina

Polina V. Lishko
Department of Molecular and Cell Biology
University of California
Berkeley, California

Alejandro López-Requena
Department of Cellular and Molecular
 Medicine
University of Leuven
Leuven, Belgium

Jialie Luo
Department of Anesthesiology
Center for the Study of Itch
Washington University School of Medicine
St. Louis, Missouri

Mack Madison
Department of Dermatology
Center for the Study of Itch
Washington University School of Medicine
St. Louis, Missouri

André Marx
Pharmakologisches Institut
Ruprecht-Karls-Universität Heidelberg
Heidelberg, Germany
and
DZHK (German Centre for Cardiovascular
    Research)
Heidelberg/Mannheim, Germany

Ilka Mathar
Pharmakologisches Institut
Ruprecht-Karls-Universität Heidelberg
Heidelberg, Germany
and
DZHK (German Centre for Cardiovascular
    Research)
Heidelberg/Mannheim, Germany

Rebekka Medert
Pharmakologisches Institut
Ruprecht-Karls-Universität Heidelberg
Heidelberg, Germany
and
DZHK (German Centre for Cardiovascular
    Research)
Heidelberg/Mannheim, Germany

Baruch Minke
Departments of Medical Neurobiology
The Institute of Medical Research Israel-
    Canada (IMRIC)
The Edmond and Lily Safra Center for Brain
    Sciences (ELSC)
Faculty of Medicine of the Hebrew University
Jerusalem, Israel

Lorenz Mittermeier
Ludwig-Maximilian University of Munich
DZHK (German Center for Cardiovascular
    Research) and DZL (German Center for
    Lung Research)
Munich, Germany

Carlene Moore
Department of Neurobiology, Department of
    Medicine/Division of Neurology—
    Duke Pain Clinics
Center for Translational Neuroscience
Duke University Medical Center
Durham, North Carolina

Yasuo Mori
Department of Synthetic Chemistry and
    Biological Chemistry
Graduate School of Engineering
Kyoto University
Kyoto, Japan
and
Department of Technology and Ecology
Hall of Global Environmental Studies
Kyoto University
Kyoto, Japan

Takayuki Nakagawa
Department of Clinical Pharmacology and
    Therapeutics
Kyoto University Hospital
Kyoto, Japan

Richard Payne
Department of Biology
University of Maryland
College Park, Maryland

Koenraad Philippaert
Department of Cellular and Molecular
    Medicine
University of Leuven
Leuven, Belgium

Celine Emmanuelle Riera
Department of Biomedical Sciences
Cedars-Sinai Diabetes and Obesity Research
    Institute
Cedars-Sinai Medical Center
Los Angeles, California

Mario G. Rosasco
Department of Physiology and Biophysics
University of Washington
Seattle, Washington

Seishiro Sawamura
Department of Molecular Physiology
Niigata University School of Medicine
Niigata, Japan

Alexander Schürger
Pharmakologisches Institut
Ruprecht-Karls-Universität Heidelberg
Heidelberg, Germany
and
DZHK (German Centre for Cardiovascular
    Research)
Heidelberg/Mannheim, Germany

Sebastian Segin
Pharmakologisches Institut
Ruprecht-Karls-Universität Heidelberg
Heidelberg, Germany
and
DZHK (German Centre for Cardiovascular
    Research)
Heidelberg/Mannheim, Germany

Hisashi Shirakawa
Department of Molecular Pharmacology
Graduate School of Pharmaceutical Sciences
Kyoto University
Kyoto, Japan

Sidney A. Simon
Department of Neurobiology
Duke University
Durham, North Carolina

Karel Talavera
Department of Cellular and Molecular
    Medicine
University of Leuven
Leuven, Belgium

Laura Van Gerven
Department of Microbiology and Immunology
University of Leuven
Leuven, Belgium
and
Clinical Department of Otorhinolaryngology
University Hospitals Leuven
Leuven, Belgium

Rudi Vennekens
Department of Cellular and Molecular
    Medicine
University of Leuven
Leuven, Belgium

Helen Wallace
Department of Cellular and Molecular
    Physiology
University of Liverpool
Liverpool, United Kingdom

Pu Yang
Department of Anesthesiology
Center for the Study of Itch
Washington University School of Medicine
St. Louis, Missouri

1 TRP Channels
What Do They Look Like?

Mario G. Rosasco and Sharona E. Gordon

CONTENTS

1.1 INTRODUCTION

The first transient receptor potential (TRP) ion channel was identified as a *Drosophila* locus that gave rise to a phenotype in which the photoreceptor light response decayed to baseline during prolonged illumination (Cosens and Manning, 1969; Minke et al., 1975). The identification of the TRP fly in 1969 and the molecular identification of the *trp* gene in 1975 set the stage for the subsequent explosion of discoveries that continue even today.

The mid-1990s through the early 2000s were a particularly productive time for identification of many new TRP subfamilies and subfamily members. This is apparent from the rapid increase in the number of publications on TRP channels listed in PubMed (Figure 1.1); once many new TRP channels were identified, work on understanding their physiology progressed rapidly. Six subfamilies of TRP channels have now been identified in mammals, with an additional subfamily found in invertebrates and nonmammalian vertebrate animals (Figure 1.1, right panel). Because of diverse nomenclature for any given TRP channel, only the characters used to describe each column in Figure 1.1 were used as search terms. Although this approach clearly underestimates the work on all TRP channels, it likely underestimates those TRP channels with primarily clinical publications more than others.

TRP channels are members of the voltage-gated superfamily of ion channels that includes the voltage gated K^+, Na^+, and Ca^{2+} channels as well as related cyclic nucleotide-gated channels. They form as tetramers of identical subunits (Figure 1.2), although heterotetramers of TRP channel subunits have been reported (reviewed in Cheng et al., 2010). Like other members of the voltage-gated superfamily, each subunit includes six membrane-spanning helices with a reentrant pore loop between the fifth and six transmembrane helices, and intracellular amino- and carboxy-terminali. The first four transmembrane segments (Figure 1.2, blue) form the voltage-sensing or voltage-sensing-like domain. The remaining two transmembrane segments, along with the reentrant pore loop (Figure 1.2, yellow), form the ion-conducting pore of the channel.

Although some TRP channels (e.g., TRPM8) show voltage-dependent activation (Voets et al., 2007); others show little or no voltage-dependent gating (e.g., TRPV1) (Liu et al., 2009). This

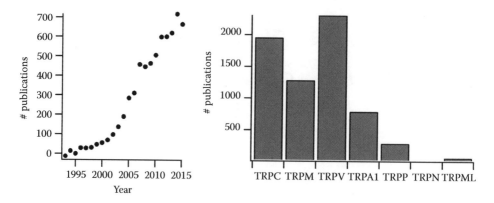

FIGURE 1.1   Search terms used were (TRP OR [transient receptor potential] AND channel). Numbers were corrected by subtracting 20 publications/year—our estimate of off-target publications identified by these search terms.

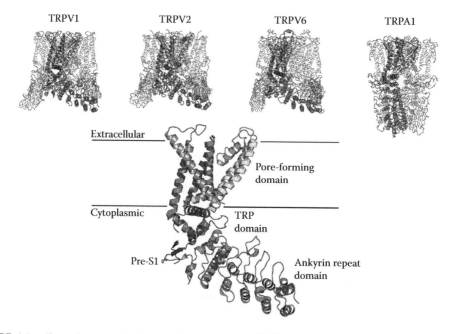

FIGURE 1.2   (See color insert.) Structural arrangement of TRP channels whose structures have recently been elucidated. In the top row are structures of the indicated TRP channels, with one subunit shown in color. Below is a zoomed-in view of one subunit from TRPV1 to highlight the structure of the voltage sensor-like domain (blue), pore domain (yellow), the pre-S1 domain (pink), and the ankyrin repeat domain (green) for those channels.

variability in function is likely due to variability in the amino acid sequence in the fourth transmembrane helix, which for voltage-gated channels includes a number of positively charged residues and for voltage-independent channels does not (Figure 1.3). It is worth noting that the macroscopic current-voltage relationship of TRPV1 shows significant outward rectification. However, this is due almost exclusively to rectification in the unitary conductance (Liu et al., 2009).

A hallmark of many TRP channels is the TRP domain following the sixth transmembrane helix (Figures 1.2 (purple) and 1.4). This can be recognized based on primary sequence in TRPC, TRPM, and TRPV channels. Although it was not obvious from the primary sequence of TRPA1 channels that they included a TRP domain, structural homology in this region was revealed by the recent

rKv1.2 290 - LAILRVIRLVRVFRIFKLSRH - 310
rTRPV1 536 - EYVASMVFSLAMGWTNMLYYT - 556
rTRPM8 832 - FCLDYIIFTLRLIHIFTVSRN - 852

FIGURE 1.3 (See color insert.) Sequence alignment of S4 membrane-spanning helices, with positive "gating charge" residues indicated in red. Histidine is colored pink to indicate its ability to be protonated and thus carry charge at near-physiological pH. The alignment shows that no charges are present in TRPV1, whereas TRPM8 retains at least some of the R-X-X-R motifs.

S6 helix   TRP Box

rTRPA1   970 - VQKHASLKRIAMQVELHTNLEKK - 992
rTRPC1   618 - IANHEDKEWKFARAKLWLSYFDD - 640
rTRPC2   658 - IEDDADVEWKFARSKLYLSYFRE - 680
rTRPC3   667 - IEDDSDVEWKFARSKLWLSYFDD - 689
rTRPC4   627 - IADHADIEWKFARTKLWMSYFEE - 649
rTRPC5   631 - IADHADIEWKFARTKLWMSYFDE - 653
rTRPC6   733 - IEDDADVEWKFARAKLWFSYFEE - 755
rTRPC7   679 - IEEDADVEWKFARAKLWLSYFDE - 701
rTRPM1   1133 - VKSISNQVWKFQRYQLIMTFH-D - 1154
rTRPM2   1053 - VQEHTDQIWKFQRHDLIEEYH-G - 1074
rTRPM3   1133 - VKSISNQVWKFQRYQLIMTFH-E - 1154
rTRPM4   1044 - VHGNSDLYWKAQRYSLIREFH-S - 1065
rTRPM5   982 - VQGNADMFWKFQRYHLIVEYH-G - 1003
rTRPM6   1079 - EVSVHEPLGT-QSMLLKRTLHIS - 1100
rTRPM7   1103 - VKAISNIVWKYQRYHFIMAYH-E - 1124
rTRPM8   986 - VQENNDQVWKFQRYFLVQEYC-N - 1007
rTRPV1   689 - IAQESKNIWKLQRAITILDTEKS - 711
rTRPV2   652 - VADNSWSIWKLQKAISVLEMENG - 674
rTRPV3   684 - VSKESERIWRLQRARTILEFEKM - 706
rTRPV4   725 - VSKESKHIWKLQWATTILDIERS - 747
rTRPV5   578 - VAQERDELWRAQVVATTVMLERK - 600
rTRPV6   584 - VAHERDELWRAQVVATTVMLERK - 606

FIGURE 1.4 Sequence alignment of the region following the S6 transmembrane helix and including the TRP domain or TRP box. TRP channel family members are indicated to the left. The character "r" preceding each protein name indicates all sequences are from the species *Rattus norvegicus*. Numbers flanking the sequences indicate the positions of the first and last residues displayed. TRP channel sequences containing a region with greater than 50% identity to the TRP box sequence "EWKFAR" have the homologous region highlighted in gray.

cryoEM structures of TRPV1 (Liao et al., 2013) and TRPA1 (Paulsen et al., 2015) and is shown in Figure 1.2. The TRP domain consists of an alpha helical segment parallel and in close proximity to the plasma membrane. Although a definitive function for the TRP domain has not been established, it is positioned well to interact with both the membrane and the amino-terminal region.

Three TRP subfamilies, TRPC, TRPV, and TRPA, have amino-terminal ankyrin repeat domains of varying lengths (Figures 1.2 and 1.5). TRPA1's domain is the longest, although we do not yet know how many ankyrin repeats may be present in TRPA1 channels—only that it is a large number. The function of these domains is not fully understood, but in some channels this structural element appears to influence gating. For example, in TRPV1 the ankyrin repeats contain a reactive cysteine that promotes channel opening (Salazar et al., 2008), and the region has been proposed to be a functionally important binding site for ATP (Lishko et al., 2007) and calmodulin (Rosenbaum et al., 2004).

The TRP channel superfamily can be subdivided into seven separate subfamilies: TRPA, TRPC, TRPM, TRPML, TRPN, TRPP, and TRPV. The individual subunits of all seven subfamilies' members are thought to contain six transmembrane segments that assemble as tetramers to form functional TRP channels (Figure 1.2). However, the tissue distribution, function, and even in which species each subfamily can be found vary wildly (Figure 1.6), representing the myriad roles that TRP channels play in neurobiology.

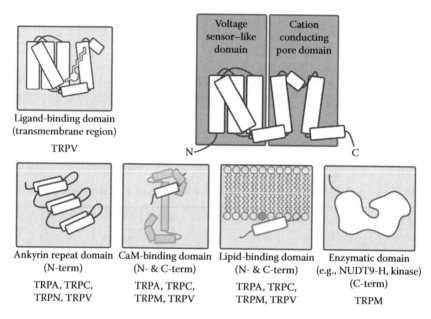

FIGURE 1.5 A cartoon representation of various functional domains found in TRP channels. In all TRP channels, single subunits contain six transmembrane segments, S1–S6 (top). S1–S4 form a voltage sensor-like domain (left), whereas S5 and S6 form the ion-conducting pore (right). In TRPV1, these transmembrane domains also form a ligand-binding domain (far left). In this cartoon, the channel subunit is oriented with the cytoplasm at the bottom and the amino terminus of the protein on left. The cytoplasmic N- and C-termini of the channel vary significantly between and within TRP subfamilies. Shown are examples of domains identified in the cytoplasmic domains of various TRP channels: an ankyrin repeat domain, a calcium-calmodulin (CaM)-binding domain (black, with CaM shown in gray), a lipid-binding domain, and a catalytic enzyme domain. Below each domain is a list of TRP subfamilies where the functional domains have been identified. Note that not all members of a given TRP subfamily may contain these domains; for example, only TRPM6 and TRPM7 have been found to contain kinase domains.

FIGURE 1.6 (See color insert.) Expression profiles of TRP channels in humans. Colored boxes under each TRP channel name correspond to the colors of various organs on the left, indicating in which tissues each channel has been identified. Evidence for expression at the RNA level is indicated by a box with a white border, and evidence for expression at the protein level is indicated by a box with a black border. Tissue systems corresponding to each color are: orange (brain/nervous system), yellow (pulmonary system), red (cardiovascular system), purple (renal system), dark blue (liver), light blue (remainder of digestive system), pink (integumentary system), and green (reproductive system).

1.2 TRPA

The TRPA subfamily contains only one member, TRPA1, so named for its extensive N-terminal ankyrin repeat domain (ARD) (Figure 1.2). Whereas many TRP channels contain an ARD, with at least 14 ankyrin repeats TRPA1 has the most of any identified mammalian TRP. The gene encoding TRPA1 is conserved from flies to mammals, and in humans the protein is expressed in sensory nerves, as well as other tissues. TRPA1's activity can be regulated by temperature, as with members of the TRPV and TRPM family; however, TRPA1's temperature sensitivity is strongly species dependent, with different species demonstrating wildly different temperature activation profiles for TRPA1 (reviewed in Laursen et al., 2015).

1.3 TRPC

The TRPC (for "classical" or "canonical") subfamily contains the TRP channels most closely related to the first identified TRP channel (Cosens and Manning, 1969; Minke et al., 1975). Since the discovery and cloning of the original *trp* locus in *Drosophila*, seven TRPC genes have been reported. Humans contain genes encoding TRPC1, TRPC3, TRPC4, TRPC5, TRPC6, and TRPC7, but lack the TRPC2 gene found in mice and other species.

1.4 TRPM

In many ways, the TRPM subfamily is the most enigmatic and diverse of the TRPs. Its founding member, TRPM1, derives its name ("TRP melastatin 1") from its putative role as a melanoma suppressor (Duncan et al., 1998). Whereas currents sensitive to TRPM1 siRNA have been recorded from melanocytes, attempts to record activity directly from recombinant TRPM1 have been so far unsuccessful (Oancea et al., 2009). This subfamily also contains the unique class of "chanzymes," in which an enzymatic domain has been evolutionarily fused to the TRP channel (Figure 1.5). Such chanzymes include TRPM2, which contains a NUDT9 homology domain that hydrolyzes adenosine diphosphate ribose (Kuhn and Luckhoff, 2004), and TRPM6 and TRPM7, and each contain a functional protein kinase (Runnels et al., 2001). TRPM8 is perhaps the best-studied member of the TRPMs, and is activated by noxious cold, and chemical compounds such as menthol and eucalyptus (McKemy et al., 2002). Recent work has also demonstrated that TRPM8 can serve as a physiologically relevant receptor for testosterone (Asuthkar et al., 2015a; Asuthkar et al., 2015b).

1.5 TRPML

The three mucolipins, TRPML1–3 (also known as MCOLN1–3) are named after loss of function mutations in TRPML1 that result in the neurodegenerative disease type IV mucolipidosis (reviewed in Slaugenhaupt, 2002). TRPMLs are the most distantly connected subfamily and lack the extensive cytoplasmic N- and C-termini found in many other TRPs. This may be related to their subcellular localization; whereas many TRPs are known to traffic to the plasma membrane where they exert their physiological functions, TRPMLs predominantly localize to intracellular membranes of the endo- and exocytosis pathways (reviewed in Venkatachalam et al., 2015). Although the endogenous regulation of TRPMLs is not well understood, it is known that the activity of these channels is regulated by phosphoinositides (Dong et al., 2010), as is true for most other TRP channels (Zheng and Trudeau, 2015). Recently, synthetic agonists for the TRPMLs have been developed to facilitate further study of these channels (Shen et al., 2012; Grimm et al., 2010).

1.6 TRPN

Absent from mammals, the TRPN ("No mechanoreceptor potential C," also called NOMPC) protein forms a mechanosensitive channel (Walker et al., 2000). Mechanosensitivity is imparted by the N-terminal ARD, which dwarfs even TRPA1's, with its 29 ankyrin repeats. These ankyrin repeats have been shown to interact with cellular microtubules, and this association is required to impart mechanosensitivity to the channels (Zhang et al., 2015).

1.7 TRPP

The TRPPs, or polycystins, are similar to the TRPMLs in that they were identified from disease-causing mutations, and bear little similarity to the rest of the TRP superfamily (Mochizuki et al., 1996). Although TRPPs are among the most widely expressed TRPs (Figure 1.6), the effects of human mutations in TRPP1 are relatively limited to the kidney, where such mutations cause autosomal dominant polycystic kidney disease (ADPKD) (reviewed in Ong and Harris, 2015). TRPP1 interacts with PKD1, an 11 transmembrane domain protein also mutated in ADPKD (Hanaoka et al., 2000). This association was originally suggested to be required for TRPP1 to form functional channels (Hanaoka et al., 2000); however, recent studies have questioned this interpretation (DeCaen et al., 2016; Shen et al., 2016). Two other TRPPs—TRPP2 and TRPP3 (also called PKD2L2 and TRPP5)—have been identified, and TRPP2 has been suggested to be the receptor for sour tastes (Huang et al., 2006; Ishimaru et al., 2006; LopezJimenez et al., 2006). TRPPs are often found localized to primary cilia, where it has been proposed that they have a role in regulating flow-induced calcium transients (Nauli et al., 2003), although recent pressure-clamp studies of primary cilia indicate that TRPPs in primary cilia are only mechanosensitive at high pressures (DeCaen et al., 2013).

The nomenclature of TRPPs can often be confusing, because of the original inclusion of the PKD1 family in the TRPPs. For this reason, TRPP1 was originally designated TRPP2, and so on. However, because the PKD1 family proteins contain 11 transmembrane domains and have not been found to form conductive channels, it has been suggested that a consistent nomenclature should leave the PKDs out of the TRPP family. For completeness, we list here the gene name for each TRPP (in italics), followed by possible aliases that are still prevalent in the literature: TRPP1 (*PKD2*, PKD2, TRPP2, polycystin-2); TRPP2 (*PKD2L1*, PKD2L1, TRPP3, polycystin-L); TRPP3 (*PKD2L2*, PKD2L2, TRPP5).

1.8 TRPV

TRPVs are named for the role of the founding member, TRPV1, as the receptor for the pungent vanilloid capsaicin (Caterina et al., 1997). Despite this nomenclature, only TRPV1 has so far been shown to be gated by vanilloids. TRPVs are by far the best-studied TRP subfamily (Figure 1.1), in large part due to the pharmacological accessibility and clear knockout phenotype of TRPV1 (Caterina et al., 1997). TRPV1 is highly expressed in sensory nerves (Figure 1.6), where it acts as a multimodal integrator of noxious stimuli, gating in response to noxious compounds, heat, lipids, and protons (Caterina et al., 1997, 2000; Lukacs et al., 2007, 2013; Rohacs et al., 2008; Klein et al., 2008; Senning et al., 2014; Ufret-Vincenty et al., 2015). TRPV1 was the first TRP for which a near-atomic structural model was determined (Liao, 2013), confirming that TRPs form tetramers of subunits with six transmembrane domains each (Figure 1.2). The TRPV family also includes the mechanosensitive channels TRPV4 and TRPY1 (the yeast homologue to TRPV channels) (Palmer et al., 2001; Strotmann et al., 2000; Liedtke et al., 2000). Alternative splice variants of TRPV1 are also thought to be mechanosensitive, responding to changes in osmotic pressure (Sharif Naeini et al., 2006; Zaelzer et al., 2015). Most TRPV channels display calcium-dependent modulation in addition to being permeable to calcium (reviewed in Gordon-Shaag et al., 2008), and TRPV5 and TRPV6 have been shown to be directly modulated by CaM (Lambers et al., 2004).

1.9 CONCLUSION

The number of TRP channel structures that have been solved has grown rapidly since the first TRPV1 structure was solved only a few short years ago (Liao et al., 2013; Paulsen et al., 2015; Huynh et al., 2016; Zubcevic et al., 2016; Saotome et al., 2016). Most recently, the first structure of a TRPP family member (PKD2, or TRPP1) was solved (Shen et al., 2016), highlighting key structural differences between these proteins and the more closely related TRPV and TRPA structures (Figure 1.2). Although these structures have provided unprecedented insight into the elegant molecular machinery that underlies TRP channel function, they also indicate that there is much work left to do. Future studies to probe the precise conformational rearrangements that occur to produce channel gating will be required to fully understand these channels' roles in normal physiology, and to guide the design of TRP-targeted therapeutics. Furthermore, the number of TRP structures determined so far represents only a fraction of the TRP superfamily, leaving open the possibility of future structural surprises in the TRP world.

REFERENCES

Asuthkar, S. et al. 2015a. The TRPM8 protein is a testosterone receptor: II. Functional evidence for an ionotropic effect of testosterone on TRPM8. *J Biol Chem*, 290(5): 2670–2688.

Asuthkar, S. et al. 2015b. The TRPM8 protein is a testosterone receptor: I. Biochemical evidence for direct TRPM8-testosterone interactions. *J Biol Chem*, 290(5): 2659–2669.

Caterina, M.J. et al. 1997. The capsaicin receptor: A heat-activated ion channel in the pain pathway. *Nature*, 389(6653): 816–824.

Caterina, M.J. et al. 2000. Impaired nociception and pain sensation in mice lacking the capsaicin receptor. *Science*, 288(5464): 306–313.

Cheng, W., C. Sun, and J. Zheng. 2010. Heteromerization of TRP channel subunits: Extending functional diversity. *Protein Cell*, 1(9): 802–810.

Cosens, D.J. and A. Manning. 1969. Abnormal electroretinogram from a *Drosophila* mutant. *Nature*, 224(5216): 285–287.

DeCaen, P.G. et al. 2013. Direct recording and molecular identification of the calcium channel of primary cilia. *Nature*, 504(7479): 315–318.

DeCaen, P.G. et al. 2016. Atypical calcium regulation of the PKD2-L1 polycystin ion channel. *Elife*, 5.

Dong, X.P. et al. 2010. PI(3,5)P(2) controls membrane trafficking by direct activation of mucolipin Ca(2+) release channels in the endolysosome. *Nat Commun*, 1: 38.

Duncan, L.M. et al. 1998. Down-regulation of the novel gene melastatin correlates with potential for melanoma metastasis. *Cancer Res*, 58(7): 1515–1520.

Gordon-Shaag, A., W.N. Zagotta, and S.E. Gordon. 2008. Mechanism of Ca(2+)-dependent desensitization in TRP channels. *Channels (Austin)*, 2(2): 125–129.

Grimm, C. et al. 2010. Small molecule activators of TRPML3. *Chem Biol*, 17(2): 135–148.

Hanaoka, K. et al. 2000. Co-assembly of polycystin-1 and -2 produces unique cation-permeable currents. *Nature*, 408(6815): 990–994.

Huang, A.L. et al. 2006. The cells and logic for mammalian sour taste detection. *Nature*, 442(7105): 934–938.

Huynh, K.W. et al. 2016. Structure of the full-length TRPV2 channel by cryo-EM. *Nat Commun*, 7: 11130.

Ishimaru, Y. et al. 2006. Transient receptor potential family members PKD1L3 and PKD2L1 form a candidate sour taste receptor. *Proc Natl Acad Sci U S A*, 103(33): 12569–12574.

Klein, R.M. et al. 2008. Determinants of molecular specificity in phosphoinositide regulation. Phosphatidylinositol (4,5)-bisphosphate (PI(4,5)P2) is the endogenous lipid regulating TRPV1. *J Biol Chem*, 283(38): 26208–26216.

Kuhn, F.J. and A. Luckhoff. 2004. Sites of the NUDT9-H domain critical for ADP-ribose activation of the cation channel TRPM2. *J Biol Chem*, 279(45): 46431–46437.

Lambers, T.T. et al. 2004. Regulation of the mouse epithelial Ca2(+) channel TRPV6 by the Ca(2+)-sensor calmodulin. *J Biol Chem*, 279(28): 28855–28861.

Laursen, W.J. et al. 2015. Species-specific temperature sensitivity of TRPA1. *Temperature (Austin)*, 2(2): 214–226.

Liao, M. et al. 2013. Structure of the TRPV1 ion channel determined by electron cryo-microscopy. *Nature*, 504(7478): 107–112.

Liedtke, W. et al. 2000. Vanilloid receptor-related osmotically activated channel (VR-OAC), a candidate vertebrate osmoreceptor. *Cell*, 103(3): 525–535.

Lishko, P.V. et al. 2007. The ankyrin repeats of TRPV1 bind multiple ligands and modulate channel sensitivity. *Neuron*, 54(6): 905–918.

Liu, B. et al. 2009 Proton inhibition of unitary currents of vanilloid receptors. *J Gen Physiol*, 134(3): 243–258.

LopezJimenez, N.D. et al. 2006. Two members of the TRPP family of ion channels, Pkd1l3 and Pkd2l1, are co-expressed in a subset of taste receptor cells. *J Neurochem*, 98(1): 68–77.

Lukacs, V. et al. 2007. Dual regulation of TRPV1 by phosphoinositides. *J Neurosci*, 27(26): 7070–7080.

Lukacs, V. et al. 2013. Distinctive changes in plasma membrane phosphoinositides underlie differential regulation of TRPV1 in nociceptive neurons. *J Neurosci*, 33(28): 11451–11463.

McKemy, D.D., W.M. Neuhausser, and D. Julius. 2002. Identification of a cold receptor reveals a general role for TRP channels in thermosensation. *Nature*, 416(6876): 52–58.

Minke, B., C. Wu, and W.L. Pak. 1975. Induction of photoreceptor voltage noise in the dark in *Drosophila* mutant. *Nature*, 258(5530): 84–87.

Mochizuki, T. et al. 1996. PKD2, a gene for polycystic kidney disease that encodes an integral membrane protein. *Science*, 272(5266): 1339–1342.

Nauli, S.M. et al. 2003. Polycystins 1 and 2 mediate mechanosensation in the primary cilium of kidney cells. *Nat Genet*, 33(2): 129–137.

Oancea, E. et al. 2009. TRPM1 forms ion channels associated with melanin content in melanocytes. *Sci Signal*, 2(70): ra21.

Ong, A.C. and P.C. Harris. 2015. A polycystin-centric view of cyst formation and disease: The polycystins revisited. *Kidney Int*, 88(4): 699–710.

Palmer, C.P. et al. 2001. A TRP homolog in Saccharomyces cerevisiae forms an intracellular Ca(2+)-permeable channel in the yeast vacuolar membrane. *Proc Natl Acad Sci U S A*, 98(14): 7801–7805.

Paulsen, C.E. et al. 2015. Structure of the TRPA1 ion channel suggests regulatory mechanisms. *Nature*, 525(7570): 552.

Rohacs, T., B. Thyagarajan, and V. Lukacs. 2008. Phospholipase C mediated modulation of TRPV1 channels. *Mol Neurobiol*, 37(2–3): 153–163.

Rosenbaum, T. et al. 2004. Ca2+/calmodulin modulates TRPV1 activation by capsaicin. *J Gen Physiol*, 123(1): 53–62.

Runnels, L.W., L. Yue, and D.E. Clapham. 2001. TRP-PLIK, a bifunctional protein with kinase and ion channel activities. *Science*, 291(5506): 1043–1047.

Salazar, H. et al. 2008. A single N-terminal cysteine in TRPV1 determines activation by pungent compounds from onion and garlic. *Nat Neurosci*, 11(3): 255–261.

Saotome, K. et al. 2016. Crystal structure of the epithelial calcium channel TRPV6. *Nature*, 534(7608): 506–511.

Senning, E.N. et al. 2014. Regulation of TRPV1 ion channel by phosphoinositide (4,5)-bisphosphate: The role of membrane asymmetry. *J Biol Chem*, 289(16): 10999–11006.

Sharif Naeini, R. et al. 2006. An N-terminal variant of Trpv1 channel is required for osmosensory transduction. *Nat Neurosci*, 9(1): 93–98.

Shen, D. et al. 2012. Lipid storage disorders block lysosomal trafficking by inhibiting a TRP channel and lysosomal calcium release. *Nat Commun*, 3: 731.

Shen, P.S. et al. 2016. The structure of the polycystic kidney disease channel PKD2 in lipid nanodiscs. *Cell*, 167(3): 763–773 e11.

Slaugenhaupt, S.A. 2002. The molecular basis of mucolipidosis type IV. *Curr Mol Med*, 2(5): 445–450.

Strotmann, R. et al. 2000. OTRPC4, a nonselective cation channel that confers sensitivity to extracellular osmolarity. *Nat Cell Biol*, 2(10): 695–702.

Ufret-Vincenty, C.A. et al. 2015. Mechanism for phosphoinositide selectivity and activation of TRPV1 ion channels. *J Gen Physiol*, 145(5): 431–442.

Venkatachalam, K., C.O. Wong, and M.X. Zhu. 2015. The role of TRPMLs in endolysosomal trafficking and function. *Cell Calcium*, 58(1): 48–56.

Voets, T. et al. 2007. TRPM8 voltage sensor mutants reveal a mechanism for integrating thermal and chemical stimuli. *Nat Chem Biol*, 3(3): 174–182.

Walker, R.G., A.T. Willingham, and C.S. Zuker. 2000. A *Drosophila* mechanosensory transduction channel. *Science*, 287(5461): 2229–2234.

Zaelzer, C. et al. 2015. DeltaN-TRPV1: A molecular co-detector of body temperature and osmotic stress. *Cell Rep*, 13(1): 23–30.

Zhang, W. et al. 2015. Ankyrin repeats convey force to gate the NOMPC mechanotransduction channel. *Cell*, 162(6): 1391–1403.

Zheng, J. and M. Trudeau. 2015. *Handbook of Ion Channels*. Boca Raton, FL: CRC Press.

Zubcevic, L. et al. 2016. Cryo-electron microscopy structure of the TRPV2 ion channel. *Nat Struct Mol Biol*, 23(2): 180–186.

2 Molecular Mechanisms of Temperature Gating in TRP Channels

León D. Islas

CONTENTS

2.1 INTRODUCTION

Transient receptor potential (TRP) channels belong to a varied superfamily of membrane proteins with multiple functions. Most TRP channels are permeable to Ca^{2+} ions and are thus involved in physiological processes related to Ca^{2+}-mediated homeostasis and signaling.

Every year, thousands of research articles are published describing new roles for TRP channels, both in normal and pathological physiological conditions.

This functional diversity probably stems from the presence of various structural motifs in each one of its members, some of which have been identified and characterized, at least at the level of their primary sequence (Latorre et al., 2009).

Among this tremendous diversity, a few TRP channels are the only ion channels that have been unequivocally characterized as being directly gated by changes in temperature (Islas and Qin, 2014).

These are generically grouped into the category of thermoTRP channels and include channels that are gated by cold or hot temperatures. The following members of three subfamilies of TRP channels are recognized as thermoTRPs: TRPA1, TRPM2, TRPM8, and TRPV1–4. These channels and their mechanisms of gating will be the focus of the present chapter.

These key proteins provide the molecular substrate for the detection of temperature and temperature changes (Islas and Qin, 2014), in organisms as varied as worms (Chatzigeorgiou et al., 2010) and humans (Gavva, 2008; Gavva et al., 2008) and play important roles in pain, inflammation, and other physiological processes.

2.2 STRUCTURAL PRELUDE

In a recent burst of research activity, mostly spurred by the modern revolution in cryo-electron microscopy (Cao et al., 2013b; Liao et al., 2013; Paulsen et al., 2015; Huynh et al., 2016; Saotome et al., 2016; Zubcevic et al., 2016), several atomic resolution structures of TRP channels have become available. These include structures of the TRPV1 channel in putative closed and open conformations, structures of TRPV2 and TRPA1, and an x-ray diffraction structure of TRPV6.

Several previously existing x-ray diffraction structures of amino (N)-terminal or carboxyl (C)-terminal isolated fragments of some TRP channels are also at our disposal (Jin et al., 2006; McCleverty et al., 2006; Lishko et al., 2007; Phelps et al., 2008; Inada et al., 2012).

These structures demonstrate that at least for the TRPVs and TRPA1, some basic architectural motifs are common to these channels.

As discussed in Chapter 1, they are assembled as tetramers where each subunit is based on the fold of the voltage-gated potassium (Kv) channels, with six transmembrane domains of which S1–S4 are located in the periphery and the pore is located in the center of the tetramer, formed by segments S5 and S6 and the P-loop (Figure 2.1).

The first four transmembrane domains (S1–S4) are organized similarly to the same domains in voltage-gated ion channels and are analogous to the voltage-sensing domains (VSDs) found in those channels but do not seem to have a role in voltage gating (Palovcak et al., 2015).

These structures also revealed fundamental differences between the structure of Kv channels and TRPVs. The N-terminus and the bulk of the C-terminus are located intracellularly. While the C-terminus structure has not been resolved in its entirety, the N-terminus is assembled into a well-formed specific domain. The N-terminus of TRPV channels of known structure is formed partially by six ankyrin repeat domains, which show remarkable structural conservation among members of the family (Figure 2.2). The sequence of ~120 amino acids before the ankyrin repeats seems to be highly disorganized and is not resolved in any of the published structures.

These ankyrin repeats seem to be of prime importance for the mechanisms of activation by chemical modifiers and for the binding of some ligands, as well as being part of the temperature gating mechanism, as will be discussed later.

The distal C-terminal, part of the S6 segment, forms a bent alpha helix that runs almost parallel to the membrane and transforms into a loop and a beta strand that interacts with some of the ankyrin repeats of an adjacent subunit. This helix is formed by the TRP domain, a signature sequence of amino acids that is commonly found in TRP channels of varied families and which seems to play

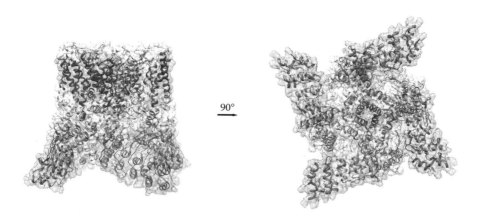

90°

FIGURE 2.1 **(See color insert.)** Structure of the TRPV1 ion channel as determined by cryo-electron microscopy. The channel is a homotetramer of subunits with six transmembrane domains and large intracellular regions formed by the N- and C-termini. The amino terminal ankyrin repeats interact between adjacent subunits at a specific region, which contains an unusually long alpha helix. This interaction also includes a beta strand from the C-terminus and part of the TRP box domain and might be determinant of heat activation. The coloring of regions in this figure was done according to b-factors, with blue hues being low values and red and orange high values. This coloring gives an indication of the regions with likely low and high mobility, respectively. The structure depicted 3J5R entry in the PDB database. The left panel is the side view and the right panel depicts a 90° rotation showing a view from the intracellular side.

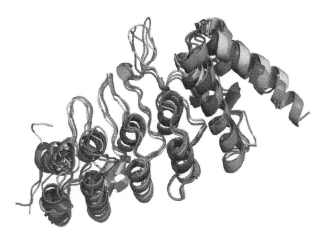

FIGURE 2.2 (See color insert.) Comparison of the known structures of ankyrin repeat domains of the amino termini of TRV1, TRPV2, and TRPV6 channels. Structural alignment of the structures of the amino terminal ankyrin-repeat domains of some TRP channels. The coloring is as follows: pink, human TRPV2; green, rat TRPV2; cyan, rat TRPV1; yellow, TRPV6.

an important role in gating and lipid modulation (García-Sanz et al., 2007; Gregorio-Teruel et al., 2015). The rest of the C-terminus is probably nested between the N-terminal ankyrin domains of two adjacent subunits (De-la-Rosa et al., 2013).

This arrangement is different from the one present in the C-terminus of Kv channels, where it is continuous with the S6 segment and does not have the same interactions with the N-terminus (Long et al., 2005).

The S5–S6 region has been demonstrated to perform as an intracellular gate that can regulate the access of small intracellular ions to the selectivity filter. Cysteine-modifying reagent accessibility experiments (SCAM) have shown that this region undergoes an opening conformational change similar to the opening of Kv channels and thus constitutes an activation gate. It was shown that both ligand gating and temperature gating promote the same conformational change in this gate region (Salazar et al., 2009).

2.3 LIGAND AND VOLTAGE GATING IN THERMOTRP CHANNELS

ThermoTRP channels can be activated by binding of very diverse ligands, and show marked promiscuity. For example, the funding member of the TRPV subfamily, TRPV1, is distinctly activated by capsaicin, although TRPV2 seems to have a vestigial binding site for this molecule (Yang et al., 2016; Zhang et al., 2016). Another promiscuous activator of TRPV2, TRPV3, TRPM8, and TRPV4 channels is 2-aminoethyl diphenylborinate (2-APB) (Hu et al., 2004). The binding site or sites of 2-APB in each of these channels has not been delineated accurately and thus the best-characterized binding site is that of capsaicin for TRPV1 channels.

This vanilloid-binding pocket (VBP) is formed by residues in S3, S4, and part of the S5 transmembrane alpha helices as well as the S4–S5 linker. Not all orthologues of TRPV1 can be activated by capsaicin, and this property has played a role in identifying important residues for capsaicin binding (Jordt and Julius, 2002). Currently the picture we have of capsaicin binding and activation is that this molecule, once bound in the VBP in a "head-down" orientation, stabilizes the S4–S5 linker in a position that allows the S6 gate to be in the open conformation (Darre and Domene, 2015; Hanson et al., 2015; Yang et al., 2015).

An interesting phenomenon in TRPV1 is the observation that capsaicin and capsazepine bind to the same site, the VBP (Phillips et al., 2004), but induce quite opposite conformational changes. While capsaicin is an activator, capsazepine is a potent competitive antagonist (Walpole et al., 1994). The molecular basis for these opposed actions remains unknown. Another recently described inhibitor of TRPV1 that binds at the VBP is oleic acid, a naturally occurring fatty acid. Oleic acid acts as an inhibitor by promoting stabilization of the closed state, perhaps by destabilizing the same conformational change that is promoted by capsaicin (Morales-Lázaro et al., 2016).

A unique mechanism of gating present in thermoTRP channels is activation by covalent modification of intracellular cysteines, which could be considered a ligand-induced mechanism (Bautista et al., 2005; Macpherson et al., 2007; Salazar et al., 2008; Chuang and Lin, 2009; Koizumi et al., 2009). Electrophilic molecules can oxidize the SH group in cysteine, and this in turn promotes opening of the channel. This form of activation is responsible for the response of TRPV1 and TRPA1 to pungent compounds found as components in extracts in several plants, such as horseradish, garlic, and onion (Bautista et al., 2005; Salazar et al., 2008). It should be noted that this is a direct gating phenomenon, since it has been shown that activation can happen in the absence of other stimuli, such as voltage, other ligands, or changes in temperature, and produces an increment of the single-channel open probability. Although the individual cysteine residues have been identified in many cases, the allosteric mechanism of coupling to pore opening remains unknown (Salazar et al., 2008).

Besides activation by ligand binding and other stimuli, thermoTRPs also show voltage dependence. The voltage activation of thermoTRPs has been mostly characterized from macroscopic current recordings, although a few single-channel studies exist (Brauchi et al., 2004; Matta and Ahern, 2007; Fernández et al., 2011).

A simple conceptualization of a voltage-dependent mechanism is a channel that opens when a charged particle with valence z is displaced through the electrical field generated by the potential difference V across the plasma membrane. This charged particle thus acts as a voltage sensor that surmounts an energy barrier, which has two components, one of which is the electrostatic energy given by the zV product, and another is an intrinsic conformational energy. If there is a single energy barrier, the mechanism involves a single closed and a single open state and the equilibrium probability of finding the channel in the open state, $p(V)$ is given by a simple Boltzmann equation:

$$p(V) = \frac{1}{1 + K_o \exp^{-zV/K_B T}}$$

where K_B is Boltzmann's constant, and K_o is related to the size of the energy barrier in the absence of voltage (0 mV). In this simple model, the charge of the voltage sensor, z, is derived from a fit of this function to the measured $p(V)$ or a proportional parameter such as the macroscopic conductance G(V). In all cases it has been found that the value of z is rather low for thermoTRPs, in the order of 1 e_o, which greatly contrasts with an equivalent z in the order of 9–13 e_o found in *bona fide* voltage-gated ion channels (Schoppa et al., 1992; Hirschberg et al., 1995; Noceti et al., 1996; Islas and Sigworth, 1999; Ishida et al., 2015).

Consistently, the S4 segment of TRP channels lacks the repeating R/KXXR/K (X = hydrophobic) motifs found in voltage-sensing domains (Long et al., 2005).

Some attempts have been made to define the possible role of charged amino acids at or near the S4 of thermoTRPs, but the results are consistent with the S4 not contributing to voltage sensing, perhaps with the exception of TRPM8 (Nilius et al., 2005; Voets et al., 2007). It is thus possible that the function of the S1–S4 domain is to provide a more or less rigid scaffold for ligand binding and energy coupling to open the pore (Steinberg et al., 2014). Consistent with this view, a bioinformatics comparative analysis of the structure of the transmembrane domains of TRPV1 and Kv channels has shown that the main structural determinants of voltage sensing are lacking in TRPV1 and possibly in all thermoTRPs (Palovcak et al., 2015).

Gating in thermoTRPs is not only different in voltage dependence but shows other important disparities when compared with the Kv channels. Experimentally, it is observed that the heat or cold activation mechanisms interact with voltage-dependent openings. Initially, it was proposed that temperature and voltage controlled the same conformational transition, by virtue of this transition having unusually large enthalpic (ΔH) and entropic changes (ΔS) in thermoTRPs (Voets et al., 2004). Formally, this can be understood from the simplified two-state gating formalism as the voltage-independent constant K_o being given by

$$K_o = e^{-(\Delta H - T\Delta S)}$$

However, a growing body of evidence now indicates that voltage and temperature contribute to different and distinct conformational changes that are likely coupled via allosteric mechanisms (Yao et al., 2010a) and that more than one transition is needed to account for the details of channel gating (Hui et al., 2003).

So, what is the mechanism of voltage sensing in thermoTRPs? Several, still unexplored mechanisms are possible. The movement in the electrical field of the partial charges of aromatic residues can afford voltage sensing. This type of voltage sensor has been recently described in G-protein coupled receptors (Ben-Chaim et al., 2006; Barchad-Avitzur et al., 2016). Interestingly, aromatic residues are found in the S4 segment of most thermoTRPs, and thus they could potentially contribute to a voltage sensor through limited outward movement of the S4 helix at depolarized voltages.

Voltage-dependent activation is also observed in potassium channels formed by only the pore domain, which also lack canonical voltage sensors. In these channels, it is found that conformational changes in the selectivity filter are coupled to the flux of permeant ions, and since ions traverse the electric field located at the selectivity filter, this conformational change is intrinsically voltage dependent (Spassova and Lu, 1998; Kurata et al., 2010). A similar mechanism also gives rise to the voltage dependence of other channels lacking voltage sensors, such as some chloride channels (Pusch et al., 1995; Chen and Miller, 1996). The voltage dependence conferred by the existence of a pore gate could be substantial, being equivalent to the translocation of up to 2.5 e_o (Schewe et al., 2016).

2.4   MECHANISM OF GATING BY TEMPERATURE

ThermoTRP channels are unique among ion channels in that they can be opened by changes in temperature. Opening by increased temperature in TRPVs can be accomplished at negative voltages, at neutral pH, and in the complete absence of any other activating ligand. Cool activated channels, like TRPM8 and some orthologues of TRPA1, can also be directly activated by reduced temperature (Brauchi et al., 2004; Chen et al., 2013).

Direct gating by temperature changes implies that the channel protein is able to absorb heat and convert it to a conformational change efficiently.

Thermodynamically, heat absorption at constant pressure is defined by the change in enthalpy (ΔH), which in thermoTRPs is much larger than in other channels (Voets et al., 2004; Yao et al., 2010a, 2011).

Most thermoTRPs are activated by heating; thus, these heat-activated channels are the better understood. Each thermoTRP channel has a distinct range of temperatures in which it is best activated; these different ranges have given rise to the concept of activation threshold, which each research group defines in different ways.

Strictly, and in a manner similar to what happens with activation by voltage, there is no activation threshold. The open probability of the channel continuously changes as heat is absorbed. Since most experiments in thermoTRPs are carried out by measuring macroscopic currents, a general operative definition of threshold is the temperature at which currents are just discernible (Tominaga et al., 1998; Vyklický et al., 1999). Other definitions of threshold are commonly in use, although none of them has a strict theoretical meaning.

The dependence of current (or open probability) increase as a function of temperature is more often quantified by the temperature coefficient or Q_{10}. This factor describes the fold-change in current magnitude produced by a 10°C change in temperature:

$$Q_{10} = \left(\frac{I_2}{I_1}\right)^{\frac{10}{(T_2 - T_1)}}$$ (2.1)

The higher value of Q_{10} implies a larger enthalpy change associated to the change of current magnitude or open probability.

Several authors have lucidly presented the quantitative thermodynamic formalism associated with the simple two-state model (Yao et al., 2010a; Baez et al., 2014). In this case, it can be shown that temperature activation can be quantitatively expressed by a simple equation of the form:

$$I = I_L e^{-\frac{\Delta H_L}{RT}} + \frac{I_{max} e^{\frac{-\Delta H}{RT}}}{1 + e^{\frac{-\Delta G}{RT}}}$$ (2.2)

This equation relates the magnitude of the ionic current, I, to the temperature change, T. In this equation $\Delta G = \Delta H - T\Delta S$ is the free energy change of the open-close transition, which is a function of the enthalpy or heat involved in the transition (ΔH) and the change in order, given by the entropic component, ΔS. It is important to note that in this equation the temperature dependence of the leak current, I_L—that is, the non-heat activated current—has been explicitly dealt with, using the enthalpy ΔH_L.

It should be noted that the Q_{10} of the conductance of a single channel is between 1.5–2 and 6–7 kCal/mol (Lee and Deutsch, 1990; Correa et al., 1992; Rodríguez et al., 1998; Vyklický et al., 1999). This means that the size of the thermoTRP current can increase more than four times for a 20°C stimulus, just because the resistance of the channel pore decreases. This is an important fact that is not always taken into account in quantitative descriptions of gating by temperature in thermoTRPs.

In this review, I would like to explicitly discuss the use of allosteric models to describe gating in thermoTRPs. There is abundant evidence that gating mechanisms are separable in these channels, making allosteric coupling models more appropriate (Liu et al., 2003; Brauchi et al., 2004; Yao et al., 2010a; Cao et al., 2013b; Jara-Oseguera and Islas, 2013). The simplest such model for the behavior of an activation module is shown is Figure 2.3a. One of the main characteristics of these models is that channels are able to open (C_o–O_2) in the absence of a stimulus, generally with a low probability given by the equilibrium constant L_o as

$$\frac{L_o}{1 + L_o}$$

Although the stimulus-independent open probability has not been measured explicitly in thermoTRPs, there is evidence of voltage-independent and temperature-independent openings (Yao et al., 2010a). For example, several mutations increase the magnitude of the current at negative voltages and room temperature in TRV1 (Boukalova et al., 2010).

The stimulus (heat, voltage, or ligand binding) acts through the C_o–C_1 transition that is governed by the equilibrium constant, K, in such a way that the opening, C_1–O_3 is increased by an allosteric factor D to DL_o. In the presence of the stimulus, the open probability is thus increased to

$$\frac{DL_o}{1 + DL_o}$$

In the case of temperature activation, the equilibrium constant $K(T)$ is given by the ratio of forward and backward rate constants:

$$\frac{k_1}{k_{-1}} = \frac{\theta_1 \cdot \exp^{-\left(\Delta H_1^{\ddagger} - T\Delta S_1^{\ddagger}\right)/RT}}{\theta_{-1} \cdot \exp^{-\left(\Delta H_{-1}^{\ddagger} - T\Delta S_{-1}^{\ddagger}\right)/RT}}$$

Here, the θ_i are the diffusion-limited rate constants at 0 K; $\Delta H_i^{\pm}$ and $\Delta S_i^{\pm}$ are the activation enthalpies and entropies, respectively; R is the gas constant; and T is the temperature in Kelvin.

For this simplest allosteric model, the steady-state open probability is given by Equation 2.3:

$$p(T) = \frac{L_o + KDL_o}{1 + K + L_o + KDL_o} \tag{2.3}$$

In general, all parameters can be temperature dependent, but for the sake of simplicity, the temperature dependence will be assumed to happen only for the equilibrium constant $K(T)$, which then represents a specific conformational change that is driven by the change in temperature.

The allosteric gating mechanism encoded in Equation 2.3 predicts that the steepness of the open probability change is related to the heat (ΔH) involved in the transition. The value of ΔS sets the temperature range over which the transition takes place. It would seem like the value of ΔH can be extracted from the steepness of this function. However, the value of the allosteric coupling factor between the temperature-sensitive transition and the pore module, D, also serves to set the steepness of the activation curve. Figure 2.3b illustrates how for a fixed value of ΔH, increasing values of D make the steepness more pronounced. This fact is stressed even more in Figure 2.3c, where it can be seen that a temperature-sensing transition with smaller associated enthalpy can give rise to a steeper activation curve if it is more strongly coupled to pore opening (larger value of the factor D). Experimentally, a measurement of the absolute open probability at a saturating temperature, together with the limiting open probability at low temperature, $\frac{L_o}{1 + L_o}$, can provide a determination of the value of D.

It should be stressed that because the strength of coupling between the temperature-dependent reaction and pore opening, D, contributes to the steepness of the activation curve and also, as discussed before, the single-channel conductance increases with temperature, it is implied that estimates of the enthalpy from current activation curves tend to be overestimates.

As a general rule, the sign of the entropic and enthalpic changes in the K transition determines whether the channel is activated by increases or decreases of the temperature. If ΔH and $\Delta S > 0$, the channel is a heat-activated channel, and if ΔH and $\Delta S < 0$, it will be activated by cold temperatures (Voets et al., 2004). This quantitative fact has led to the notion that there will be two types of transitions associated with the temperature sensor, one endergonic and one exergonic.

Any other activation module can be modeled in the same way as the heat-sensitive transition, and each module can interact with each other through an allosteric coupling factor. For example, the voltage- and temperature-sensing modules can be coupled to produce heat and voltage activation, in the manner shown in Figure 2.4a. For this model, the steady-state dependence of the open probability on voltage and temperature is given by Equation 2.4:

$$p(T) = \frac{L_o\left(KD + E\Omega + DECK\Omega\right)}{1 + K + \Omega + CK\Omega + L_o\left(1 + DK + E\Omega + DECK\Omega\right)} \tag{2.4}$$

In this new equation, Ω is the equilibrium constant of the voltage-dependent transition, E is the coupling factor between the voltage sensor and the pore, and C is the coupling between the voltage sensor and the heat sensor. Figure 2.4b shows the behavior predicted by Equation 2.4 for the

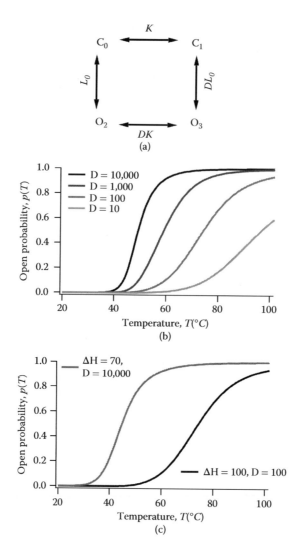

FIGURE 2.3 The contribution of coupling factors to the heat activation curves in allosteric gating models of thermoTRP channels. (a) The simplest allosteric gating model for thermal activation. The transition given by the equilibrium L_o represents pore opening, while K is the transition representing activation of a thermal sensor. The factor D represents the strength of coupling between the two modules. This model predicts the existence of several open and closed states. (b) The steepness and threshold of the activation curve can be greatly affected by the strength of coupling, given by the value of D. This result is further stressed in (c), where a simulated channel with smaller enthalpy of activation can be made to have a steeper activation curve if the value of D is increased (gray).

activation by voltage and temperature. It can be seen that the effect of temperature on voltage activation is more complex than a simple shift of the voltage of half activation. At low temperatures, large depolarization is unable to elicit the maximum open probability (Matta and Ahern, 2007), while opening the channel with voltage at high temperature increases both the maximum and minimum open probabilities.

The more complete and complex models such as this are difficult to constrain experimentally, although some researches have tried to provide empirical support at least for some sections of the more complex models (Cao et al., 2014; Jara-Oseguera et al., 2016).

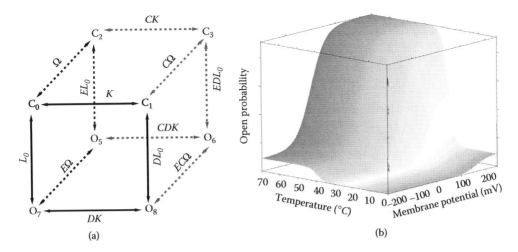

FIGURE 2.4 A general allosteric activation model for thermoTRPs. This model considers the coupling between a temperature-induced and a voltage-induced conformational change. (a) Eight possible states the channel can visit and the respective equilibrium constants. The voltage-dependent transition is given by the equilibrium constant: $\Omega = 0.001e^{(1 \cdot F \cdot V \cdot /RT)}$. The values of the other constants are $L_o = 0.001$, $D = 100$, $E = 100$, $C = 100$. The activation enthalpy, $\Delta H = 100$ kcal/mol and the entropy, $\Delta S = 0.32$ kcal·K/mol. (b) A surface plot showing that the model parameter's values yield activation curves as a function of temperature and voltage comparable to experimental results. (From Yao, J. et al., *Biophys J*, 99, 1743–1753, 2010a; Jara-Oseguera, A., and L.D. Islas, *Biophys J*, 104, 2160–2169, 2013.)

2.5 PHYSICAL THEORIES OF ACTIVATION BY TEMPERATURE

One of the main aims in thermoTRP channel research has been finding a "temperature sensor." This sensor is conceptualized as a specific region of the protein that would be particularly amenable to undergo temperature-induced changes in conformation and transmit those to an activation gate. Mutagenic manipulations have identified several channel regions that affect gating by temperature. Unfortunately, while a binding site can be identified by ligand-binding experiments (Blount and Merlie, 1989) and a conformational change in a voltage sensor can give rise to gating currents (Sigworth, 1994), a heat sensor would have to be identified by its differential absorption (or production) of heat, which is experimentally out of reach for now.

2.5.1 ROLE OF PORE REGION

Several early mutagenesis experiments identified mutations in the pore region that differentially stabilize or disrupt temperature-induced opening in TRPV1, TRPV3, and TRPA1 (Grandl et al., 2008; Grandl et al., 2010; Wang et al., 2013). It was also shown that some residues near the extracellular pore region undergo temperature-dependent conformational changes in TRPV channels (Kim et al., 2013). The importance of the channel pore in gating is also underscored by the finding that deletions and other manipulations of the long linker between S5 and the selectivity filter, known as the turret, also disrupt heat activation (Yang et al., 2010; Cui et al., 2012). Although there is good evidence that the turret modulates heat activation in TRPV1, the notion that it can act as a heat sensor is tempered by the finding that TRPV1 channels completely lacking the turret can be heat activated with high sensitivity (Yao et al., 2010b; Cao et al., 2013a; Liao et al., 2013). The pore region seems to be a hotbed for gating, since a lot of mono- and divalent ions that sensitize or produce channel activation exert their actions through binding to residues in the extracellular region of the pore (Ahern et al., 2005; Ryu et al., 2007; Aneiros et al., 2011; Cao et al., 2014; Jara-Oseguera et al., 2016).

2.5.2 ROLE OF TRANSMEMBRANE DOMAINS

The transmembrane regions of thermoTRP channels seem to be nondeterminant of heat activation. Chimera construction experiments show that thermal activation properties can be transferred between TRPV1 and TRPV2 only by transplantation of the N- and C-terminal domains (Yao et al., 2011). This is also true for chimeras between TRPA1 channels of different species (Cordero-Morales et al., 2011) and splice isoforms of *Drosophila* TRPA1 (Zhong et al., 2012).

However, it was recently suggested that the Kv2.1 voltage-gated ion channel and the ILT mutant of the *Shaker* Kv channel, are intrinsically temperature sensitive, implying that transmembrane domains are sufficient for temperature gating (Yang and Zheng, 2014). In contrast, previous studies of the temperature dependence of voltage-gated ion channels are inconsistent with these results, showing that activating conformational changes have a $Q_{10} \sim 3$ (Starace et al., 1997; Rodríguez et al., 1998; Gagnon and Bezanilla, 2010). Furthermore, it has been shown in voltage-gated channels with complex inactivation and recovery from inactivation that the temperature dependences of activation as well as inactivation are small ($Q_{10} \sim 2$) but are affected by temperature in inverse fashion, giving rise to an apparently large heat-activated open occupancy (Vandenberg et al., 2006).

2.5.3 ROLE OF INTRACELLULAR REGIONS

The role of intracellular regions was first explored in TRPV1 and TRPM8, heat- and cold-activated channels, respectively. It was shown that the C-terminus of both channels contains a transplantable region that could interchange the temperature-activation properties of these channels (Brauchi et al., 2006). This result does not automatically mean that the C-terminus is a heat sensor, especially since the chimeras showed a marked reduction in the steepness of activation, measured by the Q_{10}. It is possible that the complex allosteric interactions involved in gating could mean that the C-terminus plays a role in coupling of the stimulus to channel opening and not sensing. Evidence that the C-terminus is important in modulating allosteric activation by heat is provided by the finding that alternative splice variants of TRPV1 in vampire bats can produce channels with large truncations of the C-terminus which remain heat activated, although with significant shifts in activation threshold (Gracheva et al., 2011).

An intriguing structural feature of thermoTRPs is the presence of the multiple ankyrin repeats in the N-terminus. Ankyrin repeats are a common scaffold for protein–protein interactions (Li et al., 2006; Gaudet, 2008); they have interesting elastic properties (Sotomayor et al., 2005; Lee et al., 2006) and can bind several molecules (Grove et al., 2008). The function of the ankyrin repeats (ANKr) in thermoTRP channels is not completely understood. In TRPV channels, they can bind ATP (Lishko et al., 2007; Phelps et al., 2010), calmodulin (Rosenbaum et al., 2004; Lau et al., 2012), and are the substrate for cysteine modifications that can produce channel activation (Bautista et al., 2005; Salazar et al., 2008). As mentioned previously, the ANKr domains of TRPV1, TRPV2, TRPV4, and TRPV6, for which we have crystal structures, show remarkable structural similarity (Figure 2.2). These channels, with the exception TRPV6, are heat activated.

Several studies in thermoTRPs are consistent with the idea that the ANKr domains are important for thermal sensitivity. It was recently shown that a segment comprising at least some ANKr and a linker region just before the first transmembrane domain are capable of transplanting the enthalpy of activation between TRPV1 and TRPV2 channels (Yao et al., 2011). It can be argued that a hallmark of a region that might act as a temperature sensor should be its capacity to determine the activation enthalpy. So this region could potentially be regarded as a true heat sensor. In the recent atomic resolution structures of these two channels, this region is seen to make several contacts between adjacent subunits and also with the TRP domain, which might be indicative of its relevance in temperature-dependent gating.

The importance of the ANKr domains for temperature gating in TRPV1 has been recently stressed by the finding that single amino acid residue mutations can fine-tune the activation threshold of heat-activated currents. Remarkably, this fine-tuning seems to be important for the functionality of TRPV1 as a thermal receptor in different species adapted to different thermal environments (Laursen et al., 2016).

An important result was reported in snake and human orthologues of TRPA1 channels, where it was shown that the seven or eight proximal (to the transmembrane region) ANKr are sufficient to confer high-sensitivity heat activation to the temperature-insensitive human variant (Cordero-Morales et al., 2011). Importantly, in this study only the ANKr domains were capable of providing heat activation.

ANKr domains are also important for TRPV4 temperature activation. It has been shown that a deletion mutant that gets rid of the three proximal ANKr domains in mouse TRPV4 stops being activated by heat (Watanabe et al., 2002). This region is equivalent to the region that was demonstrated to be able to determine activation enthalpy in TRPV1 (Yao et al., 2011).

What kind of conformational changes can account for temperature-dependent activation? It was assumed that a large conformational change was the only way to account for the large enthalpic and entropic changes observed during thermoTRP gating; however, attempts based on fluorescence spectroscopy methods have failed to detect large conformational changes (De-la-Rosa et al., 2013). Initial molecular dynamics simulations of the whole TRPV1 structure also suggest that the opening conformational change is accompanied by modest rearrangements in the amino terminal and transmembrane domains (Zheng and Qin, 2015).

A stimulating recent proposal considers the physical-chemical foundations of a large enthalpy change in protein conformational dynamics. Buried hydrophobic residues in the core of a protein tend to satisfy energetic constrains by establishing contacts with nonpolar residues. Likewise, polar residues tend to reside in the periphery of the protein, or if present in the core, interact with other polar residues by forming salt bridges or by being surrounded by high dielectric solvent accessible regions (Dorairaj and Allen, 2007). When these interactions are perturbed, usually the energetic cost is very high, both in terms of heat evolved (enthalpy) and order/disorder (entropy). This energetic cost gives rise to the temperature dependence of the heat capacity of a protein during denaturalization transitions that tend to perturb the interactions mentioned above (Privalov et al., 1989).

Clapham and Miller (2011) point out that a change in heat capacity between closed and open states can explain the temperature dependence of activation in thermoTRPs. The exposure of a few nonpolar amino acids to the solvent is sufficiently energetic that if it was coupled to channel opening, it could explain the high enthalpies and entropies observed during thermoTRP activation. This explanation has been exploited to increase the temperature dependence of voltage-dependent gating in *Shaker* channels. The S4 alpha helix in the voltage sensor undergoes conformational changes that change its solvation (Jogini and Roux, 2007). By carefully engineering changes in solvent exposure of certain amino acids, it was possible to generate channels that display very high Q_{10} values of the rates of voltage-dependent gating (Chowdhury et al., 2014), lending experimental evidence that changes in heat capacity can affect the temperature dependence of channel heating. It remains to be seen if thermoTRPs do indeed use a similar mechanism.

An important prediction of the heat capacity change theory is that channel activation should be bimodal. That is, the same channel should display activation at both low and high temperatures. Although this behavior has never been observed in any thermoTRP, it should be pointed out that this prediction is not unique. In the previously discussed allosteric mechanisms, only one equilibrium constant was considered as dependent on temperature; however, the coupling factors between gating modules can also have an associated enthalpy change. If coupling is made temperature dependent, as has been previously suggested (Yao et al., 2010a), then channels can display the bimodal activation behavior. More importantly, temperature-dependent coupling predicts that a single type of heat sensor, activated by increases in temperature and thus a single type of conformational change, could be differentially coupled to either closing or opening the pore, giving rise to channels activated by cold or channels activated by heat (Jara-Oseguera and Islas, 2013). In support of these ideas, it has been found that the directionality of heat activation in TRPA1 channels can be reversed by single point mutations (Jabba et al., 2014).

Finally, just as denaturation of proteins by heat is highly dependent on temperature through perturbations of hydrophobic and polar interactions, it is still possible that at least part of the mechanism of temperature gating in thermoTRPs will turn out to be due to partial denaturation of selective regions of the channel structure.

ACKNOWLEDGMENTS

Research in the author's laboratory is financed through grant no. IN209515 from DGAPA-PAPIIT-UNAM, grant no. 248499 from CONACyT-Mexico, and grant no. 77 from Fronteras de la Ciencia-CONACyT-Mexico.

REFERENCES

Ahern, G.P. et al. 2005. Extracellular cations sensitize and gate capsaicin receptor TRPV1 modulating pain signaling. *J Neurosci*, 25: 5109–5116.

Aneiros, E. et al. 2011. The biophysical and molecular basis of TRPV1 proton gating. *EMBO J*, 30: 994–1002.

Baez, D. et al. 2014. Gating of thermally activated channels. *Curr Top Membr*, 74: 1–87.

Barchad-Avitzur, O. et al. 2016. A novel voltage sensor in the orthosteric binding site of the M2 muscarinic receptor. *Biophys J*, 111: 1396–1408.

Bautista, D.M. 2005. Pungent products from garlic activate the sensory ion channel TRPA1. *Proc Natl Acad Sci U S A*, 102: 12248–12252.

Ben-Chaim, Y. 2006. Movement of "gating charge" is coupled to ligand binding in a G-protein-coupled receptor. *Nature*, 444: 106–109.

Blount, P. and J.P. Merlie. 1989. Molecular basis of the two nonequivalent ligand binding sites of the muscle nicotinic acetylcholine receptor. *Neuron*, 3: 349–357.

Boukalova, S. et al. 2010. Conserved residues within the putative S4–S5 region serve distinct functions among thermosensitive vanilloid transient receptor potential (TRPV) channels. *J Biol Chem*, 285: 41455–41462.

Brauchi, S., P. Orio, and R. Latorre. 2004. Clues to understanding cold sensation: Thermodynamics and electrophysiological analysis of the cold receptor TRPM8. *Proc Natl Acad Sci U S A*, 101: 15494–15499.

Brauchi, S. et al. 2006. A hot-sensing cold receptor: C-terminal domain determines thermosensation in transient receptor potential channels. *J Neurosci*, 26: 4835–4840.

Cao, E. et al. 2013a. TRPV1 channels are intrinsically heat sensitive and negatively regulated by phosphoinositide lipids. *Neuron*, 77: 667–679.

Cao, E. et al. 2013b. TRPV1 structures in distinct conformations reveal activation mechanisms. *Nature*, 504: 113–118.

Cao, X. et al. 2014. Divalent cations potentiate TRPV1 channel by lowering the heat activation threshold. *J Gen Physiol*, 143: 75–90.

Chatzigeorgiou, M. et al. 2010. Specific roles for DEG/ENaC and TRP channels in touch and thermosensation in *C. elegans* nociceptors. *Nature Neurosci*, 13: 861–868.

Chen, J. et al. 2013. Species differences and molecular determinant of TRPA1 cold sensitivity. *Nature Commun*, 4.

Chen, T.-Y. and C. Miller. 1996. Nonequilibrium gating and voltage dependence of the ClC-0 Cl-channel. *J Gen Physiol*, 108: 237–250.

Chowdhury, S. et al. 2014. A molecular framework for temperature-dependent gating of ion channels. *Cell*, 158: 1148–1158.

Chuang, H.-h. and S. Lin. 2009. Oxidative challenges sensitize the capsaicin receptor by covalent cysteine modification. *Proc Natl Acad Sci*, 106: 20097–20102.

Clapham, D.E., and C. Miller. 2011. A thermodynamic framework for understanding temperature sensing by transient receptor potential (TRP) channels. *Proc Natl Acad Sci*, 108: 19492–19497.

Cordero-Morales, J.F., E.O. Gracheva, and D. Julius. 2011. Cytoplasmic ankyrin repeats of transient receptor potential A1 (TRPA1) dictate sensitivity to thermal and chemical stimuli. *Proc Natl Acad Sci*, 108: E1184–E1191.

Correa, A., F. Bezanilla, and R. Latorre. 1992. Gating kinetics of batrachotoxin-modified Na+ channels in the squid giant axon. Voltage and temperature effects. *Biophys J*, 61: 1332.

Cui, Y. et al. 2012. Selective disruption of high sensitivity heat activation but not capsaicin activation of TRPV1 channels by pore turret mutations. *J Gen Physiol*, 139: 273–283.

Darre, L. and C. Domene. 2015. Binding of capsaicin to the TRPV1 ion channel. *Mol Pharm*, 12: 4454–4465.

De-la-Rosa, V. et al. 2013. Coarse architecture of the transient receptor potential vanilloid 1 (TRPV1) ion channel determined by fluorescence resonance energy transfer. *J Biol Chem*, 288: 29506–29517.

Dorairaj, S. and T.W. Allen. 2007. On the thermodynamic stability of a charged arginine side chain in a transmembrane helix. *Proc Natl Acad Sci*, 104: 4943–4948.

Fernández, J.A. et al. 2011. Voltage-and cold-dependent gating of single TRPM8 ion channels. *J Gen Physiol*, 137: 173–195.

Gagnon, D.G. and F. Bezanilla. 2010. The contribution of individual subunits to the coupling of the voltage sensor to pore opening in Shaker K channels: Effect of ILT mutations in heterotetramers. *J Gen Physiol*, 136: 555–568.

García-Sanz, N. et al. 2007. A role of the transient receptor potential domain of vanilloid receptor I in channel gating. *J Neurosci*, 27: 11641–11650.

Gaudet, R. 2008. A primer on ankyrin repeat function in TRP channels and beyond. *Mol Biosyst*, 4: 372–379.

Gavva, N.R. 2008. Body-temperature maintenance as the predominant function of the vanilloid receptor TRPV1. *Trends Pharmacol Sci*, 29: 550–557.

Gavva, N.R. et al. 2008. Pharmacological blockade of the vanilloid receptor TRPV1 elicits marked hyperthermia in humans. *Pain*, 136: 202–210.

Gracheva, E.O. et al. 2011. Ganglion-specific splicing of TRPV1 underlies infrared sensation in vampire bats. *Nature*, 476: 88–91.

Grandl, J. et al. 2008. Pore region of TRPV3 ion channel is specifically required for heat activation. *Nature Neurosci*, 11: 1007–1013.

Grandl, J. et al. 2010. Temperature-induced opening of TRPV1 ion channel is stabilized by the pore domain. *Nat Neurosci*, 13: 708–714.

Gregorio-Teruel, L. et al. 2015. The integrity of the TRP domain is pivotal for correct TRPV1 channel gating. *Biophys J*, 109: 529–541.

Grove, T.Z., A.L. Cortajarena, and L. Regan. 2008. Ligand binding by repeat proteins: Natural and designed. *Curr Opin Struct Biol*, 18: 507–515.

Hanson, S.M. et al. 2015. Capsaicin interaction with TRPV1 channels in a lipid bilayer: Molecular dynamics simulation. *Biophys J*, 108: 1425–1434.

Hirschberg, B. et al. 1995. Transfer of twelve charges is needed to open skeletal muscle Na+ channels. *J Gen Physiol*, 106: 1053–1068.

Hu, H.-Z. et al. 2004. 2-Aminoethoxydiphenyl borate is a common activator of TRPV1, TRPV2, and TRPV3. *J Biol Chem*, 279: 35741–35748.

Hui, K., B. Liu, and F. Qin. 2003. Capsaicin activation of the pain receptor, VR1: Multiple open states from both partial and full binding. *Biophys J*, 84: 2957–2968.

Huynh, K.W. et al. 2016. Structure of the full-length TRPV2 channel by cryo-EM. *Nat Commun*, 7: 1-8.

Inada, H. et al. 2012. Structural and biochemical consequences of disease-causing mutations in the ankyrin repeat domain of the human TRPV4 channel. *Biochemistry*, 51: 6195–6206.

Ishida, I.G. et al. 2015. Voltage-dependent gating and gating charge measurements in the Kv1. 2 potassium channel. *J Gen Physiol*, 145: 345–358.

Islas, L., and F. Qin. 2014. *Thermal Sensors*. New York, NY: Academic Press.

Islas, L.D. and F.J. Sigworth. 1999. Voltage sensitivity and gating charge in Shaker and Shab family potassium channels. *J Gen Physiol*, 114: 723–742.

Jabba, S. et al. 2014. Directionality of temperature activation in mouse TRPA1 ion channel can be inverted by single-point mutations in ankyrin repeat six. *Neuron*, 82: 1017–1031.

Jara-Oseguera, A., C. Bae, and K.J. Swartz. 2016. An external sodium ion binding site controls allosteric gating in TRPV1 channels. *eLife*, 5: e13356.

Jara-Oseguera, A. and L.D. Islas. 2013. The role of allosteric coupling on thermal activation of thermo-TRP channels. *Biophys J*, 104: 2160–2169.

Jin, X., J. Touhey, and R. Gaudet. 2006. Structure of the N-terminal ankyrin repeat domain of the TRPV2 ion channel. *J Biol Chem*, 281: 25006–25010.

Jogini, V. and B. Roux. 2007. Dynamics of the Kv1. 2 voltage-gated K+ channel in a membrane environment. *Biophys J*, 93: 3070–3082.

Jordt, S.-E. and D. Julius. 2002. Molecular basis for species-specific sensitivity to "hot" chili peppers. *Cell*, 108: 421–430.

Kim, S.E., A. Patapoutian, and J. Grandl. 2013. Single residues in the outer pore of TRPV1 and TRPV3 have temperature-dependent conformations. *PLOS ONE*, 8:e59593.

Koizumi, K. et al. 2009. Diallyl sulfides in garlic activate both TRPA1 and TRPV1. *Biochem Biophys Res Commun*, 382: 545–548.

Kurata, H.T. et al. 2010. Voltage-dependent gating in a "voltage sensor-less" ion channel. *PLoS Biol*, 8: e1000315.

Latorre, R., C. Zaelzer, and S. Brauchi. 2009. Structure–functional intimacies of transient receptor potential channels. *Q Rev Biophys*, 42: 201–246.

Lau, S.-Y., E. Procko, and R. Gaudet. 2012. Distinct properties of Ca2+–calmodulin binding to N- and C-terminal regulatory regions of the TRPV1 channel. *J Gen Physiol*, 140: 541–555.

Laursen, W.J. et al 2016. Low-cost functional plasticity of TRPV1 supports heat tolerance in squirrels and camels. *Proc Natl Acad Sci*, 113: 11342–11347.

Lee, G. et al. 2006. Nanospring behaviour of ankyrin repeats. *Nature*, 440: 246–249.

Lee, S. and C. Deutsch. 1990. Temperature dependence of K (+)-channel properties in human T lymphocytes. *Biophys J*, 57: 49.

Li, J., A. Mahajan, and M.-D. Tsai. 2006. Ankyrin repeat: A unique motif mediating protein-protein interactions. *Biochemistry*, 45: 15168–15178.

Liao, M. et al. 2013. Structure of the TRPV1 ion channel determined by electron cryo-microscopy. *Nature*, 504: 107–112.

Lishko, P.V. et al. 2007. The ankyrin repeats of TRPV1 bind multiple ligands and modulate channel sensitivity. *Neuron*, 54: 905–918.

Liu, B., K. Hui, and F. Qin. 2003. Thermodynamics of heat activation of single capsaicin ion channels VR1. *Biophys J*, 85: 2988–3006.

Long, S.B., E.B. Campbell, and R. MacKinnon. 2005. Crystal structure of a mammalian voltage-dependent Shaker family K+ channel. *Science*, 309: 897–903.

Macpherson, L.J. et al. 2007. Noxious compounds activate TRPA1 ion channels through covalent modification of cysteines. *Nature*, 445: 541–545.

Matta, J.A. and G.P. Ahern. 2007. Voltage is a partial activator of rat thermosensitive TRP channels. *J Physiol*, 585: 469–482.

McCleverty, C.J. et al. 2006. Crystal structure of the human TRPV2 channel ankyrin repeat domain. *Protein Sci*, 15: 2201–2206.

Morales-Lázaro, S.L. et al. 2016. Inhibition of TRPV1 channels by a naturally occurring omega-9 fatty acid reduces pain and itch. *Nat Commun*, 7 : 1-12.

Nilius, B. et al. 2005. Gating of TRP channels: A voltage connection? *J Physiol*, 567: 35–44.

Noceti, F. et al. 1996. Effective gating charges per channel in voltage-dependent K+ and Ca2+ channels. *J Gen Physiol*, 108: 143–155.

Palovcak, E. et al. 2015. Comparative sequence analysis suggests a conserved gating mechanism for TRP channels. *J Gen Physiol*, 146: 37–50.

Paulsen, C.E. et al. 2015. Structure of the TRPA1 ion channel suggests regulatory mechanisms. *Nature*, 520: 511–517.

Phelps, C.B. et al. 2008. Structural analyses of the ankyrin repeat domain of TRPV6 and related TRPV ion channels. *Biochemistry*, 47: 2476–2484.

Phelps, C.B. et al. 2010. Differential regulation of TRPV1, TRPV3, and TRPV4 sensitivity through a conserved binding site on the ankyrin repeat domain. *J Biol Chem*, 285: 731–740.

Phillips, E. et al. 2004. Identification of species-specific determinants of the action of the antagonist capsazepine and the agonist PPAHV on TRPV1. *J Biol Chem*, 279: 17165–17172.

Privalov, P. et al. 1989. Heat capacity and conformation of proteins in the denatured state. *J Mol Biol*, 205: 737–750.

Pusch, M. et al. 1995. Gating of the voltage-dependent chloride channel CIC-0 by the permeant anion. *Nature*, 373: 527–531.

Rodríguez, B.M., D. Sigg, and F. Bezanilla. 1998. Voltage gating of Shaker K+ channels the effect of temperature on ionic and gating currents. *J Gen Physiol*, 112: 223–242.

Rosenbaum, T. et al. 2004. Ca2+/calmodulin modulates TRPV1 activation by capsaicin. *J Gen Physiol*, 123: 53–62.

Ryu, S. et al. 2007. Uncoupling proton activation of vanilloid receptor TRPV1. *J Neurosci*, 27: 12797–12807.

Salazar, H. et al. 2009. Structural determinants of gating in the TRPV1 channel. *Nat Struct Mol Biol*, 16: 704–710.

Salazar, H. et al. 2008. A single N-terminal cysteine in TRPV1 determines activation by pungent compounds from onion and garlic. *Nat Neurosci*, 11: 255–261.

Saotome, K. et al. 2016. Crystal structure of the epithelial calcium channel TRPV6. *Nature*, 534: 506–511.

Schewe, M. et al. 2016. A non-canonical voltage-sensing mechanism controls gating in K2P K+ channels. *Cell*, 164: 937–949.

Schoppa, N. et al. 1992. The size of gating charge in wild-type and mutant Shaker potassium channels. *Science*, 255: 1712–1715.

Sigworth, F.J. 1994. Voltage gating of ion channels. *Q Rev Biophys*, 27: 1–40.

Sotomayor, M., D.P. Corey, and K. Schulten. 2005. In search of the hair-cell gating spring: Elastic properties of ankyrin and cadherin repeats. *Structure*, 13: 669–682.

Spassova, M., and Z. Lu. 1998. Coupled ion movement underlies rectification in an inward-rectifier K+ channel. *J Gen Physiol*, 112: 211–221.

Starace, D.M., E. Stefani, and F. Bezanilla. 1997. Voltage-dependent proton transport by the voltage sensor of the Shaker K+ channel. *Neuron*, 19: 1319–1327.

Steinberg, X., C. Lespay-Rebolledo, and S. Brauchi. 2014. A structural view of ligand-dependent activation in thermoTRP channels. *Front Physiol*, 5: 171.

Tominaga, M. et al. 1998. The cloned capsaicin receptor integrates multiple pain-producing stimuli. *Neuron*, 21: 531–543.

Vandenberg, J.I. et al. 2006. Temperature dependence of human ether-a-go-go-related gene K+ currents. *Am J Physiol Cell Physiol*, 291: C165–C175.

Voets, T. et al. 2004. The principle of temperature-dependent gating in cold- and heat-sensitive TRP channels. *Nature*, 430: 748–754.

Voets, T. et al. 2007. TRPM8 voltage sensor mutants reveal a mechanism for integrating thermal and chemical stimuli. *Nat Chem Biol*, 3: 174–182.

Vyklický, L. et al. 1999. Temperature coefficient of membrane currents induced by noxious heat in sensory neurones in the rat. *J Physiol*, 517: 181–192.

Walpole, C.S. et al. 1994. The discovery of capsazepine, the first competitive antagonist of the sensory neuron excitants capsaicin and resiniferatoxin. *J Med Chem*, 37: 1942–1954.

Wang, H. et al. 2013. Residues in the pore region of *Drosophila* transient receptor potential A1 dictate sensitivity to thermal stimuli. *J Physiol*, 591: 185–201.

Watanabe, H. et al. 2002. Heat-evoked activation of TRPV4 channels in a HEK293 cell expression system and in native mouse aorta endothelial cells. *J Biol Chem*, 277: 47044–47051.

Yang, F. et al. 2010. Thermosensitive TRP channel pore turret is part of the temperature activation pathway. *Proc Natl Acad Sci*, 107: 7083–7088.

Yang, F. et al. 2015. Structural mechanism underlying capsaicin binding and activation of the TRPV1 ion channel. *Nat Chem Biol*, 11: 518–524.

Yang, F. et al. 2016. Rational design and validation of a vanilloid-sensitive TRPV2 ion channel. *Proc Natl Acad Sci*, 113: E3657–E3666.

Yang, F. and J. Zheng. 2014. High temperature sensitivity is intrinsic to voltage-gated potassium channels. *Elife*, 3: e03255.

Yao, J., B. Liu, and F. Qin. 2010a. Kinetic and energetic analysis of thermally activated TRPV1 channels. *Biophys J*, 99: 1743–1753.

Yao, J., B. Liu, and F. Qin. 2010b. Pore turret of thermal TRP channels is not essential for temperature sensing. *Proc Natl Acad Sci U S A*, 107: E125.

Yao, J., B. Liu, and F. Qin. 2011. Modular thermal sensors in temperature-gated transient receptor potential (TRP) channels. *Proc Natl Acad Sci*, 108: 11109–11114.

Zhang, F. et al. 2016. Engineering vanilloid-sensitivity into the rat TRPV2 channel. *eLife*, 5: e16409.

Zheng, W., and F. Qin. 2015. A combined coarse-grained and all-atom simulation of TRPV1 channel gating and heat activation. *J Gen Physiol*, 145: 443–456.

Zhong, L. et al. 2012. Thermosensory and nonthermosensory isoforms of *Drosophila melanogaster* TRPA1 reveal heat-sensor domains of a thermoTRP Channel. *Cell Rep*, 1: 43–55.

Zubcevic, L. et al. 2016. Cryo-electron microscopy structure of the TRPV2 ion channel. *Nat Struct Mol Biol*, 23(2): 180–186.

3 TRP Channels in Vision

Ben Katz, Richard Payne, and Baruch Minke

CONTENTS

3.1 INTRODUCTION

The transient receptor potential (TRP) field began (for reviews see Minke, 2010; Montell, 2011; Hardie, 2011) with the analysis of a spontaneously formed *Drosophila* mutant showing transient, rather than sustained, responses to prolonged intense illumination in electroretinogram (ERG) measurements, rendering the flies effectively blind (Cosens and Manning, 1969). Cosens and Manning (1969) isolated this mutant, designated this strain the "A-type" mutant, and attributed its phenotype to a failure of photopigment regeneration (Cosens, 1971). The isolation of this mutant, though potentially interesting, raised a number of concerns at the time. One main concern was that the results were based on a single spontaneously occurring mutant with no description of its genetic background. It was thus difficult to know what genetic alterations this strain represented. For example, the results could have been due to additive effects of alterations in several genes mapping to the same chromosome (Pak, 2010). The isolation of multiple mutated alleles from a baseline stock of known genetic background, conducted by Pak and colleagues (Pak, 2010), was important in establishing that the observed phenotype was indeed due to mutation in a single gene. Another concern at the time was that the cellular origins of ERG components were not well established. One could not be certain whether the lack of a sustained response seen in the ERG of this strain originated from the photoreceptors or from other retinal cells (Pak, 2010). This question was settled by performing intracellular recordings from the mutant photoreceptors (Minke et al., 1975). Only after determining that the defect arose from the photoreceptors, was it safe to conclude that this mutant is defective in phototransduction (Pak, 2010; Minke et al., 1975). Extensive studies of this mutant (Minke et al., 1975; Minke, 1977, 1982; Minke and Selinger, 1992a; Barash and Minke, 1994; Barash et al., 1988) provided a more descriptive name, "transient receptor potential" or *trp* (Minke et al., 1975) (Figure 3.1a) by Minke and colleagues, which was ultimately adopted by the scientific community to designate the entire gene family (Montell et al., 2002). These studies revealed that the mutant photopigment cycle was not altered and concluded that the defect was at an intermediate stage of the phototransduction cascade. A combination of electrophysiological, biochemical (Devary et al., 1987), and direct Ca^{+2} measurements in other invertebrates (Minke and Tsacopoulos, 1986) supported an hypothesis that the TRP encodes for a novel phosphoinositide-activated and Ca^{+2}-permeable channel/transporter protein, which is defective in the *trp* mutant (Devary et al., 1987; Minke and Selinger, 1991; Selinger and Minke, 1988). When the *trp* gene was cloned, its sequence indicated a transmembrane (TM) protein with eight TM helices, a topology reminiscent of known receptor/transporter/channel proteins (Montell and Rubin, 1989; Wong et al., 1989).

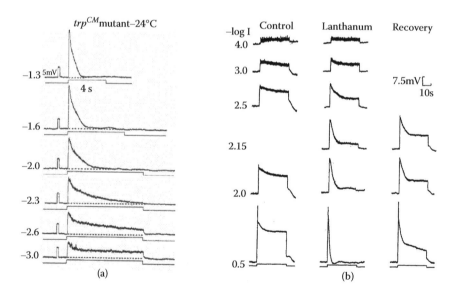

FIGURE 3.1 The phenotype of the *trp* mutant is mimicked by lanthanum (La^{3+}) in wild-type fly. (a) Intracellular recordings from single photoreceptor cell of white-eyed *trp*^{CM} raised at 24°C showing voltage responses to increasing intensities of orange lights (in relative log scale). (b) Intracellular recordings from a single photoreceptor cell of white-eyed *Musca domestica* in response to increasing intensities of orange lights, as indicated. The left column shows control responses; the middle column shows the responses of the same cell 1 minute after La^{3+} injection to the extracellular space, and the right column shows partial recovery of the response 20 minutes after La^{3+} injection. (From Minke, B., *J Gen Physiol.*, 79, 361–385, 1982; Suss Toby, E. et al., *J Gen Physiol.*, 98, 849–868, 1991. With permission.)

Immunofluorescent measurements of TRP localized the protein to the signaling compartment, the *rhabdomere*, further supporting its participation in phototransduction. However, due to the lack of homologous proteins to the TRP protein in available databases and results showing that in a presumably null *trp* alleles (Montell and Rubin, 1989; Wong et al., 1989), a sustained receptor potential persists under dim light stimulation (Minke et al., 1975; Minke, 1977), it was concluded that the *trp* gene does not encode for the light-sensitive channel (Montell and Rubin, 1989; Wong et al., 1989). Following later studies, based on the ability of La^{3+} to mimic the *trp* phenotype (Suss Toby et al., 1991; Hochstrate, 1989) (Figure 3.1b), it was proposed that the TRP might encode for an inositide-activated Ca^{2+} channel/transporter required for Ca^{2+} stores refilling (Minke and Selinger, 1992b). Consequently, using whole-cell voltage-clamp recordings to determine ionic selectivity, it was shown that the primary defect in the *trp* mutant was a drastic reduction in the Ca^{2+} permeability of the light-sensitive channels themselves (Hardie and Minke, 1992). This conclusion was further supported by studies using microfluorimetry (Peretz et al., 1994a) and Ca^{2+}-selective microelectrodes (Peretz et al., 1994b) (Figure 3.2a and b). The identification of another protein similar to the *trp* gene product, designated TRP-like (TRPL), using a Ca^{2+}/calmodulin binding assay, allowed for a reinterpretation of the phenotype of the *trp* mutation and suggested that the light response of *Drosophila* is mediated by channels composed from the TRP and TRPL gene products (Hardie and Minke, 1992; Phillips et al., 1992). Later, a third TRP homologue channel of *Drosophila* with similarity to TRP and TRPL was discovered by Montell and colleagues and was designated TRPγ (Xu et al., 2000). Using a TRPγ-antibody the authors showed that the protein is highly expressed in the retina and interacts with both the TRP and TRPL channels. Although the lack of a light response in the *trpl;trp* double null mutant (Scott et al., 1997) indicates that TRPγ cannot by itself form a light-sensitive channel, the authors suggested that TRPγ-TRPL heteromers may form an additional light-sensitive channel complex. In other insects, the role of TRPγ is even less certain. TRPγ expression was not

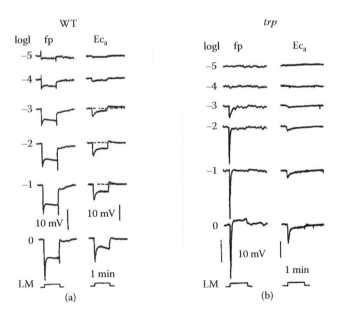

FIGURE 3.2 A comparison of light-induced Ca^{2+} influx in WT and the trp^{CM} mutant. (a) Measurements of Ca^{2+} influx from white-eyed WT *Drosophila*, in response to increasing intensities of orange light as indicated, using Ca^{2+}-selective microelectrodes that measured the reduction in extracellular Ca^{2+} as a function of time in the intact *Drosophila* retina. E_{Ca} (right column, see Peretz et al., 1994b) was measured simultaneously with the ERG response to light (fp, left column). LM indicates the light monitor. (b) The measurements similar to those of panel (a) were repeated in white-eyed trp^{CM} mutant at 24°C. (From Peretz, A. et al., *J Gen Physiol.*, 104, 1057–1077, 1994b. With permission.)

detectable in the compound eyes of the moth, *Spodoptera littoralis* (Chouquet et al., 2009), nor in those of the cockroach, *Periplaneta americana* (French et al., 2015). Hence, there is no evidence so far of a functional role for *trpγ* in phototransduction in any species, although roles in olfaction, cardiac function, and mechanosensation have been reported in insects (Chouquet et al., 2009; Wicher et al., 2006; Akitake et al., 2015).

In conclusion, the normal light-sensitive current comprises two distinct conductances: one is highly Ca^{2+} selective and is encoded by the *trp* gene, and the second is a channel responsible for the residual light-sensitive current in the *trp* mutants. We now know that the latter conductance is encoded by the homologous gene *trpl,* while the involvement of *trpγ*, if any, is unclear (Phillips et al., 1992; Niemeyer et al., 1996; Reuss et al., 1997).

3.2 DROSOPHILA PHOTOTRANSDUCTION

The physiological activation of the *Drosophila* TRP and TRPL channels by light is characterized by an outstanding performance. This includes fast response activation, sensitivity to single photons, high signal-to-noise ratio, adaptation to a wide range of light intensities, an unusually high frequency response to modulated lights, and fast response termination. These remarkable features most likely stem from the highly compartmentalized structure of the photoreceptor cells as well as the activation and regulation of the light-activated channels by a cascade of enzymatic reactions with positive feed-forward and negative feedback loops, collectively known as the phototransduction cascade. Elucidation and understanding of the molecular details underlying the phototransduction cascade have mainly stemmed from the power of *Drosophila* genetics and detailed studies conducted over many years, which are reviewed below.

3.2.1 COUPLING OF PHOTOEXCITED RHODOPSIN TO INOSITOL PHOSPHOLIPID HYDROLYSIS

TRP is an illuminating example of a novel signaling protein of prime importance whose physiological function and interactions with other signaling proteins have been elucidated due to the powerful genetic tools and functional tests that have been developed in *Drosophila*. These studies have led to the identification and characterization of TRP as a light-sensitive and Ca^{2+}-permeable channel (Minke, 2010; Montell, 2011; Hardie, 2011).

Illumination of fly photoreceptors induces a cascade of enzymatic reactions, which result in activation of the light-sensitive TRP channels (Minke, 2010; Devary et al., 1987). To function as a reliable light monitor, each stage of the phototransduction cascade needs an efficient mechanism of activation as well as an equally efficient mechanism of termination, ensuring that, at the cessation of the light stimulus, the photoreceptor potential will rapidly reach dark baseline. The use of specific nonphysiological conditions and mutated *Drosophila* strains has assisted in uncovering conditions in which specific stages underlying the light response failed to activate or terminate, and the TRP channels remain inactive or active in the dark. These experiments have revealed important mechanisms that regulate the enzymatic cascade controlling the activation and termination of the *Drosophila* TRP channel. These mechanisms are likely to also control the activation and regulation of nonvisual mammalian TRP channels.

3.2.2 RHODOPSIN AND THE PHOTOCHEMICAL CYCLE

The G protein-coupled receptor (GPCR), rhodopsin (R), is composed of a seven-transmembrane (TM) protein, opsin and the chromophore, 11-*cis* 3 hydroxyretinal (Vogt and Kirschfeld, 1984). Isomerization of the chromophore by photon absorption to all-*trans* retinal induces a conformational change in the opsin resulting in a dark stable and physiologically active photoproduct, metarhodopsin (M), that is stable in the dark (Figure 3.3). The action spectrum of this reaction depends on the specific R type and spans a wavelength range between ultraviolet (UV) and green light. To

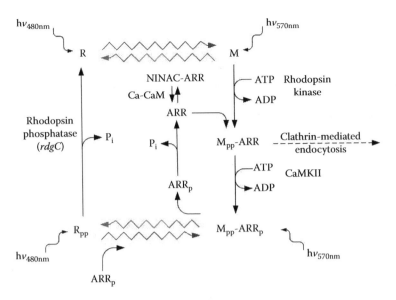

FIGURE 3.3 The photochemical cycle: the "turn-on" and "turn-off" of the photopigment. Upon photo-conversion of rhodopsin (R) to metarhodopsin (M), by illuminating with blue light, M is phosphorylated at multiple sites by rhodopsin kinase and the fly ARR2 binds to phosphorylated M (Mpp-ARR). ARR2 is then phosphorylated by Ca^{2+} calmodulin-dependent kinase (CaMK II). Photoconversion of phosphorylated M (Mpp) back to phosphorylated R (Rpp) is achieved by illuminating with orange light. Upon photoregeneration of Mpp to Rpp, phosphorylated ARR2 is released (ARRp), and the phosphorylated rhodopsin is exposed to phosphatase activity by rhodopsin phosphatase (encoded by the rdgC gene). Dephosphorylated ARR2 also binds to myosin III (NINAC) in a Ca^{2+} calmodulin (Ca-CaM) dependent manner. (Modified from Selinger, Z. et al., *Biochim Biophys Acta*, 1179, 283–299, 1993; Liu, C.H. et al., *Neuron*, 59, 778–789, 2008.)

ensure high sensitivity, high temporal resolution, and low dark noise of the photoresponse, the active M has to be quickly inactivated and recycled. The latter requirement is achieved, in invertebrates, by two means; the absorption of an additional photon by dark-stable M, which photoconverts M back to R, or by a multistep photochemical cycle (Byk et al., 1993; Selinger et al., 1993). The photochemical cycle begins by M phosphorylation (Mpp) at multiple sites by rhodopsin kinase (Doza et al., 1992) and consequently the binding of fly Arrestin 2 (ARR2) protein (Byk, 1993; Alloway et al., 2000), which is accelerated by Ca^{2+} influx acting via calmodulin (CaM) and the NINAC (see Section 3.3) (Liu et al., 2008). ARR2 is then phosphorylated by Ca^{2+} calmodulin-dependent kinase (CaMK II) (Matsumoto et al., 1994). Absorption of a photon, by the phosphorylated M-ARR2 complex (Mpp-ARRp), results in the release of the phosphorylated ARR2 (ARRp) from Mpp and the conversion into phosphorylated R (Rpp). Upon photoregeneration of Mpp to Rpp, Rpp is exposed to phosphatase activity by the rhodopsin phosphatase RDGC (Steele et al., 1992; Steele and O'Tousa, 1990), which converts it back to R (Byk et al., 1993; Selinger et al., 1993). Alternatively, in the dark, Mpp-ARR is internalized and degraded by a clathrin-mediated endocytosis (Byk et al., 1993; Kiselev et al., 2000).

3.2.3 PHOTORECEPTOR POTENTIAL AND PROLONGED DEPOLARIZING AFTERPOTENTIAL: A TOOL FOR DISCOVERY OF PHOTOTRANSDUCTION COMPONENTS

Failure of response termination at the stage of R activation was designated the prolonged depolarizing after (PDA) potential by Hillman, Hochstein, and Minke (Hillman et al., 1983; Minke, 2012). The PDA, like the light coincident receptor potential, arises from light-induced opening of the TRP channels in the plasma membrane. However, in contrast to the light coincident receptor potential, which quickly declines to baseline after the cessation of the light stimulus, the PDA is a depolarization that continues long after light offset (Figure 3.4) (see Hillman et al., 1983; Minke, 2012 for reviews). The PDA has been a major tool to screen for visual mutants of *Drosophila* and a powerful experimental tool to unravel phototransduction (Pak, 1991; Pak et al., 2012), because it drives the phototransduction cascade to the upper limit of its activation. The PDA is obtained due to (1) the bi-stable property of fly photopigment with separated peak absorption spectra of R and M states, in *Drosophila* this is obtained by a blue absorbing R spectrum and orange absorbing M spectrum, and (2) the low expression level of ARR2 relative to M, while ARR2 is required to terminate M activity (Byk et al., 1993; Selinger et al., 1993; Dolph et al., 1993). Thus, massive R to M photoconversion by intense blue light induces a PDA, while M to R photoconversion by intense orange light suppresses the PDA. Therefore, the major cause of termination for the active photopigment is either the absorption of an additional photon by M that converts it to R, or the binding of arrestin to M. In summary, the PDA is observed only when a considerable amount of photopigment (>20%) is converted from R to M. The larger the net amount of R to M conversion, the longer the PDA. The PDA can be depressed at any time by photopigment conversion from M to R. The degree of PDA depression depends also on the net amount of M to R conversion (Hillman et al., 1983). After the depression of maximal PDA, additional PDA can be induced immediately by R to M conversion (Figure 3.4). Indeed, the PDA screen yielded a plethora of novel and interesting visual mutants. One group of mutants exhibited a loss in several features of the PDA, and they were termed *nina* mutants, which stands for "neither inactivation nor afterpotential." Most *nina* mutants were found to be caused by reduced levels of R (Figure 3.4). The second group of PDA mutants lost the ability to produce the voltage response associated with the PDA but were still

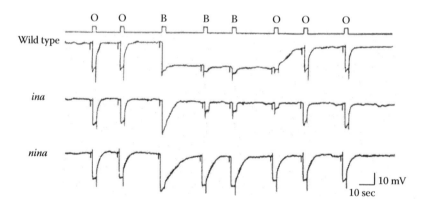

FIGURE 3.4 The prolonged depolarizing afterpotential (PDA) response of wild-type *Drosophila* and its modifications in the *nina* and *ina* mutants. *Upper trace*: ERG recordings from a wild-type white-eyed fly in response to a series of intense blue (B) and orange (O) light pulses used for induction and suppression of the PDA. This paradigm included two intense orange light pulses followed by an intense blue light pulse that converted ~80% of the Rh1 photopigment from R to M and resulted in prolonged corneal negative response that continued in the dark. Two additional intense blue lights elicited small responses that originated from the central cells (R7, R8) in which PDA was not induced, due to their UV and green absorption spectra, respectively, while the R1–6 cells were nonresponsive ("inactivated") due to maximal activation of the light-sensitive channels. The following orange light suppressed the PDA after the light is turned off. *Middle trace*: The paradigm of the upper trace was repeated in an *ina* mutant. In contrast to wild type, the response to the intense blue light declined to baseline. However, the R1–6 cells remained inactivated and additional blue lights elicited responses only in R7, R8 cells, while the following orange light removed the inactivation and allowed recovery of the R1–6 cells' response to light. *Bottom trace*: The paradigm of the upper traces was repeated in a *nina* mutant. The first blue light elicited a short PDA that quickly declined to baseline, and additional blue lights elicited responses in all photoreceptor cells and allowed additional activation of R1–6 cells by orange lights. (From Pak, W.L., *Neurogenetics, Genetic Approaches to the Nervous System,* Elsevier North-Holland, New York, 1979. With permission.)

inactivated by strong blue light, and the inactivation could be relieved by orange light. These mutants, consisting of seven allelic groups, were termed *inaA–G*, which stands for "inactivation but no afterpotential" (Figure 3.4). The *ina* mutants were found to have normal R levels but deficiencies in proteins associated with the function of the TRP channel. The *nina* and *ina* mutants have led to the identification of many of the crucial components of *Drosophila* phototransduction, some of which are novel proteins of general importance for many cells and tissues (Pak, 1991; Pak et al., 2012).

3.2.4 Light-Activated G Protein

Heterotrimeric G proteins mediate a variety of signaling processes by coupling heptahelical receptors to their downstream effectors. It has been well established in photoreceptors of several invertebrate species that photoexcited rhodopsin activates a heterotrimeric G protein (Blumenfeld et al., 1985; Fein, 1986). Direct observations of G protein–mediated chemically induced noise have suggested that there is a three- to fivefold gain at the G protein stage (Minke and Stephenson, 1985; Fein and Corson, 1979), and this was confirmed and extended by mutation analysis (Hardie et al., 2002). Direct demonstration of light-dependent G protein activity in fly eye was first demonstrated in membrane preparations using an α-^{32}P-labeled azidoanilido analog of GTP (Devary et al., 1987; Minke and Selinger, 1991). Polyacrylamide gel electrophoresis and autoradiography revealed in the blue-illuminated membranes a labeled 41-kDa protein band that was not observed when illuminated with red light (Figure 3.5a). Binding assays, which show strict dependence of GTPγS binding to the eye membrane preparation upon production of metarhodopsin with blue light (Figure 3.5b), also revealed the involvement of a G protein in fly phototransduction (Devary et al., 1987). Later studies in *Drosophila*, using genetic screens, isolated three genes encoding eye-specific G protein subunits; DG$_q\alpha$ (Lee et al., 1990), G$_q\beta_e$ (Dolph et al., 1994), and Gγ$_e$ (Schulz et al., 1999). The isolated eye-specific DG$_q\alpha$, showed ~75% identity to mouse G$_q\alpha$, known to activate phospholipase C (PLC) (Lee et al., 1990). The most direct demonstration that DG$_q\alpha$ participates in the phototransduction cascade came from studies of mutants defective in DG$_q\alpha$ which showed highly reduced sensitivity to light. In the isolated $G\alpha_q^1$ mutant, DG$_q\alpha$ protein levels are reduced to ~1%, while rhodopsin, G$_q\beta$, and PLC protein levels are virtually normal. The $G\alpha_q^1$ mutant exhibits ~1000-fold reduced sensitivity to light and slow response termination (Scott et al., 1995), thus strongly suggesting that there is no parallel pathway mediated by G proteins, as proposed for the *Limulus* eye (Dorlochter et al., 1997). Manipulations of the DG$_q\alpha$ protein level by an inducible heat-shock promoter made it possible to show a strong correlation between DG$_q\alpha$ protein levels and sensitivity to light, further establishing its major role in *Drosophila* phototransduction (Scott et al., 1995). Thus, the reason for the reduced sensitivity in G$_q\alpha$ mutants is its role as mediator of the downstream effector PLC. In addition to the reduced sensitivity to light, the $G\alpha_q^1$ mutant also shows slow response termination. This phenotype was first explained as a result of reduced Ca^{2+}-dependent receptor inactivation. Accordingly, it was suggested that in the $G\alpha_q^1$ mutant fly, the imbalance between the number of activated receptors and the light-activated currents and Ca^{2+} influx result in attenuation of receptor inactivation (Liu et al., 2008; Scott et al., 1995). A more recent study has suggested an explanation in which M/G$_q$ interaction affects M/ARR2 binding. According to this hypothesis, the reduced level of G$_q\alpha$ in the $G\alpha_q^1$ mutant fly attenuates the binding of ARR2 to the active M state, resulting in slow receptor termination and hence the slow termination of the light response (Hu et al., 2012).

The eye-specific G$_q\beta$ (G$_q\beta_e$) shares 50% amino acid identity with other Gβ homologue proteins. Two defective *Drosophila* G$_q\beta_e$ (*Gβ$_e^1$* and *Gβ$_e^2$*) mutants were isolated, exhibiting ~100-fold reduced sensitivity to light and slow response termination (Dolph et al., 1994). The reduced light sensitivity and large reduction in light-stimulated GTPγ^{35}S binding in the Gβe mutant was first interpreted as indicating a major role for G$_q\beta$ in the coupling of G$_q\alpha$ with metarhodopsin (Dolph et al., 1994). However, later studies conducted on these mutants revealed that G$_q\alpha$ is dependent on G$_q\beta$ for both membrane attachment and targeting to the rhabdomere, suggesting that the decreased light sensitivity of these mutants may result from the mislocalization of G$_q\alpha$ (Elia et al., 2005). Many studies

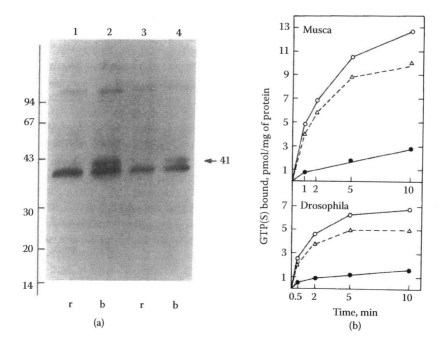

FIGURE 3.5 The first direct biochemical evidence of blue light–activated G protein in fly photoreceptors. (a) Photoaffinity labeling of light-activated G protein in *Musca* eye membranes. Eye membranes were preilluminated with either red (r) or blue (b) lights followed by incubation in binding solution containing 10 nM α^{32}-labeled azidoanilido analog of GTP for 20 minutes at 4°C in the dark. Washing and membrane absorption to nitrocellulose filter was followed by 30 second intense UV illumination followed by insertion of the filter paper into wells of 10% polyacrylamide gel, electrophoresis, and autoradiography. Lanes 1 and 2 are from untreated membranes, and lanes 3 and 4 are from membranes washed twice in order to remove GTPase activity. Strikingly, a 41 kDa band appeared after blue and not after red preillumination, and this band could be partially eliminated by treatment that reduced light-induced GTPase activity. Light-independent additional G protein that is not blue light dependent, also appeared on the gel. (b) Measurements of blue light (circles) and red light (black circles) induced binding of GTPγ^{35}S to *Musca* and *Drosophila* eye membranes (triangle represent blue-red values). (From Devary, O. et al., *Proc Natl Acad Sci U.S.A.*, 84, 6939–6943, 1987. With permission.)

have shown that G$\beta\gamma$ binds to Gα-GDP switch regions, suggesting that this interaction stabilizes the binding of GDP and suppresses spontaneous receptor-independent activation by preventing spontaneous GDP-GTP exchange (Itoh and Gilman, 1991). However, a physiologically relevant *in vivo* demonstration of this mechanism was lacking. Analysis of the stoichiometry between G$_q\alpha$ and G$_q\beta$ in *Drosophila* photoreceptors, revealed a twofold excess of G$_q\beta$ over G$_q\alpha$. Genetic elimination of the G$_q\beta$ excess by using the G$_q\beta$ heterozygote mutant (;;*Gβ_e^1/+*) led to an increase in the spontaneous activation of the visual cascade in the dark, while reestablishing the excess of G$_q\beta$ over G$_q\alpha$ in a double G$_q\alpha$, G$_q\beta$ heterozygote mutant fly (;*Gα_q^1/+;Gβ_e^1/+*) reduced the dark spontaneous activity back to its WT (wild-type) levels. The increase of dark spontaneous activity in the G$_q\beta$ heterozygote was found to be rhodopsin independent. These results suggested that G$_q\beta$ excess is essential for the suppression of dark electrical activity produced by spontaneous GDP-GTP exchange on G$_q\alpha$ (Elia et al., 2005).

Although Gγ_e has been genetically isolated and both Gγ_e hypomorph mutants and RNAi targeted against Gγ_e exist, the effect of reduced Gγ_e level has never been examined in *Drosophila* photoreceptor cells (Schulz et al., 1999). However, reduced sensitivity to light has been observed in flies overexpressing a mutated Gγ_e in which the prenylation site was modified, further establishing the participation of heterotrimeric G proteins in *Drosophila* phototransduction (Schillo et al., 2004). Lipidation, the covalent lipid modifications of proteins, play a major role in targeting heterotrimeric ($\alpha\beta\gamma$) G proteins to cellular membranes. Indeed, the three main types of lipid modifications of proteins (myristoylation, prenylation, and palmitoylation-thioacylation) are all found among different G proteins and their functional subunits (Chen and Manning, 2001). $G_q\alpha$ has two palmitoylation sites at the N-terminal end, while the $G\beta\gamma$ complex is modified by prenylation at the C-terminal of $G\gamma$. No lipid modification has been found on $G_q\beta$. However, overexpression of $G\gamma_e$ in which the prenylation site *CAAX* motif at the C-terminal was modified by replacing a cysteine with a glycine (C69G) resulted in interference with attachment of the β-subunit to the membrane (Schillo et al., 2004). These results support the notion that the peripheral membrane localization of $G_q\beta$ depends on $G\gamma_e$.

Little is known about the regulation of $G_q\alpha$ localization within specialized signaling cells. Studies using *Drosophila* show that prolonged illumination causes massive and reversible translocation of $G_q\alpha$ from the signaling compartment, the rhabdomere, to the cytosol, associated with marked architectural changes in the rhabdomere (Kosloff et al., 2003). Epistatic analysis showed that $G_q\alpha$ is necessary but not sufficient to bring about the morphological changes in the signaling organelle (Kosloff et al., 2003). Long exposure to light (light raised) followed by minutes of darkness resulted in a large reduction of the efficiency with which each absorbed photon elicited single-photon responses, while the size and shape of each single-photon response were unchanged (Frechter et al., 2007). To dissect the physiological significance of $G_q\alpha$ translocation by light, a series of *Drosophila* mutants were used. Genetic dissection showed that light-induced translocation of $G_q\alpha$ between the signaling membrane and the cytosol induced long-term adaptation. Physiological and biochemical studies revealed that the sensitivity to light depends on membrane $G_q\alpha$ levels, which were modulated either by light or by mutations that impaired membrane targeting. Thus, long-term adaptation is mediated by the movement of $G_q\alpha$ from the signaling membrane to the cytosol, thereby reducing the probability of each photon to elicit a single-photon response (bump) (Frechter et al., 2007). However, the molecular mechanism of light-dependent $G_q\alpha$ translocation is still unknown.

3.2.5 Role of PLC in Light Excitation and Adaptation

Evidence for a light-dependent $G_q\alpha$-mediated PLC activity in fly photoreceptors came from combined biochemical and electrophysiological experiments. These experiments conducted on membrane preparations and intact *Musca* and *Drosophila* eyes showed coupling of photoexcited rhodopsin to phosphatidylinositol 4,5-biphosphate (PIP_2) hydrolysis (Devary et al., 1987). They furthermore showed that light-activated PIP_2 hydrolysis was inhibited in the *norpA* mutant allele, which suppressed the response to light (Selinger and Minke, 1988). In this study illumination and $G_q\alpha$-dependent activation result in accumulation of $InsP_3$ and inositol-bisphosphate ($InsP_2$), derived from PIP_2 hydrolysis by PLC (Devary et al., 1987).

The key evidence for the participation of PLC in visual excitation of the fly was achieved by the isolation and analysis of *Drosophila* PLC gene, designated no receptor potential A (*norpA*). The *norpA* mutant has long been a strong candidate for a transduction defective mutant because of its drastically reduced receptor potential. The *norpA* gene encodes a β-class PLC, predominately expressed in the rhabdomeres, which has high amino acid homology to a PLC extracted from bovine brain (Bloomquist et al., 1988). Transgenic *Drosophila*, carrying the *norpA* gene on null *norpA* background, rescued the transformant flies from all the physiological, biochemical, and morphological defects, which are associated with the *norpA* mutants (Shortridge et al., 1991). The *norpA* mutant thus provides essential evidence for the critical role of inositol-lipid signaling in phototransduction, by showing that no excitation takes place in the absence of functional PLC. However, the events required for light excitation downstream of PLC activation remain unresolved.

In general, the cytoplasmic GTP concentration in cells is much higher than that of GDP, making the inactivation process of $G_q\alpha$ by hydrolysis of $G_q\alpha$ -GTP to $G_q\alpha$ -GDP unfavorable. In order to accelerate the GTPase reaction and terminate $G_q\alpha$ and PLC activities, specific GTPase-activating proteins (GAPs) exist. *In vitro* studies of mammalian PLCβ1 reconstituted into phospholipid vesicles with recombinant M1 muscarinic receptor and $G_{q/11}$ (Berstein et al., 1992) had shown that upon receptor stimulation the addition of PLCβ1 increases the rate at which G_q hydrolyses GTP by three orders of magnitude, suggesting its action as GAP. A reduction in the levels of PLC in mutant flies affects the amplitude and activation kinetics of the light response (Pearn et al., 1996), but also mysteriously slows response termination. Biochemical and physiological studies conducted in *Drosophila* have revealed the requirement for PLC in the induction of GAP activity *in vivo*. Using several *Drosophila* *norpA* mutant flies, a high correlation between PLC protein level, GAP activity, and response termination was observed (Cook et al., 2000). The virtually complete dependence of GAP activity on PLC (Cook et al., 2000) provides an efficient mechanism for ensuring the one photon, one bump relationship (Yeandle and Spiegler, 1973), which is critical for the fidelity of phototransduction at dim light. The apparent inability to hydrolyze GTP without PLC ensures that every activated G protein eventually encounters a PLC molecule and thereby produces a response by the downstream mechanisms. The instantaneous inactivation of the G protein by its target, the PLC, guarantees that every G protein produces no more than one bump (Cook et al., 2000). This apparently complete dependence of GTPase activity on its activator PLC, in flies, differs from the partial dependence of GTPase activity on additional GAP factors in vertebrate phototransduction (Chen et al., 2000). Vertebrate phototransduction depends on specific GAP proteins named regulators of G protein signaling (RGSs) (Arshavsky and Pugh, 1998). Accordingly, genetic elimination of RGS proteins reduces and slows down GAP activity and leads to slow termination of the light response (Chen et al., 2000).

PLC activity is known to be regulated by Ca^{2+} (Rhee, 2001). It has been shown that Ca^{2+} is bound at the catalytic site of PLC and is required as a co-factor for the catalytic reaction (Essen et al., 1996). These studies show that the positive charge of Ca^{2+} is used to counterbalance local negative charges formed in the active site during the course of the catalytic reaction. Accordingly, Ca^{2+} performs electrostatic stabilization of both the substrate and the transition state, thus providing a twofold contribution to lower the energy of the enzyme reaction (Essen et al., 1997). In *Drosophila*, both *in vitro* (Running Deer et al., 1995) and *in vivo* (Hardie, 2005) measurements revealed Ca^{2+} dependence of PLC activity. This activity shows a bell-shaped dependency of NORPA catalytic activity as a function of Ca^{2+} concentration ($[Ca^{2+}]$), with maximal basal activity in the range of 10^{-7}–10^{-5} M $[Ca^{2+}]$. This complex dependency affects both excitation and adaptation of the photoresponse. While at physiological conditions cellular Ca^{2+} levels are sufficient for the activation of NORPA, it was demonstrated that highly reduced Ca^{2+} levels eliminated excitation completely (Minke and Agam, 2003). Furthermore, it was shown that during the light response, rhabdomere Ca^{2+} concentration can reach mM levels (Oberwinkler and Stavenga, 2000; Postma et al., 1999). At this high Ca^{2+} level, NORPA activity is attenuated raising the possibility of physiological relevance (Hardie, 2005). Later studies have shown that the inhibition of NORPA activity by high Ca^{2+} concentration participates in the mechanism of fast light adaptation and prevents depletion of membrane PIP_2. Accordingly, the high Ca^{2+} concentration, reached at the peak response, attenuates NORPA activity preventing PIP_2 depletion and adapting the cells by reducing excitation (Gu et al., 2005). However, this hypothesis still needs to be substantiated by specific mutations of the NORPA, decoupling its activity from its Ca^{2+} regulation.

3.2.6 Phosphoinositide (PI) Cycle

In the phototransduction cascade of *Drosophila*, light triggers the activation of PLCβ. This catalyzes the hydrolysis of the membrane phospholipid PIP_2 into water-soluble $InsP_3$ and membrane-bound diacylglycerol (DAG) (Berridge, 1993). Genetic elimination of the single *Drosophila* $InsP_3$ receptor had no effect on light excitation (Acharya et al., 1997; Raghu et al., 2000b), questioning the role of $InsP_3$ in phototransduction (but see Kohn et al., 2015). The continuous functionality of the photoreceptors during illumination is maintained by rapid regeneration of PIP_2 in a cyclic enzymatic pathway (Figure 3.6).

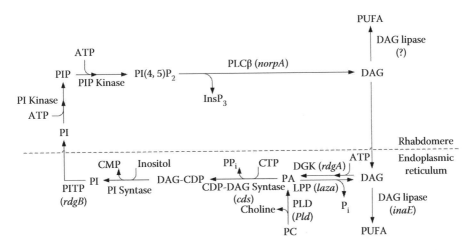

FIGURE 3.6 The phosphoinosite (PI) cycle. In the phototransduction cascade, light triggers the activation of phospholipase Cβ (PLCβ, encoded by *norpA*). This catalyzes hydrolysis of the membrane phospholipid PI(4,5) P_2 (PIP$_2$) into InsP$_3$ and DAG. DAG is transported by endocytosis to the endoplasmic reticulum and inactivated by phosphorylation converting it into phosphatidic acid (PA) via DAG kinase (DGK, encoded by *rdgA*) and to CDP-DAG via CDP-DAG synthase. Subsequently, CDP-DAG is converted into phosphatidyl inositol (PI), which is transferred back to the microvillar membrane, by the PI transfer protein (encoded by *rdgB*). PIP and PIP$_2$ are produced at the microvillar membrane by PI kinase and PIP kinase, respectively. PA can also be converted back to DAG by lipid phosphate phosphohydrolase (Lpp, encoded by laza). PA is also produced from phosphatidyl choline (PC) by phospholipase D (PLD). DAG is also converted in two enzymatic stages, one of them is by DAG lipase (encoded by *inaE*), into polyunsaturated fatty acids (PUFAs). (From Katz, B. and B. Minke, *Front Cell Neurosci.*, 3, 2, 2009. With permission.)

The phospholipid branch of the phosphatidylinositol (PI) cycle following PLC activation, begins by DAG transport through endocytosis to the endoplasmic reticulum (submicrovillar cisternae [SMC]), and subsequently, inactivation by phosphorylation into phosphatidic acid (PA) via DAG kinase (DGK; encoded by the *rdgA*) (Masai et al., 1993; Masai et al., 1997) and to cytidine diphosphate diacylglycerol (CDP-DAG) via CDP-DAG synthase (encoded by the *cds*) (Wu et al., 1995). Both retinal degeneration A (RDGA, the gene product of the retinal degeneration A gene) and CDS are located in extension of the smooth endoplasmic reticulum called SMC (Figure 3.6). PA can be reconverted back to DAG by lipid phosphate phosphohydrolase (Kwon and Montell, 2006) (LPP, also designated phosphatidic acid phosphatase, PAP, encoded by *laza*) (Kwon and Montell, 2006; Garcia-Murillas et al., 2006) or produced from phosphatidylcholine (PC) by phospholipase D (PLD, encoded by *Pld*) (LaLonde et al., 2005). DAG is also hydrolyzed by DAG lipase (encoded by *inaE*) (Leung et al., 2008) predominantly localized outside the rhabdomeres, into polyunsaturated fatty acids (Figure 3.6). Subsequently, CDP-DAG is converted into PI, which is transferred back to the microvillar membrane, by the PI transfer protein (PITP; encoded by the *rdgB*) (Vihtelic et al., 1991) located in the smooth endoplasmic reticulum. PIP and PIP$_2$ are produced at the microvillar membrane by PI kinase and PIP kinase, respectively (Raghu et al., 2012).

Mutations in most proteins of the PI pathway result in retinal degeneration. The retinal degeneration phenotypes in these mutants are thought to occur due to Ca^{2+} influx through the light-activated TRP and TRPL channels, making the PI pathway crucial for understanding phototransduction and TRP channel activation. Although it is possible to partially rescue the degeneration phenotypes by genetically eliminating the TRP channels (Raghu et al., 2000a), it is unclear whether these mutations promote channel opening directly or through indirect changes in the photoreceptor leading to their activation. Recent detailed studies by Raghu and colleagues (reviewed in Raghu et al., 2012) have shown that the principles of PI signaling seem largely conserved between *Drosophila* and other metazoan models. They concluded that with a fully sequenced, relatively compact genome and with sophisticated molecular genetic technology, studies in *Drosophila* will be influential in understanding the details and physiological implications of PI signaling in general (Raghu et al., 2012).

3.2.7 SINGLE-PHOTON RESPONSES AND SPONTANEOUS DARK BUMPS

In *Drosophila* photoreceptors, light activates PLC, which hydrolyzes PIP_2 into DAG and $InsP_3$ promoting the opening of the TRP and TRPL channels and resulting in membrane depolarization. The absorption of a single photon promotes the opening of ~15 TRP channels generating currents of ~12 pA (at a holding potential of –70 mV) (Henderson et al., 2000). At total darkness, spontaneous unitary events of ~3 pA, called dark bumps, are also observed at a rate of ~2 per second. The phenomenon of discrete current fluctuation as a result of dark bumps and single-photon absorption raises many questions on the underlying transduction mechanisms, channel regulation, and anatomical structures that enable it.

3.2.7.1 Spontaneous Dark Bumps

The dark bumps originate from spontaneous GDP to GTP exchange on the $G_q\alpha$ subunits. This has been demonstrated by showing that spontaneous bump frequency is highly correlated with the level of membrane attached $G\alpha_q$, by eliminating dark bumps in the $G\alpha_q{}^1$ mutant fly (Hardie et al., 2002; Katz and Minke, 2012), by the high rate of spontaneous bumps in the $G\beta_e{}^1/+$ heterozygote mutant fly (Elia et al., 2005), and by the reduction of spontaneous bumps as a result of cellular GDP elevation (Katz and Minke, 2012). The spontaneous bumps in *Drosophila* have a mean amplitude of ~3 pA under physiological conditions, which correspond to ~4 open TRP channels at the peak amplitude (Hardie et al., 2002; Katz and Minke, 2012). However, their amplitude can be enhanced to ~9 pA by removing Mg^{2+} from the extracellular solution, which when present, reduces the single-channel conductance of the TRP channel by an open-channel block mechanism (Hardie and Mojet, 1995; Parnas et al., 2007). The waveform and time to peak of the spontaneous bumps are only slightly faster than that of the light-induced bump supporting the notion that the kinetic parameters of both dark and quantum bumps are determined downstream of PLC (Figure 3.7) (Katz and Minke, 2012). A surprising feature of the spontaneous bumps is their sensitivity to cellular ATP level. Accordingly, reduction in cellular ATP results in enhancement of both bump rate of occurrence (frequency) and amplitude (Katz and Minke, 2012). These two phenotypes are also observed in the $rdgA^1/+$ mutant fly with reduced DAG kinase activity (Raghu et al., 2000a). Hence, accumulation of putative PLC products due to the reduction in cellular ATP or reduced DAG kinase activity by mutations facilitates spontaneous bump production and affects channel opening. Several mutants have been shown to alter dark bump frequency— these are *retinophilin* and *ninaC*—however, the underlying mechanism is still unknown (Chu et al., 2013).

In general, the dark spontaneous bumps are considered as the noise of the transduction system constituting a limiting factor for reliable single-photon detection that should be reduced to a minimum (Katz and Minke, 2012).

3.2.8 SINGLE-PHOTON RESPONSES: QUANTUM BUMPS

Dim light stimulation induces discrete voltage (or current) fluctuations in most invertebrate species, which are called quantum bumps (Yeandle and Spiegler, 1973). Each bump is assumed to be evoked by the absorption of a single photon and obeys a stochastic process described by Poisson statistics (Yeandle and Spiegler, 1973). In *Drosophila* the mean amplitude of the light-induced bumps is ~12 pA under physiological conditions, which is approximately fourfold larger than that of the dark spontaneous bumps. This has led to the suggestion that light-induced bumps are a result of synchronous activation of ~3–5 $G_q\alpha$ molecules by a single activated rhodopsin, which synchronously activates ~5 PLC molecules and promotes the opening of ~15 TRP channels (Hardie et al., 2002).

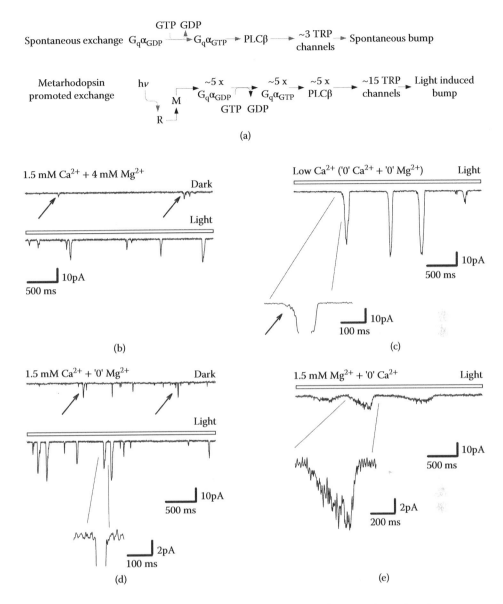

FIGURE 3.7 Initiation and properties of single-photon responses (quantum bumps) and spontaneous dark bumps. (a) A scheme of the initial stages of phototransduction underlying dark bump production (top) and quantum bump production (bottom). (b) A demonstration of dark bumps (top, arrows) and quantum bumps (bottom) under physiological conditions, obtained by whole cell recordings. (c) Quantum bumps under low Ca^{2+} (few μM) conditions; inset: magnification of the initial phases of the bump. Note the two "on" phases: an initial slow phase followed by a faster phase. (d) Dark bumps (top, arrows) and quantum bumps (bottom) under low Mg^{2+} conditions. Note that bump sizes are larger compared to dark and quantum bumps in (b); inset: magnification of a quantum bump. Note the single fast "on" phase. (e) Quantum bumps under low Ca^{2+} in the presence of 1.5 mM Mg^{2+}; inset: magnification of a quantum bump. Note the single slow "on" phase.

The light-induced bumps vary in latency, time course, and amplitude for identical stimulation with half-width duration of ~20 ms and a characteristic variable latency of between 20 and 100 ms.

The discrete nature of the bumps is not due to the quantized nature of light. This has been demonstrated by the application of nonquantized stimulus such as GTPγS (nonreversible activation of G$_q$α), eliciting quantum bump–like events (Fein and Corson, 1981). Further genetic evidence for the discrete nature of the bumps came from two *Drosophila* mutants, *arr2* and *ninaC*, in which rhodopsin inactivation is attenuated. These mutants respond to the absorption of a single photon with a train of bumps, which do not overlap. This phenotype was shown to be the consequence of failure in the rhodopsin inactivation process and by the assumption that a "refractory period" exists in bump production (Liu et al., 2008; Dolph et al., 1993; Song et al., 2012). Hence, continuous rhodopsin activation elicits trains of discrete events.

A detailed study in *Limulus* and *Locusta* photoreceptors has indicated that the latency of the bump is not correlated with the bump waveform and amplitude, thus strongly suggesting that the triggering mechanism of the bump arises from different molecular processes than that determining the bump waveform (Dorlochter and Stieve, 1997; Howard, 1983). This finding was further confirmed by experiments conducted in *Drosophila*, demonstrating that in *norpA*[H43] mutant fly, in which PLC catalytic activity is highly reduced, the bump triggering process is attenuated, while little effect is observed on bump waveform (Katz and Minke, 2012). However, both the triggering process and bump waveform are highly modulated by Ca^{2+}. At low external Ca^{2+} (100 μM), bump kinetics becomes slow with two distinct "on" phases (an initial slow phase) accompanied by a faster phase (Henderson et al., 2000), while little effect is observed on bump frequency. Further reduction in Ca^{2+} (10 μM) reduces bump kinetics to a single slow phase, and the excitation efficiency of bump production is reduced (Henderson et al., 2000). These results have led to the hypothesis that Ca^{2+} participates in both positive and negative feedback of bump production (Henderson et al., 2000) and bump triggering (Katz and Minke, 2012), but the molecular targets of Ca^{2+}-dependent positive and negative feedback in bump production are still unknown.

The single photon–single bump relationship, typical for *Drosophila* photoreceptors, requires that each step in the visual cascade must have both "turn on" and "turn-off" mechanisms (Figure 3.7). This principle has been demonstrated for rhodopsin which is inactivated by the binding of arrestin and for $G_q\alpha$ and PLC, which are inactivated by the GAP activity of PLC and promote dissociation of $G_q\alpha$ from PLC (Byk et al., 1993; Dolph et al., 1993; Cook et al., 2000). However, the still unknown mechanism by which TRP channels are inactivated, although highly dependent on PLC activity, PKC (protein kinase C; encoded by *InaC*), and Ca^{2+} concentration, may require an additional molecular component (Gu et al., 2005; Parnas et al., 2007; Parnas et al., 2009). The functional advantage of such a transduction mechanism is obvious; it produces a sensitive photon counter, very well suited for both the sensitivity and the temporal resolution required by the visual system.

A bump is thought to represent the activation of most of the light-activated channels within a single microvillus, thus representing a functional anatomical element. This has been proposed by comparing the estimated channel number underlying each bump (calculated from the division of the single-channel conductance with the mean bump amplitude) (Henderson et al., 2000) and the estimated number of channels in each microvillus (Huber et al., 1996). Moreover, it has been shown that reduced cellular ATP increases low-amplitude bumps of several mutant flies but does not significantly increase the amplitude of WT bumps (Hardie et al., 2002), demonstrating an upper limit for bump size. Further demonstration has been given by showing that conditions that highly reduce bump production by PLC mutation cannot be overcome by increasing the light intensity to a level that activates all the microvilli at once (Katz and Minke, 2012). Hence, the biochemical reactions that promote channel openings do not sum up across different microvilli.

A recent study has presented a quantitative model explaining how bumps emerge from stochastic nonlinear dynamics of the signaling cascade. Three essential "modules" govern the production of bumps in this model: (1) an "activation module" downstream of PLC but upstream of the channels, (2) a "bump-generation module" including channels and Ca^{2+}-mediated positive feedback, and (3) a Ca^{2+}-dependent "negative-feedback module." The model shows that the cascade acts as an "integrate and fire" device conjectured formerly by Henderson and Hardie (Henderson et al., 2000), much like the generation of the action potential. The model explains both the reliability of bump formation, low background noise in the dark, were able to capture mutant bump behavior and explains the dependence on external Ca^{2+}, which controls feedback regulation (Pumir et al., 2008).

This view was extended by showing experimentally that a critical level of PLC activity (threshold) is necessary to trigger bump production (Katz and Minke, 2012). Hence, the synchronization of ~5 PLC by rhodopsin results in sufficient PLC product to surpass the threshold in bump production, while single spontaneous $G_q\alpha$ molecules activating single PLC molecules have low probability of surpassing this threshold and producing a bump. This mechanism of spontaneous dark bump suppression maintains the fidelity of single-photon detection. Moreover, it was demonstrated that the Ca^{2+}-dependent bump kinetics are not due to effects on the catalytic activity of PLC but rather to actions downstream, probably at the channel level (Katz and Minke, 2012).

3.3 EVOLUTION OF TRPC CHANNELS IN PHOTORECEPTOR CELLS

The essential role of *Drosophila trp* in vision was discovered by the isolation of a spontaneous mutation and large-scale genetic screening (see Introduction). The only other protostome where the functional roles of TRPC channels have been studied by genetic screening is the blind nematode worm, *Caenorhabditis elegans*, where mutants of *trp-1* and *trp-2* are superficially WT, and mutants of *trp-3* are infertile (Xiao and Xu, 2009). Thus progress in understanding the functions of TRPC channels in vision across species has been very limited. It is not known whether phototransduction in microvillar photoreceptors in all species is mediated solely by orthologs of TRP and TRPL, or whether the mechanism that couples phospholipase C to TRPC channel activation has evolved due to selection for speed, sensitivity, and accuracy. Certainly, there are differences in the ionic permeability of the light-sensitive current and in the physiology of its activation, leading to proposals that the molecular identity of the light-sensitive channels in microvillar photoreceptors and their activation mechanism may vary across species (Bacigalupo et al., 1991; Chen et al., 2001; Garger et al., 2004; Fein and Cavar, 2000).

3.3.1 COMPLEX PHYLOGENY OF TRPC CHANNELS IN INVERTEBRATE SPECIES

Identification of genes orthologous to *Drosophila trp* and *trpl* may be helpful in identifying candidate genes in other species that mediate phototransduction. This is because orthologous genes are more likely to perform similar functions across species (Conant and Wolfe, 2008; Altenhoff et al., 2012; Rogozin et al., 2014; but see Nehrt et al., 2011). In order to identify orthologs, TRP family phylogenetic trees containing representative sequences from protostome genomes have been constructed (Matsuura et al., 2009; Rivera et al., 2010; Peng et al., 2014; Speiser et al., 2014). The tree shown in Figure 3.8a is an extension of those efforts, displaying 174 TRPC protein sequences mined from 43 animal genomes and transcriptomes. The tree predicts four major clades of protostome TRPC channels, three of which have deuterostome sister clades.

3.3.2 TRPC CHANNELS FROM TWO MAJOR CLADES SHOWN TO BE EXPRESSED IN MICROVILLAR PHOTORECEPTORS

Drosophila TRP, TRPL, and TRPγ fall within just one major clade, *trpγ-like* (Figure 3.8b), which also contains *c-elegans* TRP-2 and is a sister to that containing mammalian TRPC1, TRPC4, and TRPC5 (Venkatachalam and Montell, 2007; Denis and Cyert, 2002). The *trpγ-like* clade contains sequences from all of the protostome genomes sampled for the tree and comprises a set of orthologs of TRP and TRPL. Although *Drosophila melanogaster* is the only species for which there is functional evidence for members of the *trpγ*-like clade having a role in phototransduction, the expression of orthologs of *trp* or *trpl* has been found to be enriched in microvillar photoreceptors in the eyespots of the platyhelminth, *Schmidtea mediterranea* (*trpC-2*) (Lapan and Reddien, 2012), the simple eyes of the mollusc, *Loligo forbesii* (*strp*) (Monk et al., 1996), and the ventral eye of the chelicerate, *Limulus polyphemus* (*Lptrpl*; R. Payne, personal communication). The localization of a

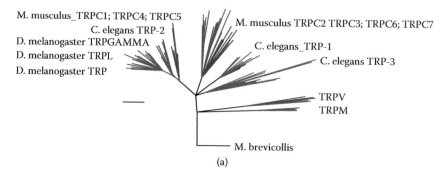

(a)

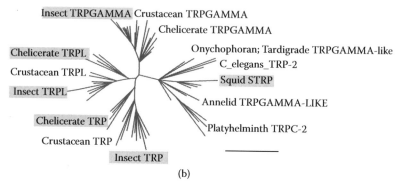

(b)

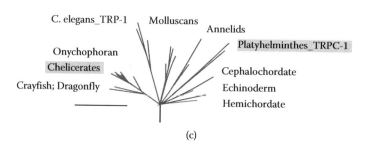

(c)

FIGURE 3.8 **(See color insert.)** Evolution of TRPC channels: The complex phylogeny of TRPC channels. (a) Phylogram of TRPC, TRPV, and TRPM channel protein species mined from 43 animal genomes or transcriptomes released to GenBank. (b) Expanded phylogram of the clade, *trpγ-like*, containing *Drosophila* TRP and *C. elegans* TRP-2. (c) Expanded phylogram of the clade, *trp-1-like,* containing *C. elegans* TRP-1. Positions of proteins referred to in the text and of proteins belonging to major phyla of protostome (red) and deuterostome (blue) lineages are indicated. The protein sequences were aligned using MAFFT (Katoh et al., 2002; Katoh et al., 2005). Trees were calculated using the maximum likelihood method, GARLI (Bazinet et al., 2014; Zwickl, 2006), running on CIPRES (Miller et al., 2010), and plotted using Dendroscope (Huson and Scornavacca, 2012). Nodes with less than 90% bootstrap support were collapsed into polytomies. The scale bars indicate one substitution per site.

trpγ-like channel, *trpC-2*, to the eyespot of a platyhelminth is illustrated in Figure 3.9a. Orthologs of *trp* and *trpl* therefore constitute a set of candidate genes for future functional studies of light-sensitive channels with a role in vision in a very broad range of protostome species.

Trp-1-like (Figure 3.8c) is a second clade of TRPC channels whose expression pattern provides evidence for a possible role in phototransduction, although physiological evidence is lacking. *Trp-1-like* clade members are limited to protostome species basal to the neoptera and to deuterostome species basal to vertebrates. Members of the *trp-1-like* clade are expressed in microvillar photoreceptors

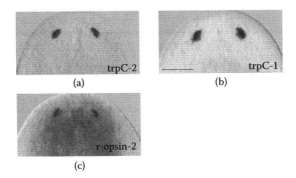

FIGURE 3.9 **(See color insert.)** *In situ* hybridization patterns of probes directed against trpC-2 (a), trpC-1 (b), and rhabdomeral r-opsin mRNA sequences in the head of the platyhelminth, *S. mediterranea.* Densest areas of r-opsin localization indicate the photoreceptor cell bodies. (From Lapan, S.W. and P.W. Reddien, *Cell Rep.,* 2, 294–307, 2012. With permission.) Scale bar indicates 200 μm.

in the platyhelminth *S. mediterranea* (*trpC-1*) (Lapan and Reddien, 2012) (Figure 3.9b) and in *Limulus* ventral photoreceptors (Bandyopadhyay and Payne, 2004), although whether they have a function in phototransduction is unknown. Localizing members of this clade to the microvillar photoreceptors of more species as well as demonstrating a functional role in at least one of those species may open up new avenues of research. Deuterostome members of this clade may also be of interest as potential mediators of the response of microvillar photoreceptors in the cephalochordate, amphioxus (Gomez et al., 2009; Angueyra et al., 2012; Ferrer et al., 2012; Pulido et al., 2012).

The gene trees of Figure 3.8 provide evidence of an evolving palette of TRPC channel subtypes available to be coupled to phototransduction in different species. For example, *trp, trpl,* and *trpγ* evolved following a double duplication of an ancestral *trpγ-like* gene at the base of the arthropods, accompanying the evolution of the compound eye. Following these duplications, *trp, trpl,* and *trpγ* may have further evolved to mediate phototransduction and other sensory processes to differing extents across species (French et al., 2015). In addition, *trp-1-like* channels appear to have been lost in neopterous insects, accompanying the evolution of fast, complex flight behaviors to track mates and capture prey.

3.3.3 A Variety of Putative Pore Selectivity Filter Sequences Have Evolved

One possible reason for the duplication and loss of channel clades within phyla may be the adaptive significance of channel ionic selectivity. In microvillar photoreceptors of several species, light-induced elevation of intracellular calcium (Ca^{2+}_i) regulates the amplification and speed of phototransduction. The relative contributions of light-induced Ca^{2+} release from intracellular stores versus Ca^{2+} influx from the extracellular space appear to vary across species (Peretz et al., 1994a; Fein and Cavar, 2000; Gomez and Nasi, 1996; Ziegler and Walz, 1990). The expression of *TRPC* channel clade members with different ionic selectivities could, in principle, provide a mechanism for this variation. An aspartate, D621, in the channel selectivity filter of *Drosophila* TRP is critical for that channel's high Ca^{2+} selectivity (Liu et al., 2007). Thus, it is interesting that the acidity of the amino acid residue that aligns with *Drosophila* TRP D621 varies across channel clades (Figure 3.10), the consensus being acidic (aspartate) in insect TRP, insect TRPGAMMA, and in the *trpγ-like* channels of species basal to the arthropods but neutral in most insect TRPL channels (glycine) (Liu et al., 2007) and neutral or basic in members of the *trp-1-like* clade (proline or lysine). By expressing a different mix of clade members, these substitutions may allow an adjustment of relative light-induced Ca^{2+} fluxes in the photoreceptors of different species.

In summary, the gene family of protostome TRPC channels is as complex as that of the deuterostomes, with the potential to couple clade members with different ionic selectivity and gating

		T	L F/L W	S	L		F G/S	I/L/V	I/T	P/Q	I/P	☐	☐
All	TRP-1-like	T	L F/L W	S	L		F G/S	I/L/V	I/T	P/Q	I/P	☐	☐
Basal	TRPGAMMA-like	S/T	L Y W	A/S	I		F/Y G	L	I/V	☐	L	☐/T	☐/☐
Arthropod	TRPGAMMA	T	L F W	A	A/S/V		F G	L	I/V	☐	L	☐	☐/S
Arthropod	TRP	S	L F W	A	S		F G	L	V	☐	L	T	S
Arthropod	TRPL	S	L F W	A	S		F G	L/M	I/V	☐/G	I/L	☐/S	☐/☐/S

D621

_____   – – – –   _____
          Pore helix   selectivity filter

FIGURE 3.10 (See color insert.) Aligned consensus sequences of portions of the pore loop and selectivity filter of selected TRPC channel groups included in the trees of Figure 3.8. Inclusion in the consensus requires more than 20% presence of an amino acid at that position. Amino acid residue symbols are colored according to Lesk (2008); small (yellow); hydrophobic (green); polar (magenta); negatively charged (red); positively charged (blue).

properties to phototransduction. While little is currently known and even the channel sequence and phylogeny are preliminary, the abundance of predicted TRPC channel sequences now available and improved techniques for heterologous expression, editing, and targeted knockdown of genes in nonmodel organisms may lead to a future understanding of the evolution of phototransduction mechanisms with possible insight into the still-unknown gating mechanism of TRP channels in all physiologically investigated species.

3.4 PROPERTIES AND ORGANIZATION OF TRP CHANNELS

3.4.1 GENERAL STRUCTURAL FEATURES OF TRP CHANNELS

The amino acid sequence of TRP channels in general, distantly relates them to the superfamily of the voltage-gated channel proteins, particularly to the voltage-gated K^+ channels and the bacterial Na^+ channel (Kelley and Sternberg, 2009). On the basis of their homology to voltage-gated K^+, bacterial Na^+ and cyclic nucleotide–gated (CNG) channels, crystal and cryo-EM solved structures a tetrameric assembly of TRP channels has been suggested (Liao et al., 2013; Saotome et al., 2016). The tetrameric channels may be composed of four identical subunits (homomultimers) or of different subunits (heteromultimers). Each subunit contains cytosolic N- and C-terminals and six transmembrane segments (S1–S6). The ion-conducting pore is formed by S5, the S5–S6 linker, and S6 (Figure 3.11). The voltage-sensing domain of voltage-gated channels, which is located at the S4 transmembrane domain and contains positive charged residues, is missing in the TRP channels (reviewed in Minke and Cook, 2002). The N- and C-terminals of TRP channels interact with various proteins, lipids, and signaling molecules and vary greatly between the different members of the TRP superfamily. There are, however, several common domains to most TRP channels (Figure 3.11):

1. *Ankyrin repeats:* The ankyrin repeat is a 33-residue motif that mediates specific protein-protein interactions with a diverse repertoire of macromolecular targets (Sedgwick and Smerdon, 1999). In TRP channels they are located at the N-terminal. The number of ankyrin repeats varies among the different TRP channels ranging from four repeats in the case of the *Drosophila* TRP, TRPL, and TRPC1(Minke and Cook, 2002); six repeats in the TRPV subfamily (Jin et al., 2006; Schindl and Romanin, 2007); 14–18 repeats for TRPA1 (Story et al., 2003); and 28 repeats for NompC (Walker et al., 2000; Sidi et al., 2003). The ankyrin repeat domain was proposed to participate in channel assembly; however, an interesting hypothesis for a functional role of long ankyrin chains (e.g., for the TRPA1) is that this repeated structure forms the gating spring of mechanoreceptors (Corey et al., 2004; Howard and Bechstedt, 2004). It was shown that the ankyrin repeats of TRP channels can bind ATP, CaM, and PIP_2. The role of the ankyrin repeats in the TRP channels is not clear. They have been suggested to function as dimerization domains (Lepage et al., 2009), but isolated

ankyrin repeats from TRP channels behave as monomers in every biochemical assay (Jin et al., 2006; Phelps et al., 2010). For TRPV1 the ankyrin repeat domain was shown to bind ATP and induces channel desensitization in the presence of the agonist capsaicin (Lishko et al., 2007) and channel activation in the presence of allicin (Salazar et al., 2008). In the case of TRPA1, which has 14 ankyrin repeats, the use of point mutations (Jabba et al., 2014) and swapping of ankyrin repeats from different species (Cordero-Morales et al., 2011) has established a role of ankyrin repeats in thermosensitivity. Whether this holds true for other TRP channels is currently unknown. Mutations in the ankyrin repeats of TRPV or TRPC channels are associated with genetic diseases, for example, TRPV4 with Charcot-Marie-Tooth disease type 2C (Landoure et al., 2010) and TRPC6 with familial focal segmental glomerulosclerosis (Winn et al., 2005). All known disease-causing TRP-channel mutations induce higher basal channel activity by an unknown mechanism. Although the structure of several mammalian TRPV ankyrin repeat domains is known, there is no knowledge about the structural and biochemical properties of TRPC ankyrin repeat domains. In the case of TRPM, TRPP, and TRPML subfamilies, there are no known ankyrin repeats in either the N- or the C-termini (Venkatachalam and Montell, 2007). Hence, the function of the ankyrin repeat domain in TRP channels is still an open question.

2. *Coiled-coil motif*: The hallmark structural feature is a heptad repeat, denoted $(abcdefg)_n$, with hydrophobic residues at the "a" and "d" positions forming a nonpolar stripe along the helical surface that is used for multimerization (Crick, 1952). The identities of the "a" and

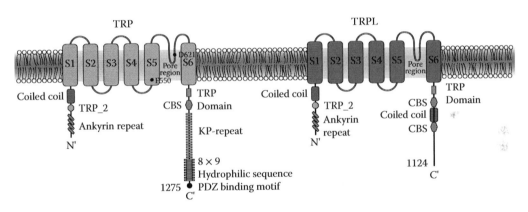

FIGURE 3.11 Structural features of the *Drosophila* TRP and TRPL channels. The *trp* and *trpl* genes encode a 1275– and 1124–amino acid long protein, respectively. The TRPL channel contains one to three putative CC domains and two CaM-binding sites (CBS) at the C-terminal of the channel. One of these is unconventional in the sense that it can bind CaM in the absence of Ca^{2+}. The C-terminus fragment of the TRP sequence also has been reported to bind CaM in a Ca^{2+}-dependent manner. Both channels have a designated TRP domain located adjacent to the S6 with a EWKFAR motif found in many members of the TRP family. At the C-terminal region of the TRP there is a proline-rich sequence with 27 KP repeats, which overlap with a multiple repeat sequence, DKDKKP (G/A) D termed 8 × 9. Such proline-rich motifs occur widely and are predicted to form a structure involved in binding interactions with other proteins, including cytoskeletal elements such as actin. This region is unique to TRP and has not been found in any other member of the TRP family. The last 14 amino acids in the C-terminal of TRP are essential for the binding of INAD and form a PDZ binding domain. The N-terminal regions of the TRP and TRPL proteins contain four ankyrin repeats and a CC domain. Both domains are believed to mediate protein-protein interactions and function in channel assembly. The N-terminal regions also contain a TRP_2 domain with a largely unclear function, predicted recently as involved in lipid binding and trafficking. Replacement of Asp621 (D621) with glycine or asparagine reduces Ca^{2+} permeability of the TRP channel. Substitution of Phe550 to isoleucine (F550I) forms a constitutively active TRP channel leading to extremely fast light-independent retinal degeneration. (From Katz, B. and B. Minke, *Front Cell Neurosci.*, 3, 2, 2009. With permission.)

"d" amino acids provide the dominant feature that determines whether a given coiled-coil (CC) helix will associate into a two-, three-, four-, or five-stranded bundle (Harbury et al., 1993; Malashkevich et al., 1996). The CC domain has been suggested to participate in channel assembly and channel activation (Gaudet, 2008; Lepage et al., 2006). The function and structures of the CC domain of TRP channels is still not clear.

3. *TRP domain*: The TRP domain is a highly conserved 23–25 amino acid region, located at the C-terminal adjacent to the S6. The TRP box is highly conserved among the TRPV, TRPC, TRPN, and TRPM subfamilies. In TRPM8 the TRP box was suggested to mediate PIP_2 sensitivity (Rohacs et al., 2005), while in TRPV1 it was proposed to participate in channel assembly and determine the activation energy needed for channel opening (Garcia-Sanz et al., 2007). The recently solved atomic structure of rat TRPV1 suggests that the TRP domain assumes an α-helical structure that runs parallel to the inner leaflet of the membrane and functions as an integrator of various channel domains facilitating allosteric coupling (Cao et al., 2013b). However, it is unclear if the role of the TRP domains of the TRPV subfamily could be generalized for all members of the TRP family.

4. *The calmodulin (CaM) binding site (CBS domain)*: The number of CBS sites varies among the different TRP channels, ranging from one for the *Drosophila* TRP, to three for the TRPC4. There are CBS domains at the N- and C-termini. All TRPCs and some of the TRPVs have CBS domains (Lepage et al., 2006). These CBS domains are suggested to mediate Ca^{2+}-dependent desensitization of TRP channels and tachyphylaxis (Lepage et al., 2006).

3.4.2 STRUCTURAL FEATURES OF DROSOPHILA TRP AND TRPL CHANNELS

The *Drosophila trp* and *trpl* genes, which constitutes the founding member of the TRPC subfamily (Minke and Cook, 2002), encode a 1275– and 1124–amino acid long protein, respectively (Figure 3.11). Several studies have helped understand the structural domains and amino acids participating in specific channel functions and properties. For example, the replacement of Asp621 (D621) with glycine or asparagine reduced Ca^{2+} permeability of the TRP channel and is critical for the selectivity filter in the pore domain of the channel (Liu et al., 2007). Another is the F550I substitution of TRP, which forms a constitutively active channel leading to extremely fast light-independent retinal degeneration (Yoon et al., 2000). This site located at the beginning of S5 (Figure 3.11) was also found to increase channel activity of other TRP members, pointing to the importance of this amino acid in channel opening. The N-terminal regions of the TRP and TRPL proteins contain four ankyrin repeats and a CC domain. Both domains are believed to mediate protein-protein interactions. The N-terminal region also contains a TRP_2 domain with unknown function, predicted recently as involved in lipid binding and trafficking (van Rossum et al., 2008). At the C-terminal, the TRPL channel contains two CaM-binding sites (CBSs) (Phillips et al., 1992). One of these is unconventional in the sense that it can bind CaM in the absence of Ca^{2+}. The C-terminus fragment of the TRP sequence also has been reported to bind CaM in a Ca^{2+}-dependent manner. Both channels have a designated TRP domain located adjacent to the S6 with a EWKFAR motif found in many members of the TRP family. At the C-terminal region of the TRP there is a proline-rich sequence with 27 KP repeats, which overlap with a multiple repeat sequence, DKDKKP(G/A)D termed 8 × 9 (Montell and Rubin, 1989). Such proline-rich motifs occur widely and are predicted to form a structure involved in binding interactions with other proteins, including cytoskeletal elements such as actin. This region is unique to TRP and has not been found in any other member of the TRP family. The last 14 amino acids in the C-terminal of TRP are essential for binding to the INAD scaffold protein and form a PDZ binding domain. This has been demonstrated by truncation experiments (Shieh et al., 1997; Tsunoda et al., 1997; Chevesich et al., 1997). Using Web-based servers for the identification of putative coiled-coil (CC) domains in TRP and TRPL (Predictor of CC-domains in proteins [Fariselli et al., 2007], PairCoil2 [McDonnell et al., 2006], and COILS [Lupas et al., 1991]), it was suggested that in the TRP channel a CC domain in the N-terminal exists with high confidence, while at the C-terminal, one CC domain was identified with low confidence. For the TRPL, one CC domain at the N-terminal were identified with high confidence. At the C-terminal, three CC domains, one with high confidence and two with low confidence, were identified (Katz et al., 2013).

3.4.3 ORGANIZATION IN SUPRAMOLECULAR SIGNALING COMPLEX VIA SCAFFOLD PROTEIN INAD

An important step toward understanding *Drosophila* phototransduction has been achieved by the finding that some of the key proteins of the phototransduction cascade are incorporated into supramolecular signaling complexes via a scaffold protein, INAD. The INAD protein was discovered by a PDA screen of defective *Drosophila* mutant (*inaD*). The first discovered *inaD* mutant, the *InaD^{P215}*, was isolated by Pak and colleagues (Pak, 1995) and subsequently was cloned and sequenced by Shieh and Zhu (Shieh and Niemeyer, 1995). Studies in *Calliphora* have shown that INAD binds not only TRP but also PLC (NORPA) and ePKC (INAC) (Huber et al., 1996). The interaction of INAD with TRP, NORPA, and INAC was later confirmed in *Drosophila*. It was further found that *inaD* is a scaffold protein, which consists of five ~90 amino acid (aa) protein interaction motifs called PDZ (PSD95, DLG, ZO1) domains (Tsunoda et al., 1997). These domains are recognized as protein modules that bind to a diversity of signaling, cell adhesion, and cytoskeletal proteins by specific binding to target sequences typically, though not always, in the final three residues of the carboxy-terminal. The PDZ domains of INAD bind to the signaling molecules as follows: PDZ1 and PDZ5 bind PLC, PDZ2 or PDZ4 bind ePKC, and PDZ3 binds TRP (Figure 3.12). This binding pattern is still in debate due to several contradictory reports (Shieh et al., 1997; Chevesich et al., 1997; Adamski et al., 1998; Tsunodaet al., 2001; Mishra et al., 2007; Xu et al., 1998). TRPL appears not to be a member of the complex, since unlike INAC, NORPA, and TRP, it remains strictly localized to the microvilli in the *inaD^I* null (Tsunoda et al., 1997). Several studies have suggested that in addition to PLC, PKC, and TRP, other signaling molecules such as CaM, rhodopsin, TRPL, and NINAC bind to the INAD signaling complex (Chevesich et al., 1997). Such binding, however, must be dynamic. Biochemical studies conducted in *Calliphora* and later in *Drosophila* have revealed that both INAD and TRP are targets for phosphorylation by the nearby PKC (Huber et al., 1998). Accordingly, the association of TRP in a signaling complex together with its activator, PLC, and possible regulator, PKC, could be

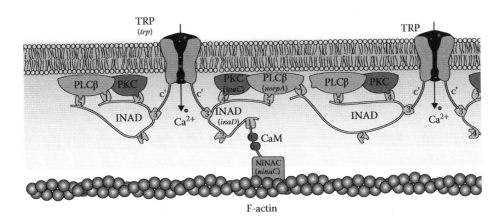

FIGURE 3.12 The INAD protein complex. The INAD sequence contained five consensus PDZ domains (indicated by numbers 1–5) and identified specific interactions between PKC and PDZ2 (or PDZ4), TRP and PDZ3, and PLC with PDZ1 and PDZ5. This binding pattern is still in debate due to several contradictory reports. It was also reported that the INAD contains a Ca^{2+}-calmodulin binding site that may be involved in its regulation. (From Katz, B. and B. Minke, *Front Cell Neurosci.*, 3, 2, 2009. With permission.)

related to increasing speed and efficiency of transduction. However, the experimental support for this notion is rather limited.

TRP plays a major role in localizing the entire INAD multimolecular complex. Association between TRP and INAD is essential for correct localization of the complex in the rhabdomeres as found in other signaling systems. This conclusion was derived using *Drosophila* mutants in which the signaling proteins, which constitute the INAD complex, were removed genetically, and by deletions of the specific binding domains, which bind TRP to INAD (Tsunoda et al., 2001; Xu et al., 1998). These experiments show that INAD is correctly localized to the rhabdomeres in *inaC* mutants (where ePKC is missing) and in *norpA* mutants (where PLC is missing), but severely mislocalized in null *trp* mutants, thus indicating that TRP but not PLC or PKC is essential for localization of the signaling complex to the rhabdomere. To demonstrate that specific interaction of INAD with TRP is required for the rhabdomeric localization of the complex, the binding site at the carboxyl terminal of TRP was removed or three conserved residues in PDZ3, which are expected to disrupt the interaction between PDZ domains and their targets, were modified. As predicted, both TRP and INAD were mislocalized in these mutants. The study of the above mutants was also used to show that TRP and INAD do not depend on each other to be targeted to the rhabdomeres; thus, INAD-TRP interaction is not required for targeting but for anchoring and retention of the signaling complex. Additional experiments on TRP and INAD further showed that INAD has other functions in addition to anchoring the signaling complex. One important function is to preassemble the proteins of the signaling complex. Another important function, at least for the case of PLC, is to prevent degradation of the unbound signaling protein (Xu et al., 1998).

A structural study of INAD has suggested that the binding of signaling proteins to INAD may be a dynamic process that constitutes an additional level of phototransduction regulation (Mishra et al., 2007). Their study showed two crystal structural states of isolated the INAD PDZ5 domain, differing mainly by the presence of a disulfide bond. This conformational change has light-dependent dynamics that was demonstrated by the use of transgenic *Drosophila* flies expressing an INAD having a point mutation that disrupts the formation of the disulfide bond. In this study a model was proposed in which ePKC phosphorylation at a still unknown site promotes the light-dependent conformational change of PDZ5, distorting its ligand binding groove to PLC and thus regulating phototransduction. Further studies showed that the redox potential of PDZ5 is allosterically regulated by its interaction with PDZ4 (Liu et al., 2011). Whereas isolated PDZ5 is stable in the oxidized state, formation of a PDZ4-5 "supramodule" locks PDZ5 in the reduced state by raising the redox potential of a disulfide bond. Acidification, potentially mediated via light-dependent PLCβ hydrolysis of PIP_2, disrupts the interaction between PDZ4 and PDZ5, leading to PDZ5 oxidation and dissociation from the TRP channel (Liu et al., 2011). However, demonstration of the physiological significance of these light-dependent changes in INAD is still lacking.

3.4.4 Assembly of TRP Channels

Several mechanisms regulate TRP channel activity and determine its biophysical properties. Examples of such mechanisms include channel assembly and translocation, which participate in the modulation of cellular currents mediated by TRP channels. Channel translocation results in a change in the number of channels present at the signaling membrane and affects the total conductance, partly utilizing the cellular transport and membrane fusion machinery. Channel assembly diversifies the functional properties of channels by assemblage of different subunits into heteromultimeric channels. The later mechanism can affect the gating, activity, and biophysical properties of the channels.

By analogy to other channels with a similar transmembrane structure that have been more extensively studied (e.g., voltage-gated K^+ channels and CNG channels), TRP channels are most likely tetramers (e.g., TRPV1 [Liao et al., 2013] and TRPV6 [Saotome et al., 2016]). Given that the mammalian TRPC subfamily consists of seven members and the mammalian TRP superfamily consists of a total of 28 isoforms, an important question arises as to the ability of TRP channels to form hetero-multimeres and which members of a specific subfamily can combine to form heteromeric channels.

The first suggestion that TRP channels are composed of heteromultimers came from studies on the *Drosophila* channels TRP, TRPL, and TRPγ (Xu et al., 2000; Xu et al., 1997). According to one study, TRP and TRPL channels co-immunoprecipitated (co-IP) both in the native tissue and in a heterologous expression system (HEK293 cells). This was also shown for TRPL and TRPγ channels (Xu et al., 2000). In studies on the effects of heteromultimerization on channel activity, whole-cell current measurements were carried out in HEK293 cells expressing the *Drosophila* TRP, TRPL, or both. Expression of TRPL channels resulted in a robust outwardly rectifying current, which resembled the native TRPL current, consistent with other studies (Parnas et al., 2007; Hardie et al., 1997; Kunze et al., 1997; Zimmer et al., 2000). However, TRP channel expression resulted in a nearly linear current-voltage relationship (Xu et al., 1997), different from the strongly rectifying *in vivo* current carried by TRP channels (Hardie and Minke, 1994). Moreover, coexpression of TRP and TRPL channels resulted in currents almost indistinguishable from the TRPL currents (Xu et al., 1997). Accordingly, while functional TRPL channels can be easily expressed, functional expression of the *Drosophila* TRP channels could not be reproduced (Minke and Parnas, 2006). Consistent with the notion that TRP and TRPL channels do not form heteromultimers, it was shown that both TRP and TRPL can form functional, light-activated ion channels in photoreceptor cells of *Drosophila* mutants in isolation, clearly showing that each channel can function independently of the other channel (Niemeyer et al., 1996; Reuss et al., 1997). Moreover, it was shown that the light-dependent conductance of WT flies could be attained by a weighted sum of the individual TRP and TRPL conductance, further supporting a solely homomultimeric TRP and TRPL channel assembly (Reuss et al., 1997). Further support for homomultimerization of TRP and TRPL channels includes the interaction of TRP but not TRPL with the scaffold protein, INAD (Tsunoda et al., 1997; Huber et al., 1996), and the light-dependent translocation of TRPL but not TRP (Bahner et al., 2002). Thus, although many types of TRP channels can assemble as heteromultimers, it was unclear for the *Drosophila* TRP and TRPL channels. This has led to further investigation of the issue of *Drosophila* TRP and TRPL channels assembly and its structural determinants. To this end a series of transgenic *Drosophila* flies expressing GFP-tagged TRP-TRPL chimeric channels were generated. The use of these chimeras indicated that TRP and TRPL assemble exclusively as homomultimeric channels in their native environment. The above chimeras revealed that the transmembrane regions of TRP and TRPL do not determine assembly specificity of these channels. However, the C-terminal regions of both TRP and TRPL predominantly specify the assembly of homomeric TRP and TRPL channels (Figure 3.13).

Various molecular determinants have been suggested for the assembly of TRPC family members. Using a yeast two-hybrid, protein-immobilization, and co-IP assays, it was shown that the *Drosophila* TRP and TRPL associate through their N-terminal regions (Xu et al., 1997). Later study refined this view by showing that TRPL and TRPγ channels assemble through the coiled-coil (CC) domain and by a 33-residue fragment located adjacent to this CC domain, located at the N-terminal (Xu et al., 2000). The involvement of the CC domain in TRPC channels assembly was supported by findings in other TRPC group members (Lepage et al., 2006). Accordingly, using yeast two-hybrid assays, it was shown that TRPC1 channels homotetramerize through their N-terminal CC domains (Engelke et al., 2002). The ankyrin repeat domain, located at the N-terminal of TRPC channels, was also implicated in channel assembly. Accordingly, it was shown that the first ankyrin repeat of TRPC1 is involved in the interaction with the N-terminus of TRPC3 (Liu et al., 2005). A later study utilizing TRPC channels from two distinct functional groups (i.e., TRPC4 and TRPC6) showed that swapping the N-terminal of TRPC4 with that of TRPC6, which contains the CC domain and ankyrin repeat by constructing a TRPC4–TRPC6 chimeric channel, enabled the interaction with TRPC4. However, the replacement of the CC domain or ankyrin repeats of TRPC6 by that of TRPC4 alone did not result in TRPC4 interaction, demonstrating that both domains are necessary for channel assembly (Lepage et al., 2009; Lepage et al., 2006).

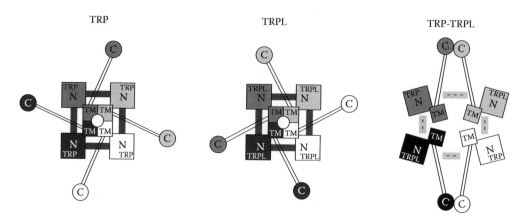

FIGURE 3.13 A model for the assembly of *Drosophila* TRP and TRPL channels. We adapted a previously proposed model in which the N-terminal interaction through the N-terminal coiled-coil (CC) domains (Engelke et al., 2002) and the ankyrin repeat domain, assembles the channels (see four squares marked by "N"). However, we extended this model by including participation of the C-terminal in subunit interactions via repulsion (see circles marked by "C"). Accordingly, the N-terminal fragments of TRP and TRPL are able to homo- and heterooligomerize. However, the TRP and TRPL do not interact because the C-terminal pattern of their CC domains differs. This model fits well with previous *in vitro* results from *Drosophila* (Xu et al., 1997, 2000), as well as mammalian TRPC channels, showing that the N-terminal but not the C-terminal fragments of these channel subunits interact. (From Lepage, P.K. et al., *Cell Calcium.*, 4, 251–259, 2009; Lepage, P.K. et al., *J Biol Chem*, 281, 30356–30364, 2006. With permission.)

3.4.5 Biophysical Properties of TRP Channels

The *Drosophila* light-sensitive channels, TRP and TRPL, could be studied separately by utilizing the most useful *trpl³⁰²* and *trp^P343* null mutants (Niemeyer et al., 1996; Pak and Leung, 2003). The channels are permeable to a variety of monovalent and divalent ions including Na^+, K^+, Ca^{2+}, and Mg^{2+}, and even to large organic cations such as TRIS and TEA (Reuss et al., 1997; Ranganathan et al., 1991). The reversal potential of the light-induced current (LIC) shows a marked dependence on extracellular Ca^{2+} indicating a high permeability for this ion. Permeability ratio measurements for a variety of divalent and monovalent ions were determined under bi-ionic conditions and confirmed a high Ca^{2+} permeability of ~57:1 = $Ca^{2+}:Cs^+$ in the *trpl* mutant (lacking the TRPL channel) and ~4.3:1 = $Ca^{2+}:Cs^+$ for the *trp* mutant (lacking the TRP channel) (Reuss et al., 1997). The TRP and TRPL channels show voltage-dependent conductance during illumination. An early study revealed that the light response could be blocked by physiological concentrations of Mg^{2+} ions (Hardie and Mojet, 1995). The block mainly influenced the TRP channel and affected its voltage dependence (Hardie and Mojet, 1995). Later, detailed analysis described the voltage dependence of heterologously expressed TRPL in S2 cells and of the native TRPL channel, using the *Drosophila trp* null mutant (Parnas et al., 2007). These studies indicated that the voltage dependence of the TRPL channel is not an intrinsic property, as is thought for other members of the TRP family, but arises from a divalent open-channel block that can be removed by depolarization. The open-channel block by divalent cations is thought to play a role in improving the signal-to-noise ratio of the response to intense light and may function in response termination (Parnas et al., 2007).

3.4.6 Stimulus-Dependent TRP Channel Translocation

The physiological properties of cells are largely determined by the specific set of ion channels at the plasma membrane. Besides regulation at the gene expression level, trafficking of ion channels into and out of the plasma membrane has been established as an important mechanism for manipulating the number of channels at a specific cellular site (for reviews see Lai and Jan, 2006; Sheng and Lee, 2001). Furthermore, stimulus-dependent trafficking (translocation) has emerged as an important regulatory mechanism with high physiological significance. Translocation has also been found for several mammalian TRP channels and for the *Drosophila* TRPL channels. The first study, which demonstrated *in vivo* translocation of a TRP channel and its physiological implications, came from studies of the light-activated *Drosophila* TRPL channel. Direct visualization of intracellular movements of TRPL in photoreceptors upon illumination was observed by immunolabeling of cross sections through *Calliphora* and *Drosophila* eyes. Accordingly, in dark-raised flies TRPL specific immunofluorescence was confined to the rhabdomere while, in light-raised flies the TRPL specific immunofluorescence was distributed over the cell body of the photoreceptor cells and was not detected in the rhabdomeres. This study further revealed that unlike TRPL, TRP is confined to the rhabdomeres, regardless of the illumination regime (Bähner et al., 2002). Since the translocation of TRPL depended on the illumination regime, a question arose as to whether or not the response to light through activation of the TRP and TRPL channels is the trigger for TRPL movement to the cell body. Using green fluorescent protein (GFP) tagged TRPL and direct fluorescent visualization of intracellular movements of TRPL in intact photoreceptors upon illumination revealed TRPL translocation from the rhabdomere to the cell body (Figure 3.14) (Meyer et al., 2006). This method was used to test TRPL translocation in the nearly null PLC mutant, *norpA*[P24] and the null mutant of TRP, *trp*[P343]. Light-induced translocation of TRPL to the cell body did not occur in both *norpA*[P24] and *trp*[P343] mutant flies indicating that TRP activation is necessary for TRPL translocation (Meyer

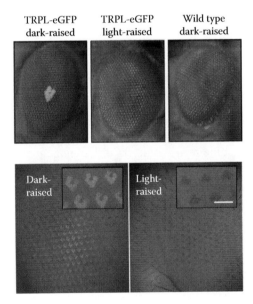

FIGURE 3.14 **(See color insert.)** Light-dependent translocation of TRPL-eGFP between the rhabdomere and cell body *in vivo*. Upper panel: Deep pseudopupil (Franceschini and Kirschfeld, 1971) of TRPL-eGFP expressing flies (*trpl-eGFP*) raised in the dark (upper left panel). A fluorescence signal of the deep pseudopupil was observed in dark-raised flies and disappeared in light-raised flies (upper middle panel) and in wild-type eyes that do not express TRPL-eGFP (upper right panel), indicating that in the dark, TRPL-eGFP are predominantly localized in the rhabdomeres. Lower panel: Subcellular localization of TRPL-eGFP in dark- and light-raised transgenic flies (*trpl-eGFP*). Fluorescence was detected in intact eyes after optical neutralization of the cornea by water immersion. Flies were kept in the dark (lower left panel) or under continuous orange light for 16 hours (lower right panel). The insets show the central area of the eye at higher magnification showing localization of TRPL-eGFP in the rhabdomeres of dark-raised flies (green dots) and its absence in the rhabdomeres of light-raised flies (black dots). Scale bar, 15 μm. (From Meyer, N.E. et al., *Cell Sci.*, 119, 2592–2603, 2006. With permission.)

et al., 2006). This result has led to the hypothesis that Ca^{2+} influx through the TRP channels is the necessary trigger for TRPL channel translocation. This expectation was directly demonstrated when light-dependent movement of TRPL was blocked by EGTA or by overexpression of the Na^+/Ca^{2+} exchanger CALX (Meyer et al., 2006; Richter et al., 2011).

Further research has shown that translocation of TRPL occurs in two distinct stages, first to the neighboring stalk membrane then to the basolateral membrane. In the first stage, light-induced translocation occurs within 5 minutes, whereas the second stage is much slower and takes over 6 hours. These two distinct translocation stages suggest that two distinct transport processes occur. The rapid first stage of translocation suggests that channels are released from the rhabdomere and diffuse laterally through the membrane into the adjoining stalk membrane. The slow second stage suggests an active internalization mechanism of the channels from the plasma membrane into vesicles (Cronin et al., 2006). A recent study has shown that TRPL and rhodopsin co-localize in endocytic particles revealing that TRPL is internalized by a vesicular transport pathway that is also utilized, at least partially, by rhodopsin endocytosis. In line with a canonical vesicular transport pathway, it was found that Rab proteins which are member of the Ras superfamily of monomeric G proteins, Rab5 and RabX4, are required for the internalization of TRPL into the cell body (Oberegelsbacher et al., 2011).

In the initial study it was shown that TRPL, but not the TRP channels undergo light-dependent translocation between the rhabdomere and cell body. Therefore, a question arises as to which of the TRPL channel segments are essential for translocation. Using transgenic flies expressing chimeric TRP and TRPL proteins that formed functional light-activated channels, translocation was induced only in the chimeric channel containing both the N- and C-terminal segments of TRPL. These results indicate that motifs present at both the N- and C-termini are required for proper channel translocation (Richter et al., 2011).

3.4.7 TRP CHANNEL REGULATION BY PHOSPHORYLATION

The activity of many proteins is controlled by phosphorylation and dephosphorylation reactions. Protein kinases and phosphatases that are activated during neuronal activity orchestrate cellular events that ultimately reshape the neuronal events via phosphorylation and dephosphorylation of various ion channels, including many members of the mammalian TRP channel superfamily (Voolstra et al., 2010; Por et al., 2013; Voolstra et al., 2013) (for a comprehensive review see Voolstra and Huber, 2014). However, the roles of phosphorylation and dephosphorylation in controlling TRP channel activity are still unclear (Cao et al., 2013a).

TRP phosphorylation: Since ePKC (INAC) is a member of the INAD signaling complex, it was assumed that this protein kinase might phosphorylate other members of the INAD complex. Indeed, ePKC was shown to phosphorylate INAD as well as TRP *in vitro* (Huber et al., 1996; Liu et al., 2000). Using quantitative mass spectrometry, 28 TRP differential phosphorylation sites from light- and dark-adapted flies (Voolstra et al., 2010; Voolstra et al., 2013) were identified. Twenty-seven phosphorylation sites resided in the predicted intracellular C-terminal region and a single site resided near the N-terminus. Fifteen of the C-terminal phosphorylation sites exhibited enhanced phosphorylation in the light, whereas a single site, Ser936, exhibited enhanced phosphorylation in the dark. To further investigate TRP phosphorylation at light-dependent phosphorylation sites, phosphor-specific antibodies were generated to specifically detect TRP phosphorylation at Thr849, at Thr864, which become phosphorylated in the light, and at Ser936, which becomes dephosphorylated in the light. To identify the stage of the phototransduction cascade that is necessary to trigger dephosphorylation of Ser936 or phosphorylation of Thr849 and Th864, phototransduction-defective *Drosophila* mutants and the phosphospecific antibodies were used. Strong phosphorylation of Ser936 in dark-adapted WT flies was observed but no phosphorylation in light-adapted WT flies was detected. Conversely, weak phosphorylation of Thr849 and Thr864 was observed in dark-adapted WT flies and strong phosphorylation was observed in light-adapted wild flies. Additionally, in phototransduction-defective mutants, strong phosphorylation of Ser936, and weak phosphorylation of Thr849 and Thr864 were observed, regardless of the light conditions. Conversely, a mutant expressing a constitutively active TRP channel (trp^P365) (Hong et al., 2002) exhibited weak phosphorylation of Ser936 and strong phosphorylation of Thr849 and Thr864 regardless of illumination. These data indicate that *in vivo*, TRP dephosphorylation at Ser936 and phosphorylation at Thr849 and Thr864 depend on the phototransduction cascade, but activation of the TRP channel alone is sufficient to trigger this process (Voolstra et al., 2013).

To identify kinases and phosphatases of Thr849 and Thr864, a candidate screen using available mutants of kinases and phosphatases that are expressed in the eye was applied. It was found that Thr849 phosphorylation was compromised in light-adapted *inaC* null mutants. Diminished phosphorylation in light-adapted *PKC^53e* mutants was also found, suggesting that these two protein kinases C synergistically phosphorylate TRP at Thr849. AMP-activated protein kinase was not involved in TRP phosphorylation at Thr864. The physiological effects of phosphorylation/dephosphorylation at the above Thr849 and Thr864 sites are still lacking However, the physiological role of Ser936 phosphorylation has been recently elucidated (Voolstra et al., 2017).

Phosphorylation of TRPL: Using mass spectrometry nine phosphorylated serine and threonine residues were identified in the TRPL channel (Cerny et al., 2013). Eight of these phosphorylation sites resided within the predicted cytosolic C-terminal region and a single site, Ser20, was located close to the TRPL N-terminus. Relative quantification revealed that Ser20 and Thr989 exhibited enhanced phosphorylation in the light, whereas Ser927, Ser1000, Ser1114, Thr1115, and Ser1116 exhibited enhanced phosphorylation in the dark. Phosphorylation of Ser730 and Ser931 was not light dependent. To further investigate the function of the eight C-terminal phosphorylation sites, these serine and threonine residues were mutated either to alanine, eliminating phosphorylation (TRPL8x), or to aspartate, mimicking phosphorylation (TRPL8xD). The mutated TRPL channels were transgenically expressed in R1-6 photoreceptor cells of flies as *trpl-eGFP* fusion constructs. The mutated channels formed multimeres with WT TRPL and produced electrophysiological responses when expressed in *trpl;trp* double-mutant background, indistinguishable from responses produced by WT TRPL. These findings indicated that TRPL channels devoid of their C-terminal phosphorylation sites form fully functional channels, and they argue against a role of TRPL phosphorylation for channel gating or regulation of its biophysical properties. Since TRPL undergoes light-dependent translocation, subcellular localization of the phosphorylation-deficient, as well as the phosphomimetic TRPL-eGFP fusion proteins by water immersion microscopy were analyzed. After initial dark adaptation, WT TRPL-eGFP was located in the rhabdomeres. After 16 hours of light adaptation, TRPL-eGFP was translocated to the cell body and successively returned to the rhabdomeres within 24 hours of dark adaptation. eGFP fluorescence obtained from the TRPL8x-eGFP displayed marked differences to the WT. After initial dark adaptation, a faint eGFP signal was observed in the cell body, but no eGFP signal was present in the rhabdomeres. After 16 hours of light adaptation, a strong eGFP signal was observed in the cell body akin to that observed in the WT. This indicated that TRPL8x-eGFP fusion construct was newly synthesized during light adaptation. After 4 hours of dark adaptation, TRPL8x-eGFP was present in the rhabdomere, but 20 hours later, only faint eGFP fluorescence was observable in the cell bodies and none in the rhabdomeres (Cerny et al., 2013) This result indicates that the phosphorylation of TRPL participates in TRPL translocation and recycling..

In conclusion, the localization of light and dark phosphorylation sites in both TRP and TRPL is well established. However, the physiological roles of these posttranslational modifications of the channels are unclear.

3.5   GATING MECHANISM OF *DROSOPHILA* TRP CHANNELS

3.5.1   Lipids Activate Light-Sensitive Channels in THE Dark

The activation of PLC results in the hydrolysis of PIP$_2$ into DAG and InsP$_3$ (Figure 3.6). The pathway, which recycles DAG back to PIP$_2$, the phosphoinositide (PI) pathway, has emerged to be highly relevant for activation of the TRP and TRPL channels, although the detailed mechanism is still unclear. The most familiar action of DAG is to activate the classical protein kinase C (PKC) synergistically with Ca^{2+}. However, null mutation of the eye-specific PKC (ePKC, *inaC*) leads to defects in response termination with no apparent effects on response activation. A second role for DAG in *Drosophila* photoreceptors is to act as a precursor for regeneration of PIP_2 via conversion to phosphatidic acid (PA) by DGK (Figure 3.6). DGK was identified after the *retinal degeneration A* (*rdgA*) mutant was isolated by Benzer and colleagues in a screen searching for defects in eye morphology. The *rdgA* gene encodes for an eye-specific DAG kinase (Masai et al., 1993) localized to the endoplasmic reticulum (ER) close to the base of the rhabdomere (Masai et al., 1997). The *rdgA* mutation leads to a severe form of light-independent photoreceptor degeneration. Electrophysiological studies showed that the light-activated TRP and TRPL channels are constitutively active in the *rdgA* mutant fly and that crossing the *rdgA* mutant into a TRP null mutant background partially rescued the retinal degeneration (Raghu et al., 2000a). Thus, the suggested mechanism of *rdgA* mutant degeneration was toxic increase in cellular Ca^{2+}, due to the constitutive activity of the TRP channels, via involvement of the phosphoinositide pathway in channel activation (Raghu et al., 2000a).

Another mutation in the PI pathway affects the CDP-diacylglycerol synthase (CDP-DAG synthase) encoded by the *cds* gene (Wu et al., 1995) (Figure 3.6). The *cds* mutant was isolated by using a screen based on a collection of P-element enhancer trap *Drosophila* lines searched by defects in ERG (Wu et al., 1995). An additional mutation in the PI pathway inactivates a PITP encoded by the *retinal degeneration B* (*rdgB*) gene (Vihtelic et al., 1991) (Figure 3.6). This mutation, like the *cds* mutation, led to severe forms of light-dependent photoreceptor degeneration. The *rdgB* gene product, the PITP protein, is essential for transferring PI from the ER to the rhabdomere (Vihtelic et al., 1993). To account for the effects of the above mutations in the PI pathway and of the *rdgA* mutation in particular, a hypothesis has been put forward whereby DAG acts as intracellular messenger, leading directly to TRP and TRPL channel activation (Hardie, 2003). However, application of DAG to isolated *Drosophila* ommatidia either did not activate the channels, or induced single-channel activity after a long delay in an isolated photoreceptor preparation (Delgado and Bacigalupo, 2009). Furthermore, application of DAG to recombinant TRPL expressed in *Drosophila* S2 cells did not result in channel opening (unpublished observations), thus questioning the DAG hypothesis.

3.5.2 PUFA AS SECOND MESSENGER OF EXCITATION

DAG is also a precursor for the formation of PUFAs via DAG lipases (Figure 3.6). Studies conducted by Hardie and colleagues showed that application of PUFAs to isolated *Drosophila* ommatidia, as well as to recombinant TRPL channels expressed in *Drosophila* S2 cells, reversibly activated the TRP and TRPL channels, respectively (Chyb et al., 1999). These results are consistent with results showing that in the *rdgA* mutant, TRP and TRPL channels are constitutively active, due to the elimination of DAG kinase, by the *rdgA* mutation. The suggested accumulation of DAG leads to a still undemonstrated accumulation of PUFAs that constitutively activate the TRP and TRPL channels, leading to a toxic increase in cellular Ca^{2+} and, thereby, degeneration. Recently, Pak and colleagues cloned and sequenced the *inaE* gene, following the isolation of the *inaE* mutant by the PDA screen (Leung et al., 2008). The *inaE* gene encodes a homologue of mammalian *sn-1*-type DAG lipase that produces the 2-MAG lipid and showed that it is expressed predominantly in the cell body of *Drosophila* photoreceptors. Mutant flies, expressing low levels of the *inaE* gene product, have an abnormal light response, while the activation of the light-sensitive channels was not prevented. The 2-MAG lipid is a precursor of PUFA that requires MAG lipase; however, this enzyme was not found in *Drosophila* (Leung et al., 2008). Thus, the participation of DAG or PUFAs in TRP and TRPL activation *in vivo* needs to be further explored.

3.5.3 Is Drosophila TRP a Mechanosensitive Channel?

It is important to realize that PLC activation also generates protons that reduce the pH and affects the light response (Huang et al., 2010). In addition, PLC activation converts PIP_2, a charged molecule, containing a large hydrophilic head-group, into DAG, devoid of the hydrophilic head-group, and this conversion causes major changes in membrane lipid packing and lipid-channel interactions. It was therefore hypothesized by Parnas, Katz, and Minke that neither PIP_2 hydrolysis nor DAG or PUFA production affect the TRP and TRPL channel as second messengers but rather act as modifiers of membrane lipid-channel interactions acting as mechanotransducers (Parnas et al., 2009). Indeed, Hardie and Franze have demonstrated by combined patch clamp recordings and atomic force microscopy that by cleaving PIP_2, PLC generates rapid physical changes in the lipid bilayer that lead to contractions of the *Drosophila* microvilli, and suggest that the resultant mechanical forces that exhibit the kinetics of the light response and are eliminated by the *norpA* mutation contribute, together with pH changes to the gating of the TRP and TRPL channels (Hardie and Franze, 2012). Clearly, the mechanosensitivity of the *Drosophila* light-sensitive channels needs further exploration.

3.6 CONCLUDING REMARKS

Channel members of the TRP superfamily are involved in fundamental mechanisms common to many cells and tissues, especially where cellular Ca^{2+} plays an important role. This characteristic of TRP channels is also reflected in the unusually large variety of stimuli that activate these channels. The light-sensitive *Drosophila* channels TRP and TRPL, which are activated via a cascade of enzymatic reactions, are unique. Thus, unlike mammalian TRPC and TRPV channels, which are also activated via a cascade of enzymatic reactions under physiological conditions, activation by light dictates special molecular mechanisms to cope with the need to follow accurately light on and off as well as a fast response to activation/termination of the external signal. Therefore, features such as sensitivity to single photons, gain, adaptation to a wide range of light intensities, and an unusually wide range of frequency responses to modulated lights characterize activation and termination of TRP and TRPL channels. The lack of structural information on TRPC channels is an obvious obstacle for future progress. However, homology modeling with the solved atomic structures of TRPV1 (Liao et al., 2013; Cao et al., 2013b), TRPV2 (Huynh et al., 2016; Zubcevic et al., 2016), and TRPV6 (Saotome et al., 2016) is expected to be a useful tool for TRPC channels because of the general similarity in the protein modules that characterize both channel subfamilies. Thus, both channel subfamilies have an ankyrin repeat domain, pore domain of relatively large sequence homology, conserved TRP domain at the end of the pore region, calmodulin binding sites, and many phosphorylation sites at the C-terminal. Therefore, a large volume of existing information on TRPV1 can guide future studies on TRP/TRPL. Similarly, vast information on activation by the inositol lipid signaling cascade and on the identity of channel interacting proteins can guide future studies of TRPV and mammalian TRPC channel activation via enzymatic cascade.

   This review emphasizes not only the importance of recognizing all the molecular components related to TRP/TRPL activation and termination mechanisms, but also the importance of accurate cellular localization and stoichiometry among the various molecular components required for proper channel function, which are essential for the high fidelity of the visual system. The power of the *Drosophila* molecular genetics, allowing the generation of mutations in virtually every important signaling protein in the living animal, allows a rather detailed functional insight that has been exploited to understand light activation of TRP/TRPL that does not exist for other TRP channels. Thus, studies of the *Drosophila* light-activated channels can serve as a guideline for investigating physiological activation mechanisms of mammalian TRP channels via the inositol lipid signaling cascade.

ACKNOWLEDGMENTS

Research parts of this review were supported by grants from the U.S.-Israel Bi National Science Foundation, the Israel Science Foundation (ISF), and the Deutsch-Israelische Projektkooperation (DIP) (to B.M.).

REFERENCES

Acharya, J.K. et al. 1997. InsP 3 receptor is essential for growth and differentiation but not for vision in *Drosophila*. *Neuron*, 18: 881–887.

Adamski, F.M. et al. 1998. Interaction of eye protein kinase C and INAD in *Drosophila*. Localization of binding domains and electrophysiological characterization of a loss of association in transgenic flies. *J Biol Chem*, 273: 17713–17719.

Akitake, B. et al. 2015. Coordination and fine motor control depend on *Drosophila* TRPγ. *Nat Commun*, 6: 7288.

Alloway, P.G., L. Howard, and P.J. Dolph. 2000. The formation of stable rhodopsin-arrestin complexes induces apoptosis and photoreceptor cell degeneration. *Neuron*, 28: 129–138.

Altenhoff, A.M. et al. 2012. Resolving the ortholog conjecture: Orthologs tend to be weakly, but significantly, more similar in function than paralogs. *PLoS Comput Biol*, 8: e1002514.

Angueyra, J.M. et al. 2012. Melanopsin-expressing amphioxus photoreceptors transduce light via a phospholipase C signaling cascade. *PLOS ONE*, 7: e29813.

Arshavsky, V.Y. and E.N. Pugh, Jr. 1998. Lifetime regulation of G protein-effector complex: Emerging importance of RGS proteins. *Neuron*, 20: 11–14.

Bacigalupo, J. et al. 1991. Light-dependent channels from excised patches of *Limulus* ventral photoreceptors are opened by cGMP. *Proc Natl Acad Sci U S A*, 88: 7938–7942.

Bähner, M. et al. 2002. Light-regulated subcellular translocation of *Drosophila* TRPL channels induces long-term adaptation and modifies the light-induced current. *Neuron*, 34: 83–93.

Bandyopadhyay, B.C. and R. Payne. 2004. Variants of TRP ion channel mRNA present in horseshoe crab ventral eye and brain. *J Neurochem*, 91: 825–835.

Barash, S. and B. Minke. 1994. Is the receptor potential of fly photoreceptors a summation of single-photon responses? *Comments Theor Biol*, 3: 229–263.

Barash, S. et al. 1988. Light reduces the excitation efficiency in the nss mutant of the sheep blowfly *Lucilia*. *J Gen Physiol*, 92: 307–330.

Bazinet, A.L., D.J. Zwickl, and M.P. Cummings. 2014. A gateway for phylogenetic analysis powered by grid computing featuring GARLI 2.0. *Syst Biol*, 63: 812–818.

Berridge, M.J. 1993. Inositol trisphosphate and calcium signalling. *Nature*, 361: 315–325.

Berstein, G. et al. 1992. Reconstitution of agonist-stimulated phosphatidylinositol 4,5-bisphosphate hydrolysis using purified m1 muscarinic receptor, G$_q$/11, and phospholipase C-á1. *J Biol Chem*, 267: 8081–8088.

Bloomquist, B.T. et al. 1988. Isolation of a putative phospholipase C gene of *Drosophila*, *norpA*, and its role in phototransduction. *Cell*, 54: 723–733.

Blumenfeld, A. et al. 1985. Light-activated guanosinetriphosphatase in *Musca* eye membranes resembles the prolonged depolarizing afterpotential in photoreceptor cells. *Proc Natl Acad Sci U S A*, 82: 7116–7120.

Byk, T. et al. 1993. Regulatory arrestin cycle secures the fidelity and maintenance of the fly photoreceptor cell. *Proc Natl Acad Sci U S A*, 90: 1907–1911.

Cao, E. et al. 2013a. TRPV1 channels are intrinsically heat sensitive and negatively regulated by phosphoinositide lipids. *Neuron*, 77: 667–679.

Cao, E. et al. 2013b. TRPV1 structures in distinct conformations reveal activation mechanisms. *Nature*, 504: 113–118.

Cerny, A.C. et al. 2013. Mutation of light-dependent phosphorylation sites of the *Drosophila* transient receptor potential-like (TRPL) ion channel affects its subcellular localization and stability. *J Biol Chem*, 288: 15600–15613.

Chen, C.A. and D.R. Manning. 2001. Regulation of G proteins by covalent modification. *Oncogene*, 20: 1643–1652.

Chen, C.K. et al. 2000. Slowed recovery of rod photoresponse in mice lacking the GTPase accelerating protein RGS9-1. *Nature*, 403: 557–560.

Chen, F.H. et al. 2001. A cGMP-gated channel subunit in *Limulus* photoreceptors. *Vis Neurosci*, 18: 517–526.

Chevesich, J., A.J. Kreuz, and C. Montell. 1997. Requirement for the PDZ domain protein, INAD, for localization of the TRP store-operated channel to a signaling complex. *Neuron*, 18: 95–105.

Chouquet, B. et al. 2009. A TRP channel is expressed in *Spodoptera littoralis* antennae and is potentially involved in insect olfactory transduction. *Insect Mol Biol*, 18: 213–222.

Chu, B. et al. 2013. Common mechanisms regulating dark noise and quantum bump amplification in *Drosophila* photoreceptors. *J Neurophysiol*, 109: 2044–2055.

Chyb, S., P. Raghu, and R.C. Hardie. 1999. Polyunsaturated fatty acids activate the *Drosophila* light-sensitive channels TRP and TRPL. *Nature*, 397: 255–259.

Conant, G.C. and K.H. Wolfe. 2008. Turning a hobby into a job: How duplicated genes find new functions. *Nat Rev Genet*, 9: 938–950.

Cook, B. et al. 2000. Phospholipase C and termination of G-protein-mediated signalling in vivo. *Nat Cell Biol*, 2: 296–301.

Cordero-Morales, J.F., E.O. Gracheva, and D. Julius. 2011. Cytoplasmic ankyrin repeats of transient receptor potential A1 (TRPA1) dictate sensitivity to thermal and chemical stimuli. *Proc Natl Acad Sci U S A*, 108: E1184–E1191.

Corey, D.P. et al. 2004. TRPA1 is a candidate for the mechanosensitive transduction channel of vertebrate hair cells. *Nature*, 432: 723–730.

Cosens, D. 1971. Blindness in a *Drosophila* mutant. *J Insect Physiol*, 17: 285–302.

Cosens, D.J. and A. Manning. 1969. Abnormal electroretinogram from a *Drosophila* mutant. *Nature*, 224: 285–287.

Crick, F.H. 1952. Is alpha-keratin a coiled coil? *Nature*, 170: 882–883.

Cronin, M.A., M.H. Lieu, and S. Tsunoda. 2006. Two stages of light-dependent TRPL-channel translocation in *Drosophila* photoreceptors. *J Cell Sci*, 119: 2935–2944.

Delgado, R. and J. Bacigalupo. 2009. Unitary recordings of TRP and TRPL channels from isolated *Drosophila* retinal photoreceptor rhabdomeres: Activation by light and lipids. *J Neurophysiol*, 101: 2372–2379.

Denis, V. and M.S. Cyert. 2002. Internal Ca(2+) release in yeast is triggered by hypertonic shock and mediated by a TRP channel homologue. *J Cell Biol*, 156: 29–34.

Devary, O. et al. 1987. Coupling of photoexcited rhodopsin to inositol phospholipid hydrolysis in fly photoreceptors. *Proc Natl Acad Sci U S A*, 84: 6939–6943.

Dolph, P.J. et al. 1993. Arrestin function in inactivation of G protein-coupled receptor rhodopsin in vivo. *Science*, 260: 1910–1916.

Dolph, P.J. et al. 1994. An eye-specific Gα subunit essential for termination of the phototransduction cascade. *Nature*, 370: 59–61.

Dorlochter, M., M. Klemeit, and H. Stieve. 1997. Immunological demonstration of G_q-protein in *Limulus* photoreceptors. *Vis Neurosci*, 14: 287–292.

Dorloechter, M. and H. Stieve. 1997. The Limulus ventral photoreceptor: Light response and the role of calcium in a classic preparation. *Prog Neurobiol*, 53: 451–515.

Doza, Y.N. et al. 1992. Characterization of fly rhodopsin kinase. *Eur J Biochem*, 209: 1035–1040.

Elia, N. et al. 2005. Excess of G βe over G qαe in vivo prevents dark, spontaneous activity of *Drosophila* photoreceptors. *J Cell Biol*, 171: 517–526.

Engelke, M. et al. 2002. Structural domains required for channel function of the mouse transient receptor potential protein homologue TRP1β. *FEBS Lett*, 523: 193–199.

Essen, L.O. et al. 1996. Crystal structure of a mammalian phosphoinositide-specific phospholipase C delta. *Nature*, 380: 595–602.

Essen, L.O. et al. 1997. Structural mapping of the catalytic mechanism for a mammalian phosphoinositide-specific phospholipase C. *Biochemistry*, 36: 1704–1718.

Fariselli, P. et al. 2007. Prediction of structurally-determined coiled-coil domains with hidden markov models. In *Bioinformatics Research and Development*, edited by S. Hochreiter and R. Wagner, 292–302. Berlin/Heidelberg: Springer.

Fein, A. 1986. Blockade of visual excitation and adaptation in *Limulus* photoreceptor by GDP-α-S. *Science*, 232: 1543–1545.

Fein, A. and D.W. Corson. 1979. Both photons and fluoride ions excite *Limulus* ventral photoreceptors. *Science*, 204: 77–79.

Fein, A. and D.W. Corson. 1981. Excitation of Limulus photoreceptors by vanadate and by a hydrolysis-resistant analog of guanosine triphosphate. *Science*, 212: 555–557.

Fein, A. and S. Cavar. 2000. Divergent mechanisms for phototransduction of invertebrate microvillar photoreceptors. *Vis Neurosci*, 17: 911–917.

Ferrer, C. et al. 2012. Dissecting the determinants of light sensitivity in amphioxus microvillar photoreceptors: Possible evolutionary implications for melanopsin signaling. *J Neurosci*, 32: 17977–17987.

Franceschini, N. and K. Kirschfeld. 1971. Pseudopupil phenomena in the compound eye of *Drosophila*. *Kybernetik*, 9: 159–182.

Frechter, S. et al. 2007. Translocation of $G_q\alpha$ mediates long-term adaptation in *Drosophila* photoreceptors. *J Neurosci*, 27: 5571–5583.

French, A.S. et al. 2015. Transcriptome analysis and RNA interference of cockroach phototransduction indicate three opsins and suggest a major role for TRPL channels. *Front Physiol*, 6: 207.

Garcia-Murillas, I. et al. 2006. lazaro encodes a lipid phosphate phosphohydrolase that regulates phosphatidylinositol turnover during *Drosophila* phototransduction. *Neuron*, 49: 533–546.

Garcia-Sanz, N. et al. 2007. A role of the transient receptor potential domain of vanilloid receptor I in channel gating. *J Neurosci*, 27: 11641–11650.

Garger, A.V., E.A. Richard, and J.E. Lisman. 2004. The excitation cascade of *Limulus* ventral photoreceptors: Guanylate cyclase as the link between InsP3-mediated Ca^{2+} release and the opening of cGMP-gated channels. *BMC Neurosci*, 5: 7.

Gaudet, R. 2008. TRP channels entering the structural era. *J Physiol*, 586: 3565–3575.

Gomez, M.D. and E. Nasi. 1996. Ion permeation through light-activated channels in rhabdomeric photoreceptors. Role of divalent cations. *J Gen Physiol*, 107: 715–730.

Gomez, M.E.P., J.M. Angueyra, and E. Nasi. 2009. Light-transduction in melanopsin-expressing photoreceptors of amphioxus. *Proc Natl Acad Sci U S A*, 106: 9081–9086.

Gu, Y. et al. 2005. Mechanisms of light adaptation in *Drosophila* photoreceptors. *Curr Biol*, 15: 1228–1234.

Harbury, P.B. et al. 1993. A switch between two-, three-, and four-stranded coiled coils in GCN4 leucine zipper mutants. *Science*, 262: 1401–1407.

Hardie, R.C. 1997. Functional equivalence of native light-sensitive channels in the *Drosophila* trp 301 mutant and TRPL cation channels expressed in a stably transfected *Drosophila* cell line. *Cell Calcium*, 21: 431–440.

Hardie, R.C. 2003. Regulation of trp channels via lipid second messengers. *Annu Rev Physiol*, 65: 735–759.

Hardie, R.C. 2005. Inhibition of phospholipase C activity in *Drosophila* photoreceptors by 1,2-bis(2-aminophenoxy)ethane N,N,N',N'-tetraacetic acid (BAPTA) and di-bromo BAPTA. *Cell Calcium*, 38: 547–556.

Hardie, R.C. 2011. A brief history of trp: Commentary and personal perspective. *Pflugers Arch*, 461: 493–498.

Hardie, R.C. and B. Minke. 1992. The trp gene is essential for a light-activated Ca^{2+} channel in *Drosophila* photoreceptors. *Neuron*, 8: 643–651.

Hardie, R.C. and B. Minke. 1994. Calcium-dependent inactivation of light-sensitive channels in *Drosophila* photoreceptors. *J Gen Physiol*, 103: 409–427.

Hardie, R.C. and K. Franze. 2012. Photomechanical responses in *Drosophila* photoreceptors. *Science* 338: 260–263.

Hardie, R.C. and M.H. Mojet. 1995. Magnesium-dependent block of the light-activated and trp-dependent conductance in *Drosophila* photoreceptors. *J Neurosci*, 74: 2590–2599.

Hardie, R.C. et al. 2002. Molecular basis of amplification in *Drosophila* phototransduction. Roles for G protein, phospholipase C, and diacylglycerol kinase. *Neuron*, 36: 689–701.

Henderson, S.R., H. Reuss, and R.C. Hardie. 2000. Single photon responses in *Drosophila* photoreceptors and their regulation by Ca^{2+}. *J Physiol Lond*, 524(Pt 1): 179–194.

Hillman, P., S. Hochstein, and B. Minke. 1983. Transduction in invertebrate photoreceptors: Role of pigment bistability. *Physiol Rev*, 63: 668–772.

Hochstrate, P. 1989. Lanthanum mimicks the trp photoreceptor mutant of *Drosophila* in the blowfly *Calliphora*. *J Comp Physiol A*, 166: 179–187.

Hong, Y.S. et al. 2002. Single amino acid change in the fifth transmembrane segment of the TRP Ca^{2+} channel causes massive degeneration of photoreceptors. *J Biol Chem*, 277: 33884–33889.

Howard, J. 1983. Variations in the voltage response to single quanta of light in the photoreceptor of *Locusta migratoria*. *Biophys Struct Mech*, 9: 341–348.

Howard, J. and S. Bechstedt. 2004. Hypothesis: A helix of ankyrin repeats of the NOMPC-TRP ion channel is the gating spring of mechanoreceptors. *Curr Biol*, 14: R224–R226.

Hu, W. et al. 2012. Protein G_q modulates termination of phototransduction and prevents retinal degeneration. *J Biol Chem*, 287: 13911–13918.

Huang, J. et al. 2010. Activation of TRP channels by protons and phosphoinositide depletion in *Drosophila* photoreceptors. *Curr Biol*, 20: 189–197.

Huber, A. et al. 1996. The transient receptor potential protein (Trp), a putative store-operated Ca^{2+} channel essential for phosphoinositide-mediated photoreception, forms a signaling complex with NorpA, InaC and InaD. *EMBO J*, 15: 7036–7045.

Huber, A. et al. 1998. The TRP Ca^{2+} channel assembled in a signaling complex by the PDZ domain protein INAD is phosphorylated through the interaction with protein kinase C (ePKC). *FEBS Lett*, 425: 317–322.

Huber, A., P. Sander, and R. Paulsen. 1996. Phosphorylation of the InaD gene product, a photoreceptor membrane protein required for recovery of visual excitation. *J Biol Chem*, 271: 11710–11717.

Huson, D.H. and C. Scornavacca. 2012. Dendroscope 3: An interactive tool for rooted phylogenetic trees and networks. *Syst Biol*, 61: 1061–1067.

Huynh, K.W. et al. 2016. Structure of the full-length TRPV2 channel by cryo-EM. *Nat Commun*, 7: 11130.

Itoh, H. and A.G. Gilman. 1991. Expression and analysis of Gs alpha mutants with decreased ability to activate adenylylcyclase. *J Biol Chem*, 266: 16226–16231.

Jabba, S. et al. 2014. Directionality of temperature activation in mouse TRPA1 ion channel can be inverted by single-point mutations in ankyrin repeat six. *Neuron*, 82: 1017–1031.

Jin, X., J. Touhey, and R. Gaudet. 2006. Structure of the N-terminal ankyrin repeat domain of the TRPV2 ion channel. *J Biol Chem*, 281: 25006–25010.

Katoh, K. et al. 2002. MAFFT: A novel method for rapid multiple sequence alignment based on fast Fourier transform. *Nucleic Acids Res*, 30: 3059–3066.

Katoh, K. et al. 2005. MAFFT version 5: Improvement in accuracy of multiple sequence alignment. *Nucleic Acids Res*, 33: 511–518.

Katz, B. and B. Minke. 2009. *Drosophila* photoreceptors and signaling mechanisms. *Front Cell Neurosci*, 3: 2.

Katz, B. and B. Minke. 2012. Phospholipase C-mediated suppression of dark noise enables single-photon detection in *Drosophila* photoreceptors. *J Neurosci*, 32: 2722–2733.

Katz, B. et al. 2013. The *Drosophila* TRP and TRPL are assembled as homomultimeric channels in vivo. *J Cell Sci*, 126: 3121–3133.

Kelley, L.A. and M.J. Sternberg. 2009. Protein structure prediction on the Web: A case study using the Phyre server. *Nat Protoc*, 4: 363–371.

Kiselev, A. et al. 1990. A molecular pathway for light-dependent photorecptor apoptosis in *Drosophila*. *Neuron*, 28: 139–152.

Kohn, E. et al. 2015. Functional cooperation between the IP3 receptor and phospholipase C secures the high sensitivity to light of *Drosophila* photoreceptors in vivo. *J Neurosci*, 35: 2530–2546.

Kosloff, M. et al. 2003. Regulation of light-dependent $G_q\alpha$ translocation and morphological changes in fly photoreceptors. *EMBO J*, 22: 459–468.

Kunze, D.L. et al. 1997. Properties of single *Drosophila* Trpl channels expressed in Sf9 insect cells. *Am J Physiol*, 272: C27–C34.

Kwon, Y. and C. Montell. 2006. Dependence on the Lazaro phosphatidic acid phosphatase for the maximum light response. *Curr Biol*, 16: 723–729.

Lai, H.C. and L.Y. Jan. 2006. The distribution and targeting of neuronal voltage-gated ion channels. *Nat Rev Neurosci*, 7: 548–562.

LaLonde, M.M. et al. 2005. Regulation of phototransduction responsiveness and retinal degeneration by a phospholipase D-generated signaling lipid. *J Cell Biol*, 169: 471–479.

Landoure, G. et al. 2010. Mutations in TRPV4 cause Charcot-Marie-Tooth disease type 2C. *Nat Genet*, 42: 170–174.

Lapan, S.W. and P.W. Reddien. 2012. Transcriptome analysis of the planarian eye identifies ovo as a specific regulator of eye regeneration. *Cell Rep*, 2: 294–307.

Lee, Y.J. et al. 1990. dgq: A *Drosophila* gene encoding a visual system-specific G alpha molecule. *Neuron*, 5: 889–898.

Lepage, P.K. et al. 2006. Identification of two domains involved in the assembly of transient receptor potential canonical channels. *J Biol Chem*, 281: 30356–30364.

Lepage, P.K. et al. 2009. The self-association of two N-terminal interaction domains plays an important role in the tetramerization of TRPC4. *Cell Calcium*, 4: 251–259.

Lesk, A.M. 2008. *Introduction to Bioinformatics*. Oxford: Oxford University Press.

Leung, H.T. et al. 2008. DAG lipase activity is necessary for TRP channel regulation in *Drosophila* photoreceptors. *Neuron*, 58: 884–896.

Liao, M. et al. 2013. Structure of the TRPV1 ion channel determined by electron cryo-microscopy. *Nature*, 504: 107–112.

Lishko, P.V. et al. 2007. The ankyrin repeats of TRPV1 bind multiple ligands and modulate channel sensitivity. *Neuron*, 54: 905–918.

Liu, C.H. et al. 2007. In vivo identification and manipulation of the Ca^{2+} selectivity filter in the *Drosophila* transient receptor potential channel. *J Neurosci*, 27: 604–615.

Liu, C.H. et al. 2008. Ca²⁺-dependent metarhodopsin inactivation mediated by calmodulin and NINAC myosin III. *Neuron*, 59: 778–789.

Liu, M. et al. 2000. Reversible phosphorylation of the signal transduction complex in *Drosophila* photoreceptors. *J Biol Chem*, 275: 12194–12199.

Liu, W. et al. 2011. The INAD scaffold is a dynamic, redox-regulated modulator of signaling in the *Drosophila* eye. *Cell*, 145: 1088–1101.

Liu, X. et al. 2005. Molecular analysis of a store-operated and 2-acetyl-sn-glycerol-sensitive non-selective cation channel. Heteromeric assembly of TRPC1-TRPC3. *J Biol Chem*, 280: 21600–21606.

Lupas, A., M. Van Dyke, and J. Stock. 1991. Predicting coiled coils from protein sequences. *Science*, 252: 1162–1164.

Malashkevich, V.N. et al. 1996. The crystal structure of a five-stranded coiled coil in COMP: A prototype ion channel? *Science*, 274: 761–765.

Masai, I. et al. 1993. *Drosophila* retinal degeneration: A gene encodes an eye-specific diacylglycerol kinase with cysteine-rich zinc-finger motifs and ankyrin repeats. *Proc Natl Acad Sci U S A*, 90: 11157–11161.

Masai, I. et al. 1997. Immunolocalization of *Drosophila* eye-specific diacylgylcerol kinase, rdgA, which is essential for the maintenance of the photoreceptor. *J Neurobiol*, 32: 695–706.

Matsumoto, H. et al. 1994. Phosrestin I undergoes the earliest light-induced phosphorylation by a calcium/calmodulin-dependent protein kinase in *Drosophila* photoreceptors. *Neuron*, 12: 997–1010.

Matsuura, H. et al. 2009. Evolutionary conservation and changes in insect TRP channels. *BMC Evol Biol*, 9: 228.

McDonnell, A.V. et al. 2006. Paircoil2: Improved prediction of coiled coils from sequence. *Bioinformatics*, 22: 356–358.

Meyer, N.E. et al. 2006. Subcellular translocation of the eGFP-tagged TRPL channel in *Drosophila* photoreceptors requires activation of the phototransduction cascade. *J Cell Sci*, 119: 2592–2603.

Miller, M.A., W. Pfeiffer, and T. Schwartz. 2010. Creating the CIPRES Science Gateway for inference of large phylogenetic trees. *Proceedings of the Gateway Computing Environments Workshop* (GCE), 14 November 2010, New Orleans, LA, pp. 1–8.

Minke, B. 1977. *Drosophila* mutant with a transducer defect. *Biophys Struct Mechanism*, 3: 59–64.

Minke, B. 1982. Light-induced reduction in excitation efficiency in the trp mutant of *Drosophila*. *J Gen Physiol*, 79: 361–385.

Minke, B. 2010. The history of the *Drosophila* TRP channel: The birth of a new channel superfamily. *J Neurogenet*, 24: 216–233.

Minke, B. 2012. The history of the prolonged depolarizing afterpotential (PDA) and its role in genetic dissection of *Drosophila* phototransduction. *J Neurogenet*, 26: 106–117.

Minke, B. and K. Agam. 2003. TRP gating is linked to the metabolic state and maintenance of the *Drosophila* photoreceptor cells. *Cell Calcium*, 33: 395–408.

Minke, B. and B. Cook. 2002. TRP channel proteins and signal transduction. *Physiol Rev*, 82: 429–472.

Minke, B. and M. Parnas. 2006. Insights on TRP channels from in vivo studies in *Drosophila*. *Annu Rev Physiol*, 68: 649–684.

Minke, B. and Selinger, Z. 1991. Inositol lipid pathway in fly photoreceptors: Excitation, calcium mobilization and retinal degeneration. In *Progress in Retinal Research*, edited by N.A. Osborne and G.J. Chader, 99–124. Oxford: Pergamon Press.

Minke, B. and Z. Selinger. 1992a. Intracellular messengers in invertebrate photoreceptors studied in mutant flies. In *Neuromethods*, edited by A. Boulton, G. Baker, and C. Taylor, 517–563. Clifton, NJ: Humana Press.

Minke, B. and Z. Selinger. 1992b. The inositol-lipid pathway is necessary for light excitation in fly photoreceptors. In *Sensory Transduction*, edited by D. Corey and S.D. Roper, 202–217. New York, NY: Rockefeller University Perss.

Minke, B. and R.S. Stephenson. 1985. The characteristics of chemically induced noise in *Musca* photoreceptors. *J Comp Physiol*, 156: 339–356.

Minke, B. and M. Tsacopoulos. 1986. Light induced sodium dependent accumulation of calcium and potassium in the extracellular space of bee retina. *Vision Res*, 26: 679–690.

Minke, B., C. Wu, and W.L. Pak. 1975. Induction of photoreceptor voltage noise in the dark in *Drosophila* mutant. *Nature*, 258: 84–87.

Mishra, P. et al. 2007. Dynamic scaffolding in a G protein-coupled signaling system. *Cell*, 131: 80–92.

Monk, P.D. et al. 1996. Isolation, cloning, and characterisation of a trp homologue from squid (*Loligo forbesi*) photoreceptor membranes. *J Neurochem*, 67: 2227–2235.

Montell, C. 2011. The history of TRP channels, a commentary and reflection. *Pflugers Arch*, 461: 499–506.

Montell, C. and G.M. Rubin. 1989. Molecular characterization of the *Drosophila* trp locus: A putative integral membrane protein required for phototransduction. *Neuron*, 2: 1313–1323.

Montell, C. et al. 2002. A unified nomenclature for the superfamily of TRP cation channels. *Mol Cell*, 9: 229–231.

Nehrt, N.L. et al. 2011. Testing the ortholog conjecture with comparative functional genomic data from mammals. *PLOS Comput Biol*, 7: e1002073.

Niemeyer, B.A. et al. 1996. The *Drosophila* light-activated conductance is composed of the two channels TRP and TRPL. *Cell*, 85: 651–659.

Oberegelsbacher, C. et al. 2011. The *Drosophila* TRPL ion channel shares a Rab-dependent translocation pathway with rhodopsin. *Eur J Cell Biol*, 90: 620–630.

Oberwinkler, J. and D.G. Stavenga. 2000. Calcium transients in the rhabdomeres of dark- and light-adapted fly photoreceptor cells. *J Neurosci*, 20: 1701–1709.

Pak, W.L. 1979. Study of photoreceptor function using *Drosophila* mutants. In *Neurogenetics, Genetic Approaches to the Nervous System*, edited by X. Breakfield, 67–99. New York, NY: Elsevier North-Holland.

Pak, W.L. 1991. Molecular genetic studies of photoreceptor function using *Drosophila* mutants. In *Molecular Biology of the Retina: Basic and Clinical Relevant Studies*, edited by J. Piatigorsky, T. Shinohara, and P.S. Zelenka, 1–32. New York, NY: Wiley-Liss.

Pak, W.L. 1995. *Drosophila* in vision research. The Friedenwald Lecture. *Invest Ophthalmol Vis Sci*, 36: 2340–2357.

Pak, W.L. 2010. Why *Drosophila* to study phototransduction? *J Neurogenet*, 24: 55–66.

Pak, W.L. and H.T. Leung. 2003. Genetic approaches to visual transduction in *Drosophila melanogaster*. *Recept Channel*, 9: 149–167.

Pak, W.L., S. Shino, and H.T. Leung. 2012. PDA (prolonged depolarizing afterpotential)-defective mutants: The story of nina's and ina's—Pinta and santa maria, too. *J Neurogenet*, 26: 216–237.

Parnas, M. et al. 2009. Membrane lipid modulations remove divalent open channel block from TRP-like and NMDA channels. *J Neurosci*, 29: 2371–2383.

Parnas, M., B. Katz, and B. Minke. 2007. Open channel block by Ca^{2+} underlies the voltage dependence of *Drosophila* TRPL channel. *J Gen Physiol*, 129: 17–28.

Pearn, M.T. et al. 1996. Molecular, biochemical, and electrophysiological characterization of *Drosophila* norpA mutants. *J Biol Chem*, 271: 4937–4945.

Peng, G., X. Shi, and T. Kadowaki. 2014. Evolution of TRP channels inferred by their classification in diverse animal species. *Mol Phylogenet Evol*, 84C: 145–157.

Peretz, A. et al. 1994a. The light response of *Drosophila* photoreceptors is accompanied by an increase in cellular calcium: Effects of specific mutations. *Neuron*, 12: 1257–1267.

Peretz, A. et al. 1994b. Genetic dissection of light-induced Ca2+ influx into *Drosophila* photoreceptors. *J Gen Physiol*, 104: 1057–1077.

Phelps, C.B. et al. 2010. Differential regulation of TRPV1, TRPV3, and TRPV4 sensitivity through a conserved binding site on the ankyrin repeat domain. *J Biol Chem*, 285: 731–740.

Phillips, A.M., A. Bull, and L.E. Kelly. 1992. Identification of a *Drosophila* gene encoding a calmodulin-binding protein with homology to the *trp* phototransduction gene. *Neuron*, 8: 631–642.

Por, E.D. et al. 2013. Phosphorylation regulates TRPV1 association with β-arrestin-2. *Biochem J*, 451: 101–109.

Postma, M., J. Oberwinkler, and D.G. Stavenga. 1999. Does Ca^{2+} reach millimolar concentrations after single photon absorption in *Drosophila* photoreceptor microvilli? *Biophys J*, 77: 1811–1823.

Pulido, C. et al. 2012. The light-sensitive conductance of melanopsin-expressing Joseph and Hesse cells in amphioxus. *J Gen Physiol*, 139, 19–30.

Pumir, A. et al. 2008. Systems analysis of the single photon response in invertebrate photoreceptors. *Proc Natl Acad Sci U S A*, 105: 10354–10359.

Raghu, P. et al. 2000a. Constitutive activity of the light-sensitive channels TRP and TRPL in the *Drosophila* diacylglycerol kinase mutant, rdgA. *Neuron*, 26: 169–179.

Raghu, P. et al. 2000b. Normal phototransduction in *Drosophila* photoreceptors lacking an InsP 3 receptor gene. *Mol Cell Neurosci*, 15: 429–445.

Raghu, P., S. Yadav, and N.B. Mallampati. 2012. Lipid signaling in *Drosophila* photoreceptors. *Biochim Biophys Acta*, 1821: 1154–1165.

Ranganathan, R. et al. 1991. A *Drosophila* mutant defective in extracellular calcium-dependent photoreceptor deactivation and rapid desensitization. *Nature*, 354: 230–232.

Reuss, H. et al. 1997. In vivo analysis of the *Drosophila* light-sensitive channels, TRP and TRPL. *Neuron*, 19: 1249–1259.

Rhee, S.G. 2001. Regulation of phosphoinositide-specific phospholipase C. *Annu Rev Biochem*, 70: 281–312.

Richter, D. et al. 2011. Translocation of the *Drosophila* transient receptor potential-like (TRPL) channel requires both the N- and C-terminal regions together with sustained Ca^{2+} entry. *J Biol Chem*, 286: 34234–34243.

Rivera, A.S. et al. 2010. Gene duplication and the origins of morphological complexity in pancrustacean eyes, a genomic approach. *BMC Evol Biol*, 10: 123.

Rogozin, I.B. et al. 2014. Gene family level comparative analysis of gene expression in mammals validates the ortholog conjecture. *Genome Biol Evol*, 6: 754–762.

Rohacs, T. et al. 2005. PI(4,5)P 2 regulates the activation and desensitization of TRPM8 channels through the TRP domain. *Nat Neurosci*, 8: 626–634.

Running Deer, J.L., J.B. Hurley, and S.L. Yarfitz. 1995. G protein control of *Drosophila* photoreceptor phospholipase C. *J Biol Chem*, 270: 12623–12628.

Salazar, H. et al. 2008. A single N-terminal cysteine in TRPV1 determines activation by pungent compounds from onion and garlic. *Nat Neurosci*, 11: 255–261.

Saotome, K. et al. 2016. Crystal structure of the epithelial calcium channel TRPV6. *Nature*, 534: 506–511.

Schillo, S. et al. 2004. Targeted mutagenesis of the farnesylation site of *Drosophila* Gγe disrupts membrane association of the G protein βγ complex and affects the light sensitivity of the visual system. *J Biol Chem*, 279: 36309–36316.

Schindl, R. and C. Romanin. 2007. Assembly domains in TRP channels. *Biochem Soc Tran*, 35: 84–85.

Schulz, S. et al. 1999. A novel Gγ isolated from *Drosophila* constitutes a visual G protein γ subunit of the fly compound eye. *J Biol Chem*, 274: 37605–37610.

Scott, K. et al. 1995. $G_q\alpha$ protein function in vivo: Genetic dissection of its role in photoreceptor cell physiology. *Neuron*, 15: 919–927.

Scott, K. et al. 1997. Calmodulin regulation of *Drosophila* light-activated channels and receptor function mediates termination of the light response in vivo. *Cell*, 91: 375–383.

Sedgwick, S.G. and S.J. Smerdon. 1999. The ankyrin repeat: A diversity of interactions on a common structural framework. *Trends Biochem Sci*, 24: 311–316.

Selinger, Z. and Minke, B. 1988. Inositol lipid cascade of vision studied in mutant flies. *Cold Spring Harb Symp Quant Biol*, 53(Pt 1): 333–341.

Selinger, Z., Y.N. Doza, and B. Minke. 1993. Mechanisms and genetics of photoreceptors desensitization in *Drosophila* flies. *Biochim Biophys Acta*, 1179: 283–299.

Sheng, M. and S.H. Lee. 2001. AMPA receptor trafficking and the control of synaptic transmission. *Cell*, 105: 825–828.

Shieh, B.H. and B. Niemeyer. 1995. A novel protein encoded by the InaD gene regulates recovery of visual transduction in *Drosophila*. *Neuron*, 14: 201–210.

Shieh, B.H. et al. 1997. Association of INAD with NORPA is essential for controlled activation and deactivation of *Drosophila* phototransduction in vivo. *Proc Natl Acad Sci U S A*, 94: 12682–12687.

Shortridge, R.D. et al. 1991. A *Drosophila* phospholipase C gene that is expressed in the central nervous system. *J Biol Chem*, 266: 12474–12480.

Sidi, S., R.W. Friedrich, and T. Nicolson. 2003. NompC TRP channel required for vertebrate sensory hair cell mechanotransduction. *Science*, 301: 96–99.

Song, Z. et al. 2012. Stochastic, adaptive sampling of information by microvilli in fly photoreceptors. *Curr Biol*, 22: 1371–1380.

Speiser, D.I. et al. 2014. Using phylogenetically-informed annotation (PIA) to search for light-interacting genes in transcriptomes from non-model organisms. *BMC Bioinformatics*, 15: 350.

Steele, F. and J.E. O'Tousa. 1990. Rhodopsin activation causes retinal degeneration in *Drosophila* rdgC mutant. *Neuron*, 4: 883–890.

Steele, F.R. et al. 1992. *Drosophila* retinal degeneration C (rdgC) encodes a novel serine/threonine protein phosphatase. *Cell*, 69: 669–676.

Story, G.M. et al. 2003. ANKTM1, a TRP-like channel expressed in nociceptive neurons, is activated by cold temperatures. *Cell*, 112: 819–829.

Suss Toby, E., Z. Selinger, and Minke, B. 1991. Lanthanum reduces the excitation efficiency in fly photoreceptors. *J Gen Physiol*, 98: 849–868.

Tsunoda, S. et al. 1997. A multivalent PDZ-domain protein assembles signalling complexes in a G-protein-coupled cascade. *Nature*, 388: 243–249.

Tsunoda, S. et al. 2001. Independent anchoring and assembly mechanisms of INAD signaling complexes in *Drosophila* photoreceptors. *J Neurosci*, 21: 150–158.

van Rossum, D.B. et al. 2008. TRP_2, a lipid/trafficking domain that mediates diacylglycerol-induced vesicle fusion. *J Biol Chem*, 283: 34384–34392.

Venkatachalam, K. and C. Montell. 2007. TRP channels. *Annu Rev Biochem*, 76: 387–417.

Vihtelic, T.S. et al. 1993. Localization of *Drosophila* retinal degeneration B, a membrane- associated phosphatidylinositol transfer protein. *J Cell Biol*, 122: 1013–1022.

Vihtelic, T.S., D.R. Hyde, and J.E. O'Tousa. 1991. Isolation and characterization of the *Drosophila* retinal degeneration B (rdgB) gene. *Genetics*, 127: 761–768.

Vogt, K. and K. Kirschfeld. 1984. Chemical identity of the chromophores of fly visual pigment. *Naturwissen*, 77: 211–213.

Voolstra, O. and A. Huber. 2014. Post-translational modifications of TRP channels. *Cells*, 3: 258–287.

Voolstra, O. et al. 2010. Light-dependent phosphorylation of the *Drosophila* transient receptor potential ion channel. *J Biol Chem*, 285: 14275–14284.

Voolstra, O. et al. 2013. Phosphorylation of the *Drosophila* transient receptor potential ion channel is regulated by the phototransduction cascade and involves several protein kinases and phosphatases. *PLOS ONE*, 8: e73787.

Voolstra O. et al. 2017. The phosphorylation state of the Drosophila TRP channel modulates the frequency response to oscillating light in vivo. *J Neurosci*, 37(15):4213-4224.

Walker, R.G., A.T. Willingham, and C.S. Zuker. 2000. A *Drosophila* mechanosensory transduction channel. *Science*, 287: 2229–2234.

Wicher, D. et al. 2006. TRPγ channels are inhibited by cAMP and contribute to pacemaking in neurosecretory insect neurons. *J Biol Chem*, 281: 3227–3236.

Winn, M.P. et al. 2005. A mutation in the TRPC6 cation channel causes familial focal segmental glomerulosclerosis. *Science*, 308: 1801–1804.

Wong, F. et al. 1989. Proper function of the *Drosophila trp* gene product during pupal development is important for normal visual transduction in the adult. *Neuron*, 3: 81–94.

Wu, L. et al. 1995. Regulation of PLC-mediated signalling in vivo by CDP-diacylglycerol synthase. *Nature*, 373: 216–222.

Xiao, R. and X.Z. Xu. 2009. Function and regulation of TRP family channels in *C. elegans*. *Pflugers Arch*, 458: 851–860.

Xu, X.Z. et al. 2000. TRPγ, a *Drosophila* TRP-related subunit, forms a regulated cation channel with TRPL. *Neuron*, 26: 647–657.

Xu, X.Z., et al. 1998. Coordination of an array of signaling proteins through homo- and heteromeric interactions between PDZ domains and target proteins. *J Cell Biol*, 142: 545–555.

Xu, X.Z.S. et al. 1997. Coassembly of TRP and TRPL produces a distinct store-operated conductance. *Cell*, 89: 1155–1164.

Yeandle, S. and J.B. Spiegler. 1973. Light-evoked and spontaneous discrete waves in the ventral nerve photoreceptor of *Limulus*. *J Gen Physiol*, 61: 552–571.

Yoon, J. et al. 2000. Novel mechanism of massive photoreceptor degeneration caused by mutations in the trp gene of *Drosophila*. *J Neurosci*, 20: 649–659.

Ziegler, A. and B. Walz. 1990. Evidence for light-induced release of Ca^{2+} from intracellular stores in bee photoreceptors. *Neurosci Lett*, 111: 87–91.

Zimmer, S. et al. 2000. Modulation of recombinant transient-receptor-potential-like (TRPL) channels by cytosolic Ca^{2+}. *Pflugers Arch*, 440: 409–417.

Zubcevic, L. et al. 2016. Cryo-electron microscopy structure of the TRPV2 ion channel. *Nat Struct Mol Biol*, 23: 180–186.

Zwickl, D.J. 2006. Genetic algorithm approaches for the phylogenetic analysis of large biological sequence datasets under the maximum likelihood criterion. Ph.D. dissertation. The University of Texas at Austin, Austin, TX.

4 Sensory Mechanotransduction and Thermotransduction in Invertebrates

Shana Geffeney

CONTENTS

4.1 STRUCTURE AND FUNCTION OF TRP CHANNELS IN INVERTEBRATES

Mechanosensory and thermosensory neurons have a diversity of tasks in invertebrate animals. Mechanosensory neurons must detect mechanical stimuli as harsh as the penetration of a parasitoid wasp ovipositor and as gentle as the brush of bacteria while an animal glides along an agar plate (Sawin et al., 2000; Hwang et al., 2007). Thermosensory neurons must detect temperatures warmer and cooler than preferred temperatures to allow animals to return to appropriate temperatures for proper growth as well as detect tissue-damaging heat to allow animals to avoid immediate danger (Kwon et al., 2008; Klein et al., 2015; Sayeed and Benzer, 1996; Lee et al., 2005; Hamada et al., 2008; Gallio et al., 2011; Tracey et al., 2003; Hwang et al., 2007; Xu et al., 2006). Within these cells, the threshold for detection of mechanical and thermal stimuli is adjusted by changes in the external and internal environment. The detection of food or the release of cytokines by dying cells can alter the cellular response of mechanosensory and thermosensory neurons (Sawin et al., 2000; Hilliard et al., 2005; Babcock et al., 2009; Babcock et al., 2011). Genes that encode transient receptor potential (TRP) channel subunits are members of a large gene family that encodes cation channels with key roles in the function of many types of sensory neurons both as transduction channels that transform a specific stimulus into an electrical signal and as regulators of the cellular response.

Work in invertebrates has provided important insights into the structure and function of TRP channels. TRP channel subunits contain long cytoplasmic N- and C-termini on either side of six transmembrane domains that form the pore portion of the channel (Wu et al., 2010; Kadowaki, 2015). Four channel subunits join together to form the channel pore, and channels are formed by homomeric and heteromeric subunits (Gong, 2004; Cao et al., 2013; Gao et al., 2016) (Figure 4.1).

This review focuses on members of the TRP channel gene family that are expressed in mechanosensory and thermosensory neurons of invertebrates and have been demonstrated to affect cellular responses to stimuli. Because the most extensive work has been done investigating the role of TRP channels in the nematode *Caenorhabditis elegans* and the fruit fly *Drosophila melanogaster*, this review focuses on the role of TRP channel subunits in detection of mechanical and thermal stimuli in these two animals. Specifically, work on invertebrate animals has highlighted the importance of the N-terminus for controlling the ability of TRP channels to function in mechanosensory and thermosensory cells as transduction channels or to control the cellular response as signal amplifiers. The N-terminus of the TRPN channel subunit, NOMPC, has the largest number of ankyrin repeats at 29 (Figure 4.1). The ankyrin repeats domain of some TRP channels may form an elastic tether that pulls open the channel during mechanotransduction (Howard and Bechstedt, 2004; Jin et al., 2006; Sotomayor et al., 2005; Gaudet, 2008). In mechanosensory neurons of *Drosophila* halteres, the N-terminus of NOMPC forms a critical component of the link between the plasma membrane and microtubules, which is required to gate the channel and may accomplish this task by pulling the channel open (Liang et al., 2013; Liang et al., 2014; Zhang et al., 2015).

In contrast to NOMPC, the gene that encodes a TRPA channel subunit, dTRPA1, is expressed as four different protein variants, each with different elements in the N-terminus. When the four protein variants are individually expressed in heterologous cells, only two of the variants form thermotansduction channels (Kwon et al., 2010a; Zhong et al., 2012). The portion of N-terminus that differs between the variants that form thermotransduction channels and those that do not is a portion of the N-terminus between the ankyrin repeats and the first transmembrane domain (Figure 4.1). Surprisingly, when expressed in polymodal neurons of the *Drosophila* larvae, it is one of the non-temperature-responsive variants that rescues the behavioral response to high heat (Zhong et al., 2012). The role of dTRPA1 channels in these cells appears to be amplification of thermotransduction signals. These results suggest that an understanding of the genes as well as the variants of those genprotein variants a neuron may be critical for defining the role of TRP channels in the cell.

The small size and well-studied morphology of nematodes and fruit flies permit researchers to examine how individual or small groups of identified neurons function. This work has allowed for precise determination of the TRP channels expressed in specific mechanosensory and thermosensory cells as well as the demonstration of which TRP channels are functioning as transduction channels. To function as a transduction channel, a protein must transform a physical stimulus into an electrical signal by allowing ions to flow through its pore. Mechanical stimuli activate currents in mechanosensory cells within milliseconds, which is faster than the fastest known second messenger-based sensory transduction pathway. These results suggest that mechanotransduction channels can be directly gated by force (Hardie, 2001; Walker et al., 2000; Kang et al., 2010; O'Hagan et al., 2004; Geffeney et al., 2011). Thermotransduction channels can be reconstituted in artificial membranes to produce a channel that opens in response to temperature changes (Cao et al., 2013; Gao et al., 2016), suggesting that these channels can also be directly gated by heat and cold. To determine the transduction channel in a cell, it is critical to be able to separate the currents generated by transduction channels from those generated by other ion channels. In invertebrates, the ability to voltage-clamp individual neurons of particular interest has identified TRPN channels as the mechanotransduction channels in several mechanosensory neurons (Walker et al., 2000; Kang et al., 2010; Li et al., 2011). Additionally, this allowed identification of demonstrated that genes from other gene families form mechanotransduction channel in these

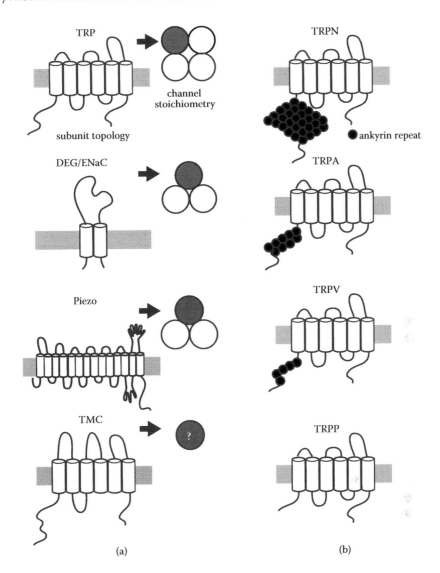

FIGURE 4.1 Topology and stoichiometry of TRP channel proteins along with three others proposed to form mechanotransduction channels in animals. (a) TRP channel protein subunits are predicted to have six transmembrane domains and assemble into tetrameric ion channels (Kahn-Kirby and Bargmann, 2006; Damann et al., 2008; Nilius and Owsianik, 2011; Kadowaki, 2015). TRP channels genes are conserved in eukaryotes (Damann et al., 2008; Kadowaki, 2015). DEG/ENaC protein channel subunits are predicted to have two transmembrane domains with a large extracellular domain. Three DEG/ENaC protein subunits form an ion channel. The DEG/ENaC channel genes are conserved in animals but absent from plants, yeast, and other microbes (Goodman and Schwarz, 2003). Both TRP and DEG/ENaC proteins can form homomeric and heteromeric channels, increasing the potential for diversity in channel structure and function. Recently, two additional classes of transmembrane proteins have been linked to mechanotransduction in animals, Piezo and TMC (Coste et al., 2012; Kawashima et al., 2015; Kim et al., 2012). Piezo proteins are huge with an increasingly better resolved topology, and they appear to form trimeric channels as do DEG/ENaC channels (Ge et al., 2015; Zhao et al., 2016). TMC channel proteins are predicted to have six transmembrane domains, but their stoichiometry is unknown (Kawashima et al., 2015). (b) TRP channel subunits have been organized into seven subfamilies based on protein homology (Kahn-Kirby and Bargmann, 2006; Damann et al., 2008; Nilius and Owsianik, 2011; Kadowaki, 2015). Four subunit family members are found in mechanosensory and thermosensory neurons of invertebrates, TRPN, TRPA, TRPV, and TRPP. Members of different TRP subfamilies can be distinguished by a number of structural features including presence or absence of ankyrin repeats in the N-terminus of the protein as well as the number of ankyrin repeats in the N-terminus.

cells (O'Hagan et al., 2004; Geffeney et al., 2011) (Figure 4.1). The ability to study identifiable cells has provided an understanding that multiple TRP channels are expressed in the same mechanosensory or thermosensory cell and function together to generate cellular responses. This review highlights work in nematodes and fruit flies that has deepened our understanding of how TRP channels work together and with other ion channels in sensory neurons.

4.2 MECHANOSENSATION AND THERMOSENSATION IN ADULT *CAENORHABDITIS ELEGANS*

4.2.1 RESPONSE TO GENTLE (AND HARSH TOUCH) IN *C. ELEGANS* ADULTS, TRPN

The CEP neurons mediate the slowing of nematodes that is the behavioral response to the gentle touch of bacteria striking the nose as the animals glide along a surface (Sawin et al., 2000). CEP neurons have ciliated endings that lie parallel to the cuticle and an extension of the cilium is embedded in the cuticle (Perkins et al., 1986). CEP neurons express at least one TRP and one DEG/ENaC channel protein (Figure 4.2). TRPN channel proteins, specifically TRP-4, are the pore-forming subunits of the mechanotransduction channel in CEPs. Recordings of mechanically evoked currents indicate that mechanoreceptor currents are eliminated by loss of TRP-4, and mutations in the pore domain of the channel subunits alter the reversal potential of mechanoreceptor currents (Kang et al., 2010).

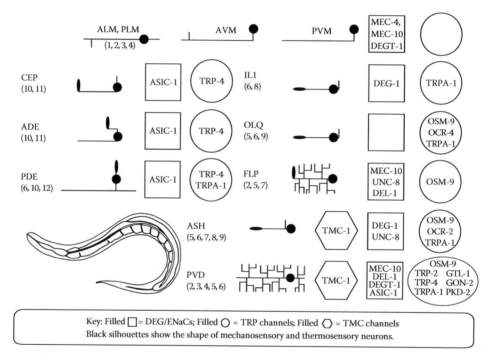

FIGURE 4.2 TRP, DEG/ENaC, and TMC channel proteins coexpressed in mechanosensory and thermosensory neurons in *C. elegans* nematodes. The gross morphology of identified mechanosensory and thermosensory neurons in *C. elegans are diagrammed*, and the ion channel subunits that are expressed in each cell are listed. Sources for *C. elegans* neuron expression are listed by numbers: (1) Driscoll, M. and M. Chalfie, *Nature,* 349, 6310, 588–593, 1991; (2) Huang, M. and M. Chalfie, *Nature,* 367, 6462, 467–470, 1994; (3) Chatzigeorgiou, M. et al., *Nature Neurosci,* 13, 7, 861–868, 2010; (4) Smith, C.J. et al., *Dev Biol,* 345, 1, 18–33, 2010; (5) Colbert, H.A. et al., *J Neurosci,* 17, 21, 8259–8269, 1997; (6) Kindt, K.S. et al., *Nature Neurosci,* 10, 5, 568–577, 2007; (7) Tavernarakis, N. et al., *Neuron,* 18, 107–119, 1997; (8) Hall, D.H. et al., *J Neurosci,* 17, 3, 1033–1045, 1997; (9) Tobin, D.M. et al., *Neuron,* 35, 307–318, 2002; (10) Voglis, G. and N. Tavernarakis, *EMBO J,* 27, 24, 3288–3299, 2008; (11) Walker, R.G. et al., *Science,* 287, 5461, 2229–2234, 2000; (12) Li, W. et al., *Nature,* 440, 7084, 684–687, 2006.

In contrast to CEPs, touch receptor neurons (ALM, PLM, AVM, and PVM) express several amiloride-sensitive Na⁺ (DEG/ENaC) channel proteins, but no TRP channel subunits (Figure 4.2).These neurons are responsible for the avoidance response to gentle touch on the sides of the body. Recordings of mechanically evoked currents indicate that two DEG/ENaC channel subunits MEC-4 and MEC-10 form mechanotransduction channels (O'Hagan et al., 2004). Neurons that are activated by gentle touch appear to function without the need for a more elaborate system of response channels.

The PDE neurons also mediate animal slowing as worms move in a bacterial lawn (Sawin et al., 2000) as well as forward movement generated by harsh touch to the sides of the animal (Li et al., 2011). These cells have a ciliated sensory ending as well as two sensory projections along the body wall muscles (White et al., 1986). These cells express a single DEG/ENaC channel protein ASIC-1 as well as two TRP channel proteins, TRP-4 and TRPA-1 (Figure 4.2). Slightly larger displacements are required to evoke mechanoreceptor currents in these cells compared to those that initiate currents in CEP, 1 μm compared to 0.5 μm (Li et al., 2011). These mechanoreceptor currents are completely eliminated in *trp-4* mutant animals suggesting that TRP-4 channel subunits form the mechanotransduction channel in these cells.Thus, currents in PDEs that are initiated by relatively small displacements are also generated by a single type of mechanotransduction channel.

4.2.2 RESPONSE TO HARSH TOUCH IN *C. ELEGANS* ADULTS, **TRPV** AND **TRPA**

The ASH neurons are polymodal and act as nociceptors in the animal, initiating avoidance behavior when the nose of the animal is stimulated by multiple aversive stimuli including touch, acid, or chemicals (Hart et al., 1999; Troemel et al., 1995; Hart et al., 1995; Maricq et al., 1995; Kaplan and Horvitz, 1993; Culotti and Russell, 1978; Hilliard et al., 2002). These neurons have ciliated endings with access to small pore openings next to the mouth of the animal (Perkins et al., 1986). As is characteristic of a nociceptor, ASH neurons require more intense forces for activation than PLMs and larger displacements for activation than CEPs and PDE (Geffeney et al., 2011). These cells express multiple DEG/ENaC and TRP channel proteins as well as a TMC channel, but the major mechanoreceptor current is carried by a mechanotransduction channel formed by the DEG/ENaC channel protein, DEG-1 (Geffeney et al., 2011) (Figure 4.2). A minor current remains in *deg-1* null mutants and is carried by an ion channel that is not selective for sodium ions (Geffeney et al., 2011). These results suggest that the remaining current is not carried by a DEG/ENaC channel, though it is possible that DEG-1 and the channel responsible for the minor current function in series with DEG-1 amplifying the minor current. However, the data suggest that the channels function in parallel because loss of DEG-1 does not alter the rise rate of mechanoreceptor currents in ASH.

Three TRP proteins are expressed in ASH neurons, two TRPV channel subunits (OSM-9, OCR-2) and one TRPA channel subunit (TRPA-1). OSM-9, OCR-2, and TRPA-1 are required for behavioral responses to nose touch that are mediated by ASH (Colbert et al., 1997; Tobin et al., 2002; Kindt et al., 2007). In contrast to TRP channel subunits, loss of the TMC channel subunit expressed in ASH, TMC-1, has no effect on the behavioral response to nose touch (Chatzigeorgiou et al., 2013). The loss of TRPA-1 has no effect on the calcium increases in ASH that can be evoked by nose touch, and its role in ASH remains unclear (Kindt et al., 2007). However, when animals are exposed to serotonin, calcium increases that are initiated by touch require OSM-9 (Hilliard et al., 2005). Yet, without serotonin exposure, loss of these TRPV proteins does not affect mechanically evoked currents or potentials in ASH (Geffeney et al., 2011). These TRPV proteins are likely to affect cell activity after mechanotransduction, and this response may be modulated by serotonin.

A conserved function of nociceptors is their regulation by biogenic amines like serotonin (Walters and Moroz, 2009). In mammals, nociceptor sensitization following injury is induced by the release of a suite of chemicals including serotonin (Basbaum et al., 2009). In nematodes, serotonin alters the cellular response to noxious chemical stimuli by decreasing the amplitude of internal calcium release and increasing the amplitude of membrane depolarization that follows these stimuli (Zahratka et al.,

2015). The behavioral response to chemical stimuli is faster when animals are exposed to serotonin or when animals are on food (Chao et al., 2004; Harris et al., 2009). Serotonin is released when nematodes are eating bacteria (Sawin et al., 2000).Together, these data suggest that changes in cellular response mediated by serotonin are linked to increased sensitivity to stimuli when animals are on food. Behavioral responses to touch that are mediated by ASH require that animals are on food or exposed to serotonin, and these responses are eliminated in animals with null mutations in *osm-9* and *ocr-2* Thus, increased sensitivity to nose touch when animals are exposed to food or serotonin may be mediated by OSM-9 and OCR-2 acting after mechanotransduction in ASH.

4.2.3 RESPONSE TO HARSH TOUCH AND LOW TEMPERATURES IN *C. ELEGANS* ADULTS, TRPV AND TRPA

The polymodal PVD neurons are activated by mechanical and thermal stimuli and may function as nociceptors. These multidendritic neurons send projections to the posterior ¾ of the cuticle (White et al., 1986) and express multiple DEG/ENaC and TRP channels (Figure 4.2). Multiple lines of evidence suggest that mechanotransduction currents are generated by DEG/ENaC channels. Mechanotransduction currents in these cells have the properties of DEG/ENaC channels and not TRP channels. The currents are amiloride sensitive and have reversal potentials that are close to the equilibrium potential for sodium ions (Li et al., 2011). The loss of MEC-10 or the reduction of DEGT-1 expression by RNAi eliminates calcium responses to harsh touch in PVDs (Chatzigeorgiou et al., 2010); however, the loss of MEC-10 alone does not eliminate mechanotransduction currents in PVDs (Li et al., 2011). The loss of two TRP channel subunits, OSM-9 or TRPA-1, has no effect on calcium responses to harsh touch (Chatzigeorgiou et al., 2010), but the loss of TRPA-1 eliminates calcium responses to cold. A 5°C drop in temperature from 20°C to 15°C initiates an increase in calcium in the cell PVD that is lost in *trpa-1* mutants. Additionally, neurons that normally do not show a calcium response to the same temperature drop, FLP and ALM, can be induced to respond to the change when they express TRPA-1. Together these data suggest that in the polymodal PVD neurons, mechanotransduction and thermotransduction are split between DEG/ENaC and TRPA channels, respectively. The function of the other six TRP channels expressed in PVD is unclear. However by examining the behavioral response to light-evoked cell activation, Husson et al. (2012) demonstrated that reduced expression of a TRPM channel GTL-1 by RNAi altered the behavioral response to light stimulation in animals that expressed Channelrhodospin-2 (ChR2). Altered GTL-2 expression reduced both the number of animals that responded to and the maximum velocity induced in animals by all light intensities. Thus, GTL-1 may be responsible for mediating cellular responses after mechanotransduction and thermotransduction, specifically amplifying responses to transduction currents.

4.2.4 RESPONSE TO HARSH TOUCH AND HIGH TEMPERATURES IN *C. ELEGANS* ADULTS, TRPV

Both noxious mechanical and thermal stimuli activate FLP neurons (Chatzigeorgiou and Schafer, 2011). These are multidendritic neurons that send projections to the anterior quarter of the animal as well as a ciliated ending to the nose (Perkins et al., 1986). These neurons respond to both large amplitude displacements of the head and smaller amplitude displacements applied to the nose of the animal (Chatzigeorgiou and Schafer, 2011). These polymodal neurons express three DEG/ENaC channels and one TRPV channel, OSM-9 (Figure 4.2). One of the DEG/ENaC channels, MEC-10, is required for calcium increases in response to harsh touch to the head, but loss of OSM-9 has no effect on this response (Chatzigeorgiou and Schafer, 2011). The, contribution of TRP channels to nose-touch response is complicated by the fact that that FLP is electrically coupled to an interneuron RIH that receives input from two other mechanoreceptors that innervate the nose, OLQs and CEPs. Though the calcium response in FLP to nose touch is affected by loss of OSM-9, the response to nose touch in FLP is rescued by expression of OSM-9 in OLQs. Mechanotransduction currents have not been recorded in FLP directly; thus, it is not clear whether the DEG/ENaC channel MEC-10 functions as a mechanotransduction channel responding to either harsh head touch or nose touch. However, because the effects of OSM-9 rescue occur when the channel is expressed in OLQs and not FLPs, these data suggest that the critical effects of OSM-9 on cell activity are limited to OLQ.

FLPs also respond to temperature changes of 15°C, from 20°C to 35°C, with an increase in internal calcium (Chatzigeorgiou and Schafer, 2011). These responses are reduced in *osm-9* mutants and can be rescued by expression of OSM-9 in FLPs directly. These results suggest two possible roles for OSM-9 in FLPs including (1) OSM-9 may act as a thermotransduction channel in parallel with another unidentified thermotransduction channel or (2) OSM-9 may be important for amplifying cellular response to thermotransduction currents that are generated by another channel. Other possibilities exist, but critical for defining the role of OSM-9 in thermotransduction is the need to record heat-evoked currents directly in FLPs.

4.3 MECHANOSENSATION IN ADULT AND LARVAL *DROSOPHILA* FLIES

4.3.1 MECHANOSENSORY NEURONS IN *DROSOPHILA* ADULT FLIES, TRPN AND TRPV

The antennae of *Drosophila* flies are multipart sensory organs that rotate in response to various mechanical stimuli including sound waves, wind, and gravity. Inside each antenna is a structure called the Johnston's organs with a collection of about 500 ciliated neurons (Johnston's organ neurons, JONs) attached to a portion of the antenna that rotates in response to mechanical stimulation (Kamikouchi et al., 2006). Classes of JONs can be distinguished by their position in the JO as well as the type of stimuli they respond to including the direction of antennal movement (e.g., anterior or posterior) and whether antennal movement in a direction is sustained or oscillatory (e.g., the push of wind or vibrations from sound waves) (Kamikouchi et al., 2006; Kamikouchi et al., 2009; Yorozu et al., 2009; Matsuo et al., 2014; Mamiya et al., 2011). At least four TRP channel subunits are expressed in JONs including two TRPV channel subunits (Nanchung and Inactive) and a TRPN channel subunit (NOMPC) (Figure 4.3).

Efforts to understand the components of mechanotransduction have focused on two classes of JONs (A and B) that respond to sound waves and are responsible for the ability of flies to hear courtship songs (Kamikouchi et al., 2006). These neurons respond best to sound waves of particular frequencies as well as amplifying the antennal oscillations in a frequency- and intensity-dependent fashion. At low intensity the best frequency of vibrations is tuned to courtship song at 150–200 Hz (Göpfert et al., 2005; Nadrowski et al., 2008; Nadrowski and Göpfert, 2009; Riabinina et al., 2011). Particular attention has been paid to the roles of NOMPC and two TRPV channel subunits, Nanchung (Nan) and Inactive (Iav) in auditory transduction. These proteins are expressed in different portions of the mechanosensory cilia of JONs and may have different roles in the cellular response to movement.

NOMPC localizes to the distal tip of mechanosensory cilia in all JONs, not just the two classes responsible for auditory transduction (Lee et al., 2010; Cheng et al., 2010; Liang et al., 2010). The loss of NOMPC affects JON function by eliminating mechanical amplification (Göpfert et al., 2006) and reducing the amplitude and sensitivity of sound-evoked afferent neuron potentials (Eberl et al., 2000; Effertz et al., 2011). Measurements of compound calcium responses from JONs indicate that NOMPC is required for the response in sound-sensitive JONs but not in other classes of JONs that respond to wind and gravity (Effertz et al., 2011). However, recordings from the giant fiber neuron that is electrically coupled to class A and B JONs demonstrate that loss of NOMPC does not eliminate sound-evoked responses (Lehnert et al., 2013).

Two different models of mechanotransduction have been developed to synthesize experimental data because there are no direct recordings of mechanotransduction currents from these neurons. One model is NOMPC dependent, proposing that NOMPC is the channel responsible for mechanotransduction in sound-sensitive JONs (Göpfert et al., 2006; Effertz et al., 2011; Effertz et al., 2012). In this model, antennal vibrations are coupled to the opening of NOMPC channels that

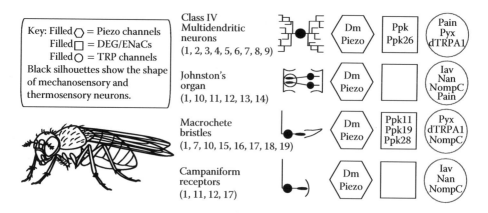

FIGURE 4.3 TRP, DEG/ENaC, and piezo channel proteins coexpressed in mechanosensory and thermosensory neurons in *Drosophila melanogaster* flies. The gross morphology of mechanosensory and thermosensory neurons in larval and adult flies are diagrammed, and the ion channel subunits that are expressed in each cell type are listed. Sources for *Drosophila* neuron expression are listed by numbers: (1) Kim, S.E. et al., Nature, 483, 7388, 209-212, 2012; (2) Ainsley, J.A et al., *Curr Biol*, 13, 17, 1557–1563, 2003; (3) Zhong, L. et al., *Curr Biol*, 20, 5, 429–434, 2010; (4) Guo, Y. et al., *Cell Rep*, 9, 4, 1183–1190, 2014; (5) Gorczyca, D.A. et al., *Cell Rep*, 9, 4, 1446–1458, 2014; (6) Tracey, W.D. et al., *Cell*, 113, 2, 261–273, 2003; (7) Lee, Y. et al., *Nature Gen*, 37, 3, 305–310, 2005; (8) Neely, G.G. et al., *PLOS ONE*, 6, 8, e24343, 2011; (9) Zhong, L. et al., *Cell Rep*, 1, 1, 43–55, 2012; (10) Lee, J. et al., *PLOS ONE*, 5, 6, e11012, 2010; (11) Kim, J. et al., *Nature*, 424, 6944, 81–84, 2003; (12) Gong, Z. et al., *J Neurosci*, 24, 41, 9059–9066, 2004; (13) Kamikouchi, A. et al., *Nature*, 457, 7235, 165–171, 2009; (14) Sun, Y. et al., *PNAS*, 106, 32, 13606–13611, 2009; (15) Chen, Z. et al., *J Neurosci*, 30, 18, 6247–6252, 2010; (16) Liu, L. et al., *Neuron*, 39, 1, 133–146, 2003; (17) Walker, R.G. et al., *Science*, 287, 5461, 2229–2234, 2000; (18) Hamada, F.N. et al., *Nature*, 454, 7201, 217–220, 2008; (19) Kim, S.H. et al., *PNAS*, 107, 18, 8440–8445, 2010.

transduce the mechanical stimulus into an electrical signal and provide mechanical amplification. In the NOMPC-independent model, NOMPC is responsible for mechanical amplification that takes place in parallel to mechanotransduction (Lehnert et al., 2013). Independently of NOMPC, mechanotransduction of sound wave vibrations requires the TRPV channel subunits, Nan and Iav.

Nan and Iav are expressed in virtually all JONs and are localized in the proximal region of the mechanosensory cilium separated from the distal attachment point of the cilia where NOMPC is expressed (Kim et al., 2003; Gong et al., 2004). Disrupting Iav expression disrupts Nan expression and vice versa, suggesting that Nan and Iav function as a heteromeric channel in JONs (Gong et al., 2004). Null mutations in either *nan* or *iav* eliminate sound-evoked nerve responses and eliminate sound-evoked responses recorded from the giant fiber neuron that receives input from multiple class A and B JONs (Kim et al., 2003; Gong et al., 2004; Lehnert et al., 2013). These results provide support for the NOMPC-independent auditory transduction model in which Nan-Iav channels are responsible for mechanotransduction. In this model, mechanical amplification is independent of transduction and provided by the activity of NOMPC channels. Surprisingly, when Nan-Iav channels are lost, mechanical feedback amplification is not eliminated but rather increases. These results provide support for both models of mechanotransduction. In both models, NOMPC is responsible for mechanical amplification of sound-evoked vibrations, and Nan-Iav negatively regulates mechanical amplification by NOMPC (Göpfert et al., 2006; Lehnert et al., 2013). In the NOMPC-dependent model, NOMPC is responsible for mechanotransduction, and Nan-Iav provides amplification of the electrical signal required to generate sound-evoked nerve affect nerve response. In the absence of individual recordings from JONs, an understanding of how the Nan-Iav channel is gated would provide a crucial test for the models. Nan and Iav channels can be activated by hypotonicity when they are expressed heterologously (Gong et al., 2004; Kim et al., 2003); however, more work will be needed to confirm whether Nan-Iav channels are mechanically gated or can be gated by changes in membrane potential.

Other sensory and chordotonal neurons with sensory cilia innervating mechanosensory organs express NOMPC in the adult fly (Walker et al., 2000) (Figure 4.3). These organs include both large and small bristles on the body (macrochaetes and microchaetes) as well as sensory bristles on the proboscis and campaniform sensilla that detect cuticle deformation of the halteres and wings. Recordings of mechanoreceptor currents from neurons innervating the large bristle sensilla indicate that NOMPC channels are responsible for the majority of the neuronal response to movement (Walker et al., 2000). Displacement of the bristle in the best direction for inducing a response initiates a large rapidly adapting current that declines to a steady state. In *nompC* mutant flies, the large peak currents are lost and only the small, steady-state current remains in response to bristle displacement. These data suggest that, though NOMPC channels may carry the major, fast-adapting mechanotransduction current, other channels are responsible for the minor, sustained response. Adult fly bristle mechanoreceptors express multiple other TRP channel as well as DEG/ENaC channel subunits, but the identity of the channel responsible for the minor current is unclear (Figure 4.3).

Examinations of the ultrastructure of the campaniform sensilla found in adult fly halteres suggest a mechanism for mechanical gating of NOMPC channels in these structures. In the mechanosensory cilia of these organs, there are filamentous connections between the plasma membrane and the microtubule cytoskeleton (Liang et al., 2013, 2014; Zhang et al., 2015). When a mutant form of NOMPC is expressed in these neurons with an N-terminus containing double the number of ankyrin repeats (Figure 4.1), the length of the filaments increases as well as the distance between the plasma membrane and microtubules in the mechanosensory cilia (Zhang et al., 2015). In *nompC* null mutants, the filamentous connectors are lost and the distance between the plasma membrane and microtubules is again increased (Zhang et al., 2015). These data suggest that the N-terminus of NOMPC is a critical component of a structure linking the plasma membrane to microtubules in the sensory cilia and that this structure could form an elastic tether responsible for opening the channel.

4.3.2 RESPONSE TO GENTLE TOUCH IN *DROSOPHILA* LARVAE, TRPN

Four classes of multidendritic (md) neurons innervate large portions of the body wall of each segment in *Drosophila* larvae (Bodmer et al., 1987; Grueber et al., 2002). Class III md neurons are responsible for the behavioral response to the gentle touch of an eyelash (Tsubouchi et al., 2012; Yan et al., 2012). When synaptic transmission in class III md neurons is inhibited with tetanus toxin, larvae do not respond to touch by an eyelash hair but will still respond if tetanus toxin is expressed in class IV md neurons (Tsubouchi et al., 2012; Yan et al., 2012). Multiple lines of evidence support the hypothesis that NOMPC functions as a mechanotransduction channel in class III md neurons. NOMPC is expressed in class III md neurons but not in class IV neurons (Yan et al., 2012). Extracellular recordings of action potentials and imaging of calcium responses from class III and class IV md neurons indicate that class III md neurons respond to small displacements of the cuticle that do not excite class IV md neurons. Action potential generation and calcium responses to small displacements in class III neurons are eliminated in *nompC* mutants and expression of NOMPC in class III md neurons of *nompC* mutants rescues the response. Additionally, ectopically expressing NOMPC in class IV causes them to generate action potentials and calcium increases in response to small displacements. Finally, when *Drosophila* S2 cells are transfected with NOMPC the protein forms a mechanically gated channel. Together these data provide evidence that NOMPC proteins may function as mechanotransduction channels in class III neurons allowing them to respond to gentle touch and small displacements of the cuticle. However, NOMPC may not be the only mechanotransduction channel in class III md neurons. Rather *ripped pocket* that encodes a DEG-ENaC channel subunit is expressed in these neurons as well. Ripped pocket (RPK) is found throughout the dendritic arbor, cell body, and axons of all multidendritic neurons (Tsubouchi et al., 2012). Null mutants for *rpk* exhibit a severely reduced behavioral response to gentle touch that can be rescued by expression of RPK (Tsubouchi et al., 2012).

4.3.3 Response to Harsh Touch in *Drosophila* Larvae, TRPA

Class IV md neurons function as nociceptors and respond to multiple types of noxious stimuli including harsh mechanical stimulation, high heat, and short-wavelength light (Tracey et al., 2003; Xiang et al., 2010). These neurons express three TRPA channel subunits, Painless, Pyrexia, and dTRPA1 (Figure 4.3). The contribution of Pyrexia to mechanosensitivity in class IV md neurons has not been tested, but deletion of *painless* and *dTRPA1* increases the threshold for aversive responses to heat and force (Tracey et al., 2003; Zhong et al., 2012). Additionally, these nociceptors express multiple genes that affect the behavioral response to mechanical stimulation including DEG/ENaC channel subunits, *pickpocket* (*ppk*) and *balboa* or *pickpocket26* (*ppk26*), as well as the sole *piezo* gene in *Drosophila*, Dm*piezo* (Kim et al., 2012; Ainsley et al., 2003; Zhong et al., 2010; Guo et al., 2014; Gorczyca et al., 2014).

Dm*piezo* is widely expressed in all types of sensory neurons as well as in several nonneuronal tissues (Kim et al., 2012); however, Dm*piezo* knockout animals only have defects in the behavioral response of larvae to harsh touch. Dm*piezo* knockout animals do not demonstrate reduced coordination or a loss of bristle mechanoreceptor potential as adult flies. In Dm*piezo* knockout animals, gentle touch and the response to high heat are unaffected. Together these results suggest that Dm*piezo* is critical for the function of class IV md neurons.

Genes that encode DEG/ENaC channel subunits, *pickpocket* and *pickpocket26*, may form protein subunits that function as mechanotransduction channels in parallel with Dm*piezo* channels. Both *ppk* and *ppk26* are expressed in class IV md neurons (Ainsley et al., 2003; Zhong et al., 2010; Guo et al., 2014; Gorczyca et al., 2014). Loss of *ppk* and *ppk26* reduces the behavioral response of larvae to harsh touch but not to high heat (Guo et al., 2014; Gorczyca et al., 2014; Zhong et al., 2010). PPK and PPK 26 are interdependent for cell surface expression of either partner suggesting that together these DEG/ENaC subunits form a heteromeric channel (Guo et al., 2014; Gorczyca et al., 2014). Dm*piezo* is expressed in *ppk*-expressing cells, and when *ppk*-expressing cells are isolated from larvae, these cells have a mechanically gated current that is eliminated by loss of Dm*piezo* (Kim et al., 2012). If both *ppk* and Dm*piezo* are knocked down by expressing RNAi for the genes in *ppk*-expressing cells, then the loss of behavioral response to harsh mechanical stimulation is more severe than that for larvae where *ppk* or Dm*piezo* expression has been reduced individually using RNAi (Kim et al., 2012). Double *ppk26::piezo* mutants have a significantly reduced response to harsh mechanical stimulation compared to either mutant alone. However, *ppk::ppk26* double mutants were not less sensitive to harsh mechanical stimulus than *ppk* mutants. Together these data suggest that Dm*piezo* and the two DEG/ENaC subunits, PPK and PPK 26, may form mechanotransduction channels that function in parallel, separate pathways. In contrast to the results for DEG/ENaC channels and DmPiezo, evidence suggests that DmPiezo and the TRPA channel Pain may function in series to transmit mechanical signals in class IV md neurons. Double *Dmpiezo::pain* mutants have the same level of behavioral response to harsh touch as individual mutants (Kim et al., 2012). Thus, Pain and DmPiezo may function in the same pathway.

4.3.4 Proprioception in *Drosophila* Larvae, TRPA

DmPiezo and the dTRPA1 channel may function in series to transmit mechanical signals in the dorsal bipolar dendritic (dbd) sensory neurons that enervate the larval body wall (Suslak et al., 2015). These cells respond to stretch of the body wall with a complex change in membrane potential composed of a fast adapting peak and a smaller sustained depolarized potential (Suslak et al., 2015). In *dTRPA1* mutants or larvae in which *dTRPA1* expression is reduced by RNAi knockdown, the fast-adapting peak response is lost, but the smaller sustained response remains. In contrast to *dTRPA1* RNAi knockdown lines, the peak response is eliminated and the minor, sustained response is severely reduced in two Dm*piezo* knockdown lines. Together these results support a model that DmPiezo proteins form channels responsible for mechanotransduction, and dTRPA1 channels amplify the response to mechanical stimuli in dbd neurons. Another mechanotransduction channel may be responsible for the remaining sustained response to mechanical stimulation.

4.4 THERMOSENSATION IN ADULT AND LARVAL *DROSOPHILA* FLIES

4.4.1 THERMOTRANSDUCTION: TRPA CHANNEL CANDIDATES

Three TRPA channels are candidates for thermotransduction channels in *Drosophila* larvae and adults: dTRPA1, Pain, and Pyrexia (Figure 4.1). All three produce thermotransduction currents when heterologously expressed (Hwang et al., 2012; Neely et al., 2011; Tracey et al., 2003; Zhong et al., 2012; Sokabe et al., 2008; Lee et al., 2005; Viswanath et al., 2003). In addition, their function as thermotransduction channels is supported by evidence that loss of the genes affects behavior of the animals as well as calcium imaging and action potential recordings from *Drosophila* cells. Though none of these experiments demonstrate the specific role of these channels in thermosensory cells, this work suggests that the channels function as both thermotransduction channels and downstream regulators of cellular response. Each of the three genes, *dTRPA1*, *painless*, and *pyrexia*, are expressed as multiple isoforms that have different cellular roles. There are at least four isoforms of dTRPA1 expressed in different *Drosophila* cells (Kwon et al., 2010b; Zhong et al., 2012), two of the isoforms dTRPA1-B and dTRPA1-C do not respond to temperature change when expressed in heterologous cells though they will form channels that respond to the agonist allyl isothiocyanate (AITC) (Zhong et al., 2012). These isoforms may function downstream of thermotransduction channels in cells, for example, dTRPA1-C is not activated by temperature changes but can rescue the behavioral response to noxious heat when expressed in larval nociceptors in *dTRPA1* mutant animals (Zhong et al., 2012). When Pain is heterologously expressed, it forms a thermotransduction channel with different thresholds for heat response depending on the presence or absence of internal calcium ions; without internal calcium, the threshold is 44°C and with internal calcium, the threshold is lowered by 2°C (Sokabe et al., 2008). In Drosophila cells, the painless gene is transcribed as three RNA variants that are predicted to encode Pain protein isoforms with variable length N-terminal domains (Hwang et al., 2012). The *painless* RNA variant that encodes a protein with the longest N-terminus contains eight ankyrin repeats and the variant that encodes a protein with the shortest N-terminus lacks the ankyrin repeat portion of the N-terminus (Hwang et al., 2012). The shortest variant cannot rescue the behavioral defect in thermal nociception of the *painless* mutant when expressed in larval nociceptor neurons; however, the longest variant with all eight ankyrin repeats can. These data suggest that amino acids in the N-terminus, possibly those within the ankyrin repeat domain, allow the protein to respond to changes in temperature. The shortest variant can rescue the behavioral response to harsh mechanical stimulation when expressed in larval nociceptors, but the longest version cannot. In contrast to pain, the two isoforms of Pyrexia, Pyx-A and Pyx-B, both form thermotransduction channels when expressed in heterologous cells, though one does not contain an ankyrin repeat domain (Lee et al., 2005). The shorter variant, Pyx-B, does not contain the nine ankyrin repeats that are found in the Pyx-A variant. The difference between the variants in a heterologous expression system is the threshold for activation. The shorter isoform, Pyx-B, forms a channel with an activation threshold that is 2.5°C cooler than the activation threshold for Pyx-A.

All of these TRPA isoforms have activation thresholds above 24°C, yet both adult and larval *Drosophila* respond to a wider range of temperatures suggesting that more thermotransduction channels function in these animals. Larval animals have a preferred temperature of 18°C (Kwon et al., 2008; Klein et al., 2015) and respond to both warmer and cooler temperatures with thermotaxis back to their preferred temperature (Kwon et al., 2008; Klein et al., 2015). Adult flies prefer 25°C (Sayeed and Benzer, 1996; Lee et al., 2005; Hamada et al., 2008) and avoid cooler and warmer temperatures (Hamada et al., 2008; Gallio et al., 2011). Additionally, both larval and adult animals have a separate behavioral response to high temperatures, larvae writhe and roll in response to temperatures above 39°C (Tracey et al., 2003; Hwang et al., 2007) and adults jump in response to temperatures above 45°C (Xu et al., 2006). Some of the cells and ion channels that are important for thermotaxis and avoidance behavior have been identified for larval and adult animals.

4.4.2 RESPONSE TO HEAT IN *DROSOPHILA* LARVAE, TRPA

The behavioral response to high heat, harsh touch as well as short wavelength light is dependent on class IV md neurons (Tracey et al., 2003; Xiang et al., 2010; Hwang et al., 2012). The rolling and writhing behavioral responses to noxious temperatures and mechanical stimuli appear to be an antipredator behavior because larvae roll in the direction of the stimulus when they are attacked by a predatory wasp as well as when they are touched by a hot probe (Hwang et al., 2012). This behavior works to dislodge the wasp but does not help larvae avoid injury from a hot probe. These cells express all three TRPA channel subunits that form thermotransduction channels when heterologously expressed, dTRPA1, Pain, and Pyx (Hwang et al., 2012; Neely et al., 2011; Tracey et al., 2003; Zhong et al., 2012; Sokabe et al., 2008; Lee et al., 2005; Viswanath et al., 2003). Mutations in *dTRPA1*, *painless* and *pyrexia* reduce larval behavioral response to noxious heat (Hwang et al., 2012; Neely et al., 2011; Tracey et al., 2003; Zhong et al., 2012). The threshold for behavioral response and cellular activation to noxious heat is ~39°C–43°C (Tracey et al., 2003; Xiang et al., 2010; Terada et al., 2016). Pain has a temperature activation threshold of ~39°C–42°C when heterologously expressed in HEK cells (Sokabe et al., 2008). Pyx is activated by temperatures above ~37.5°C–40°C when expressed in *Xenopus* oocytes and HEK cells (Lee et al., 2005). Either of these TRPA channels could function as thermotransduction channels in class IV md neurons. Two variants of dTRPA1 are activated at much lower temperatures, dTRPA1-A is activated by temperature changes between 24°C and 29°C (Viswanath et al., 2003) when expressed in *Xenopus* oocytes, and dTRPA1-D activates calcium increases at 34°C when expressed in cells from the *Drosophila* S2R+ cell line (Zhong et al., 2012). Inducing dTRPA1-A expression in class IV md neurons lowers the threshold for the behavioral response to noxious temperatures to 30°C suggesting that dTRPA1-A is not the isoform that functions in these cells. The behavioral response to heat is lowered to 34°C after short wavelength ultraviolet (UV-C) radiation suggesting that dTRPA1-D may be involved in the alteration of temperature response after tissue damage (Babcock et al., 2009). dTRPA1-C does not function as a thermotransduction channel when heterologously expressed, but it can rescue the behavioral response to high heat when expressed in class IV md neurons of *dTRPA1* mutant animals (Zhong et al., 2012). dTRPA1 isoforms may be important for amplifying the cellular response to thermotransduction as well as shifting the temperature threshold for class IV md neurons' response that follows tissue damage.

Recordings of action potentials from class IV md neurons suggest that the cellular response to high heat (45°C) is a specific firing pattern of high-frequency bursts interspersed with short pauses (Terada et al., 2016). The loss of dTRPA1 and Pain affect this response in different ways suggesting they have different roles in cell activation. In *pain* mutants, heating class IV md to 40°C induces significantly slower action potential firing frequencies during the bursts that are elicited by noxious heat. Action potential bursts in *dTrpA1* mutants are also significantly slowed and, in addition, the time between stimulus application and maximum firing rate is significantly longer. Together these data suggest that loss of Pain reduces the size of the cellular response and might be involved in transduction or amplification but does not appear to be the sole thermotransduction channel in class IV md neurons. Loss of dTPRA1 alters the timing of the response as well the size, providing support for the hypothesis that dTRPA1 is involved in rapid amplification of the cellular response.

PAIN and dTRPA1 may have separate roles in the behavior changes that are induced by tissue damage. Exposure to ultraviolet radiation causes tissue damage to larvae leaving epidermal cells with abnormal morphology and activating a marker of apoptosis, caspase 3 (Babcock et al., 2009). Following UV radiation exposure, larvae demonstrate increased probability of writhing and rolling in response to noxious temperatures (>40°C) as well as to temperatures that normally do not elicit that response (38°C). When apoptosis is blocked, larvae continued to exhibit increased response to noxious temperatures (hyperalgesia) but did not exhibit increased response to warmth (38°C, allodynia). Signaling through the TNF-α pathway induces allodynia in *Drosophila* larvae. Expressing Eigen, the *Drosophila* homolog of tumor necrosis factor alpha (TNF-α) in sensory neurons can induce allodynia in the absence of UV irradiation. Larvae with null mutations in *eiger* and larvae with the Eiger-receptor Wengen knocked down do not develop allodynia (Babcock et al., 2009). TNF-α is one of the signaling molecules that can induce nociceptor sensitization in mammals (Zhou et al., 2016). Hedgehog (Hh) signaling acts in parallel to Eiger to induce allodynia and can also induce hyperalgesia (Babcock et al., 2011). Pain is required for the development of Hh- or Eigen-induced allodynia. Knocking down dTRPA-1 reduces the development of allodynia that follows UV radiation, and dTRPA1 is required for Hh-induced thermal hyperalgesia.

4.4.3 RESPONSE TO WARM TEMPERATURES IN *DROSOPHILA* LARVAE, TRPA

As well as responding to noxious heat, larvae avoid temperatures above and below 18°C (Rosenzweig, 2005; Kwon et al., 2008). Without dTRPA1 larvae no longer avoid slightly higher temperatures (19°C–24°C) but still avoid even higher temperatures (26°C–32°C) as well as lower temperatures (14°C–16°C). When expressed heterologously, dTRPA1 is activated at higher temperatures than the larvae normally avoid (Viswanath et al., 2003); however, dTRPA1 may be activated indirectly in this lower temperature range (19°C–24°C). Thermotaxis to avoid temperatures in the 19°C–24°C range was eliminated in larvae with mutations in genes that encode phospholipase C and G_q, *norpA* and *Gα49B* respectively, suggesting that the G_q-PLC pathway may activate TRP channels during thermotaxis as it does during phototransduction (Bloomquist et al., 1988). dTRPA1 is expressed in the larval brain as well as two pairs of neurons adjacent to the mouth hooks. Antibodies to NORPA and Gα49B do not label these cells; however, expressing NORPA in *dTRPA1*-expressing cells rescues *norpA* mutant larvae. Thus, a G protein-coupled receptor may initiate thermosensory signaling cascades in dTRPA1-expressing cells that activate dTRPA1 and allow it to function as a thermotransduction channel outside of the range of temperatures that activate the channel directly.

4.4.4 RESPONSE TO COOL TEMPERATURES IN *DROSOPHILA* LARVAE, TRPV

To avoid cool temperatures, larvae require the activity of chordotonal neurons because suppressing neurotransmitter release by expressing tetanus toxin in these neurons impairs the ability of larvae to choose between 14°C and 17.5°C (Kwon et al., 2010 a). The TRPV channel subunit Iav is expressed in these cells, and *iav* mutant animals are defective in their ability to discriminate between 14°C and 17.5°C. However, loss of *iav* does not affect their response to temperatures lower than 12°C and higher than 22°C. The ability to choose 17.5°C over 14°C can be rescued by expressing Iav in chordotonal neurons. These neurons also express Nanchung (Gong, 2004) and Pain (Tracey et al., 2003), but loss of pain does not affect the ability to avoid cool temperatures, and *nan* mutants have such difficulty moving that their response to cool temperatures cannot be assayed (Kwon et al., 2010 a). Though Iav is important for the behavioral response to cool temperatures, it does not elicit a cool-activated current when expressed in *Xenopus* oocytes (Kwon et al., 2010 a). Thus, it is not clear whether Iav is directly activated by cool temperatures or downstream of another signal as dTRPA1 may be.

4.4.5 RESPONSE TO WARM TEMPERATURES IN *DROSOPHILA* ADULT FLIES, TRPA

Adult flies prefer warmer temperatures than larvae and the temperature responses of thermosensory cells reflect this difference. Both the anterior cell (AC) neurons in the brain and the hot cell (HC) neurons in the arista portion of the antennae are activated by warmth and increases in internal calcium can be measured in these cells in response to temperatures warmer than 25°C (Hamada et al., 2008; Gallio et al., 2011; Tang et al., 2013). The cell bodies of HC neurons sit at the base of the arista portion of the antennae, and the three HCs in each arista are located next to three neurons that respond to cooling temperatures, cold cell (CC) neurons. Though the activity pattern of HCs suggests that dTRPA1 might function as a thermotransduction channel in these cells, increases in internal calcium remain in *dTRPA1* null flies. The role of dTRPA1 in AC neurons is clearer.

The cell bodies of AC neurons are located near the brain surface where the antennal nerve enters the brain and AC neurons receive input from the antennae (Tang et al., 2013). These cells have a complicated response to warming temperatures with two calcium peaks: one at 25°C and another at 27°C. The AC neurons express dTRPA1, and the first calcium peak at 25°C is eliminated in *dTRPA1* mutant animals, but not the second peak at 27°C (Hamada et al., 2008; Tang et al., 2013). In contrast, removing the antennae severely reduces the size of the second peak (Tang et al., 2013). When examined in more detail, removal of the second antennal segment causes a severe reduction in the second calcium peak. The second antennal segment contains chordotonal neurons that form the Johnston's organ. This structure is required for hearing as well as the mechanosensory response to gravity and wind. Five TRP channel subunits are expressed in the cells of the second antennal segment (Sun et al., 2009) including two TRPA channel subunits Pain and Pyrexia (Tracey et al., 2003; Lee et al., 2005). Neurons expressing *pyx* appear to project from the second antennal segment to the soma of AC neurons. In *pyx* null mutants, the second response is eliminated and the first response remains (Tang et al., 2013). Increasing temperatures elicit the same calcium increases in *pain* mutant flies as those measured in wild-type flies. The dTRPA1-A variant forms a thermotransduction channel in heterologous cells that responds to temperatures in the proper temperature range (above 24°C) to generate the first response and dTRPA1 may function as a thermotransduction channel in AC neurons. However, the temperature threshold for Pyx is ~10°C higher than the temperature that elicits the second response in AC neurons. Thus, Pyx channels may not function as the thermotransduction channel in neurons that synapse onto AC neurons unless their temperature threshold is modified in the cells.

4.4.6 RESPONSE TO COOL TEMPERATURES IN *DROSOPHILA* ADULT FLIES, TRPP

The cell bodies of cold cell (CC) neurons sit at the base of the arista portion of the antennae, and the three CCs in each arista are located next to three neurons that respond to warm temperatures, HC neurons. As temperatures drop below 23°C, CCs respond with calcium increases (Gallio et al., 2011). These cells express the *brividol* gene that encodes a TRPP channel subunit. Flies with mutations in the three *brv* genes, *brv1*, *brv2*, and *brv3*, all have defects in their avoidance of temperatures below 24°C. CCs in *brv1* and *brv2* mutants have minimal calcium increases in response to cooling. It has not been possible to map the expressions of *brv2* and *brv3*. However, both the effects of a null mutation in *brv2* on calcium responses in the *brv1*-expressing CCs and the effects of knocking down *brv3* with RNAi expressed in *brv1*-expressing cells mimic the phenotype of animals with ubiquitous *brv3* RNAi expression. Thus, *brv2* and *brv3* might be expressed in CCs. It is not clear whether any of three TRPP channel subunits form thermotransduction channels or are important components of downstream amplification in the cellular response. Attempts to express any of the three genes in vertebrate and insect cell expression systems did not produce thermotransduction channels and an attempt to misexpress them in other *Drosophila* cells did not induce thermosensitivity in those cells (Gallio et al., 2011).

4.5 CONCLUSIONS

Work to understand how mechanosensory and thermosensory neurons function in invertebrates may provide insights into how these cells function in other animals. Both information about which proteins function as transduction channels and how transduction channels interact with other cellular components to control the response to mechanical and thermal stimuli may provide lessons about the role of TRP channels that are applicable to other animals. Consistently, in both nematodes and flies, researchers have found that mechanosensory cells that respond to smaller displacements or forces have TRPN channels that function as mechanotransduction channels (Figures 4.1 through 4.3). Cells that respond to larger displacements and greater forces express members of the DEG/ENaC and Piezo channel gene family and these channels form mechanotransduction channels in the cells (Figures 4.1 through 4.3). TRP channel subunits are expressed in mechanosensory cells that respond to intense stimulation, but their role may not be as transduction channels in these cells. Though no TRPN channel subunit genes remain in vertebrates beyond fish (Damann et al., 2008; Wu et al., 2010; Nilius and Owsianik, 2011; Kadowaki, 2015), these data suggest that TRP channels may have different roles in different types of mechanosensory neurons and that TRP channels may share the role of the mechanotransduction channel with members of other gene families.

TRPA channels have a clear role as transduction channels in *Drosophila* thermosensory cells and can form channels that are gated by temperature when heterologously expressed. However, the temperature ranges that gate TRPA channels in heterologous cells do not represent the full range of temperatures to which adult and larval flies respond. Specifically, the TRPV and TRPP channel subunits in cells that respond to cool temperatures appear to be important for cell function, though these channel subunits do not form thermotransduction channels when heterologously expressed. These data suggest several possibilities including TRPP and TRPV channels do form thermotransduction channels in these cells, but they must be modified to function properly as thermotransduction channels. Many TRP channels can be modulated by phosphatidylinositol-4,5-bisphosphate hydrolysis catalyzed by phospholipase-C (PLC) (Hardie, 2006; Qin, 2007). When rat TRPV1 is purified and incorporated into lipid nanodiscs or liposomes, it appears that the channels interact with membrane phosphatidylinositol lipids and that those interactions affect the ability of the channels to respond to temperature changes (Cao et al., 2013; Gao et al., 2016). These results suggest that channel gating can be affected by modification of membrane lipids and that lipid modification may only properly occur in cells in which the channels are normally expressed. These results additionally suggest a way that TRP channel activity could be modified in nociceptors to modulate the cellular response threshold to mechanical and thermal stimulation. Finally, work in invertebrate animals has provided information about important structural features of TRP channels. Altering the length of the N-terminus and the number of ankyrin repeats in TRPN channels affects interactions between the plasma membrane and microtubules in *Drosophila* mechanosensory cells. This portion of the channel may provide a tether that links the channel to the cytoskeleton and supplies a mechanism for force to pull the channel open. Though the N-terminus has important effects on the ability of TRPA channels to function as thermotransduction channels, the ankyrin repeats may not be the critical structures. Both variants of Pyrexia (Pyx-A and Pyx-B) can form thermotransduction channels in heterologous cells despite the fact that Pyx-B does not contain the ankyrin repeat domain (Lee et al., 2005). Two of the four known variants of the *Drosophila* dTRPA-1 channel subunit form channels that open in response to temperature changes, yet what distinguishes these channel variants from the variants that do not respond to temperature is a region between the ankyrin repeat domain and the first transmembrane helix (Zhong et al., 2012). Thus, work in invertebrate animals has provided multiple insights into the critical roles that TRP channels play in the function of mechanosensory and thermosensory neurons.

REFERENCES

Ainsley, J.A. et al. 2003. Enhanced locomotion caused by loss of the *Drosophila* DEG/ENaC protein pickpocket1. *Curr Biol*, 13(17): 1557–1563. doi:10.1016/S0960-9822(03)00596-7.

Babcock, D.T., C. Landry, and M. J Galko. 2009. Cytokine signaling mediates UV-induced nociceptive sensitization in *Drosophila* larvae. *Curr Biol*, 19(10): 799–806. doi:10.1016/j.cub.2009.03.062.

Babcock, D.T. et al. 2011. Hedgehog signaling regulates nociceptive sensitization. *Curr Biol*, 21(18): 1525–1533. doi:10.1016/j.cub.2011.08.020.

Basbaum, A.I. et al. 2009. Cellular and molecular mechanisms of pain. *Cell*, 139(2): 267–284. doi:10.1016/j.cell.2009.09.028.

Bloomquist, B.T. et al. 1988. Isolation of a putative phospholipase C gene of *Drosophila*, *norpA*, and its role in phototransduction. *Cell*, 54(5): 723–733.

Bodmer, R. et al. 1987. Transformation of sensory organs by mutations of the cut locus of *D. melanogaster*. *Cell*, 51(2): 293–307.

Cao, E. et al. 2013. TRPV1 channels are intrinsically heat sensitive and negatively regulated by phosphoinositide lipids. *Neuron*, 77(4): 667–679. doi:10.1016/j.neuron.2012.12.016.

Cao, E. et al. 2013. TRPV1 structures in distinct conformations reveal activation mechanisms. *Nature*, 504(7478): 113–118. doi:10.1038/nature12823.

Chao, M.Y. et al. 2004. Feeding status and serotonin rapidly and reversibly modulate a *Caenorhabditis elegans* chemosensory circuit. *Proc Natl Acad Sci U S A*, 101(43): 15512–15517. doi:10.1073/pnas.0403369101.

Chatzigeorgiou, M. and W.R. Schafer. 2011. Lateral facilitation between primary mechanosensory neurons controls nose touch perception in *C. elegans*. *Neuron*, 70(2): 299–309. doi:10.1016/j.neuron.2011.02.046.

Chatzigeorgiou, M. et al. 2010. Specific roles for DEG/ENaC and TRP channels in touch and thermosensation in *C. elegans* nociceptors. *Nature Neurosci*, 13(7): 861–868. doi:10.1038/nn.2581.

Chatzigeorgiou, M. et al. 2013. Tmc-1 encodes a sodium-sensitive channel required for salt chemosensation in *C. elegans*. *Nature*, 494(7435): 95–99. doi:10.1038/nature11845.

Chen, Z., Q. Wang, and Z. Wang. 2010. The amiloride-sensitive epithelial Na+ channel PPK28 is essential for *Drosophila* gustatory water reception. *J Neurosci*, 30(18): 6247–6252. doi:10.1523/JNEUROSCI.0627-10.2010.

Cheng, L.E. et al. 2010. The role of the TRP channel NompC in *Drosophila* larval and adult locomotion. *Neuron*, 67(3): 373–380. doi:10.1016/j.neuron.2010.07.004.

Colbert, H.A., T.L. Smith, and C.I. Bargmann. 1997. OSM-9, a novel protein with structural similarity to channels, is required for olfaction, mechanosensation, and olfactory adaptation in *Caenorhabditis elegans*. *J Neurosci*, 17(21): 8259–8269.

Coste, B. et al. 2012. Piezo proteins are pore-forming subunits of mechanically activated channels. *Nature*, 483(7388): 176–181. doi:10.1038/nature10812.

Culotti, J.G. and R.L Russell. 1978. Osmotic avoidance defective mutants of the nematode *Caenorhabditis elegans*. *Genetics*, 90(2): 243–256.

Damann, N., T. Voets, and B. Nilius. 2008. TRPs in our senses. *Curr Biol*, 18(18): R880–R889. doi:10.1016/j.cub.2008.07.063.

Driscoll, M. and M. Chalfie. 1991. The *Mec-4* gene is a member of a family of *Caenorhabditis elegans* genes that can mutate to induce neuronal degeneration. *Nature*, 349(6310): 588–593. doi:10.1038/349588a0.

Eberl, D.F., R.W. Hardy, and M.J. Kernan. 2000. Genetically similar transduction mechanisms for touch and hearing in *Drosophila*. *J Neurosci*, 20(16): 5981–5988.

Effertz, T., R. Wiek, and M.C. Göpfert. 2011. NompC TRP channel is essential for *Drosophila* sound receptor function. *Curr Biol*, 21(7): 592–597. doi:10.1016/j.cub.2011.02.048.

Effertz, T. et al. 2012. Direct gating and mechanical integrity of *Drosophila* auditory transducers require TRPN1. *Nature Neurosci*, 15(9): 1198–1200. doi:10.1038/nn.3175.

Gallio, M. et al. 2011. The coding of temperature in the *Drosophila* brain. *Cell*, 144(4): 614–624. doi:10.1016/j.cell.2011.01.028.

Gao, Y. et al. 2016. TRPV1 structures in nanodiscs reveal mechanisms of ligand and lipid action. *Nature*, 534(7607): 347–351. doi:10.1038/nature17964.

Gaudet, R. 2008. A primer on ankyrin repeat function in TRP channels and beyond. *Mol Biosyst*, 4(5): 372–379. doi:10.1039/b801481g.

Ge, J. et al. 2015. Architecture of the mammalian mechanosensitive Piezo1 channel. *Nature*, 527(7576): 64–69. doi:10.1038/nature15247.

Geffeney, S.L. et al. 2011. DEG/ENaC but not TRP channels are the major mechanoelectrical transduction channels in a *C. elegans* nociceptor. *Neuron*, 71(5): 845–857. doi:10.1016/j.neuron.2011.06.038.

Gong, Z. et al. 2004. Two interdependent TRPV channel subunits, inactive and nanchung, mediate hearing in *Drosophila*. *J Neurosci*, 24(41): 9059–9066. doi:10.1523/JNEUROSCI.1645-04.2004.

Goodman, M.B. and E.M. Schwarz. 2003. Transducing touch in *Caenorhabditis elegans*. *Ann Rev Physiol*, 65(1): 429–452. doi:10.1146/annurev.physiol.65.092101.142659.

Göpfert, M.C., A.D.L. Humphris, and J.T. Albert. 2005. Power gain exhibited by motile mechanosensory neurons in *Drosophila* ears. *Proc Natl Acad Sci U S A*, 102(2): 325–330.

Göpfert, M.C. et al. 2006. Specification of auditory sensitivity by *Drosophila* TRP channels. *Nature Neurosci*, 9(8): 999–1000. doi:10.1038/nn1735.

Gorczyca, D.A. et al. 2014. Identification of Ppk26, a DEG/ENaC channel functioning with Ppk1 in a mutually dependent manner to guide locomotion behavior in *Drosophila*. *Cell Rep*, 9(4): 1446–1458. doi:10.1016/j.celrep.2014.10.034.

Grueber, W.B, L.Y Jan, and Y.N. Jan. 2002. Tiling of the *Drosophila* epidermis by multidendritic sensory neurons. *Development*, 129(12): 2867–2878.

Guo, Y. et al. 2014. The role of PPK26 in *Drosophila* larval mechanical nociception. *Cell Rep*, 9(4): 1183–1190. doi:10.1016/j.celrep.2014.10.020.

Hall, D.H. et al. 1997. Neuropathology of degenerative cell death in *Caenorhabditis elegans*. *J Neurosci*, 17(3): 1033–1045.

Hamada, F.N. et al. 2008. An internal thermal sensor controlling temperature preference in *Drosophila*. *Nature*, 454(7201): 217–220. doi:10.1038/nature07001.

Hardie, R.C. 2001. Phototransduction in *Drosophila melanogaster*. *J Exper Biol*, 204(Pt 20): 3403–3409.

Hardie, R.C. 2006. TRP channels and lipids: From *Drosophila* to mammalian physiology. *J Physiol*, 578(1): 9–24. doi:10.1113/jphysiol.2006.118372.

Harris, G.P. et al. 2009. Three distinct amine receptors operating at different levels within the locomotory circuit are each essential for the serotonergic modulation of chemosensation in *Caenorhabditis elegans*. *J Neurosci*, 29(5): 1446–1456. doi:10.1523/JNEUROSCI.4585-08.2009.

Hart, A.C., S. Sims, and J.M. Kaplan. 1995. Synaptic code for sensory modalities revealed by *C. elegans* GLR-1 glutamate receptor. *Nature*, 378(6552): 82–85. doi:10.1038/378082a0.

Hart, A.C. et al. 1999. Distinct signaling pathways mediate touch and osmosensory responses in a polymodal sensory neuron. *J Neurosci*, 19(6): 1952–1958.

Hilliard, M.A., C.I. Bargmann, and P. Bazzicalupo. 2002. *C. elegans* responds to chemical repellents by integrating sensory inputs from the head and the tail. *Curr Biol*, 12(9): 730–734.

Hilliard, M.A. et al. 2005. In vivo imaging of *C. elegans* ASH neurons: Cellular response and adaptation to chemical repellents. *EMBO J*, 24(1): 63–72. doi:10.1038/sj.emboj.7600493.

Howard, J. and S. Bechstedt. 2004. Hypothesis: A helix of ankyrin repeats of the NOMPC-TRP ion channel is the gating spring of mechanoreceptors. *Curr Biol*, 14(6): R224–R226.

Huang, M. and M. Chalfie. 1994. Gene interactions affecting mechanosensory transduction in *Caenorhabditis elegans*. *Nature*, 367(6462): 467–470. doi:10.1038/367467a0.

Husson, S.J. et al. 2012. Optogenetic analysis of a nociceptor neuron and network reveals ion channels acting downstream of primary sensors. *Curr Biol*, 22(9): 743–752. doi:10.1016/j.cub.2012.02.066.

Hwang, R.Y., N.A. Stearns, and W.D. Tracey. 2012. The ankyrin repeat domain of the TRPA protein painless is important for thermal nociception but not mechanical nociception. *PLOS ONE*, 7(1): e30090. doi:10.1371/journal.pone.0030090.g004.

Hwang, R.Y. et al. 2007. Nociceptive neurons protect *Drosophila* larvae from parasitoid wasps. *Curr Biol*, 17(24): 2105–2116. doi:10.1016/j.cub.2007.11.029.

Jin, X., J. Touhey, and R. Gaudet. 2006. Structure of the N-terminal ankyrin repeat domain of the TRPV2 ion channel. *J Biol Chem*, 281(35): 25006–25010. doi:10.1074/jbc.C600153200.

Kadowaki, T. 2015. Evolutionary dynamics of metazoan TRP channels. *Pflügers Archiv*, 467(10): 2043–2053. doi:10.1007/s00424-015-1705-5.

Kahn-Kirby, A.H. and C.I. Bargmann. 2006. Trp channels in *C. elegans*. *Ann Rev Physiol*, 68(1): 719–736. doi: 10.1146/annurev.physiol.68.040204.100715.

Kamikouchi, A., T. Shimada, and K. Ito. 2006. Comprehensive classification of the auditory sensory projections in the brain of the fruit fly *Drosophila melanogaster*. *J Comparat Neurol*, 499(3): 317–356. doi:10.1002/cne.21075.

Kamikouchi, A. et al. 2009. The neural basis of *Drosophila* gravity-sensing and hearing. *Nature*, 457(7235): 165–171. doi:10.1038/nature07810.

Kang, L. et al. 2010. *C. elegans* TRP family protein TRP-4 is a pore-forming subunit of a native mechanotransduction channel. *Neuron*, 67(3): 381–391. doi:10.1016/j.neuron.2010.06.032.

Kaplan, J.M. and H.R. Horvitz. 1993. A dual mechanosensory and chemosensory neuron in *Caenorhabditis elegans*. *Proc Natl Acad Sci U S A*, 90(6): 2227–2231.

Kawashima, Y. et al. 2015. Transmembrane channel-like (TMC) genes are required for auditory and vestibular mechanosensation. *Pflügers Archiv*, 467(1): 85–94. doi:10.1007/s00424-014-1582-3.

Kim, J. et al. 2003. A TRPV family ion channel required for hearing in *Drosophila*. *Nature*, 424(6944): 81–84. doi:10.1038/nature01733.

Kim, S.E. et al. 2012. The role of *Drosophila* Piezo in mechanical nociception. *Nature*, 483(7388): 209–212. doi:10.1038/nature10801.

Kim, S.H. et al. 2010. *Drosophila* TRPA1 channel mediates chemical avoidance in gustatory receptor neurons. *PNAS*, 107(18): 8440–8445. doi:10.1073/pnas.1001425107.

Kindt, K.S. et al. 2007. *Caenorhabditis elegans* TRPA-1 functions in mechanosensation. *Nature Neurosci*, 10(5): 568–577. doi:10.1038/nn1886.

Klein, M. et al. 2015. Sensory determinants of behavioral dynamics in *Drosophila* thermotaxis. *PNAS*, 112(2): E220–E229. doi:10.1073/pnas.1416212112.

Kwon, Y. et al. 2008. Control of thermotactic behavior via coupling of a TRP channel to a phospholipase C signaling cascade. *Nature Neurosci*, 11(8): 871–873. doi:10.1038/nn.2170.

Kwon, Y. et al. 2010a. Fine thermotactic discrimination between the optimal and slightly cooler temperatures via a TRPV channel in chordotonal neurons. *J Neurosci*, 30(31): 10465–10471. doi:10.1523/JNEUROSCI.1631-10.2010.

Kwon, Y. et al. 2010b. *Drosophila* TRPA1 channel is required to avoid the naturally occurring insect repellent citronellal. *Curr Biol*, 20(18): 1672–1678. doi:10.1016/j.cub.2010.08.016.

Lee, J. et al. 2010. *Drosophila* TRPN(=NOMPC) channel localizes to the distal end of mechanosensory cilia. *PLOS ONE*, 5(6): e11012. doi:10.1371/journal.pone.0011012.s002.

Lee, Y. et al. 2005. Pyrexia is a new thermal transient receptor potential channel endowing tolerance to high temperatures in *Drosophila melanogaster*. *Nature Gen*, 37(3): 305–310. doi:10.1038/ng1513.

Lehnert, B.P. et al. 2013. Distinct roles of TRP channels in auditory transduction and amplification in *Drosophila*. *Neuron*, 77(1): 115–128. doi:10.1016/j.neuron.2012.11.030.

Li, W. et al. 2006. A *C. elegans* stretch receptor neuron revealed by a mechanosensitive TRP channel homologue. *Nature*, 440(7084): 684–687. doi:10.1038/nature04538.

Li, W. et al. 2011. The neural circuits and sensory channels mediating harsh touch sensation in *Caenorhabditis elegans*. *Nature Commun*, 2(May): 315–319. doi:10.1038/ncomms1308.

Liang, X., J. Madrid, and J. Howard. 2014. The microtubule-based cytoskeleton is a component of a mechanical signaling pathway in fly campaniform receptors. *Biophys J*, 107(12): 2767–2774. doi:10.1016/j.bpj.2014.10.052.

Liang, X. et al. 2010. NOMPC, a member of the TRP channel family, localizes to the tubular body and distal cilium of *Drosophila* campaniform and chordotonal receptor cells. *Cytoskeleton*, 68(1): 1–7. doi:10.1002/cm.20493.

Liang, X. et al. 2013. A NOMPC-dependent membrane-microtubule connector is a candidate for the gating spring in fly mechanoreceptors. *Curr Biol*, 23(9): 755–763. doi:10.1016/j.cub.2013.03.065.

Liu, L. et al. 2003. Contribution of *Drosophila* DEG/ENaC genes to salt taste. *Neuron*, 39(1): 133–146.

Mamiya, A. et al. 2011. Active and passive antennal movements during visually guided steering in flying *Drosophila*. *J Neurosci*, 31(18): 6900–6914. doi:10.1523/JNEUROSCI.0498-11.2011.

Maricq, A.V. et al. 1995. Mechanosensory signalling in *C. elegans* mediated by the GLR-1 glutamate receptor. *Nature*, 378(6552): 78–81. doi:10.1038/378078a0.

Matsuo, E. et al. 2014. Identification of novel vibration- and deflection-sensitive neuronal subgroups in Johnston's organ of the fruit fly. *Front Physiol*, 5(January): 179. doi:10.3389/fphys.2014.00179.

Nadrowski, B., J.T Albert, and M.C Göpfert. 2008. Transducer-based force generation explains active process in *Drosophila* hearing. *Curr Biol*, 18(18): 1365–1372. doi:10.1016/j.cub.2008.07.095.

Nadrowski, B. and M.C Göpfert. 2009. Level-dependent auditory tuning: Transducer-based active processes in hearing and best-frequency shifts. *Commun Integr Biol*, 2(1): 7–10.

Neely, G.G. et al. 2011. TrpA1 regulates thermal nociception in *Drosophila*. *PLOS ONE*, 6(8): e24343. doi:10.1371/journal.pone.0024343.s001.

Nilius, B. and G. Owsianik. 2011. The transient receptor potential family of ion channels. *Genome Biol*, 12(3): 218. doi:10.1186/gb-2011-12-3-218.

O'Hagan, R., M. Chalfie, and M.B Goodman. 2004. The MEC-4 DEG/ENaC channel of *Caenorhabditis elegans* touch receptor neurons transduces mechanical signals. *Nature Neurosci*, 8(1): 43–50. doi:10.1038/nn1362.

Perkins, L.A. et al. 1986. Mutant sensory cilia in the nematode *Caenorhabditis elegans*. *Dev Biol*, 117(2): 456–487.

Qin, F. 2007. Regulation of TRP ion channels by phosphatidylinositol-4,5-bisphosphate. In *Transient Receptor Potential (TRP) Channels*, edited by V. Flockerzi, and B. Nilius, 509–525. New York, NY: Springer.

Riabinina, O. et al. 2011. Active process mediates species-specific tuning of *Drosophila* ears. *Curr Biol*, 21(8): 658–664. doi:10.1016/j.cub.2011.03.001.

Rosenzweig, M. 2005. The *Drosophila* ortholog of vertebrate TRPA1 regulates thermotaxis. *Genes Dev*, 19(4): 419–424. doi:10.1101/gad.1278205.

Sawin, E.R., R. Ranganathan, and H.R. Horvitz. 2000. *C. elegans* locomotory rate is modulated by the environment through a dopaminergic pathway and by experience through a serotonergic pathway. *Neuron*, 26(3): 619–631. doi:10.1016/S0896-6273(00)81199-X.

Sayeed, O. and S. Benzer. 1996. Behavioral genetics of thermosensation and hygrosensation in *Drosophila*. *Proc Natl Acad Sci U S A*, 93(12): 6079–6084.

Smith, C.J. et al. 2010. Time-lapse imaging and cell-specific expression profiling reveal dynamic branching and molecular determinants of a multi-dendritic nociceptor in *C. elegans*. *Dev Biol*, 345(1): 18–33. doi:10.1016/j.ydbio.2010.05.502.

Sokabe, T. et al. 2008. *Drosophila* painless is a Ca²⁺-requiring channel activated by noxious heat. *J Neurosci*, 28(40): 9929–9938. doi:10.1523/JNEUROSCI.2757-08.2008.

Sotomayor, M., D.P. Corey, and K. Schulten. 2005. In search of the hair-cell gating spring elastic properties of ankyrin and cadherin repeats. *Structure*, 13(4): 669–682. doi:10.1016/j.str.2005.03.001.

Sun, Y. et al. 2009. TRPA channels distinguish gravity sensing from hearing in Johnston's organ. *PNAS*, 106(32): 13606–13611. doi:10.1073/pnas.0906377106.

Suslak, T.J. et al. 2015. Piezo is essential for amiloride-sensitive stretch-activated mechanotransduction in larval *Drosophila* dorsal bipolar dendritic sensory neurons. *PLOS ONE*, 10(7): e0130969. doi:10.1371/journal.pone.0130969.g007.

Tang, X. et al. 2013. Temperature integration at the AC thermosensory neurons in *Drosophila*. *J Neurosci*, 33(3): 894–901. doi:10.1523/JNEUROSCI.1894-12.2013.

Tavernarakis, N. et al. 1997. *unc-8*, a DEG/ENaC family member, encodes a subunit of a candidate mechanically gated channel that modulates *C. elegans* locomotion. *Neuron*, 18: 107–119.

Terada, S.I. et al. 2016. Neuronal processing of noxious thermal stimuli mediated by dendritic Ca²⁺ influx in *Drosophila* somatosensory neurons. *eLife*, 1–26. doi:10.7554/eLife.12959.001.

Tobin, D.M. et al. 2002. Combinatorial expression of TRPV channel proteins defines their sensory functions and subcellular localization in *C. elegans* neurons. *Neuron*, 35: 307–318.

Tracey, W.D. et al. 2003. *painless*, a *Drosophila* gene essential for nociception. *Cell*, 113(2): 261–273.

Troemel, E.R. et al. 1995. Divergent seven transmembrane receptors are candidate chemosensory receptors in *C. elegans*. *Cell*, 83(2): 207–218.

Tsubouchi, A., J.C. Caldwell, and W.D. Tracey. 2012. Dendritic filopodia, ripped pocket, NOMPC, and NMDARs contribute to the sense of touch in *Drosophila* larvae. *Curr Biol*, 22(22): 2124–2134. doi:10.1016/j.cub.2012.09.019.

Viswanath, V. et al. 2003. Opposite thermosensor in fruitfly and mouse. *Nature*, 423(6942): 822–823. doi:10.1038/423822a.

Voglis, G. and N. Tavernarakis. 2008. A synaptic DEG/ENaC ion channel mediates learning in *C. elegans* by facilitating dopamine signalling. *EMBO J*, 27(24): 3288–3299. doi:10.1038/emboj.2008.252.

Walker, R.G., A.T. Willingham, and C.S. Zuker. 2000. A *Drosophila* mechanosensory transduction channel. *Science*, 287(5461): 2229–2234.

Walters, E.T. and L.L. Moroz. 2009. Molluscan memory of injury: Evolutionary insights into chronic pain and neurological disorders. *Brain Behav Evol*, 74(3): 206–218. doi:10.1159/000258667.

White, J.G. et al. 1986. The structure of the nervous system of the nematode *Caenorhabditis elegans*. *Philos Trans R Soc Lond B Biol Sci*, 314(1165): 1–340.

Wu, L.J., T.B. Sweet, and D.E. Clapham. 2010. International union of basic and clinical pharmacology. LXXVI. Current progress in the mammalian TRP ion channel family. *Pharmacol Rev*, 62(3): 381–404. doi:10.1124/pr.110.002725.

Xiang, Y. et al. 2010. Light-avoidance-mediating photoreceptors tile the *Drosophila* larval body wall. *Nature*, 468(7326): 921–926. doi:10.1038/nature09576.

Xu, S.Y. et al. 2006. Thermal nociception in adult *Drosophila*: Behavioral characterization and the role of the *painless* gene. *Genes Brain Behav*, 5(8): 602–613. doi:10.1111/j.1601-183X.2006.00213.x.

Yan, Z. et al. 2012. *Drosophila* NOMPC is a mechanotransduction channel subunit for gentle-touch sensation. *Nature*, 493(7431): 221–225. doi:10.1038/nature11685.

Yorozu, S. et al. 2009. Distinct sensory representations of wind and near-field sound in the *Drosophila* brain. *Nature*, 457(7235): 201–205. doi:10.1038/nature07843.

Zahratka, J.A. et al. 2015. Serotonin differentially modulates Ca²⁺ transients and depolarization in a *C. elegans* nociceptor. *J Neurophysiol*, 113(4): 1041–1050. doi:10.1152/jn.00665.2014.

Zhang, W. et al. 2015. Ankyrin repeats convey force to gate the NOMPC mechanotransduction channel. *Cell*, 162(6): 1391–1403. doi:10.1016/j.cell.2015.08.024.

Zhao, Q. et al. 2016. Ion permeation and mechanotransduction mechanisms of mechanosensitive piezo channels. *Neuron*, 89(6): 1248–1263. doi:10.1016/j.neuron.2016.01.046.

Zhong, L., R.Y. Hwang, and W.D. Tracey. 2010. Pickpocket is a DEG/ENaC protein required for mechanical nociception in *Drosophila* larvae. *Curr Biol*, 20(5): 429–434. doi:10.1016/j.cub.2009.12.057.

Zhong, L. et al. 2012. Thermosensory and nonthermosensory isoforms of *Drosophila melanogaster* TRPA1 reveal heat-sensor domains of a thermoTRP channel. *Cell Rep*, 1(1): 43–55. doi:10.1016/j.celrep.2011.11.002.

Zhou, Y.Q. et al. 2016. Interleukin-6: An emerging regulator of pathological pain. *J Neuroinflammation*, 13(141): 1–9. doi:10.1186/s12974-016-0607-6.

5 Osmomechanical-Sensitive TRPV Channels in Mammals

Carlene Moore and Wolfgang B. Liedtke

CONTENTS

5.1 INTRODUCTION: THE TRPV SUBFAMILY

Within the transient receptor potential (TRP) superfamily of ion channels (Cosens and Manning, 1969; Montell and Rubin, 1989; Wong et al., 1989; Hardie and Minke, 1992; Zhu et al., 1995), the TRPV subfamily stepped into the limelight in 1997 (Colbert et al., 1997; Caterina et al., 1997), when its founding members, OSM-9 in *Caenorhabditis elegans* and TRPV1 in mammals, were first reported. OSM-9 was identified through genetic screening for worms' defects in osmotic avoidance (Colbert et al., 1997). TRPV1 was identified by an expression cloning strategy (Caterina et al., 1997). (This is also true for TRPV5 and TRPV6, which will not be discussed in this chapter because, up to now, they have not been implicated in osmotic and mechanical signaling.) TRPV2, TRPV3, and TRPV4 were identified by a candidate gene approach, respectively (Caterina and Julius, 1999; Peier et al., 2002; Gunthorpe et al., 2002; Xu et al., 2002; Kanzaki et al., 1999; Liedtke et al., 2000; Strotmann et al., 2000; Wissenbach et al., 2000). The latter strategy also led to the identification of four additional *C. elegans ocr* genes (Tobin et al., 2002) and two *Drosophila trpv* genes, Nanchung (NAN) and Inactive (IAV) (Kim et al., 2003; Gong et al., 2004). The TRPV channels can be subgrouped into four branches by sequence comparison. One branch includes four members of mammalian TRPVs, TRPV1, TRPV2, TRPV3, and TRPV4; *in vitro* whole cell recording showed that they respond to temperatures higher than 42°C, 52°C, 31°C, and 27°C, respectively, suggesting that they are involved in thermosensation, hence the term *thermo-TRPs*. Illuminating review articles on thermo-TRPs are available for in-depth reading (Clapham, 2003; Patapoutian, 2005; Tominaga and Caterina, 2004; Caterina and Julius, 1999; Caterina and Montell, 2005). The second mammalian branch includes the Ca^{2+}-selective channels, TRPV5 and TRPV6, possibly subserving Ca^{2+} uptake in the kidney and intestine (Hoenderop et al., 1999; den Dekker et al., 2003; Peng et al., 1999, 2003). One invertebrate branch includes *C. elegans* OSM-9 and *Drosophila* IAV; the other branch comprises OCR-1 to OCR-4 in *C. elegans* and *Drosophila* NAN.

This chapter elucidates the role of mammalian TRPV channels in signal transduction in response to osmotic and mechanical stimuli, as well as provides comments on selected recent insights regarding other TRP ion channels that respond to osmotic and mechanical cues. These "osmo- and mechano-TRPs" (Liedtke and Kim, 2005) are TRPV1 (Sharif-Naeini et al., 2006; Zaelzer et al., 2015), TRPV2 (Muraki et al., 2003), TRPV4 (Liedtke et al., 2000; Strotmann et al., 2000), TRPC1 (Chen and Barritt, 2003), TRPC3 (Quick et al., 2012), TRPC6 (Spassova et al., 2006), TRPA1 (Corey et al.,

2004; Nagata et al., 2005), TRPP2 (Nauli et al., 2003), TRPP3 (Murakami et al., 2005), TRPM3 (Grimm et al., 2003), TRPM4 (Earley et al., 2004), TRPM7 (Numata et al., 2007), and TPML3 (Di Palma et al., 2002). A full listing of mammalian TRPs involved in osmomechanosensation can be found in Table 5.1.

TABLE 5.1

Mammalian TRP Channels Involved in Osmo- and Mechanosensation

TRPs	Mechanoactivation	Osmoactivation	Critical expression	References
TRPA1	—	Hypertonicity	Sensory neurons	(Zhang et al., 2008)
TRPC1	Stretch	Hypotonicity	CHO cells, blood vessels, DRG, liver	(Chen and Barritt, 2003)
TRPC3	Stretch	—	Sensory neurons	(Quick et al., 2012)
TRPC4	—	—	—	—
TRPC5	Stretch	Hypotonicity	HEK293	(Gomis et al., 2008)
TRPC6	Stretch	—	CHO cells, blood vessels, DRG, kidney	(Spassova et al., 2006; Quick et al., 2012)
TRPC7	—	—	—	—
TRPM1	—	—	—	—
TRPM2	—	—	—	—
TRPM3	—	Hypotonicity	Kidney	(Grimm et al., 2003)
TRPM4	Stretch	—	Vascular smooth muscle	(Earley et al., 2004)
TRPM5	—	—	—	—
TRPM6	—	—	—	—
TRPM7	Stretch, shear stress	Hypotonicity	—	(Numata et al., 2007)
TRPML1	—	—	—	—
TRPML2	—	—	—	—
TRPML3	Shear stress, stretch	—	Hair cells of the ear	(Di Palma et al., 2002)
TRPP2	Shear stress	—	Endothelial cells, kidney	(Nauli et al., 2003)
TRPP3	—	Hypotonicity	Endoplasmic reticulum, HEK293	(Murakami et al., 2005)
TRPP5	—	—	—	—
TRPV1	Stretch	Hypertonicity	Eye epithelial, kidney, neurons	(Sharif Naeini et al., 2006; Zaelzer et al., 2015)
TRPV2	Stretch, shear stress	Hypotonicity	CHO, HEK293, skeletal and cardiac muscle, stomach, neurons	(Muraki et al., 2003; Shibasaki et al., 2010; Mihara et al., 2013)
TRPV3	—	—	—	—
TRPV4	Stretch, shear stress	Hypotonicity	CHO, HEK293, skeletal and cardiac muscle, sensory neurons, airway, bone, chondrocytes, eye	(Strotmann et al., 2000; Liedtke et al., 2000; O'Conor et al., 2014; Clark et al., 2010)
TRPV5	—	—	—	—
TRPV6	—	—	—	—

5.2 TRPV1

Sharif-Naeini et al. reported that *Trpv1*$^{-/-}$ mice failed to express an N-terminal variant of the *Trpv1* gene in magnocellular neurons, known to be osmotically sensitive, of the supraoptic and paraventricular nucleus of the hypothalamus (Sharif-Naeini et al., 2006). As these osmosensory neurons are known to secrete vasopressin, the *Trpv1*$^{-/-}$ mice were found to have an impaired ADH secretion in response to systemic hypertonic stimuli, and their magnocellular neurons did not show an appropriate electrical response to hypertonicity (Sharif-Naeini et al., 2006). These findings made Bourque and colleagues reason that this *Trpv1* N-terminal variant was very likely involved as (part of) a tonicity sensor of intrinsically osmosensitive magnocellular neurons. The respective cDNA was indeed cloned from mouse in the Liedtke-Lab, then from rat in Bourque's lab, and further characterized by Bourque's group as a TRPV1 ion channel, noncapsaicin receptor, and responsive to hypertonicity and also thermal cues in the vicinity of 37°C, published recently (Zaelzer et al., 2015). In keeping with the latter identification of osmotically and thermally sensitive nonvanilloid receptor TRPV1 channels, hyperosmolality activated TRPV1 in vasopressin neurons of rat supraoptic nucleus (SON), resulted in cell depolarization and Ca^{2+} entry through TRPV1 channels themselves and voltage-dependent Ca^{2+} channels suggesting an osmosensing role of TRPV1 in these neurons (Moriya et al., 2015).

In terms of regulation of renal function, TRPV1-mediated mechanosensation in the rat kidney has been attributed to H$_2$O$_2$ generated by NADPH oxidase 4 (Nox4). H$_2$O$_2$ augmented the release of substance P (SP) from kidney-innervating sensory neurons by enhancing the activity of Ca^{2+}-permeable TRPV1 channels (capsaicin receptors). Both TRPV1 and neurokinin-1 receptor activation contributed to increases in afferent renal nerve activity (ARNA) after mechanostimulation (Lin et al., 2015). The function of Nox4 in renal mechanosensitive innervating nerve fibers has a profound effect on the reflex control of urinary excretion, because Nox inhibition attenuates H$_2$O$_2$ production and diuretic/natriuretic responses in the renorenal reflex and in response to saline loading (Lin et al., 2015).

Hypertonic sensing TRPV1 channels found in human conjunctival epithelial cells, where TRPV1 was involved in Ca^{2+} regulation of volume (Mergler et al., 2012), are possibly based on the hypersensitivity-gated TRPV1 variant identified by Liedtke-Bourque (Zaelzer et al., 2015).

5.3 TRPV2

In heterologous cellular systems, TRPV2 was initially described as a temperature-gated ionotropic receptor for stimuli >52°C (Caterina and Julius, 1999). TRPV2 was also demonstrated to respond to hypotonicity and mechanical stimulation (Muraki et al., 2003). Heterologously expressed TRPV2 in CHO cells displayed a similar response to hypotonicity. These cells were also subjected to stretch by applying negative pressure to the patch-pipette and by stretching the cell membrane on a mechanical stimulator. Both maneuvers led to Ca^{2+} influx that was dependent on heterologous TRPV2 expression. Arterial smooth muscle cells from various arteries expressed TRPV2. These myocytes responded to hypotonic stimulation with Ca^{2+} influx. This activation could be reduced by specific downregulation of TRPV2 protein by an antisense strategy (Muraki et al., 2003).

Recently, TRPV2 was implicated in osmosensation in skeletal muscle fibers (Zanou et al., 2015). The response to hyperosmotic shock in normal muscle fibers and in muscle fibers expressing a dominant negative mutant of the TRPV2 channel (TRPV2-DN) was investigated. Hyperosmotic shock induced TRPV2 activation, which accelerated muscle cell depolarization and allowed the subsequent Ca^{2+} release from the sarcoplasmic reticulum, activation of the Na(+) −K(+) −Cl(-) cotransporter by SPAK (Ste20-related Proline/Alanine-rich Kinase), and the regulatory volume increase (RVI) response. In TRPV2-DN cells, slower membrane depolarization, loss of the Ca^{2+} transients, and RVI were reported in response to hyperosmotic shock (Zanou et al., 2015).

TRPV2 has been implicated as a candidate stretch-activated channel in myocyte intercalated discs, and for the mechanical stimulation-dependent Ca^{2+} signaling of cardiomyocytes, required in the maintenance of cardiac structure and function (Katanosaka et al., 2014). Using cardiac-specific TRPV2-knockout mice, cardiac-specific TRPV2 elimination led to a severe decline in the heart's pump function with the disorganization of the intercalated disc structure, conduction defects, and increased mortality. Loss of TRPV2 resulted in neonatal cardiomyocytes with no intercalated discs and showed no intracellular Ca^{2+} increase upon stretch stimulation. TRPV2-deficient hearts showed downregulation of insulin-like growth factor-1 (IGF-1) receptor/PI3K/Akt signaling. In TRPV2-deficient hearts, IGF-1 administration partially prevented chamber dilation and improved cardiac pump function.

In odontoblasts, TRPV2 was activated by extracellular hypotonicity, resulting in Ca^{2+} influx and inward currents, which were inhibited using TPRV2 antagonist tranilast and tetraethylammonium-chloride (TEA) (Sato et al., 2013).

TRPV2 has been implicated by Shibasaki et al. to be involved in the regulation of axonal outgrowth. TRPV2 expression was found in spinal motor neurons, dorsal root ganglia (DRG), axonal shafts, and growth cones (Shibasaki et al., 2010). When TRPV2 was activated in a membrane stretch–dependent manner, it promoted axon outgrowth. Knockdown of TRPV2 using a dominant-negative TRPV2 and TRPV2-specific shRNA inhibited axonal outgrowth (Shibasaki et al., 2010).

Mahari et al. studied the contribution of TRPV2 in the stomach as it responds to mechanical stimuli associated with food intake by looking at the contribution of TRPV2 to gastric adaptive relaxation (GAR) and gastric emptying (GE) (Mihara et al., 2013). *Trpv2* mRNA was detected throughout the mouse gastrointestinal tract and myenteric neurons in the stomach. GAR, which was expressed as the rate of decline of intragastric pressure in response to volume stimuli, was significantly enhanced by the TRPV2 activating probenecid, and the effect was inhibited by the TRPV2 inhibiting tranilast. GE was significantly accelerated by TRPV2 agonist applications, and the enhancement was significantly inhibited by inhibitor coapplication (Mihara et al., 2013). However, these compounds lack specificity for TRPV2 channels; therefore, final doubts about these data remain.

TRPV3 has not (yet) been characterized as osmomechano-TRP, neither in heterologous systems nor in live animals or human studies. The same is true for TRPV5 and TRPV6.

5.4 TRPV4

CHO cells responded to hypotonic solution when they were (stably) transfected with TRPV4 (Liedtke et al., 2000). HEK-293T cells, when maintained by the same authors, were found to harbor *Trpv4* cDNA, which was cloned from these cells. However, *trpv4* cDNA was not found in other batches of HEK 293T cells, so this cell line was used as a heterologous expression vehicle by other groups (Strotmann et al., 2000; Wissenbach et al., 2000). Notably, when comparing the two settings, it was obvious that the single-channel conductance was different (Liedtke et al., 2000; Strotmann et al., 2000). This underscores the relevance of gene expression in heterologous cellular systems for the functioning of TRPV4 in response to a basic biophysical stimulus. Also, it was found that the sensitivity of TRPV4 could be tuned by warming of the media. Peak sensitivity of gating in response to hypotonicity was recorded at core body temperature of the respective organism, and TRPV4 channels from both birds (chick, core body temperature 40°C) and mammals (rat, 37°C) were compared, again in CHO cells (Liedtke et al., 2000). Similar results were found in another investigation with mammalian TRPV4 in HEK-293T cells (Gao et al., 2003). Later work by Ching Kung's group unambiguously suggests that TRPV4 is critically and closely involved in mechano-transduction (Loukin et al., 2009). Other work by Thodeti and Ingber strongly argues along the same lines (Matthews et al., 2010).

In addition, in the earlier study by Gao and O'Neil, the cells were mechanically stretched, without a change of tonicity. At room temperature, there was no response upon mechanical stretch; however, at 37°C the isotonic response to stretch resulted in the maximum Ca^{2+} influx of all conditions tested. In two other investigations, TRPV4 was found to be responsive to changes in temperature

(Guler et al., 2002; Watanabe et al., 2002). Temperature change was accomplished by heating the streaming bath solution. However, flow is a mechanical stimulus, presenting as mechanical shear stress to the cell. Gating of TRPV4 was amplified when hypotonic solution was used as streaming bath. In one investigation, temperature stimulation could not activate the TRPV4 channel in cell-detached inside-out patches (Watanabe et al., 2002). With respect to the gating mechanism of TRPV4 in response to hypotonicity, two other investigations report conflicting results on phosphorylation sites of TRPV4 that are necessary for the response to hypotonicity. One study reported that TRPV4 was tyrosine-phosphorylated in HEK-293T cells and in distal convoluted tubule cells from mouse kidney (Xu et al., 2003; Tian et al., 2004). Tyrosine phosphorylation was sensitive to specific inhibition of the Src family tyrosine kinases. The Lyn tyrosine kinase was found to coimmunoprecipitate TRPV4 protein and to bear a prominent role in phosphorylation of TRPV4 (Y253). A point mutation of Y253 greatly reduced hypotonicity-induced gating. In another investigation, in HEK-293T cells, hypotonicity activated TRPV4 by phospholipase-A2–mediated formation of arachidonic acid via a cytochrome P450 epoxygenase pathway (Vriens et al., 2004). In HEK cells, this signaling mechanism did not apply for gating of TRPV4 by increased temperature or by the nonphosphorylating phorbol-ester 4-alpha PDD. This latter activation mechanism was reported to be dependent on phosphorylation of Y555. However, the authors of this study could not replicate the aforementioned finding, namely that tyrosine kinase phosphorylation of Y253 of TRPV4 was critical for hypotonicity-induced gating. Why this divergence? The discrepancy reiterates the pivotal role of the host cell in heterologous expression experiments.

In the mammalian oviduct, ciliary beat frequency of ciliated cells was found to be influenced by gating of TRPV4 (Andrade et al., 2005). In explanted ciliated cells, and also in heterologously transfected HeLa cells, TRPV4 could be activated (mechanically) by exposing the cells to hyperviscous, isotonic media. In follow-up studies, TRPV4 was found to function as an ionotropic receptor for particulate matter air pollution from combustion engines in human bronchial epithelia, where it regulated a proinflammatory and proteolytic phenotype, capable of enhancing chronic obstructive pulmonary disease (COPD) (Li et al., 2011).

TRPV4 also has been found to play a role in maintenance of cellular osmotic homeostasis. One particular cellular defense mechanism of cellular osmotic homeostasis is regulatory volume change, namely regulatory volume decrease (RVD) in response to hypotonicity and regulatory volume increase (RVI) in response to hypertonicity. Valverde's group published that TRPV4 mediates the cell swelling–induced Ca^{2+} influx into bronchial epithelial cells that triggers RVD via Ca^{2+}-dependent potassium channels (Arniges et al., 2004). This cell-swelling response did not work in cystic fibrosis (CFTR) bronchial epithelia, where, on the other hand, TRPV4 could be activated by 4-alpha-PDD, leading to Ca^{2+} influx. Thus, in CFTR bronchial epithelia, RVD could not be elicited by hypotonicity, but by 4-alpha-PDD. In yet another investigation, Ambudkar and colleagues found the concerted interaction of aquaporin 5 (AQP-5) with TRPV4 in hypotonic swelling-induced RVD of salivary gland epithelia (Ciura et al., 2011). These findings shed light on mechanisms operative in secretory epithelia (such as salivary, tear, sweat, pancreatic and intestinal glands, and airway) that underlie watery secretion based on a concerted interaction of TRPV4 and AQP-5.

TRPV4 was found to be a key osmosensing component in airway sensory nerve reflexes, as it induced activation of guinea pig airway-specific primary nodose ganglion cells. TRPV4 activators and hypoosmotic solutions caused depolarization of murine, guinea pig, and human vagus and firing of Aδ-fibers (not C-fibers), which was inhibited by TRPV4 and P2X3 receptor antagonists. Both antagonists blocked TRPV4-induced cough (Bonvini et al., 2016). TRPV4 was implicated as a novel therapeutic target for neuronal hyperresponsiveness in the airways and symptoms, such as chronic cough. Chronic cough can be viewed as an airway manifestation of a neural hypersensitivity.

In regard to pain, in *Trpv4−/−* mice, the response to noxious mechanical, not noxious thermal, stimulation is diminished (Liedtke and Friedman, 2003). In the absence of TRPV4, which in wild-type mice could be shown to be expressed in sensory ganglia (Liedtke and Friedman, 2003; Liedtke et al., 2000; Chen et al., 2014, 2013) and, in skin, in subcutaneous nerve fibers and keratinocytes

(Delany et al., 2001; Guler et al., 2002), the threshold to noxious mechanical stimulation was significantly elevated. Earlier follow-up studies, particularly those by Jon Levine's and Nigel Bunnett's groups, were characterized in an earlier TRP book edited by Liedtke and Heller and will not be reviewed here (Liedtke and Heller, 2007).

As for thirst and central osmoregulation, earlier work on the role of TRPV4 in thirst and central osmoregulation was reflected in the earlier TRP book, and also is not reviewed here. More recent work appears to suggest a role for TRPV4 in thirst and central osmoregulation, which needs to be resolved more clearly by tissue- or cell lineage–specific and inducible knockouts of the channel (Ciura et al., 2011; Janas et al., 2016; Sakuta et al., 2016).

In regard to TRPV4's function in the skeleton, TRPV4 is highly expressed in articular chondrocytes where the channel responds to osmotic cues (Phan et al., 2009). In humans, TRPV4 mutations that alter the function of the channel disrupt normal skeletal development and joint health (Leddy et al., 2014; Nilius and Voets, 2013). In mice, *Trpv4* pan-null deletions resulted in a lack of osmotically induced Ca^{2+} signaling in articular chondrocytes, plus a progressive, sex-dependent increase in bone density and osteoarthritic joint degeneration (Clark et al., 2010). In another recent study by the Liedtke-Guilak group, using tissue-specific, inducible *Trpv4* gene-targeted mice, it was demonstrated that loss of chondrocyte TRPV4 resulted in a loss of TRPV4-mediated cartilage mechanotransduction in adulthood, and a reduction of the severity of aging-associated osteoarthritis (OA), not posttraumatic OA. They also noted cartilage-specific deletion of *Trpv4* showed a decrease in total joint bone volume and decreased osteophytes in the joint (O'Conor et al., 2016).

In another recent study, O'Connor et al. revealed that upon mechanical stimulation, chondrocytes responded in a TRPV4-dependent manner, which involved the transcriptional enhancement of anabolic growth factor gene expression and inhibition of proinflammatory mediators. Inhibition of TRPV4 with GSK205 blocked both the compositional and functional augmentation of mechanically loaded agarose-embedded chondrocytes constructs, further supporting the role of TRPV4-mediated mechanotransduction in response to mild mechanical stress in regulating chondrocyte matrix metabolism toward an anabolic phenotype (O'Conor et al., 2014). Activation of TRPV4 with GSK101 in the absence of mechanical loading similarly enhanced anabolic and suppressed catabolic gene expression and potently enhanced matrix accumulation and functional properties of the chondrocyte-agarose constructs, and potently increased cartilage matrix biosynthesis. This opens the door for tissue engineering, following a molecular logic of TRPV4 signaling in cartilage.

In an attempt to decipher the mechanism, Kobayakawa et al. showed that in human chondrocytic cells, excessive mechanical stress loading induced the metalloproteinase ADAM10 expression and enhanced CD44 (a cell-surface glycoprotein) cleavage, which can lead to the loss of extracellular matrices in chondrocytes. Chemical inhibition of TRPV4 significantly suppressed mechanical stress–induced ADAM10 expression and CD44 cleavage. Conversely, chemical activation of TRPV4 increased ADAM10 expression and enhanced CD44 cleavage. Such a finding suggests a mechanism whereby mechanical loading significantly increases the expression of ADAM10, which in turn enhances CD44 cleavage in HCS-2/8 cells, with the TRPV4 mechanoreceptor mediating this process (Kobayakawa et al., 2016). These findings corroborate the Liedtke-Guilak concept (O'Conor et al., 2016) that TRPV4-mediated Ca^{2+} signaling in chondrocytes plays a central role in the transduction of mechanical signals to support cartilage extracellular matrix maintenance and joint health.

In the vertebrate eye, TRPV4 is involved in mechanical intraocular pressure sensing and its regulation. TRPV4 was found in human conjunctival epithelial cells, where it was involved in Ca^{2+} regulation of volume (Mergler et al., 2012). TRPV4 was observed to regulate Ca^{2+} homeostasis, cytoskeletal remodeling, conventional outflow, and intraocular pressure (IOP) in the mammalian eye (Ryskamp et al., 2016). Mechanical force was applied to trabecular meshwork (TM) cells, which resulted in sustained, stretch-dependent Ca^{2+} elevations that could be mimicked by GSK101 and suppressed by TRPV4 blocker. Systemic delivery, intraocular injection, or topical application of putative TRPV4 antagonist prodrug analogs lowered IOP in glaucomatous mouse eyes and protected retinal neurons from IOP-induced death. It was observed that phospholipase A2 (PLA2) antagonists inhibit stretch-induced Ca^{2+} signals and that TRPV4 blockers suppress AA-induced Ca^{2+} increases, suggesting that the channel is activated through the canonical pathway. A mechanism proposed was that mechanical stress (e.g., pressure, swelling, and tissue distension) stretches the plasma membrane and activates TRPV4 and a Ca^{2+}- and stretch-sensitive PLA2. The product, arachidonic acid (AA), leads to the synthesis of eicosanoid metabolites (EETs), which activates TRPV4. Stretch might activate PLA2 simultaneously with TRPV4, or alternatively, stretch-induced TRPV4 activation could stimulate Ca^{2+}-dependent PLA2s which amplify the initial TRPV4 signal. This study implicates TRPV4 as a potential IOP sensor within the conventional outflow pathway and as a novel target for treating ocular hypertension. Future studies will have to deconstruct ocular TRPV4 function using cell- and lineage-specific and inducible knockout models.

5.5   OUTLOOK FOR FUTURE APPLICATIONS INVOLVING TRP ION CHANNELS

Targeting osmomechano-TRPs for treatment of human disease has become a more compelling rationale since the field has been founded (Kanju et al., 2016; Brierley et al., 2009). Basic science progress as well as translational advances, such as those communicated in the latter study, have nourished this development. In particular the arenas of pain and inflammation (Wilson et al., 2013; D'Aldebert et al., 2011; Engel et al., 2011; Bonvini et al., 2015), skeletal pathophysiology and disease (McNulty et al., 2015), and cardiovascular pathophysiology and disease (White et al., 2016) appear to be areas of interest where nonincremental progress toward rationally guided medical diagnoses, prevention, and treatments could be imminent.

ACKNOWLEDGMENT

This work was supported in part by the U.S. National Institutes of Health (grants DE018549, AR48182, AR48182-S1), U.S. Department of Defense (W81XWH-13-1-0299), and the Harrington Discovery Institute (Cleveland, OH).

REFERENCES

Andrade, Y.N. et al. 2005. TRPV4 channel is involved in the coupling of fluid viscosity changes to epithelial ciliary activity. *J Cell Biol*, 168(6): 869–874.

Arniges, M. et al. 2004. Swelling-activated Ca2+ entry via TRPV4 channel is defective in cystic fibrosis airway epithelia. *J Biol Chem*, 279(52): 54062–54068.

Bonvini, S.J. et al. 2015. Targeting TRP channels for chronic cough: From bench to bedside. *Naunyn Schmiedebergs Arch Pharmacol*, 388(4): 401–420. doi:10.1007/s00210-014-1082-1.

Bonvini, S.J. et al. 2016. Transient receptor potential cation channel, subfamily V, member 4 and airway sensory afferent activation: Role of adenosine triphosphate. *J Allergy Clin Immunol*, 138(1): 249–261.e12. doi:10.1016/j.jaci.2015.10.044.

Brierley, S.M. et al. 2009. The ion channel TRPA1 is required for normal mechanosensation and is modulated by algesic stimuli. *Gastroenterology*, 137(6): 2084–2095.e3. doi:10.1053/j.gastro.2009.07.048.

Caterina, M.J. and D. Julius. 1999. Sense and specificity: A molecular identity for nociceptors. *Curr Opin Neurobiol*, 9(5): 525–530.

Caterina, M.J. and C. Montell. 2005. Take a TRP to beat the heat. *Genes Dev*, 19(4): 415–418.

Caterina, M.J. et al. 1997. The capsaicin receptor: A heat-activated ion channel in the pain pathway [see comments]. *Nature*, 389(6653): 816–824.

Chen, J. and G.J. Barritt. 2003. Evidence that TRPC1 (transient receptor potential canonical 1) forms a Ca(2+)-permeable channel linked to the regulation of cell volume in liver cells obtained using small interfering RNA targeted against TRPC1. *Biochem J*, 373(Pt 2): 327–336. doi:10.1042/bj20021904.

Chen, Y. et al. 2013. Temporomandibular joint pain: A critical role for Trpv4 in the trigeminal ganglion. *Pain*, 154(8): 1295–1304. doi:10.1016/j.pain.2013.04.004.

Chen, Y. et al. 2014. TRPV4 is necessary for trigeminal irritant pain and functions as a cellular formalin receptor. *Pain*, 155(12): 2662–2672. doi:10.1016/j.pain.2014.09.033.

Ciura, S., W. Liedtke, and C.W. Bourque. 2011. Hypertonicity sensing in organum vasculosum lamina terminalis neurons: A mechanical process involving TRPV1 but not TRPV4. *J Neurosci*, 31(41): 14669–14676. doi:10.1523/jneurosci.1420-11.2011.

Clapham, D.E. 2003. TRP channels as cellular sensors. *Nature*, 426(6966): 517–524.

Clark, A.L. et al. 2010. Chondroprotective role of the osmotically sensitive ion channel transient receptor potential vanilloid 4: Age- and sex-dependent progression of osteoarthritis in Trpv4-deficient mice. *Arthritis Rheum*, 62(10): 2973–2983. doi:10.1002/art.27624.

Colbert, H.A., T.L. Smith, and C.I. Bargmann. 1997. OSM-9, a novel protein with structural similarity to channels, is required for olfaction, mechanosensation, and olfactory adaptation in *Caenorhabditis elegans*. *J Neurosci*, 17(21): 8259–8269.

Cosens, D.J., and A. Manning. 1969. Abnormal electroretinogram from a *Drosophila* mutant. *Nature*, 224(216): 285–287.

D'Aldebert, E. et al. 2011. Transient receptor potential vanilloid 4 activated inflammatory signals by intestinal epithelial cells and colitis in mice. *Gastroenterology*, 140(1): 275–285. doi:10.1053/j.gastro.2010.09.045.

Delany, N.S. et al. 2001. Identification and characterization of a novel human vanilloid receptor- like protein, VRL-2. *Physiol Genomics*, 4(3): 165–174.

den Dekker, E. et al. 2003. The epithelial calcium channels, TRPV5 and TRPV6: From identification towards regulation. *Cell Calcium*, 33(5–6): 497–507.

Di Palma, F. et al. 2002. Mutations in Mcoln3 associated with deafness and pigmentation defects in varitint-waddler (Va) mice. *Proc Natl Acad Sci U S A*, 99(23): 14994–14999. doi:10.1073/pnas.222425399.

Earley, S., B.J. Waldron, and J.E. Brayden. 2004. Critical role for transient receptor potential channel TRPM4 in myogenic constriction of cerebral arteries. *Circ Res*, 95(9): 922–929. doi:10.1161/01.res.0000147311.54833.03.

Engel, M.A. et al. 2011. TRPA1 and substance P mediate colitis in mice. *Gastroenterology*, 141(4): 1346–1358. doi:10.1053/j.gastro.2011.07.002.

Gao, X., L. Wu, and R.G. O'Neil. 2003. Temperature-modulated diversity of TRPV4 channel gating: Activation by physical stresses and phorbol ester derivatives through protein kinase C-dependent and -independent pathways. *J Biol Chem*, 278: 27129–27137.

Gomis, A. et al. 2008. Hypoosmotic- and pressure-induced membrane stretch activate TRPC5 channels. *J Physiol*, 586(23): 5633–5649. doi:10.1113/jphysiol.2008.161257.

Gong, Z. et al. 2004. Two interdependent TRPV channel subunits, inactive and Nanchung, mediate hearing in *Drosophila*. *J Neurosci*, 24(41): 9059–9066.

Grimm, C. et al. 2003. Molecular and functional characterization of the melastatin-related cation channel TRPM3. *J Biol Chem*, 278(24): 21493–21501. doi:10.1074/jbc.M300945200.

Guler, A.D. et al. 2002. Heat-evoked activation of the ion channel, TRPV4. *J Neurosci*, 22(15): 6408–6414.

Gunthorpe, M.J. et al. 2002. The diversity in the vanilloid (TRPV) receptor family of ion channels. *Trends Pharmacol Sci*, 23(4): 183–191.

Hardie, R.C. and B. Minke. 1992. The *trp* gene is essential for a light-activated Ca2+ channel in *Drosophila* photoreceptors. *Neuron*, 8(4): 643–651.

Hoenderop, J.G. et al. 1999. Molecular identification of the apical Ca2+ channel in 1, 25-dihydroxyvitamin D3-responsive epithelia. *J Biol Chem*, 274(13): 8375–8378.

Janas, S. et al. 2016. TRPV4 is associated with central rather than nephrogenic osmoregulation. *Pflugers Arch*, 468(9): 1595–1607. doi:10.1007/s00424-016-1850-5.

Kanju, P. et al. 2016. Small molecule dual-inhibitors of TRPV4 and TRPA1 for attenuation of inflammation and pain. *Sci Rep*, 6:26894. doi:10.1038/srep26894.

Kanzaki, M. et al. 1999. Translocation of a calcium-permeable cation channel induced by insulin- like growth factor-I. *Nat Cell Biol*, 1(3): 165–170.

Katanosaka, Y. et al. 2014. TRPV2 is critical for the maintenance of cardiac structure and function in mice. *Nat Commun*, 5: 3932. doi:10.1038/ncomms4932.

Kim, J. et al. 2003. A TRPV family ion channel required for hearing in *Drosophila*. *Nature*, 424(6944): 81–84.

Kobayakawa, T. et al. 2016. Mechanical stress loading induces CD44 cleavage in human chondrocytic HCS-2/8 cells. *Biochem Biophys Res Commun*, 478(3): 1230–1235. doi:10.1016/j.bbrc.2016.08.099.

Leddy, H.A. et al. 2014. Follistatin in chondrocytes: The link between TRPV4 channelopathies and skeletal malformations. *FASEB J*, 28(6): 2525–2537. doi:10.1096/fj.13-245936.

Li, J. et al. 2011. TRPV4-mediated calcium influx into human bronchial epithelia upon exposure to diesel exhaust particles. *Environ Health Perspect*, 119(6): 784–793. doi:10.1289/ehp.1002807.

Liedtke, W., and J.M. Friedman. 2003. Abnormal osmotic regulation in trpv4-/- mice. *Proc Natl Acad Sci U S A*, 100:13698–13703.

Liedtke W., Heller S, editors. 2007. *TRP Ion Channel Function in Sensory Transduction and Cellular Signaling Cascades*. Boca Raton, FL: CRC Press/Taylor & Francis.

Liedtke, W., and C. Kim. 2005. Functionality of the TRPV subfamily of TRP ion channels: Add mechano-TRP and osmo-TRP to the lexicon! *Cell Mol Life Sci*, 62(24): 2985–3001.

Liedtke, W. et al. 2000. Vanilloid receptor-related osmotically activated channel (VR-OAC), a candidate vertebrate osmoreceptor. *Cell*, 103(3): 525–535.

Lin, C.S. et al. 2015. H2O2 generated by NADPH oxidase 4 contributes to transient receptor potential vanilloid 1 channel-mediated mechanosensation in the rat kidney. *Am J Physiol Renal Physiol*, 309(4): F369–F376. doi:10.1152/ajprenal.00462.2014.

Loukin, S.H., Z. Su, and C. Kung. 2009. Hypotonic shocks activate rat TRPV4 in yeast in the absence of polyunsaturated fatty acids. *FEBS Lett*, 583(4): 754–758. doi:10.1016/j.febslet.2009.01.027.

Matthews, B.D. et al. 2010. Ultra-rapid activation of TRPV4 ion channels by mechanical forces applied to cell surface beta1 integrins. *Integr Biol (Camb)*, 2(9): 435–442. doi:10.1039/c0ib00034e.

McNulty, A.L. et al. 2015. TRPV4 as a therapeutic target for joint diseases. *Naunyn Schmiedebergs Arch Pharmacol*, 388(4): 437–450. doi:10.1007/s00210-014-1078-x.

Mergler, S. et al. 2012. Calcium regulation by thermo- and osmosensing transient receptor potential vanilloid channels (TRPVs) in human conjunctival epithelial cells. *Histochem Cell Biol*, 137(6): 743–761. doi:10.1007/s00418-012-0924-5.

Mihara, H. et al. 2013. TRPV2 ion channels expressed in inhibitory motor neurons of gastric myenteric plexus contribute to gastric adaptive relaxation and gastric emptying in mice. *Am J Physiol Gastrointest Liver Physiol*, 304(3): G235–G240. doi:10.1152/ajpgi.00256.2012.

Montell, C., and G.M. Rubin. 1989. Molecular characterization of the *Drosophila trp* locus: A putative integral membrane protein required for phototransduction. *Neuron*, 2(4): 1313–1323.

Moriya, T. et al. 2015. Full-length transient receptor potential vanilloid 1 channels mediate calcium signals and possibly contribute to osmoreception in vasopressin neurones in the rat supraoptic nucleus. *Cell Calcium*, 57(1): 25–37. doi:10.1016/j.ceca.2014.11.003.

Murakami, M. et al. 2005. Genomic organization and functional analysis of murine PKD2L1. *J Biol Chem*, 280(7): 5626–5635. doi:10.1074/jbc.M411496200.

Muraki, K. et al. 2003. TRPV2 is a component of osmotically sensitive cation channels in murine aortic myocytes. *Circ Res*, 93(9): 829–838.

Nauli, S.M. et al. 2003. Polycystins 1 and 2 mediate mechanosensation in the primary cilium of kidney cells. *Nat Genet*, 33(2): 129–137. doi:10.1038/ng1076.

Nilius, B. and T. Voets. 2013. The puzzle of TRPV4 channelopathies. *EMBO Rep*, 14(2): 152–163. doi:10.1038/embor.2012.219.

Numata, T., T. Shimizu, and Y. Okada. 2007. Direct mechano-stress sensitivity of TRPM7 channel. *Cell Physiol Biochem*, 19(1–4): 1–8. doi:10.1159/000099187.

O'Conor, C.J. et al. 2014. TRPV4-mediated mechanotransduction regulates the metabolic response of chondrocytes to dynamic loading. *Proc Natl Acad Sci U S A*, 111(4): 1316–1321. doi:10.1073/pnas.1319569111.

O'Conor, C.J. et al. 2016. Cartilage-specific knockout of the mechanosensory ion channel TRPV4 decreases age-related osteoarthritis. *Sci Rep*, 6: 29053. doi:10.1038/srep29053.

Patapoutian, A. 2005. TRP channels and thermosensation. *Chem Senses*, 30(Suppl 1): i193–i194.

Peier, A.M. et al. 2002. A heat-sensitive TRP channel expressed in keratinocytes. *Science*, 296(5575): 2046–2049.

Peng, J.B., E.M. Brown, and M.A. Hediger. 2003. Epithelial Ca2+ entry channels: Transcellular Ca2+ transport and beyond. *J Physiol*, 551(Pt 3): 729–740.

Peng, J.B. et al. 1999. Molecular cloning and characterization of a channel-like transporter mediating intestinal calcium absorption. *J Biol Chem*, 274(32): 22739–22746.

Phan, M.N. et al. 2009. Functional characterization of TRPV4 as an osmotically sensitive ion channel in porcine articular chondrocytes. *Arthritis Rheum*, 60(10): 3028–3037. doi:10.1002/art.24799.

Quick, K. et al. 2012. TRPC3 and TRPC6 are essential for normal mechanotransduction in subsets of sensory neurons and cochlear hair cells. *Open Biol*, 2(5): 120068. doi:10.1098/rsob.120068.

Ryskamp, D.A. et al. 2016. TRPV4 regulates calcium homeostasis, cytoskeletal remodeling, conventional outflow and intraocular pressure in the mammalian eye. *Sci Rep*, 6:30583. doi:10.1038/srep30583.

Sakuta, H. et al. 2016. Nax signaling evoked by an increase in [Na+] in CSF induces water intake via EET-mediated TRPV4 activation. *Am J Physiol Regul Integr Comp Physiol*, 311(2): R299–R306. doi:10.1152/ajpregu.00352.2015.

Sato, M. et al. 2013. Hypotonic-induced stretching of plasma membrane activates transient receptor potential vanilloid channels and sodium-calcium exchangers in mouse odontoblasts. *J Endod*, 39(6): 779–787. doi:10.1016/j.joen.2013.01.012.

Sharif-Naeini, R., et al. 2006. An N-terminal variant of Trpv1 channel is required for osmosensory transduction. *Nat Neurosci*, 9 (1):93-98. doi: 10.1038/nn1614.

Shibasaki, K. et al. 2010. TRPV2 enhances axon outgrowth through its activation by membrane stretch in developing sensory and motor neurons. *J Neurosci*, 30(13): 4601–4612. doi:10.1523/jneurosci.5830-09.2010.

Spassova, M.A. et al. 2006. A common mechanism underlies stretch activation and receptor activation of TRPC6 channels. *Proc Natl Acad Sci U S A*, 103(44): 16586–16591. doi:10.1073/pnas.0606894103.

Strotmann, R. et al. 2000. OTRPC4, a nonselective cation channel that confers sensitivity to extracellular osmolarity. *Nat Cell Biol*, 2(10): 695–702.

Tian, W. et al. 2004. Renal expression of osmotically responsive cation channel TRPV4 is restricted to water-impermeant nephron segments. *Am J Physiol Renal Physiol*, 287(1): F17–F24.

Tobin, D. et al. 2002. Combinatorial expression of TRPV channel proteins defines their sensory functions and subcellular localization in *C. elegans* neurons. *Neuron*, 35: 307–318.

Tominaga, M. and M. J. Caterina. 2004. Thermosensation and pain. *J Neurobiol*, 61(1): 3–12.

Vriens, J. et al. 2004. Cell swelling, heat, and chemical agonists use distinct pathways for the activation of the cation channel TRPV4. *Proc Natl Acad Sci U S A*, 101 (1):396-401. doi: 10.1073/pnas.0303329101.

Watanabe, H. et al. 2002. Heat-evoked activation of TRPV4 channels in a HEK293 cell expression system and in native mouse aorta endothelial cells. *J Biol Chem*, 277(49): 47044–47051.

White, J.P. et al. 2016. TRPV4: Molecular conductor of a diverse orchestra. *Physiol Rev*, 96(3): 911–973. doi:10.1152/physrev.00016.2015.

Wilson, S.R. et al. 2013. The ion channel TRPA1 is required for chronic itch. *J Neurosci*, 33(22): 9283–9294. doi:10.1523/jneurosci.5318-12.2013.

Wissenbach, U. et al. 2000. Trp12, a novel Trp related protein from kidney. *FEBS Lett*, 485(2–3): 127–134.

Wong, F. et al. 1989. Proper function of the *Drosophila trp* gene product during pupal development is important for normal visual transduction in the adult. *Neuron*, 3(1): 81–94.

Xu, H. et al. 2002. TRPV3 is a calcium-permeable temperature-sensitive cation channel. *Nature*, 418(6894): 181–186.

Xu, H. et al. 2003. Regulation of a transient receptor potential (TRP) channel by tyrosine phosphorylation. SRC family kinase-dependent tyrosine phosphorylation of TRPV4 on TYR-253 mediates its response to hypotonic stress. *J Biol Chem*, 278(13): 11520–11527.

Zaelzer, C. et al. 2015. DeltaN-TRPV1: A molecular co-detector of body temperature and osmotic stress. *Cell Rep*, 13(1): 23–30. doi:10.1016/j.celrep.2015.08.061.

Zanou, N. et al. 2015. Osmosensation in TRPV2 dominant negative expressing skeletal muscle fibres. *J Physiol*, 593(17): 3849–3863. doi:10.1113/jp270522.

Zhang, X. F. et al. 2008. Transient receptor potential A1 mediates an osmotically activated ion channel. *Eur J Neurosci*, 27(3): 605-611. doi: 10.1111/j.1460-9568.2008.06030.x.

Zhu, X. et al. 1995. Molecular cloning of a widely expressed human homologue for the *Drosophila trp* gene. *FEBS Lett*, 373(3): 193–198.

6 A Critical Role for TRP Channels in the Skin

Pu Yang, Jing Feng, Jialie Luo,
Mack Madison, and Hongzhen Hu

CONTENTS

6.1 INTRODUCTION

The transient receptor potential (TRP) channels were first described in *Drosophila*, in which photoreceptors carrying *trp* gene mutations exhibited a transient voltage response to continuous light stimulation (Minke, 1977; Montell et al., 1985). Mammalian TRP channels have six subfamilies including TRPC, TRPV, TRPM, TRPA, TRPML, and TRPP (Clapham, 2003), with about 28 mammalian subfamily members, most of which have splicing variants. All TRP channels have six transmembrane domains with the N- and C-terminal regions located inside the cell and are assembled as tetramers to form nonselective cation-permeable pores (Liao et al., 2014). TRP channels are expressed in a wide variety of tissues, and they are commonly embedded either in the membrane surface or cytosolic organelles, such as endosomes and lysosomes. Activation of TRP channels generally promotes excitability of excitable cells and Ca^{2+} influx in many forms of cellular processes in both excitable and nonexcitable cells.

The skin is divided into three layers: (1) The epidermis, the outermost layer of skin, provides a waterproof barrier and creates the skin tone. Although the most abundant cells of the epidermis are keratinocytes, there are also nonepithelial immune cells present in the epidermis, such as Langerhans cells and dendritic epidermal T cells (DETCs). (2) The dermis, directly under the epidermis, contains tough connective tissue, hair follicles (HFs), and sweat glands. The dermis also hosts different subtypes of T cells that recirculate through skin-draining lymph nodes and are involved in normal immunity as well as inflammatory skin diseases such as psoriasis (Bos et al., 1987; Streilein, 1983). In addition to T cells, the dermis is enriched with tissue macrophages and dendritic cells that originate from the yolk sac and self-renew within the skin under inflammatory conditions (Jenkins, 2011). Together with cutaneous innate immune cells, the circulating monocytes

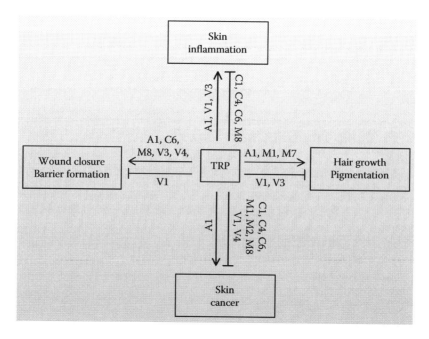

FIGURE 6.1 Expression of TRP channels in the skin. Many types of skin-resident cells besides the primary sensory nerve endings are present in the epidermis and dermis. TRP channels that were reported to be expressed by specific skin-resident cells are labeled.

traffic through the skin to survey the environment and transport antigens to the draining lymph nodes (Jakubzick et al., 2013). (3) The deeper subcutaneous tissue (hypodermis) is made of fat and connective tissue.

In the skin, TRP channels are not only expressed in sensory nerve endings but also in many nonneuronal cell populations including keratinocytes and skin-resident immune cells (Figure 6.1). Various TRP channels participate in the formation and maintenance of skin barrier, HF growth, and cutaneous immunological and inflammatory processes, thereby maintaining skin homeostasis as well as contributing to many types of skin disorders (Figure 6.2). More importantly, several skin-expressing TRP channels act as the first-order sensors of temperature, mechanical, and chemical stimuli and mediate our senses of temperature, touch, itch, and pain under both physiological and pathological conditions.

6.2 TRP CHANNELS IN SKIN PHYSIOLOGY

6.2.1 KERATINOCYTE DIFFERENTIATION AND PROLIFERATION

The epidermal keratinocytes are constantly undergoing proliferation and differentiation (Marcelo et al., 1978). The proliferation is restricted to the basal keratinocyte cell layers (Rheinwald and Green, 1975), while cells in the more superficial layers lose their ability to divide and begin terminal differentiation (Sun and Green, 1976). Interestingly, there is a Ca^{2+} gradient in the epidermis with the highest Ca^{2+} concentration in the outer layer of healthy skin, which is critical to skin homeostasis (Elsholz et al., 2014). As a result, extracellular Ca^{2+} induces terminal differentiation of keratinocytes (Watt, 1989), formation of the cornified envelope (Nemes and Steinert, 1999) as well as epidermal lipid synthesis (Watanabe et al., 1998). siRNA silencing of the TRPC6 channels in epidermal keratinocytes prevents the induction of differentiation by extracellular Ca^{2+}. Moreover, promotion of the TRPC6 function by hyperforin or triterpenes is sufficient to induce keratinocyte differentiation and inhibit keratinocyte proliferation *in vitro* (Muller et al., 2008; Woelfle et al., 2010). Besides TRPC6, several other TRPC channels including TRPC1, TRPC4, TRPC5, and TRPC7 are also expressed by differentiating epidermal keratinocytes (Cai et al., 2005; Fatherazi et al., 2007) (Figure 6.1). siRNA silencing of TRPC1 or TRPC4 channels also prevents the induction of differentiation by extracellular Ca^{2+} (Fatherazi et al., 2007; Cai et al., 2006; Beck et al., 2008), suggesting some functional redundancy among various TRPC channels.

The TRPV6 channel is involved in mediating Ca^{2+} uptake and accounts for the basal intracellular Ca^{2+} levels in human keratinocytes (Lehen'kyi et al., 2007). Extracellular Ca^{2+}-induced differentiation upregulates both mRNA and protein levels of TRPV6, and siRNA silencing of TRPV6 in human primary keratinocytes impairs extracellular Ca^{2+}-induced keratinocyte differentiation as reflected by a marked reduction of the expression of differentiation markers such as involucrin, transglutaminase-1, and keratin-10 (Lehen'kyi et al., 2007). siRNA silencing of TRPV6 not only affects cell morphology and development of intercellular contacts, but also the ability for cells to stratify, suggesting that TRPV6 may play an important role in the formation of the stratum corneum (Lehen'kyi et al., 2007). Consistent with these findings, the skin of TRPV6 knockout mice has fewer and thinner layers of the stratum corneum, decreased total Ca^{2+} content, and loss of the normal Ca^{2+} gradient. As a result, 20% of all TRPV6 knockout animals develop alopecia and dermatitis (Bianco et al., 2007).

Deviation of the ambient temperature is a constant challenge to mammals' skin. A subset of TRP channels are activated by partially overlapping temperatures ranging from noxious cold to noxious heat. For instance, TRPV1, TRPV2, and TRPM3 are activated by noxious heat, while TRPV3, TRPV4, TRPM2, TRPM4, and TRPM5 are activated by warm temperatures. TRPM8 is activated by cool temperatures, and TRPA1 is shown to be activated by noxious cold. These *thermoTRPs* serve as molecular thermal sensors in our skin and internal organs (Patapoutian et al., 2003; Dhaka et al., 2006). Activation of thermoTRPs by heat stress such as cutaneous burn injuries leads to apoptosis in the epidermis *in vivo*, and heat stimulation also results in a marked decrease in colony-forming efficiency and protracted cell death in human keratinocytes in culture, which is prevented by the TRPV1 antagonists capsazepine or ruthenium red, although the latter is not a specific inhibitor of this channel (Radtke et al., 2011). Many reports have demonstrated the expression of TRPV1 in human epidermal keratinocytes using immunohistochemistry (Denda et al., 2001; Inoue et al., 2002; Southall et al., 2003; Bodo et al., 2004; Bodo et al., 2005). Although there are no direct measurements of TRPV1-mediated membrane currents in human keratinocytes, activation of TRPV1 by capsaicin as well as acidification elevates intracellular Ca^{2+} level in cultured human epidermal keratinocytes, which is inhibited by the TRPV1 inhibitor capsazepine (Inoue et al., 2002). Further, application of glycolic acid induces epidermal proliferation and ATP release in a skin equivalent model, which is blocked by capsazepine, suggesting that TRPV1 and purinergic receptor might be involved in low pH-elicited skin proliferation (Denda et al., 2010b). It is also proposed that anandamide induces Ca^{2+} influx, causes cell death, inhibits keratinocyte proliferation, and suppresses cell growth through activation of TRPV1 in human keratinocytes (Toth et al., 2011). Primary keratinocytes from human skin biopsies display orders of magnitudes lower TRPV1 mRNA levels compared to sensory ganglia. In addition, neither capsaicin nor resiniferatoxin (RTX), an extremely potent TRPV1 activator, is able to induce cytotoxicity to human keratinocytes, suggesting that human keratinocytes are vanilloid resistant. One explanation for this result is that a dominant negative splice variant of TRPV1, TRPV1β, is also present in human keratinocytes (Pecze et al., 2008). It has been reported that the mouse TRPV1β is not functional by itself, but its coexpression inhibits the function of the full-length TRPV1α (Wang et al., 2004).

TRPV3 is abundantly expressed by the epidermal keratinocytes, especially those in the epidermal basal layer and detects thermal and chemical stimuli in mammalian skin (Xu et al., 2006; Peier et al., 2002), as depicted in Figure 6.1. TRPV3 is activated by innocuous warm temperatures (>33°C) and a variety of bioactive substances (Luo and Hu, 2014). TRPV3-deficient mice have accelerated early keratinocyte differentiation, and cultured TRPV3-deficient primary keratinocytes express more early differentiation markers such as loricrin, keratin 1, and keratin 10 than wild-type keratinocytes (Cheng et al., 2010). In addition, keratinocyte-specific ablation of the TRPV3 function impairs keratinocyte cornification resulting from dysregulation of the activities of transglutaminases, a family of Ca^{2+}-dependent cross-linking enzymes essential for keratinocyte cornification (Cheng et al., 2010) (Figure 6.1).

A new epidermal isoform of TRPM8 (eTRPM8) has been cloned recently (Bidaux et al., 2015). This TRPM8 isoform is located in the membrane of the endoplasmic reticulum and controls mitochondrial Ca^{2+} concentration, which in turn modulates ATP and superoxide synthesis in a cold-dependent manner. As a result, mild cold (25°C) induces eTRPM8-dependent superoxide-mediated necrosis of keratinocytes (Bidaux et al., 2016). Therefore, the epidermal TRPM8 might mediate the fine-tuning of ATP and superoxide levels in response to cooling, which controls the balance between keratinocyte proliferation and differentiation. Indeed, genetic ablation of the eTRPM8 decreases epidermal proliferation and increases late keratinocyte differentiation (Bidaux et al., 2015). In human skin preparations, TRPA1 immunoreactivity is also present in the basal layer of the epidermis, dermis, and HFs, and both keratinocytes and melanocytes have displayed TRPA1 immunoreactivity (Atoyan et al., 2009). Activation of TRPA1 by icilin induces expression of many genes involved in the control of keratinocyte differentiation, cornification, and cell cycle (Atoyan et al., 2009), suggesting a role of TRPA1 in epidermal keratinocyte differentiation.

6.2.2 SKIN BARRIER FUNCTION

One of the major functions of the epidermal keratinocytes is to construct and maintain a water-impermeable barrier, the stratum corneum (SC). Disruption of the skin barrier by injury substantially increases TRPC6 mRNA transcripts in wounds, and TRPC6 deficiency results in impaired dermal wound healing after injury (Davis et al., 2012). The stretch-induced ATP release through mechanosensitive hemichannels elicits Ca^{2+} waves by activating the P2Y receptors coupled to the TRPC6 activation, which accounts for the stretch-accelerated wound closure (Takada et al., 2014) (Figure 6.2).

In hairless mouse skin after tape stripping, TRPV1 activation by capsaicin delays barrier recovery, whereas capsazepine, a TRPV1 antagonist, blocks this delay (Denda et al., 2007). Consistent with this finding, PAC-14028, another TRPV1 antagonist, accelerates skin barrier repair (Yun et al., 2011). TRPV1 activation also inhibits basal and arachidonic acid–induced lipid synthesis in sebocytes (Toth et al., 2009), further suggesting that TRPV1 activation inhibits skin barrier function (Figures 6.1 and 6.2).

Absence of functional TRPV3 in the epidermal keratinocytes leads to dysregulated activities of transglutaminases, which cause defects in skin barrier formation (Cheng et al., 2010). Although activation of TRPV3 by 2-aminoethyl diphenyl borate (2-APB) or camphor does not affect barrier recovery rate after tape stripping in either hairless mouse skin or human skin (Denda et al., 2007), a gain-of-function mutation in TRPV3 (Q580P) causes focal palmoplantar keratoderma in humans, which is associated with decreased expression of keratin 10, transglutaminases, and filaggrins (He et al., 2015). A few other human TRPV3 mutations at residues G521, G568, G573, M672, L673, and W692 cause Olmsted syndrome, which is a rare genodermatosis featuring symmetric and mutilating palmoplantar keratoderma and periorificial keratotic plaques (Wilson et al., 2015), suggesting that TRPV3 plays important roles in regulating human skin barrier function (Figure 6.2).

TRPV4 is expressed in mouse keratinocytes (Chung et al., 2004), and activation of TRPV4 by either warm temperatures or selective TRPV4 agonists accelerates barrier regeneration in mice (Denda et al., 2007) (Figure 6.1). Epidermal TRPV4 protein is located at the cell-cell junctions, and TRPV4 protein directly associates with the E-cadherin complex via β-catenin, which promotes cell-cell junction development and formation of the tight barrier between skin keratinocytes (Sokabe and Tominaga, 2010; Sokabe et al., 2010; Kida et al., 2012). Consistent with this finding, activation of TRPV4 strengthens the tight junction-associated barrier of epidermal cells through upregulation of tight junction structural proteins occludin and claudin-4, and tight junction regulatory

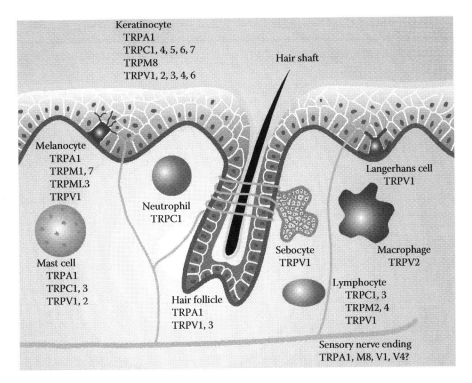

FIGURE 6.2 **(See color insert.)** Role of TRP channels in skin homeostasis and skin diseases. The skin-expressed TRP channels can either contribute to maintaining normal skin physiology or play important roles in the pathogenesis of skin diseases.

factor phospho-atypical PKCζ/ι (Akazawa et al., 2013). TRPV4 deficiency causes abnormal cell-cell junction structures, resulting in higher intercellular permeability *in vitro* (Sokabe et al., 2010) (Figure 6.2).

TRPA1 is also critically involved in epidermal barrier homeostasis. Brief cold (10°C–15°C) exposure to the skin surface of one flank for 1 minute or topical application of TRPA1 agonist allyl isothiocyanate (AITC) or cinnamaldehyde accelerates barrier recovery after tape stripping, which is blocked by a TRPA1 antagonist HC030031 (Denda et al., 2010a). Furthermore, topical application of bradykinin, an endogenous TRPA1 activator (Bautista et al., 2006; Bandell et al., 2004), accelerates barrier recovery, which is also blocked by HC030031 (Denda et al., 2010c). TRPA1 activation also enhances lamellar body secretion between stratum corneum and stratum granulosum, which promotes barrier recovery (Denda et al., 2010b). Similar to TRPA1, chemical activation of TRPM8 by WS12 accelerates skin barrier recovery after tape stripping and reduces epidermal proliferation associated with barrier disruption under low humidity, which is blocked by the selective TRPM8 antagonist BTCT (Denda et al., 2010a) (Figure 6.2).

6.2.3 HAIR GROWTH AND SKIN PIGMENTATION

The HF is a highly sensitive mini-organ cyclically transforming from phases of rapid growth (anagen), to apoptosis-driven regression (catagen), to relative quiescence (telogen) (Muller-Rover et al., 2001). Various TRP channels are expressed in the HFs and are involved in the HF cycling and hair shaft elongation. TRPV1 immunoreactivity has been identified in the outer root sheath (ORS) (Bodo et al., 2005), the inner root sheath (IRS), and the infundibulum of HFs (Stander et al., 2004) (Figure 6.1). In human skin HF organ culture or pilosebaceous units, activation of TRPV1 results in decreased hair shaft elongation, suppression of proliferation, induction of apoptosis, and premature

HF regression (Bodo et al., 2005; Toth et al., 2009). Compared with age-matched wild-type mice, TRPV1-deficient mice exhibit a significant delay in the first spontaneous transition of their HFs from the first anagen and the development of the subsequent telogen is also retarded. The development of the second anagen and subsequent HF cycling are not affected. These results suggest that TRPV1-mediated signaling is important for modulating the transition from the final stages of HF morphogenesis to regression (Biro et al., 2006) (Figure 6.2).

TRPV3 immunoreactivity is confined to epithelial compartments of both human and mouse ORS of HFs (Xu et al., 2006; Peier et al., 2002; Xu et al., 2002; Borbiro et al., 2011) (Figure 6.1). Although TRPV3 knockout mice have normal HF cycles, they display misaligned HFs compared with the wild-type mice (Cheng et al., 2010). Both global and conditional (epidermal) TRPV3 knockout mice display wavy hair coat and curly whiskers, suggesting an indispensable role for TRPV3 in normal hair morphogenesis (Cheng et al., 2010). The mouse strain DS-Nh originates from a mutant male mouse with deficient hair growth in an inbred DS strain colony at the Aburahi Laboratories (Shionogi & Co., Ltd., Shiga, Japan) (Hikita et al., 2002). Genetic studies have identified gain-of-function TRPV3 mutations causing the hairless phenotypes in the DS-Nh mice and WBN/Kob-Ht rats (Gly573 to Ser [Nh mutation] or Gly573 to Cys [Ht mutation]) (Asakawa et al., 2006; Xiao et al., 2008). DNA microarray data have revealed decreased expression of keratin-associated protein 16–1, 16–3, and 16–9 genes related to the anagen phase compared with the age-matched DS mice. Importantly, the anagen phase persists in the DS-Nh mice, but the telogen phase is not present at 21 days of age (Imura et al., 2007). Consistent with mouse studies, activation of TRPV3 by eugenol or 2-APB produces a dose-dependent inhibition of human hair shaft elongation, suppression of proliferation, and induction of apoptosis and premature HF regression (catagen), further demonstrating that TRPV3 plays important roles in modulating mammalian hair development (Asakawa et al., 2006; Imura et al., 2007; Akimoto et al., 2000; Watanabe et al., 2003) (Figure 6.2).

Human skin pigmentation is driven by melanin synthesis in epidermal melanocytes, which is a protective measure to avoid damage caused by environmental ultraviolet radiation (UVR). The melanocytes are dendritic, pigment-synthesizing cells derived from the neural crest and predominantly confined to the basal layer of the skin. Melanocytes synthesize and package melanin in melanosomes, lysosome-derived organelles that move to the melanocyte-keratinocyte junction and transfer melanin to the recipient keratinocytes (Lin and Fisher, 2007). TRPA1 mediates Ca^{2+} influx and a retinal-dependent photocurrent in human epidermal melanocytes in response to physiological doses of UVR, a process that requires G proteins and phospholipase C signaling. Activation of TRPA1 by UVR induces early increase in cellular melanin, suggesting that TRPA1 is an essential component in the phototransduction pathway in human melanocytes (Bellono et al., 2013). Although TRPV1 is also expressed by human melanocytes, the melanin content and tyrosinase activity are affected by neither activation nor inhibition of TRPV1, suggesting that TRPV1 is not directly involved in melanogenesis (Choi et al., 2009) (Figures 6.1 and 6.2).

Both TRPM1 and TRPM7 are expressed by melanocytes (Figure 6.1). The TRPM7 channel also plays a role in melanocyte survival through detoxification of the intermediates of melanogenesis (Iuga and Lerner, 2007; McNeill et al., 2007). A mutation in the TRPM1 gene is linked to Leopard complex spotting patterns in horses, highlighting the importance of TRPM1 in skin pigmentation (Bellone et al., 2010). TRPM1 expression also correlates with melanin content in neonatal human epidermal melanocytes (Oancea et al., 2009). Moreover, TRPM1 expression and intracellular Ca^{2+} levels are significantly higher in differentiating melanocytes compared with that in the proliferating melanocytes, suggesting that enhanced TRPM1 activity might be required for melanocyte differentiation. Consistent with this finding, lentiviral short hairpin RNA (shRNA)-mediated TRPM1 knockdown results in reduced intracellular Ca^{2+} and decreased intracellular melanin pigment (Devi et al., 2009) (Figure 6.2). The varitint-waddler (Va) mice have pigmentation defects with variegated/dilute coat color besides hearing loss and circling motor behavior (Cuajungco and Samie, 2008), which is caused by gain-of-function mutations (primarily an alanine to proline substitution at amino acid position 419, A419P) in the TRPML3 channels (Xu et al., 2007; Kim et al., 2007). TRPML3 is highly expressed in normal melanocytes, and the A419P mutation locks the channel in a constitutive open state resulting Ca^{2+} overload and melanocyte death (Xu et al., 2007) (Figures 6.1 and 6.2).

6.2.4 CUTANEOUS IMMUNE REGULATION

Besides serving as a physical barrier, the skin is equipped with many types of specialized immune cells to protect against invading pathogens (Nestle et al., 2009). The TRPC1 channels are involved in N-formylmethionine-leucyl-phenylalanine (fMLP)-mediated chemotaxis and migration of neutrophils, and TRPC1 deficiency results in impaired migration, transmigration, and chemotaxis of neutrophils (Lindemann et al., 2015). Both TRPC1 and TRPC3 can also modulate the Ca^{2+} release–activated Ca^{2+} channel protein (ORAI)–dependent Ca^{2+} influx in T cells and mast cells (Saul et al., 2014) (Figure 6.1).

TRPV1 immunoreactivity is present in cutaneous mast cells (Stander et al., 2004) and epidermal Langerhans cells (Bodo et al., 2004), as depicted in Figure 6.1. Activation of TRPV1 by capsaicin in a human epidermal keratinocyte cell line, HaCaT, induces a Ca^{2+}-dependent expression of cyclooxygenase-2 (COX-2) and release of many inflammatory mediators including interleukin (IL)-8, and prostaglandin E2 (PGE2), and the expression of matrix metalloproteinase (MMP)-1. The inflammatory action of capsaicin is attenuated by the TRPV1 antagonist capsazepine (Southall et al., 2003; Li et al., 2007). In cultured human ORS keratinocytes, activation of TRPV1 by capsaicin increases the production of IL-1β and transforming growth factor (TGF)-β2 (Bodo et al., 2005). In HaCaT cells, UV irradiation induces a TRPV1-dependent Ca^{2+} influx that increases the expression of many types of proinflammatory cytokines (Lee et al., 2009; Lee et al., 2011). TRPV1 is also expressed by the primary human T cells and Jurkat cells (an immortalized cell line of human T cells) as well as the primary mouse splenic T cells (Figure 6.1). TRPV1 acts as a non-store-operated Ca^{2+} channel and contributes to T-cell receptor (TCR)-mediated Ca^{2+} influx and T-cell activation (Bertin et al., 2014). Moreover, TRPV1 expression in T cells is upregulated during concanavalin A-driven mitogenic and anti-CD3/CD28 stimulated TCR activation. Accordingly, TRPV1 blockers suppress T-cell activation and production of inflammatory cytokines (Majhi et al., 2015).

TRPV2 is abundantly expressed by innate immune cells, especially macrophages, and activation of TRPV2 promotes lipopolysaccharide (LPS)-induced TNF-α and IL-6 production (Santoni et al., 2013) (Figure 6.1). TRPV2 is recruited to the nascent phagosome during macrophage phagocytosis, resulting in plasma membrane depolarization and elevated synthesis of phosphatidylinositol-4,5-bisphosphate (PI[4,5]P2), which triggers partial actin depolymerization required for occupancy-elicited phagocytic receptor clustering. As a result, TRPV2-deficient macrophages have impaired phagocytosis and chemoattractant-elicited motility (Link et al., 2010). In mast cells, activation of TRPV2 elicits Ca^{2+} influx and stimulates protein kinase A–dependent degranulation (Santoni et al., 2013).

The DS-Nh mice and WBN/Kob-Ht rats associated with TRPV3 gain-of-function mutations have more mast cells in the skin compared to their parent strains (Asakawa et al., 2006). As a result, the histamine concentration in the skin of the DS-Nh mice is significantly higher than that in the DS mice (Asakawa et al., 2006). Serum levels of several C-C motif chemokine ligands are also significantly higher in the DS-Nh mice at 30 weeks of age than that in the age-matched DS mice, suggesting that TRPV3 is an immune modulator in the skin (Yoshioka et al., 2009).

TRPA1-immunoreactivity is also present in dermal mast cells, and the expression of TRPA1 is enhanced in patients with postburn pruritus or atopic dermatitis (Yang et al., 2015; Oh et al., 2013). Cultured human primary epidermal keratinocytes produce pro-inflammatory cytokines IL-1α and IL-1β when treated with the TRPA1 agonist, icilin (Atoyan et al., 2009). Both TRPM2 and TRPM4 are expressed by lymphocytes, and TRPM4 contributes to Ca^{2+} oscillations following T-cell activation (Launay et al., 2004; Schwarz et al., 2007).

6.3 TRP CHANNELS IN SKIN DISEASES

6.3.1 SKIN INFLAMMATION

Ca^{2+}-dependent keratinocyte differentiation is impaired in both atopic dermatitis (AD) and psoriasis, which is critically associated with progression of these two skin disorders (Leuner et al., 2011; Radresa et al., 2013).

The reduced differentiation in psoriatic keratinocytes is correlated with decreased expression of TRPC channels including TRPC1, TRPC4, and TRPC6 (Leuner et al., 2011), suggesting that dysfunction of TRPC channels may trigger the process of these inflammatory skin diseases (Figure 6.2). Strikingly, activation of the TRPC6 channels can partially overcome the Ca^{2+} entry defect in the psoriatic keratinocytes, suggesting that selective TRPC activators may be useful for treating psoriasis (Leuner et al., 2011).

Rosacea is another chronic inflammatory skin disease associated with neurogenic inflammation involving activation of both TRPV1 and TRPA1 channels (Figure 6.2) and subsequent release of substance P and calcitonin gene-related peptide (CGRP) from primary sensory nerve endings, both of which promote the progression of rosacea (Steinhoff et al., 2013). Moreover, compared with healthy skin and lupus erythematosus, different types of rosacea such as erythematotelangiectatic rosacea (ETR), papulopustular rosacea (PPR), and phymatous rosacea (PhR) display increased expression of many types of TRP channels including TRPV1, TRPV2, TRPV3, and TRPV4, at either molecular or protein levels, suggesting that dysregulation of these TRPV channels might be an important mechanism underlying the initiation and maintenance of rosacea (Sulk et al., 2012).

The DS-Nh mice are commonly used as an animal model of AD because these mice develop dermatitis spontaneously when housed in a conventional environment, and *Staphylococcus aureus* (*S. aureus*) is considered the cause of the AD (Haraguchi et al., 1997). The DS-Nh mice have infiltrations of many types of inflammatory cells in the skin lesions including CD11b+ macrophages, CD4+ T cells, mast cells, and eosinophils (Hikita et al., 2002). Moreover, TRPV3 gain-of-function mutations enhance skin responses to haptens and contribute to the development of hapten-induced dermatitis (Imura et al., 2009; Yamamoto-Kasai et al., 2013). The TRPV3(Gly573Ser) transgenic mice recapitulate the symptoms of the spontaneous allergic and pruritic dermatitis when maintained under a conventional environment, further highlighting the importance of TRPV3 in skin inflammation (Yoshioka et al., 2009) (Figure 6.2).

Consistent with TRPV3 being an inflammatory mediator in the skin, activation of TRPV3 results in release of various inflammatory mediators from keratinocytes such as ATP (Mandadi et al., 2009), nitrogen oxide (NO) (Miyamoto et al., 2011), PGE2 (Huang et al., 2008), and IL-1α (Xu et al., 2006). Compared with DS mice, keratinocytes from the DS-Nh mice have enhanced production and release of thymic stromal lymphopoietin (TSLP), which promotes allergen sensitization through skin and triggers an "atopic march" in mice (Leyva-Castillo et al., 2013). In contrast to TRPV3, activation of TRPM8 reduces ultraviolet B-induced PGE2 production in keratinocytes (Park et al., 2013), suggesting that the activation of TRPM8 may reduce inflammatory responses in the skin (Figure 6.2).

6.3.2 SKIN CANCER

In basal cell carcinoma, expression of TRP channels, especially the TRPC1 and TRPC4 channels, is markedly reduced, resulting in a lack of Ca^{2+} entry through these pathways. Therefore, dysregulation of TRPC channels in keratinocytes may play a direct role in skin tumorigenesis (Beck et al., 2008). Application of triterpenes which inhibits *in vitro* growth of cancer cells to form actinic keratosis, a precancerous patch of thick and scaly skin, suppresses keratinocyte proliferation and promotes keratinocyte differentiation *in vivo*. The effect is mediated at least partly by TRPC6 because triterpenes selectively promote TRPC6 expression in cultured primary human keratinocytes and in the epidermis of human skin explants *ex vivo*, resulting in keratinocyte terminal differentiation (Woelfle et al., 2010) (Figure 6.2).

TRPV1 has a tumor-suppressing effect by downregulating epidermal growth factor receptor (EGFR) signaling through direct interaction with the EGFR, which leads to EGFR degradation (Bode et al., 2009). As a result, loss of interaction between the TRPV1 and EGFR in TRPV1 knock-out mice causes a significant increase in 12-O-tetradecanoylphorbol 13-acetate (TPA)-induced skin carcinogenesis. Similarly, topical application of AMG9810, a selective TRPV1 antagonist, enhances EGFR expression and its downstream Akt/mammalian target of rapamycin (mTOR) signaling, which promotes skin tumorigenesis (Li et al., 2011). Surprisingly, the TRPV1 activator, capsaicin, also promotes TPA-induced skin carcinogenesis *in vivo* through directly activating the EGFR/MEK signaling to increase COX-2 expression, which is independent of TRPV1 activation (Hwang et al., 2010) (Figure 6.2).

By comparing TRP channel expression patterns in healthy human skin and keratinocytic tumors, Fusi et al. have detected similar levels of TRPV1, TRPV2, and TRPV3 expression in keratino-cytes between the healthy and tumor tissues. However, both TRPV4-immunoreactivity and mRNA transcripts are abrogated in the keratinocytic tumors while maintained at high levels in healthy or inflamed skin, which seems to be caused by keratinocyte-derived IL-8, suggesting that the selective reduction of TRPV4 expression may represent an early biomarker of skin carcinogenesis. But, TRPA1 staining was increased selectively in the samples of intraepidermal proliferative disorders such as solar keratosis (Fusi et al., 2014).

A few members in the TRPM subfamily are implicated in the pathogenesis of melanomas. The TRPM1 mRNA transcripts display uniform expression in benign Spitz nevi but show variable expression in primary melanomas. More strikingly, TRPM1 mRNA transcripts are completely absent in melanoma metastasis (Duncan et al., 1998; Fang and Setaluri, 2000; Zhiqi et al., 2004; Deeds et al., 2000), suggesting that loss of TRPM1 expression might be an indication for aggressiveness of the primary cutaneous melanoma. A melanoma-enriched antisense TRPM2 (TRPM2-AS) transcript and another tumor-enriched TRPM2 transcript (TRPM2-TE) are upregulated in melanoma. Ablation of the TRPM2-TE, as well as overexpression of the wild-type TRPM2, enhance melanoma susceptibility to apoptosis and necrosis (Orfanelli et al., 2008). TRPM8 is identified in human melanoma cells and tissues (Yamamura et al., 2008; Slominski, 2008), and activation of TRPM8 in cultured human melanoma cells inhibited cell viability (Yamamura et al., 2008), suggesting that TRPM8 might be involved in the regulation of melanoma progression (Figure 6.2).

6.4 SKIN-EXPRESSED TRP CHANNELS MEDIATE SENSORY PERCEPTION

6.4.1 TEMPERATURE SENSING

Our ability to perceive temperature enables us to maintain a constant body temperature and protects us from noxious thermal stimuli. Although it is well established that sensory nerve endings mediate our sense of temperatures, emerging evidence suggests that skin cells are also equipped with many types of temperature-gated ion channels, especially the thermoTRPs, and act in concert with sensory neurons to allow us to perceive the thermal environment.

Two of these thermoTRPs, TRPV3 and TRPV4, are abundantly expressed in skin keratinocytes (Peier et al., 2002; Chung et al., 2004; Xu et al., 2002; Smith et al., 2002; Guler et al., 2002). TRPV3 is activated by warm temperatures when expressed in heterologous cells or endogenously expressed in the cultured primary keratinocytes (Peier et al., 2002; Xu et al., 2002; Smith et al., 2002; Moqrich et al., 2005; Grandl et al., 2008). High-throughput screening of random mutant TRPV3 clones has identified single point mutations that specifically abolish heat activation but do not perturb chemical activation or voltage modulation, suggesting that the temperature sensitivity of TRPV3 is separable from all other known activation mechanisms. All five TRPV3 mutations are located in the extracellular pore loop region and implicate a specific region in temperature sensing (Grandl et al., 2008). Interestingly, nuclear genome sequencing of Asian elephants and the extinct woolly mammoths has identified a hypomorphic woolly mammoth amino acid substitution in TRPV3 (N647D), which is one of the extracellular pore loop TRPV3 heat mutants identified in the original high-throughput screening (Grandl et al., 2008). This amino acid substitution causes a loss of heat activation and may contribute to evolution of cold tolerance, long hair, and large adipose stores in the mammoths (Lynch et al., 2015).

Although TRPV3 mRNA transcripts are expressed in monkey dorsal root ganglia (DRG) neurons, functional expression of TRPV3 protein in mouse DRG has not been confirmed (Xu et al., 2002; Grandl et al., 2008). Behavioral tests demonstrate that genetic ablation of TRPV3 function causes strong deficits in response to innocuous and noxious heat but not in other sensory modalities, suggesting that skin-expressed TRPV3 channels might participate in thermosensation *in vivo* (Moqrich et al., 2005). Interestingly, heat sensation is not completely abolished in single TRPV1 or TRPV3 knockout mice, which strongly suggests functional redundancies exist among various thermosensitive ion channels. Consistent with this observation, when both TRPV1 and TRPV3 are inactivated, there is a cooperative role between skin-derived TRPV3 and primary sensory neuron-enriched TRPV1 for sensing a well-defined window of acute moderate heat temperature (Marics et al., 2014).

TRPV4 is also involved in normal thermal responsiveness *in vivo* as TRPV4 knockout mice select warmer floor temperatures than wild-type littermates on a thermal gradient, and they also display prolonged withdrawal latencies during acute tail heating (Lee et al., 2005). Further, TRPV4 channel is involved in the recruitment of behavioral and autonomic warmth-defense responses to regulate core body temperature in rats through a peripheral but not a central nervous system (CNS) mechanism (Vizin et al., 2015). Moreover, the TRPV4 knockout mice display a significantly longer latency to escape from the plates at 35°C–45°C when hyperalgesia is induced by carrageenan without changes in foot volumes. TRPV4 therefore determines the sensitivity rather than the threshold of painful heat detection and plays an essential role in thermal hyperalgesia (Todaka et al., 2004). Other studies show that both TRPV3 and TRPV4 likely make limited and strain-dependent contributions to innocuous warm temperature perception or noxious heat sensation, suggesting the existence of additional significant heat perception mechanisms (Huang et al., 2011).

6.4.2 PAIN

Although great strides have been made in the past decades in identifying critical ion channels and G protein-coupled receptors in the primary afferent neurons and spinal cord interneurons that are critically involved in initiating and maintaining both inflammatory and neuropathic pain, the role of skin-resident cells in the pathogenesis of pain is not fully understood. Recent exciting studies have demonstrated that epidermal keratinocytes are the first-order sensors for sensory processing. Baumbauer et al. have used mice selectively expressing channelrhodopsin (ChR2) or halorhodopsin in keratinocytes and demonstrated that activation of ChR2-expressing keratinocytes by blue light provokes action potential firing in multiple types of cutaneous sensory neurons, while inhibition of halorhodopsin-expressing keratinocytes by yellow light suppresses action potential firing elicited by naturalistic stimuli. These findings suggest that activation of the epidermal keratinocytes is sufficient to activate the sensory pathway (Baumbauer et al., 2015). In another study, Pang et al. have selectively expressed TRPV1 under the keratin 5 promoter in the keratinocytes of TRPV1-knockout mice. Strikingly, activation of the TRPV1-expressing keratinocytes by capsaicin is sufficient to induce expression of the neuronal activation marker, c-fos, in laminae I and II of the ipsilateral spinal cord dorsal horn and to evoke acute paw-licking nocifensive behavior and conditioned place aversion, providing direct evidence that keratinocyte stimulation is sufficient to evoke acute nociception-related responses *in vivo* (Pang et al., 2015). Keratinocytes can also secrete inflammatory cytokines to sensitize the nociceptors, such as TNF-α and IL-6, which are elevated in complex regional pain syndrome (CRPS) (Birklein et al., 2014). Moreover, transplantation of human keratinocytes into a ligated and transected peripheral nerve increases nerve growth factor (NGF) levels at the transplant site and elicits axon sprouting and extreme hyperexcitability in DRG neurons whose cut axons are present near the transplanted human keratinocytes. Keratinocyte-derived NGF, especially in the presence of tissue injury, promotes membrane excitability of primary afferent neurons (Matsuda et al., 1998), which results in inflammatory hypersensitivity and spontaneous pain (Koltzenburg et al., 1999; Ma and Woolf, 1997). Consistent with these findings, animals with the keratinocyte transplants exhibit spontaneous pain behavior, suggesting that keratinocytes are critical contributors to persistent pain associated with peripheral nerve injury (Radtke et al., 2010).

Activation of TRPA1 by AITC or TRPV1 by capsaicin induces a differential secretion pattern of PGE2 and leukotriene B4 (LTB4) in human dermal fibroblasts and keratinocytes and elicits painful sensations (Jain et al., 2011). TRPV1 expression is also upregulated in the epidermal keratinocytes of herpes zoster patients, which is correlated with the degree of pain that may result from elevated levels of cytokines (Han et al., 2016).

Expression levels of both TRPV3 and TRPV4 are significantly increased in keratinocytes of patients with breast pain, which is correlated with elevated expression of NGF in keratinocytes and sensitization of nociceptors (Gopinath et al., 2005). Activation of TRPV3 in the keratinocytes can modulate thermal pain transduction through the release of PGE2 (Huang et al., 2008). In addition, recent studies suggest that the endogenous lipid molecule resolvin D1 inhibits both TRPV3 and TRPV4 and suppresses TRPV3-mediated acute pain behaviors and inflammation-induced thermal hypersensitivity (Bang et al., 2010). Other studies show that keratinocytes may play dual roles in pain sensation. For instance, keratinocyte-derived endothelin-1 (ET-1) acts on the endothelin-A receptors expressed by nociceptors to initiate pain, while ET-1 binding with the endothelin-B receptors on the keratinocytes produces analgesia (Khodorova et al., 2003). Interestingly, a recent study shows that UVB exposure directly activates TRPV4, and the subsequent Ca^{2+} influx induces upregulation of ET-1 in keratinocytes to trigger sunburn pain (Moore et al., 2013).

6.4.3 ITCH

Chronic itch is one of the major clinical symptoms of many skin disorders. TRP channels, especially the TRPA1 and TRPV1, are critical initiators of itching by directly activating the peripheral pruriceptors (Luo et al., 2015; Akiyama and Carstens, 2013). Itch-sensitive neurons are activated by a variety of exogenous pruritogens (itch-causing compounds), as well as pruritogens produced endogenously by epithelial and immune cells, suggesting that skin keratinocytes and resident immune cells may also play roles in the genesis of itch sensation. Recent studies have shown that several TRP channels in the skin are important mediators of itch in both mice and humans.

Mouse and human genetic studies have provided compelling evidence that gain-of-function TRPV3 mutations are associated with enhanced itching in addition to hair loss and dermatitis. The DS-Nh mice develop allergic and pruritic dermatitis under conventional conditions, which is correlated with elevated inflammatory factors resulted from constitutive TRPV3 activation (Yoshioka et al., 2009). For instance, these mice have higher levels of histamine and NGF in serum and/or skin tissues than age-matched DS mice without dermatitis. In addition, prolonged treatment with an H1 receptor antagonist reduces the production of keratinocyte-derived NGF and spontaneous scratching in the DS-Nh mice (Yoshioka et al., 2006).

Although both the hairless and dermatitis phenotypes in the DS-Nh mice are inherited in an autosomal-dominant fashion and are not separable, the spontaneous scratching behavior caused by the TRPV3(Gly573Ser) mutation is independent of the development of dermatitis as revealed by using the TRPV3(Gly573Ser) transgenic mice (Yoshioka et al., 2009). TRPV3 null mice also have less protein gene product 9.5 (PGP9.5)-positive nerve endings in the skin compared with wild-type mice in a mouse model of chronic dry skin–associated itch (Miyamoto et al., 2002). More importantly, the spontaneous scratching in the mouse dry skin model is lost in the TRPV3 null mice, suggesting that TRPV3 may be required for the development of some forms of chronic itch (Yamamoto-Kasai et al., 2012). A few human TRPV3 mutations including the TRPV3(Gly573Ser) are associated with Olmsted syndrome (OS) (Lin et al., 2012; Duchatelet and Hovnanian, 2015). All affected individuals complain of severe itching in the lesions, resulting in frequent scratching and sleep disturbances. Therefore, TRPV3 seems to have a conserved itch-mediating function in both humans and rodents.

Although it was recently reported that TRPV4 is involved in histamine- and serotonin-induced acute itch, the precise role of TRPV4 in acute itch remains controversial. Akiyama et al. showed that TRPV4 null mice exhibited significantly fewer serotonin-evoked scratching bouts compared with wild-type mice. The serotonin-induced itching was also suppressed by pharmacological inhibition of TRPV4 or 5-HT2 receptor, suggesting that the TRPV4-dependent pathway is likely downstream of 5-HT2-mediated itch. Alternatively, the number of scratch bouts elicited by histamine and SLIGRL (a protease-activated receptor 2/mas-related G protein-coupled receptor C11 agonist) is not different between the TRPV4 null mice and wild-type mice (Akiyama et al., 2016). Moreover, the serotonin-elicited scratching in TRPA1 null mice was not significantly different from that in the wild-type mice, which is inconsistent with another study showing that serotonin acts on 5-HT7 receptors coupled to TRPA1 channels generating itch sensation (Morita et al., 2015).

In contrast, Chen et al. reported that the scratching behaviors evoked by all histaminergic pruritogens including histamine, compound 48/80, and ET-1, were significantly attenuated in the keratinocyte-specific, tamoxifen-inducible *Trpv4* knockdown mice (TRPV4 cKO). Topical application of a TRPV4 specific antagonist, GSK205, on wild-type mice also blocked the scratching responses elicited by all these histaminergic pruritogens (Chen et al., 2016). These results suggested that TRPV4 in keratinocytes is responsible for the histaminergic itch. Further pharmacological studies showed that histamine induced TRPV4-dependent Ca^{2+} influx into the keratinocyte through the H1, H3, or H4 histamine receptors, which results in phosphorylation of MAP-kinase, ERK, and triggers the itch signaling (Chen et al., 2016). However, it is not clear if topically applied inhibitors for the TRPV4 channels, the histamine receptors, or MAPK/ERK signaling pathways also act on cutaneous sensory nerve endings in addition to skin-resident cells. Besides the phenotypic differences, whether TRPV4 expression in the skin or the DRG neurons is the main mediator of itch is still an open question (Akiyama et al., 2016; Chen et al., 2016). Akiyama et al. proposed that TRPV4 is functionally expressed by DRG neurons to mediate serotonin-induced scratching, whereas Chen et al. showed that TRPV4 expressed by the epidermal keratinocytes mediates the histaminergic itch (Akiyama et al., 2016; Chen et al., 2016).

Chloroquine (CQ) is another well-established pruritogen-mediating histamine-independent itch in mice. Interestingly, these two studies also had different outcomes in CQ-elicited itching in the TRPV4 null mice: one study showed that the CQ-elicited itching was significantly increased in the TRPV4 global knockout mice compared with the wild-type mice, whereas the other study found that CQ-evoked scratching was not affected in the TRPV4 global knockout mice or skin-specific TRPV4 cKO mice (Akiyama et al., 2016; Chen et al., 2016). Therefore, the exact role of TRPV4 in mediating acute itching warrants further investigation.

Although TRPA1 expressed in DRG neurons is a well-established itch receptor mediating the CQ-induced nonhistaminergic itch, recent studies also found that TRPA1 is expressed in dermal mast cells and that TRPA1 expression level was increased in the scar tissues of patients with post-burn pruritus or atopic dermatitis, suggesting that TRPA1 might also contribute to the genesis of itch through activation of the skin-resident cells (Yang et al., 2015; Oh et al., 2013). In contrast to itch-initiating TRP channels, activation of TRPM8 inhibits both histaminergic and lichenification-associated itch in humans (Bromm et al., 1995), suggesting that activation of TRPM8 may inhibit itch signaling, which is consistent with the fact that activation of TRPM8 by menthol or cooling skin in a water bath are commonly used in antipruritic therapies (Hong et al., 2011).

ACKNOWLEDGMENTS

This work was supported by grants from the National Institutes of Health R01DK103901 and R01GM101218, and the Center for the Study of Itch of the Department of Anesthesiology at the Washington University School of Medicine to HH.

REFERENCES

Akazawa, Y. et al. 2013. Activation of TRPV4 strengthens the tight-junction barrier in human epidermal keratinocytes. *Skin Pharmacol Physiol*, 26: 15–21.

Akimoto, T. et al. 2000. Locus of dominant hairless gene (Ht) causing abnormal hair and keratinization maps to rat chromosome 10. *Exp Anim*, 49: 137–140.

Akiyama, T. and E. Carstens. 2013. Neural processing of itch. *Neuroscience*, 250: 697–714.

Akiyama, T. et al. 2016. Involvement of TRPV4 in serotonin-evoked scratching. *J Invest Dermatol*, 136: 154–160.

Asakawa, M. et al. 2006. Association of a mutation in TRPV3 with defective hair growth in rodents. *J Invest Dermatol*, 126: 2664–2672.

Atoyan, R., D. Shander, and N.V. Botchkareva. 2009. Non-neuronal expression of transient receptor potential type A1 (TRPA1) in human skin. *J Invest Dermatol*, 129: 2312–2315.

Bandell, M. et al. 2004. Noxious cold ion channel TRPA1 is activated by pungent compounds and bradykinin. *Neuron*, 41: 849–857.

Bang, S. et al. 2010. Resolvin D1 attenuates activation of sensory transient receptor potential channels leading to multiple anti-nociception. *Br J Pharmacol*, 161: 707–720.

Baumbauer, K.M. et al. 2015. Keratinocytes can modulate and directly initiate nociceptive responses. *Elife*, 4.

Bautista, D.M. et al. 2006. TRPA1 mediates the inflammatory actions of environmental irritants and proalgesic agents. *Cell*, 124: 1269–1282.

Beck, B. et al. 2008. TRPC channels determine human keratinocyte differentiation: New insight into basal cell carcinoma. *Cell Calcium*, 43: 492–505.

Bellone, R.R. et al. 2010. Fine-mapping and mutation analysis of TRPM1: A candidate gene for leopard complex (LP) spotting and congenital stationary night blindness in horses. *Brief Funct Genomics*, 9: 193–207.

Bellono, N.W. et al. 2013. UV light phototransduction activates transient receptor potential A1 ion channels in human melanocytes. *Proc Natl Acad Sci U S A*, 110: 2383–2388.

Bertin, S. et al. 2014. The ion channel TRPV1 regulates the activation and proinflammatory properties of CD4(+) T cells. *Nat Immunol*, 15: 1055–1063.

Bianco, S.D. et al. 2007. Marked disturbance of calcium homeostasis in mice with targeted disruption of the *Trpv6* calcium channel gene. *J Bone Miner Res*, 22: 274–285.

Bidaux, G. et al. 2015. Epidermal TRPM8 channel isoform controls the balance between keratinocyte proliferation and differentiation in a cold-dependent manner. *Proc Natl Acad Sci U S A*, 112: E3345–E3354.

Bidaux, G. et al. 2016. Fine-tuning of eTRPM8 expression and activity conditions keratinocyte fate. *Channels (Austin)*, 10: 320–331.

Birklein, F. et al. 2014. Activation of cutaneous immune responses in complex regional pain syndrome. *J Pain*, 15: 485–495.

Biro, T. et al. 2006. Hair cycle control by vanilloid receptor-1 (TRPV1): Evidence from TRPV1 knockout mice. *J Invest Dermatol*, 126: 1909–1912.

Bode, A.M. et al. 2009. Transient receptor potential type vanilloid 1 suppresses skin carcinogenesis. *Cancer Res*, 69: 905–913.

Bodo, E. et al. 2004. Vanilloid receptor-1 (VR1) is widely expressed on various epithelial and mesenchymal cell types of human skin. *J Invest Dermatol*, 123: 410–413.

Bodo, E. et al. 2005. A hot new twist to hair biology: Involvement of vanilloid receptor-1 (VR1/TRPV1) signaling in human hair growth control. *Am J Pathol*, 166: 985–998.

Borbiro, I. et al. 2011. Activation of transient receptor potential vanilloid-3 inhibits human hair growth. *J Invest Dermatol*, 131: 1605–1614.

Bos, J.D. et al. 1987. The skin immune system (SIS): Distribution and immunophenotype of lymphocyte subpopulations in normal human skin. *J Invest Dermatol*, 88: 569–573.

Bromm, B. et al. 1995. Effects of menthol and cold on histamine-induced itch and skin reactions in man. *Neurosci Lett*, 187: 157–160.

Cai, S. et al. 2005. TRPC channel expression during calcium-induced differentiation of human gingival keratinocytes. *J Dermatol Sci*, 40: 21–28.

Cai, S. et al. 2006. Evidence that TRPC1 contributes to calcium-induced differentiation of human keratinocytes. *Pflugers Arch*, 452: 43–52.

Chen, Y. et al. 2016. Transient receptor potential vanilloid 4 ion channel functions as a pruriceptor in epidermal keratinocytes to evoke histaminergic itch. *J Biol Chem*, 291(19): 10252–10262.

Cheng, X. et al. 2010. TRP channel regulates EGFR signaling in hair morphogenesis and skin barrier formation. *Cell*, 141: 331–343.

Choi, T.Y. et al. 2009. Endogenous expression of TRPV1 channel in cultured human melanocytes. *J Dermatol Sci*, 56: 128–130.

Chung, M.K. et al. 2004. TRPV3 and TRPV4 mediate warmth-evoked currents in primary mouse keratinocytes. *J Biol Chem*, 279: 21569–21575.

Clapham, D.E. 2003. TRP channels as cellular sensors. *Nature*, 426: 517–524.

Cuajungco, M.P. and M.A. Samie. 2008. The varitint-waddler mouse phenotypes and the TRPML3 ion channel mutation: Cause and consequence. *Pflugers Arch*, 457: 463–473.

Davis, J. et al. 2012. A TRPC6-dependent pathway for myofibroblast transdifferentiation and wound healing in vivo. *Dev Cell*, 23: 705–715.

Deeds, J., F. Cronin, and L.M. Duncan. 2000. Patterns of melastatin mRNA expression in melanocytic tumors. *Human Pathol*, 31: 1346–1356.

Denda, M. et al. 2001. Immunoreactivity of VR1 on epidermal keratinocyte of human skin. *Biochem Biophys Res Commun*, 285: 1250–1252.

Denda, M. et al. 2007. Effects of skin surface temperature on epidermal permeability barrier homeostasis. *J Invest Dermatol*, 127: 654–659.

Denda, M., M. Tsutsumi, and S. Denda. 2010a. Topical application of TRPM8 agonists accelerates skin permeability barrier recovery and reduces epidermal proliferation induced by barrier insult: Role of cold-sensitive TRP receptors in epidermal permeability barrier homoeostasis. *Exp Dermatol*, 19: 791–795.

Denda, M. et al. 2010b. Topical application of TRPA1 agonists and brief cold exposure accelerate skin permeability barrier recovery. *J Invest Dermatol*, 130: 1942–1945.

Denda, S. et al. 2010c. Glycolic acid induces keratinocyte proliferation in a skin equivalent model via TRPV1 activation. *J Dermatol Sci*, 57: 108–113.

Devi, S. et al. 2009. Calcium homeostasis in human melanocytes: Role of transient receptor potential melastatin 1 (TRPM1) and its regulation by ultraviolet light. *Am J Phys*, 297: C679–C687.

Dhaka, A., V. Viswanath, and A. Patapoutian. 2006. Trp ion channels and temperature sensation. *Annu Rev Neurosci*, 29: 135–161.

Duchatelet, S. and A. Hovnanian. 2015. Olmsted syndrome: Clinical, molecular and therapeutic aspects. *Orphanet J Rare Dis*, 10: 33.

Duncan, L.M. et al. 1998. Down-regulation of the novel gene melastatin correlates with potential for melanoma metastasis. *Cancer Res*, 58: 1515–1520.

Elsholz, F. et al. 2014. Calcium—A central regulator of keratinocyte differentiation in health and disease. *Eur J Dermatol*, 24: 650–661.

Fang, D. and V. Setaluri. 2000. Expression and up-regulation of alternatively spliced transcripts of melastatin, a melanoma metastasis-related gene, in human melanoma cells. *Biochem Biophys Res Commun*, 279: 53–61.

Fatherazi, S. et al. 2007. Evidence that TRPC4 supports the calcium selective I(CRAC)-like current in human gingival keratinocytes. *Pflugers Arch*, 453: 879–889.

Fusi, C. et al. 2014. Transient receptor potential vanilloid 4 (TRPV4) is downregulated in keratinocytes in human non-melanoma skin cancer. *J Invest Dermatol*, 134: 2408–2417.

Gopinath, P. et al. 2005. Increased capsaicin receptor TRPV1 in skin nerve fibres and related vanilloid receptors TRPV3 and TRPV4 in keratinocytes in human breast pain. *BMC Women's Health*, 5: 2.

Grandl, J. et al. 2008. Pore region of TRPV3 ion channel is specifically required for heat activation. *Nature Neurosci*, 11: 1007–1013.

Guler, A.D. et al. 2002. Heat-evoked activation of the ion channel, TRPV4. *J Neurosci*, 22: 6408–6414.

Han, S.B. et al. 2016. Transient receptor potential vanilloid-1 in epidermal keratinocytes may contribute to acute pain in herpes zoster. *Acta Dermato Venereol*, 96: 319–322.

Haraguchi, M. et al. 1997. Naturally occurring dermatitis associated with *Staphylococcus aureus* in DS-Nh mice. *Exp Anim*, 46: 225–229.

He, Y. et al. 2015. A gain-of-function mutation in TRPV3 causes focal palmoplantar keratoderma in a Chinese family. *J Invest Dermatol*, 135: 907–909.

Hikita, I. et al. 2002. Characterization of dermatitis arising spontaneously in DS-Nh mice maintained under conventional conditions: Another possible model for atopic dermatitis. *J Dermatol Sci*, 30: 142–153.

Hong, J. et al. 2011. Management of itch in atopic dermatitis. *Semin Cutan Med Surg*, 30: 71–86.

Huang, S.M. et al. 2008. Overexpressed transient receptor potential vanilloid 3 ion channels in skin keratinocytes modulate pain sensitivity via prostaglandin E2. *J Neurosci*, 28: 13727–13737.

Huang, S.M. et al. 2011. TRPV3 and TRPV4 ion channels are not major contributors to mouse heat sensation. *Mol Pain*, 7: 37.

Hwang, M.K. et al. 2010. Cocarcinogenic effect of capsaicin involves activation of EGFR signaling but not TRPV1. *Cancer Res*, 70: 6859–6869.

Imura, K. et al. 2007. Influence of TRPV3 mutation on hair growth cycle in mice. *Biochem Biophys Res Commun*, 363: 479–483.

Imura, K. et al. 2009. Role of TRPV3 in immune response to development of dermatitis. *J Inflamm (Lond)*, 6: 17.

Inoue, K. et al. 2002. Functional vanilloid receptors in cultured normal human epidermal keratinocytes. *Biochem Biophys Res Commun*, 291: 124–129.

Iuga, A.O. and E.A. Lerner. 2007. TRP-ing up melanophores: TRPM7, melanin synthesis, and pigment cell survival. *J Invest Dermatol*, 127: 1855–1856.

Jain, A. et al. 2011. TRP-channel-specific cutaneous eicosanoid release patterns. *Pain*, 152: 2765–2772.

Jakubzick, C. et al. 2013. Minimal differentiation of classical monocytes as they survey steady-state tissues and transport antigen to lymph nodes. *Immunity*, 39: 599–610.

Jenkins, S.J. et al. 2011. Local macrophage proliferation, rather than recruitment from the blood, is a signature of TH2 inflammation. *Science*, 332: 1284–1288.

Khodorova, A. et al. 2003. Endothelin-B receptor activation triggers an endogenous analgesic cascade at sites of peripheral injury. *Nat Med*, 9: 1055–1061.

Kida, N. et al. 2012. Importance of transient receptor potential vanilloid 4 (TRPV4) in epidermal barrier function in human skin keratinocytes. *Pflugers Arch*, 463: 715–725.

Kim, H.J. et al. 2007. Gain-of-function mutation in TRPML3 causes the mouse Varitint-Waddler phenotype. *J Biol Chem*, 282: 36138–36142.

Koltzenburg, M. et al. 1999. Neutralization of endogenous NGF prevents the sensitization of nociceptors supplying inflamed skin. *Eur J Neurosci*, 11: 1698–1704.

Launay, P. et al. 2004. TRPM4 regulates calcium oscillations after T cell activation. *Science*, 306: 1374–1377.

Lee, H. et al. 2005. Altered thermal selection behavior in mice lacking transient receptor potential vanilloid 4. *J Neurosci*, 25: 1304–1310.

Lee, Y.M. et al. 2009. A novel role for the TRPV1 channel in UV-induced matrix metalloproteinase (MMP)-1 expression in HaCaT cells. *J Cell Physiol*, 219: 766–775.

Lee, Y.M. et al. 2011. Inhibitory effects of TRPV1 blocker on UV-induced responses in the hairless mice. *Arch Dermatol Res*, 303: 727–736.

Lehen'kyi, V. et al. 2007. TRPV6 is a Ca^{2+} entry channel essential for Ca^{2+}-induced differentiation of human keratinocytes. *J Biol Chem*, 282: 22582–22591.

Leuner, K. et al. 2011. Reduced TRPC channel expression in psoriatic keratinocytes is associated with impaired differentiation and enhanced proliferation. *PLOS ONE*, 6: e14716.

Leyva-Castillo, J.M. et al. 2013. TSLP produced by keratinocytes promotes allergen sensitization through skin and thereby triggers atopic march in mice. *J Invest Dermatol*, 133: 154–163.

Li, S. et al. 2011. TRPV1-antagonist AMG9810 promotes mouse skin tumorigenesis through EGFR/Akt signaling. *Carcinogenesis*, 32: 779–785.

Li, W.H. et al. 2007. Transient receptor potential vanilloid-1 mediates heat-shock-induced matrix metalloproteinase-1 expression in human epidermal keratinocytes. *J Invest Dermatol*, 127: 2328–2335.

Liao, M. et al. 2014. Single particle electron cryo-microscopy of a mammalian ion channel. *Curr Opin Struct Biol*, 27: 1–7.

Lin, J.Y. and D.E. Fisher. 2007. Melanocyte biology and skin pigmentation. *Nature*, 445: 843–850.

Lin, Z. et al. 2012. Exome sequencing reveals mutations in TRPV3 as a cause of Olmsted syndrome. *Am J Hum Genet*, 90: 558–564.

Lindemann, O. et al. 2015. TRPC1 regulates fMLP-stimulated migration and chemotaxis of neutrophil granulocytes. *Biochim Biophys Acta*, 1853: 2122–2130.

Link, T.M. et al. 2010. TRPV2 has a pivotal role in macrophage particle binding and phagocytosis. *Nat Immunol*, 11: 232–239.

Luo, J. et al. 2015. Molecular and cellular mechanisms that initiate pain and itch. *Cell Mol Life Sci*, 72: 3201–3223.

Luo, J. and H. Hu. 2014. Thermally activated TRPV3 channels. *Curr Top Membr*, 74: 325–364.

Lynch, V.J. et al. 2015. Elephantid genomes reveal the molecular bases of woolly mammoth adaptations to the arctic. *Cell Rep*, 12: 217–228.

Ma, Q.P. and C.J Woolf. 1997. The progressive tactile hyperalgesia induced by peripheral inflammation is nerve growth factor dependent. *Neuroreport*, 8: 807–810.

Majhi, R.K. et al. 2015. Functional expression of TRPV channels in T cells and their implications in immune regulation. *FEBS J*, 282: 2661–2681.

Mandadi, S. et al. 2009. TRPV3 in keratinocytes transmits temperature information to sensory neurons via ATP. *Pflugers Arch*, 458: 1093–1102.

Marcelo, C.L. et al. 1978. Stratification, specialization, and proliferation of primary keratinocyte cultures: Evidence of a functioning in vitro epidermal cell system. *J Cell Biol*, 79: 356–370.

Marics, I. et al. 2014. Acute heat-evoked temperature sensation is impaired but not abolished in mice lacking TRPV1 and TRPV3 channels. *PLOS ONE*, 9: e99828.

Matsuda, H. et al. 1998. Role of nerve growth factor in cutaneous wound healing: Accelerating effects in normal and healing-impaired diabetic mice. *J Exp Med*, 187: 297–306.

McNeill, M.S. et al. 2007. Cell death of melanophores in zebrafish trpm7 mutant embryos depends on melanin synthesis. *J Invest Dermatol*, 127: 2020–2030.

Minke, B. 1977. *Drosophila* mutant with a transducer defect. *Biophys Struct Mech*, 3: 59–64.

Miyamoto, T. et al. 2002. Itch-associated response induced by experimental dry skin in mice. *Jpn J Pharmacol*, 88: 285–292.

Miyamoto, T. et al. 2011. TRPV3 regulates nitric oxide synthase-independent nitric oxide synthesis in the skin. *Nat Commun*, 2: 369.

Montell, C. et al. 1985. Rescue of the *Drosophila* phototransduction mutation trp by germline transformation. *Science*, 230: 1040–1043.

Moore, C. et al. 2013. UVB radiation generates sunburn pain and affects skin by activating epidermal TRPV4 ion channels and triggering endothelin-1 signaling. *Proc Natl Acad Sci U S A*, 110: E3225–E3234.

Moqrich, A. et al. 2005. Impaired thermosensation in mice lacking TRPV3, a heat and camphor sensor in the skin. *Science*, 307: 1468–1472.

Morita, T. et al. 2015. HTR7 mediates serotonergic acute and chronic itch. *Neuron*, 87: 124–138.

Muller, M. et al. 2008. Specific TRPC6 channel activation: A novel approach to stimulate keratinocyte differentiation. *J Biol Chem*, 283: 33942–33954.

Muller-Rover, S. et al. 2001. A comprehensive guide for the accurate classification of murine hair follicles in distinct hair cycle stages. *J Invest Dermatol*, 117: 3–15.

Nemes, Z. and P.M. Steinert. 1999. Bricks and mortar of the epidermal barrier. *Exp Mol Med*, 31: 5–19.

Nestle, F.O. et al. 2009. Skin immune sentinels in health and disease. *Nat Rev*, 9: 679–691.

Oancea, E. et al. 2009. TRPM1 forms ion channels associated with melanin content in melanocytes. *Sci Signal*, 2: ra21.

Oh, M.H. et al. 2013. TRPA1-dependent pruritus in IL-13-induced chronic atopic dermatitis. *J Immunol*, 191: 5371–5382.

Orfanelli, U. et al. 2008. Identification of novel sense and antisense transcription at the TRPM2 locus in cancer. *Cell Res*, 18: 1128–1140.

Pang, Z. et al. 2015. Selective keratinocyte stimulation is sufficient to evoke nociception in mice. *Pain*, 156: 656–665.

Park, N.H. et al. 2013. Activation of transient receptor potential melastatin 8 reduces ultraviolet B-induced prostaglandin E2 production in keratinocytes. *J Dermatol*, 40: 919–922.

Patapoutian, A. et al. 2003. ThermoTRP channels and beyond: Mechanisms of temperature sensation. *Nat Rev Neurosci*, 4: 529–539.

Pecze, L. et al. 2008. Human keratinocytes are vanilloid resistant. *PLOS ONE*, 3: e3419.

Peier, A.M. et al. 2002. A heat-sensitive TRP channel expressed in keratinocytes. *Science*, 296: 2046–2049.

Radresa, O., M. Pare, and J.S. Albert. 2013. Multiple roles of transient receptor potential (TRP) channels in inflammatory conditions and current status of drug development. *Curr Top Med Chem*, 13: 367–385.

Radtke, C. et al. 2010. Keratinocytes acting on injured afferents induce extreme neuronal hyperexcitability and chronic pain. *Pain*, 148: 94–102.

Radtke, C. et al. 2011. TRPV channel expression in human skin and possible role in thermally induced cell death. *J Burn Care Res*, 32: 150–159.

Rheinwald, J.G. and H. Green. 1975. Serial cultivation of strains of human epidermal keratinocytes: The formation of keratinizing colonies from single cells. *Cell*, 6: 331–343.

Santoni, G. et al. 2013. The role of transient receptor potential vanilloid type-2 ion channels in innate and adaptive immune responses. *Front Immunol*, 4: 34.

Saul, S. et al. 2014. How ORAI and TRP channels interfere with each other: Interaction models and examples from the immune system and the skin. *Eur J Pharmacol*, 739: 49–59.

Schwarz, E.C. et al. 2007. TRP channels in lymphocytes. In *Handbook of Experimental Pharmacology*, edited by V. Flockerzi and B. Nilius, 445–456. Berlin: Springer-Verlag.

Slominski, A. 2008. Cooling skin cancer: Menthol inhibits melanoma growth. Focus on "TRPM8 activation suppresses cellular viability in human melanoma." *Am J Physiol*, 295: C293–C295.

Smith, G.D. et al. 2002. TRPV3 is a temperature-sensitive vanilloid receptor-like protein. *Nature*, 418: 186–190.

Sokabe, T. et al. 2010. The TRPV4 channel contributes to intercellular junction formation in keratinocytes. *J Biol Chem*, 285: 18749–18758.

Sokabe, T. and M. Tominaga. 2010. The TRPV4 cation channel: A molecule linking skin temperature and barrier function. *Commun Integr Biol*, 3: 619–621.

Southall, M.D. et al. 2003. Activation of epidermal vanilloid receptor-1 induces release of proinflammatory mediators in human keratinocytes. *J Pharmacol Exp Ther*, 304: 217–222.

Stander, S. et al. 2004. Expression of vanilloid receptor subtype 1 in cutaneous sensory nerve fibers, mast cells, and epithelial cells of appendage structures. *Exp Dermatol*, 13: 129–139.

Steinhoff, M., J. Schauber, and J.J. Leyden. 2013. New insights into rosacea pathophysiology: A review of recent findings. *J Am Acad Dermatol*, 69: S15–S26.

Streilein, J.W. 1983. Skin-associated lymphoid tissues (SALT): Origins and functions. *J Invest Dermatol*, 80(Suppl): 12s–16s.

Sulk, M. et al. 2012. Distribution and expression of non-neuronal transient receptor potential (TRPV) ion channels in rosacea. *J Invest Dermatol*, 132: 1253–1262.

Sun, T.T. and H. Green. 1976. Differentiation of the epidermal keratinocyte in cell culture: Formation of the cornified envelope. *Cell*, 9: 511–521.

Takada, H., K. Furuya, and M. Sokabe. 2014. Mechanosensitive ATP release from hemichannels and Ca^{2+} influx through TRPC6 accelerate wound closure in keratinocytes. *J Cell Sci*, 127: 4159–4171.

Todaka, H. et al. 2004. Warm temperature-sensitive transient receptor potential vanilloid 4 (TRPV4) plays an essential role in thermal hyperalgesia. *J Biol Chem*, 279: 35133–35138.

Toth, B.I. et al. 2009. Transient receptor potential vanilloid-1 signaling as a regulator of human sebocyte biology. *J Invest Dermatol*, 129: 329–339.

Toth, B.I. et al. 2011. Endocannabinoids modulate human epidermal keratinocyte proliferation and survival via the sequential engagement of cannabinoid receptor-1 and transient receptor potential vanilloid-1. *J Invest Dermatol*, 131: 1095–1104.

Vizin, R.C. et al. 2015. TRPV4 activates autonomic and behavioural warmth-defence responses in Wistar rats. *Acta Physiol (Oxf)*, 214: 275–289.

Wang, C. et al. 2004. An alternative splicing product of the murine trpv1 gene dominant negatively modulates the activity of TRPV1 channels. *J Biol Chem*, 279: 37423–37430.

Watanabe, A. et al. 2003. Role of the Nh (Non-hair) mutation in the development of dermatitis and hyperproduction of IgE in DS-Nh mice. *Exp Anim*, 52: 419–423.

Watanabe, R. et al. 1998. Up-regulation of glucosylceramide synthase expression and activity during human keratinocyte differentiation. *J Biol Chem*, 273: 9651–9655.

Watt, F.M. 1989. Terminal differentiation of epidermal keratinocytes. *Curr Opin Cell Biol*, 1: 1107–1115.

Wilson, N.J. et al. 2015. Expanding the phenotypic spectrum of Olmsted syndrome. *J Invest Dermatol*, 135: 2879–2883.

Woelfle, U. et al. 2010. Triterpenes promote keratinocyte differentiation in vitro, ex vivo and in vivo: A role for the transient receptor potential canonical (subtype) 6. *J Invest Dermatol*, 130: 113–123.

Xiao, R. et al. 2008. The TRPV3 mutation associated with the hairless phenotype in rodents is constitutively active. *Cell Calcium*, 43: 334–343.

Xu, H. et al. 2002. TRPV3 is a calcium-permeable temperature-sensitive cation channel. *Nature*, 418: 181–186.

Xu, H. et al. 2006. Oregano, thyme and clove-derived flavors and skin sensitizers activate specific TRP channels. *Nat Neurosci*, 9: 628–635.

Xu, H. et al. 2007. Activating mutation in a mucolipin transient receptor potential channel leads to melanocyte loss in varitint-waddler mice. *Proc Natl Acad Sci U S A*, 104: 18321–18326.

Yamamoto-Kasai, E. et al. 2012. TRPV3 as a therapeutic target for itch. *J Invest Dermatol*, 132: 2109–2112.

Yamamoto-Kasai, E. et al. 2013. Impact of TRPV3 on the development of allergic dermatitis as a dendritic cell modulator. *Exp Dermatol*, 22: 820–824.

Yamamura, H. et al. 2008. TRPM8 activation suppresses cellular viability in human melanoma. *Am J Physiol*, 295: C296–C301.

Yang, Y.S. et al. 2015. Increased expression of three types of transient receptor potential channels (TRPA1, TRPV4 and TRPV3) in burn scars with post-burn pruritus. *Acta Dermato Venereol*, 95: 20–24.

Yoshioka, T. et al. 2006. Spontaneous scratching behaviour in DS-Nh mice as a possible model for pruritus in atopic dermatitis. *Immunology*, 118: 293–301.

Yoshioka, T. et al. 2009. Impact of the Gly573Ser substitution in TRPV3 on the development of allergic and pruritic dermatitis in mice. *J Invest Dermatol*, 129: 714–722.

Yun, J.W. et al. 2011. Antipruritic effects of TRPV1 antagonist in murine atopic dermatitis and itching models. *J Invest Dermatol*, 131: 1576–1579.

Zhiqi, S. et al. 2004. Human melastatin 1 (TRPM1) is regulated by MITF and produces multiple polypeptide isoforms in melanocytes and melanoma. *Melanoma Res*, 14: 509–516.

7 TRP Channels at the Periphery of the Taste and Trigeminal Systems

Sidney A. Simon and Ranier Gutierrez

CONTENTS

7.1   INTRODUCTION

The mammalian taste system consists of taste buds, which are groups of 50–100 taste cells that are found throughout the oral cavity. On the tongue, which is the focus of this chapter, taste buds are located on circumvallate, foliate, and fungiform papillae (Figure 7.1a). Taste cells synapse with afferent fibers from branches of the facial (CN VII), glossopharyngeal (CN IX), and vagus (CN X) cranial nerves (Figure 7.1b) that, in turn, transmit information to the central nervous system (CNS) attributes of tastant quality, intensity, and hedonic nature (Gutierrez and Simon, 2011; Carleton et al., 2010; Vincis and Fontanini, 2016). The list includes several classes of chemical stimuli such as sugars, salts, acids, proteins, and organic compounds that are perceived as bitter tasting (Simon et al., 2006). Taste buds are embedded in a stratified squamous epithelium, which contains somatosensory branches of the trigeminal (CN V), glossopharyngeal (CN IX), and vagus (CN X) cranial nerves. The terminals of these somatosensory fibers often surround taste buds (Figures 7.1b and 7.4), indicating the close association of the taste and somatosensory systems. Information from these general sensory nerves provides information to the CNS about mechanical, thermal, and painful stimuli (Julius, 2013; Kaneko and Szallasi, 2014). The painful stimuli can arise from strong or sharp mechanical stimuli, abnormally high or low temperatures, or chemical stimuli such as capsaicin, which is found in chili peppers and causes a burning taste sensation. As both the peripheral taste and somatosensory systems contain transient receptor potential (TRPs) (Ramsey et al., 2006; Julius, 2013), here we will initially review general properties of TRPs, and then describe their roles in the peripheral taste and somatosensory systems. We do not, however, discuss their presence in keratinocytes, but refer the reader to Chapter 5. Finally, for additional details regarding their roles in taste and as condiments used for cooking, we refer the reader to several excellent reviews on this topic (Vriens et al., 2008; Roper, 2013; Talavera, 2015; Roper, 2014).

Since their discovery in 1989, as seen in Figure 7.2, numerous TRP channels have been identified. What they have in common is that they are all cation selective (most for Ca^{2+}), and they are composed of four subunits with six transmembrane spans (S1–S6), with a pore region between S5 and S6 (Morales-Lázaro et al., 2013). Moreover, they often have ankyrin repeats in their N-terminus, exhibit

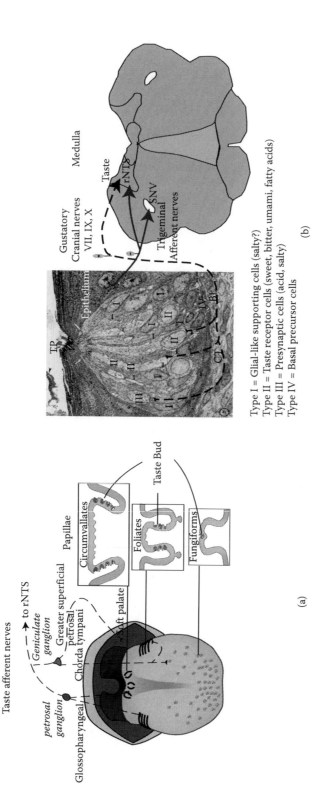

Type I = Glial-like supporting cells (salty?)
Type II = Taste receptor cells (sweet, bitter, umami, fatty acids)
Type III = Presynaptic cells (acid, salty)
Type IV = Basal precursor cells

FIGURE 7.1 The gustatory taste system and its association with lingual trigeminal perigemmal fibers. (a) Dorsal surface of a human tongue showing the location of the three types of papillae containing taste buds and their associated afferent nerves. In fungiform papillae, taste buds are on the dorsal surface, whereas in foliate and circumvallate papillae they are in trenches on the side and posterior parts of the tongue. Note that the chorda tympani nerve innervates both the fungiform papillae and the anterior part of foliate papillae, whereas the posterior part of foliate papillae and the circumvallate papillae are innervated by a branch of the glossopharyngeal nerve. The soft palate is innervated by the greater superficial petrosal nerve. (Modified from Yarmolinsky, D.A. et al., *Cell*, 139, 234–244, 2009. With permission.) (b) The left panel shows an electron micrograph of a section of a taste bud through the taste pore showing four different types of taste cells. Below are their responses to selective tastants. Also illustrated are general sensory perigemmal neurons that project to the trigeminal complex in the medulla and send branches to the rostral nucleus of the solitary tract (NTS). (From Royer, S.M. and J.C. Kinnamon, *J Comp Neurol.*, 306, 49–72, 1991. With permission.)

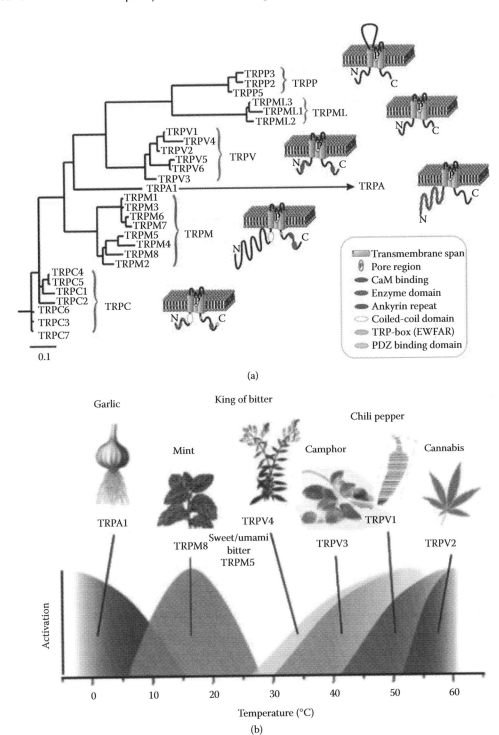

FIGURE 7.2 (a) Dendrogram showing the different groups comprising the family of TRP channels. In each TRP only two of the six subunits that flank the pore region are shown. TRP channels are composed of tetramers of subunits. (From Nilius, B. et al., *Sci STKE.*, 2005, re8, 2005. With permission.) (b) A selected group of thermal TRPs (and their temperature ranges) that are activated by tastants (TRPM5) or various spices (TRPA1, TRPV1) or other compounds (TRPM8, TRPV2). (Modified from McKemy, D.D., *Pflügers Archiv.*, 454, 777–791, 2007. With permission.)

inwardly rectifying current-voltage curves, and are modulated by calmodulin and phospholipases and kinases (Ramsey et al., 2006; Julius, 2013). As seen in Figure 7.2a, TRPs fall into six subfamilies: TRPC for "canonical" (TRPC1–7), TRPM for "melastatin" (TRPM1–8), TRPA for "ankyrin" (TRPA1), TRPV for "vanilloid" (TRPV1–6), TRPML for "mucolipin" (TRPML1–3), and TRPP for "polycystin" (TRPP2, TRPP3, TRPP5). Other interesting properties of many TRP channels, illustrated in Figure 7.2b and elaborated below, are that they are important in gustatory processing, are very sensitive to changes in temperature, and are activated by many compounds found in plants that are often used as spices (Figure 7.2b).

7.2 TRPM5 AND TRPM4 IN TASTE CELLS

Taste buds are located throughout the oral cavity and are found in specialized regions called papillae (Figure 7.1a). As noted, here we focus on those taste buds on the dorsal tongue. In the front (anterior) of the tongue, taste buds are in the fungiform papillae (innervated by chorda tympani afferent fibers), on the sides in foliate papillae, and in the posterior region in circumvallate papillae (papillae are innervated by the facial [foliate] and glossopharyngeal afferent nerves; Figure 7.1a). Most taste buds (~80%) are located in trenches in the foliate and circumvallate papillae. However, independent of their location, taste buds are composed of four distinct types (I-IV) of taste cells (Figure 7.1a). Type I cells, the most prevalent, are "glial-like" cells; type IV are basal precursor cells. Neither of these types appear to contain TRP channels (Roper, 2014; Shigemura and Ninomiya, 2016), and despite their importance in maintaining healthy taste cells, they are not further discussed.

As described by Roper (2014), type II taste cells make up about 20%–30% of cells in the taste bud. These cells are called taste receptor cells because they possess G protein-coupled receptors (GCPRs) for sweet, bitter, and umami (i.e., monosodium glutamate) tastants. The receptors for sweet (T1R2/T1R3) and umami (T1R1/T1R3) are heterodimers, whereas there are about 30 monomeric T2Rs that associate with molecules perceived to be bitter (Yarmolinsky et al., 2009). The primary transduction mechanism in type II taste receptor cells is as follows (Figure 7.3). Sweet, umami, or bitter tastants interact with their appropriate GPCRs that, in turn, activate PLC-β2 (phospholipase C beta 2) which produces diacylglycerol (DAG) and inositol triphosphate (IP3) from phosphatidylinositol-4,5-bisphosphate (PIP2). In turn, IP3 activates inositol 1,4,5-trisphosphate receptor type 3 (IP3R3) receptors in the endoplasmic reticulum producing an increase in intracellular Ca^{2+}, which then activates TRPM5 channels (Perez et al., 2002) on the basolateral surface of the taste receptor cells. The activation of TRPM5 by intracellular Ca^{2+} gates opens the channel to let Na^+ ions diffuse into these taste cells to depolarize them. In this regard, TRPM5 is unusual among the TRP channels in that it is selective for Na^+ (and other monovalent cations) but is not permeable to Ca^{2+} (Liman, 2007). The entry of Na^+ will depolarize taste cells, evoking action potentials that will via the Ca^{2+}-activated and voltage-dependent Ca^{2+} homeostasis modulator 1 (CALHM1) channels (Taruno et al., 2013) release the transmitter, ATP, which binds to $P2X_2/P2X_3$ receptors on the afferent neurons associated with the particular tastants and activates them to transmit their information to the CNS (Finger et al., 2005).

Each subset of type II cells is usually (but not exclusively as seen with artifical sweeteners [see Winnig et al., 2008]) activated by only one of the following: bitters, sweets, umami, or fatty acid tastants. These activate a transduction cascade that terminates with the release of ATP through CALMH1 channels (Taruno et al., 2013) that subsequently activate $P2X_2/P2X_3$ receptors of primary afferent neurons from branches from cranial nerves VII, IX, and supposedly X. Also shown are receptors for peptides that have been shown to modulate taste responses (see below). There are other important characteristics of TRPM5 channels that have been suggested to influence their role in taste processing (Roper, 2014; Talavera, 2015; Liman, 2007). First, at least in human embryonic kidney (HEK) cells, TRPM5 channels are inhibited by acid added from the extracellular surface; hence, in type II cells, acid might be expected to diminish responses to tastants. However, given that these channels are located on the basolateral surface of type II taste cells (i.e., below the tight junctions

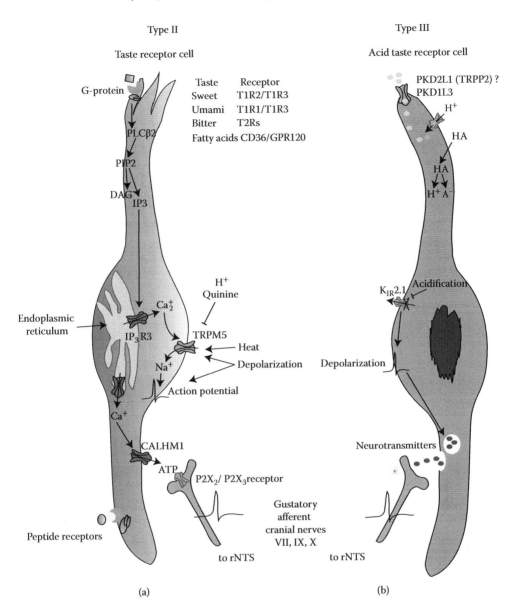

FIGURE 7.3 **(See color insert.)** (a) The transduction machinery of a type II taste receptor cell. (b) A type III taste cell that responds to acidic stimuli. The subset of type III cells responsive to salt are not shown, because it is not known whether they contain TRP channels.

[Dando et al., 2015; Holland et al., 1989]), and thus are protected to a large extent from extracellular pH changes, it is not evident that this property is important in psychophysical studies of taste processing (Keast and Breslin, 2003). Second, as seen in Figure 7.2b, TRPM5 is temperature dependent in that it is activated over a physiological temperature range of ~15°C–35°C. Obviously, cooling will have the opposite effect. In this regard, Talavera et al. (2005) showed that warming taste stimuli to 35°C selectively enhanced responses to sweet tastants, which they attributed to TRPM5 activation. However, they were unable to explain why it is only selective for sweet tastants. Third, the bitter tastant, quinine, which because of its more hydrophobic nature can diffuse to basolateral TRPM5 channels (Peri et al., 2000; Naim et al., 1994), has been shown to inhibit them (Talavera et al., 2008) and reduce responses to sweet tastants, thereby rationalizing the psychophysical and physiological results regarding how a bitter tastant like quinine may suppress sweet taste (Keast and Breslin, 2003). However, as discussed with temperature, they were unable to explain the selectivity for sweet tastants. Fourth, and finally, in a subset of mouse type II cells, mostly long-chain polyunsaturated fatty acids, such as linoleic acid, depolarize taste cells and increase intracellular Ca^{2+} that in turn activates TRPM5 (Besnard et al., 2016; Keast and Costanzo, 2015). However, whether humans perceive fatty acids as tastants remains controversial (Besnard et al., 2016).

Like TRPM5, TRPM4 is a member of the TRPM subfamily of transient receptor potential ion channels (Figure 7.2a). There is about a 40% homology between TRPM4 and TRPM5. Moreover, both channels are voltage dependent, Ca^{2+} impermeable, and are activated by increases in intracellular Ca^{2+}. However, as pointed out by Roper (2014), there is only very indirect evidence that TRPM4 channels are present in taste buds. That is, in taste cells neither the transcripts for TRPM4 nor the protein itself have been detected. The only evidence, albeit indirect, was performing patch clamp recordings on mouse $Trpm5^{-/-}$ taste cells where Ca^{2+}-activated currents were recorded with properties suggestive of TRPM4 (Zhang et al., 2007).

7.3 TRPP CHANNELS IN TASTE TRANSDUCTION TYPE III CELLS

Members of the polycystic kidney disease–like ion channel division (TRPP) of the TRP superfamily, namely PKD2L1 (TRPP2) and its associated transmembrane protein PKD1L3 (polycystic kidney disease 1-like 3") have been identified in the apical region of type III taste cells (i.e., see Figure 7.3b and Ishimaru et al. [2006]). Type III cells are called presynaptic cells because they have much of the same machinery to release synaptic vesicles onto associated nerve terminals that are found in like cells throughout the animal kingdom (Roper, 2014; Shigemura and Ninomiya, 2016). They represent about 15% of the cells in taste buds and, importantly, respond directly to acidic (sour) taste stimuli (Roper, 2014). With regard to PKD2L1 and PKD1L3 it was found that these proteins dimerize to form cation-selective channels and that mice lacking PKD2L1-expressing cells lack taste responses to acidic stimuli (LopezJimenez et al., 2006; Huang et al., 2006; Ishimaru et al., 2006). However, Horio et al. (2011) found that compared to wild-type mice, responses from primary gustatory nerves in mice lacking the $Pkd2L1$ gene were markedly reduced, whereas the responses in Pkd1L3-knockout mice were similar to those evoked from wild-type mice. These authors concluded that in mice, PKD2L1, but not PKD1L3, could partly contribute to acid (sour) taste responses. Although there is no doubt that the subset of type III taste cells containing PKD2L1 contribute to the responses to acidic stimuli, the physiological roles of PKD2L1 and PKD1L3 remain to be definitively elucidated. In this regard, recent studies on acid responses in type III cells showed that for strong acids, like hydrochloric acid, the protons diffuse through a Zn^{2+}-sensitive proton channel to depolarize these cells. Moreover, for weak acids, like citric acid, the protonated form will diffuse into the cell, dissociate to acidify the cytoplasm, which then closes an inward-rectifying potassium channel, $K_{IR}2.1$, thereby depolarizing PKD2L1 containing type III expressing cells, which then triggers the release of neurotransmitters (Ye et al., 2016) (see Figure 7.3b).

7.4 TRPV1 IN TASTE CELLS

TRPV1 is a member of the vanilloid group of TRP channels (Caterina et al., 1997) (Figure 7.2). It is activated by a wide variety of stimuli (Tominaga et al., 1998) including spices such as capsaicin, pain-producing temperatures (>42°C), acidic stimuli, voltage, and a variety of organic and inorganic molecules (for reviews see Julius, 2013; Morales-Lázaro et al., 2014). Despite the fact that there have been many studies suggesting its role in taste processing, as pointed out by Roper (2014), there is a paucity of evidence suggesting that it is actually present in taste cells. There is no question that it is present in somatosensory nerves innervating the oral cavity (see Figure 7.4), as we have all experienced the burning sensation on our tongue when eating chili peppers. It is also possible that it is located in rat geniculate ganglion neurons (Katsura et al., 2006), but that has not

Trigeminal TRPV1 + perigemmal fibers

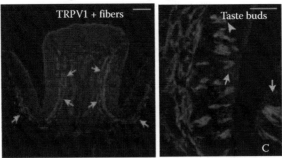

FIGURE 7.4   (See color insert.) Left panel: Perigemmal expression of TRPV1 immunoreactive (TRPV1-IR) nerve fibers in rodent circumvallate papillae (in red). Right panel: The co-localization of TRPV1-IR immunoreactivity (red) and anti-gustducin alpha-subunit (green); the latter is present in type II taste receptor cells. Arrows indicate perigemmal TRPV1+ fibers. (From Ishida, Y. et al., *Mol Brain Res.*, 107, 17–22, 2002. With permission.)

been demonstrated beyond a doubt (see discussion by Roper, 2004). For these reasons we posit that the effects of TRPV1 on taste cells likely occur via its activation in perigemmal fibers that then indirectly affect taste cells.

There are two reasons for doubting its role in taste processing. First, immunostaining for TRPV1 was not found in taste cells (Moon et al., 2010; Gu et al., 2009; Ishida et al., 2002). Second, no significant differences in $Trpv1^{-/-}$ and wild-type mice were found in behavioral studies performed on amiloride-insensitive salt taste (Lyall et al., 2004), which has been attributed to TRPV1 or a variant of this channel (Treesukosol et al., 2007; Ruiz et al., 2006; Smith et al., 2012). More recently, it has been found that amiloride-insensitive salt taste is mediated by two distinct types of type III cells (Lewandowski et al., 2016). It is established that at least in rodents, salt (NaCl) taste is mediated through amiloride-sensitive ENaC channels most likely in a subset of type III taste cells (Roper, 2015).

We note that capsaicin, especially at high concentrations, can have a variety of nonspecific effects (Lundbæk et al., 2005; Costa et al., 2005) that can alter the normal function of proteins. In this regard, in $Trpv1^{-/-}$ mice it was shown that capsaicin can decrease preference to sucrose and inhibit voltage-dependent sodium channels in taste cells (Costa et al., 2005). There are, however, additional ways in which the activation of TRPV1 in perigemmal neurons can affect taste processing. Specifically, this could occur via capsaicin inducing the release of peptides such as calcitonin gene-related peptide (CGRP) or substance P (SP) from TRPV1, containing nerve terminals surrounding taste cells (Grant, 2012; Huang and Wu, 2015) that can modulate taste responses by vasodilatation (Zygmunt et al., 1999), thereby increasing the tongue's temperature (Wang et al., 1995), and/or by interacting with their receptors on taste cells that will modulate taste responses (see Figures 7.3 and 7.5b) (Yamasaki et al., 1985).

7.5   TRP CHANNELS IN GENERAL SENSORY NERVES SURROUNDING TASTE BUDS

The nerve terminals surrounding taste buds and others located throughout the oral cavity are branches of the trigeminal (TG), glossopharyngeal, and vagus nerves that give to sensations involving touch (mechanosensors), temperature (thermosensors), and pain (nociceptors). The responses of these neurons to chemical stimuli and the resulting perception are called chemesthesis (Green, 1996). For brevity, and also to focus on the taste and associated oral responses to chemical stimuli, we only consider those TRPs associated with terminals in the lingual branch of the trigeminal nerve that surround taste buds called perigemmal nerves (Figure 7.5a). For reviews that are more extensive

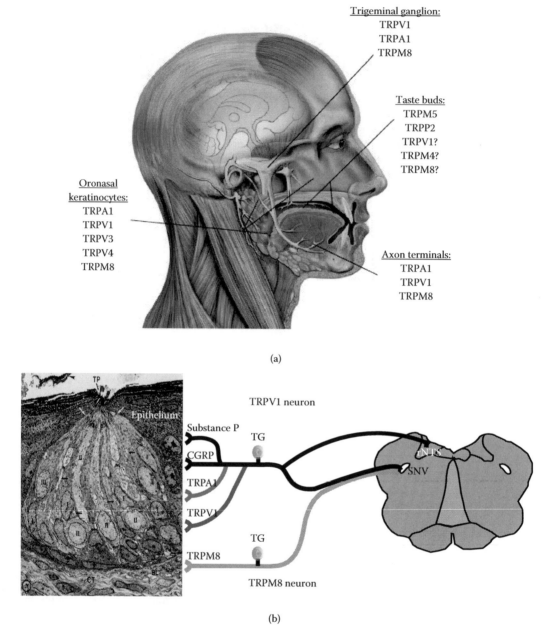

(a)

(b)

FIGURE 7.5   (See color insert.) The oronasal cavity and its innervation, showing TRP channels involved in taste and chemesthesis. (a) TRP channels are expressed in sensory ganglion neurons, their axon terminals, and in epithelial keratinocytes throughout the oronasal cavity. The TRP channels illustrated here have known or implied functions in gustation and chemesthesis. Other TRP channels are expressed in these sites, but their functions in taste and chemesthesis are not well understood. (b) A cartoon of two types of trigeminal ganglion cells and their projection to medulla. The fiber containing TRPV1 projects both to the nucleus of the solitary tract (NST) where it can influence responses from taste fibers and the SNV where it can produce responses that will be perceived as pain. The activation of these fibers can also release peptides such as substance P and CGRP as well as other neuropeptides that can bind to receptors on taste cells as seen in Figure 7.3. Also shown is a perigemmal fiber containing TRPM8 (Abe et al., *Mol Brain Res,* 136(1-2): 91-98, 2005) that projects to SNV where it provides information about cooling.

see Roper (2014), Nilius and Appendino (2013), and Viana (2011). The best characterized and those most abundant of the family of TRPs are TRPV1, TRPM8, and TRPA1, and for this reason, we focus on them.

In addition to them being activated by chemical stimuli, these three TRP channels are also activated by thermal stimuli (Figure 7.2). For TRPV1, TRPM8, and TRPA1, the activation temperatures are 42°C, 28°C, and < 17°C, respectively (Morales-Lázaro et al., 2013). TRPV1 is activated by heating, whereas the other two are activated by cooling. (We note that the basal temperature in the mouth is about 34°C.) When the subset of TRPV1 containing nociceptors that are activated by compounds, like capsaicin (even at temperatures <42°C), they evoke, depending on the concentration, a warm or burning oral sensation (Green, 1996). Similarly, neurons in the mouth that express TRPM8 when activated by menthol, for example, produce a cooling sensation. In this regard, it has been recently shown that TRPV1 tracts the absolute temperature, whereas TRPM8 tracks the relative temperature (Dhaka et al., 2007; Ran et al., 2016). In TG neurons, TRPA1 is often co-expressed with TRPV1 (see Roper, 2014) (Figure 7.5b) and thus its activation produces similar sensations to the activation of TRPV1. This perhaps may rationalize why people frequently report that garlic, which also activates TRPV1 (Salazar et al., 2008), induces a similar "hot" sensation as chili peppers. In other TRPA1 containing neurons, it may evoke an irritating sensation (Lanosa et al., 2010).

There is a wide range of molecules, many of which are quite familiar to us, that activate one or more of these three TRPs. These include spices, such as black pepper, chili pepper, garlic, onions, horseradish, mustard oil, as well as other molecules such as ethanol, acid, carbon dioxide, menthol, and high concentrations of salts. The activation of these neurons produces a variety of pungent sensations including cooling, warmth, burning, numbing, stinging, tingling, creamy, metallic, and astringency, to name a few (Prescott and Stevenson, 1995). Although not all of these sensations are the result of the activation of TRP channels, many are.

Here, we elaborate on the types of chemical stimuli that are commonly eaten or placed in the mouth that activate these three TRP channels. Suffice to say that these molecules depolarize nerve terminals that contain these channels (Ramsey et al., 2006; Julius, 2013) and that evoke both afferent responses involving the transmission of action potentials to the CNS and efferent responses involving the release of peptides and other compounds including ATP (Figures 7.3 and 7.5b).

Several of the compounds that activate TRPV1 include capsaicin (chili pepper), piperine (black pepper), 6-gingerol (ginger), euganol (cloves), acids (vinegar, extracellular), ethanol, allicin (garlic), gingerol (ginger), 6-shogal (ginger), ferric sulfate (metallic taste), high concentrations of monovalent divalent salts (aversive), and endocannabinoids (both natural compounds, such as anandamide, [N-arachidonoylethanolamine] found in the body, as well as tetrahydrocannabinol [THC] found in marijuana plants) (Roper 2014; Talavera, 2015; Nilius and Appendino, 2013; Salazar et al., 2008). It is important to note that not all these compounds produce the same pungent sensation (Green and Hayes, 2004; Prescott and Stevenson, 1995). This is because they are not completely selective for TRPV1, and with most being hydrophobic, they will diffuse into the lingual epithelium at different rates (depending on their membrane/water partition coefficient and size) so that they will first activate and then desensitize the perigemmal terminals at different times (Liu et al., 2000).

Several of the compounds found to activate TRPM8 include menthol (cough drops and vapor rubs); 1,8-cineole (eucalyptus oil [Takaishi et al., 2012]), wintergreen, eugenol (cloves), geraniol (in lemon-grass), linalool (in floral scents of Onagraceae species), menthyl lactate (peppermint oil), and L-carvone (from spearmint oil) (Nilius and Appendino, 2013; Vriens et al., 2008).

TRPA1 is activated by many different types of chemicals (Nilius and Appendino, 2013). Many of the well-studied taste compounds that activate TRPA1 are electrophiles (i.e., compounds that accept a pair of electrons to form a new covalent bond). Isothiocyanate, a compound found in mustard, is a good example of an electrophile. Isothiocyanates are also found in horseradish, wasabi, and Brussels sprouts. Other TRPA1 agonists include allicin (Bautista et al., 2005), which is found in garlic and activates TRPV1 (Salazar et al., 2008). Similarly, eugenol, which is found in cloves, also activates both channels as do shogaols, which are in ginger (Morera et al., 2012). Another spice that activates TRPA1 is curcumin, which is found in turmeric. In addition, it is activated by thymol, a major component of thyme and oregano. Finally, TRPA1 is activated by both acidic and basic stimuli (Dhaka et al., 2009).

7.6 IN SUMMARY

Here, we briefly reviewed the fundamental role that peripheral TRP channels play for both the gustatory and trigeminal sensory systems. It is clear that the intimate relationship—at least at the peripheral level—between both sensory systems is in fact an important determinant of many of our eating habits and the incorporation of condiments to our culinary cuisines (Gutierrez and Simon, 2016). The use of several spices that activate TRP channels that are solely present in the trigeminal system (TRPV1, TRPA1, TRPM8) and their thermal sensitivity, certainly complement and enrich our repertoire of chemosensory sensations way beyond that induced by activation of TRP channels solely present in the gustatory system (i.e., TRPM5, TRPP2). In other words, TRP channels are not only important components of the intracellular signaling cascades necessary to perceive the quality attributes evoked by the gustatory system (e.g., sweet, bitter, umami, and partially sour), they also potentiate (directly, but most frequently indirectly) our eating experience by evoking pungent sensations via compounds that activate trigeminal TRPs channels.

ACKNOWLEDGMENTS

Research was supported by CONACyT Fronteras de la Ciencia (Grant 63) and Problemas Nacionales (Grant 464) to R.G.

REFERENCES

Abe, J. et al. 2005. TRPM8 protein localization in trigeminal ganglion and taste papillae. *Mol Brain Res*, 136(1–2): 91–98.

Bautista, D.M. et al. 2005. Pungent products from garlic activate the sensory ion channel TRPA1. *Proc Natl Acad Sci U S A*, 102(34): 12248–12252.

Besnard, P., P. Passilly-Degrace, and N.A. Khan. 2016. Taste of fat: A sixth taste modality? *Physiol Rev*, 96(1): 151–176.

Carleton, A., R. Accolla, and S.A. Simon. 2010. Coding in the mammalian gustatory system. *Trends Neurosci*, 33(7): 326–334.

Caterina, M.J. et al. 1997. The capsaicin receptor: A heat-activated ion channel in the pain pathway. *Nature*, 389(6653): 816–824.

Costa, R.M. et al. 2005. Gustatory effects of capsaicin that are independent of TRPV1 receptors. *Chem Senses*, 30(Suppl 1): i198–i200.

Dando, R. et al. 2015. A permeability barrier surrounds taste buds in lingual epithelia. *Am J Physiol Cell Physiol*, 308(1): C21–C32.

Dhaka, A. et al. 2007. TRPM8 is required for cold sensation in mice. *Neuron*, 54(3): 371–378.

Dhaka, A. et al. 2009. TRPV1 is activated by both acidic and basic pH. *J Neurosci*, 29(1): 153–158.

Finger, T.E. et al. 2005. ATP signaling is crucial for communication from taste buds to gustatory nerves. *Science*, 310(5753): 1495–1499.

Grant, J. 2012. Tachykinins stimulate a subset of mouse taste cells. *PLOS ONE*, 7(2): e31697.

Green, B.G. 1996. Chemesthesis: Pungency as a component of flavor. *Trends Food Sci Technol*, 7(12): 415–420.

Green, B.G., and J.E. Hayes. 2004. Individual differences in perception of bitterness from capsaicin, piperine and zingerone. *Chem Senses*, 29(1): 53–60.

Gu, X.F. et al. 2009. Intra-oral pre-treatment with capsaicin increases consumption of sweet solutions in rats. *Nutr Neurosci*, 12(4): 149–154.

Gutierrez, R., and S.A. Simon. 2011. Chemosensory processing in the taste-reward pathway. *Flavour Fragr J*, 26(4): 231–238.

Gutierrez, R., and S.A. Simon. 2016. Why do people living in hot climates like their food spicy? *Temperature (Austin)*, 3(1): 48–49.

Holland, V.F., G.A. Zampighi, and S.A. Simon. 1989. Morphology of fungiform papillae in canine lingual epithelium: Location of intercellular junctions in the epithelium. *J Comp Neurol*, 279(1): 13–27.

Horio, N., et al. 2011. Sour taste responses in mice lacking PKD channels. *PLoS One*, 6(5): e20007.

Huang, A.L. et al. 2006. The cells and logic for mammalian sour taste detection. *Nature*, 442(7105): 934–938.

Huang, A.Y., and S.Y. Wu. 2015. Calcitonin gene-related peptide reduces taste-evoked ATP secretion from mouse taste buds. *J Neurosci*, 35(37): 12714–12724.

Ishida, Y. et al. 2002. Vanilloid receptor subtype-1 (VR1) is specifically localized to taste papillae. *Mol Brain Res*, 107(1): 17–22.

Ishimaru, Y. et al. 2006. Transient receptor potential family members PKD1L3 and PKD2L1 form a candidate sour taste receptor. *Proc Natl Acad Sci U S A*, 103(33): 12569–12574.

Julius, D. 2013. TRP channels and pain. *Annu Rev Cell Dev Biol*, 29: 355–584.

Kaneko, Y., and A. Szallasi. 2014. Transient receptor potential (TRP) channels: A clinical perspective. *Br J Pharmacol*, 171(10): 2474–2507.

Katsura, H. et al. 2006. Differential expression of capsaicin-, menthol-, and mustard oil-sensitive receptors in naive rat geniculate ganglion neurons. *Chem Senses*, 31(7): 681–688.

Keast, R.S.J., and P.A.S. Breslin. 2003. An overview of binary taste-taste interactions. *Food Qual Prefer*, 14(2): 111–124.

Keast, R.S.J., and A. Costanzo. 2015. Is fat the sixth taste primary? Evidence and implications. *Flavour*, 4(1): 1–7.

Lanosa, M.J. et al. 2010. Role of metabolic activation and the TRPA1 receptor in the sensory irritation response to styrene and naphthalene. *Toxicol Sci*, 115(2): 589–595.

Lewandowski, B.C. et al. 2016. Amiloride-insensitive salt taste is mediated by two populations of type III taste cells with distinct transduction mechanisms. *J Neurosci*, 36(6): 1942–1953.

Liman, E.R. 2007. TRPM5 and taste transduction. *Handb Exp Pharmacol*, (179): 287–298.

Liu, L. et al. 2000. Different responses to repeated applications of zingerone in behavioral studies, recordings from intact and cultured TG neurons, and from VR1 receptors. *Physiol Behav*, 69(1–2): 177–186.

LopezJimenez, N.D. et al. 2006. Two members of the TRPP family of ion channels, Pkd1l3 and Pkd2l1, are co-expressed in a subset of taste receptor cells. *J Neurochem*, 98(1): 68–77.

Lundbæk, J.A. et al. 2005. Capsaicin regulates voltage-dependent sodium channels by altering lipid bilayer elasticity. *Mol Pharmacol*, 68(3): 680–689.

Lyall, V. et al. 2004. The mammalian amiloride-insensitive non-specific salt taste receptor is a vanilloid receptor-1 variant. *J Physiol*, 558(1): 147–159.

McKemy, D.D. 2007. Temperature sensing across species. *Pflügers Archiv*, 454(5): 777–791.

Moon, Y.W. et al. 2010. Capsaicin receptors are colocalized with sweet/bitter receptors in the taste sensing cells of circumvallate papillae. *Genes Nutr*, 5(3): 251–255.

Morales-Lázaro, S., T. Rosenbaum, and S. Simon. 2014. A portal to pain: The transient receptor potential (TRP) vanilloid 1 channel. Physiol News, (94): 26–31.

Morales-Lázaro, S.L., S.A. Simon, and T. Rosenbaum. 2013. The role of endogenous molecules in modulating pain through transient receptor potential vanilloid 1 (TRPV1). *J Physiol*, 591(13): 3109–3121.

Morera, E. et al. 2012. Synthesis and biological evaluation of [6]-gingerol analogues as transient receptor potential channel TRPV1 and TRPA1 modulators. *Bioorg Med Chem Lett*, 22(4): 1674–1677.

Naim, M. et al. 1994. Some taste substances are direct activators of G-proteins. *Biochem J*, 297(Pt 3): 451–454.

Nilius, B., and G. Appendino. 2013. Spices: The savory and beneficial science of pungency. *Rev Physiol Biochem Pharmacol*, 164: 1–76.

Nilius, B., T. Voets, and J. Peters. 2005. TRP channels in disease. *Sci STKE*, 2005(295): re8.

Perez, C.A. et al. 2002. A transient receptor potential channel expressed in taste receptor cells. *Nat Neurosci*, 5(11): 1169–1176.

Peri, I. et al. 2000. Rapid entry of bitter and sweet tastants into liposomes and taste cells: Implications for signal transduction. *Am J Physiol Cell Physiol*, 278(1): C17–C25.

Prescott, J., and R.J. Stevenson. 1995. Pungency in food perception and preference. *Food Rev Int*, 11(4): 665–698.

Ramsey, I.S., M. Delling, and D.E. Clapham. 2006. An introduction to TRP channels. *Annu Rev Physiol*, 68: 619–647.

Ran, C., M.A. Hoon, and X. Chen. 2016. The coding of cutaneous temperature in the spinal cord. *Nat Neurosci*, 19(9): 1201–1209.

Roper, S.D. 2014. TRPs in taste and chemesthesis. *Handb Exp Pharmacol*, 223: 827–871.

Roper, S.D. 2015. The taste of table salt. *Pflugers Arch*, 467(3): 457–463.

Roper, S.D. 2013. Taste buds as peripheral chemosensory processors. *Semin Cell Dev Biol*, 24(1): 71–79.

Royer, S.M., and J.C. Kinnamon. 1991. HVEM serial-section analysis of rabbit foliate taste buds: I. Type III cells and their synapses. *J Comp Neurol*, 306(1): 49–72.

Ruiz, C. et al. 2006. Detection of NaCl and KCl in TRPV1 knockout mice. *Chem Senses*, 31(9): 813–820.

Salazar, H. et al. 2008. A single N-terminal cysteine in TRPV1 determines activation by pungent compounds from onion and garlic. *Nat Neurosci*, 11(3): 255–261.

Shigemura, N., and Y. Ninomiya. 2016. Recent advances in molecular mechanisms of taste signaling and modifying. *Int Rev Cell Mol Biol*, 323: 71–106.

Simon, S.A. et al. 2006. The neural mechanisms of gustation: A distributed processing code. *Nat Rev Neurosci*, 7(11): 890–901.

Smith, K.R. et al. 2012. Contribution of the TRPV1 channel to salt taste quality in mice as assessed by conditioned taste aversion generalization and chorda tympani nerve responses. *Am J Physiol Regul Integr Comp Physiol*, 303(11): R1195–R1205.

Takaishi, M. et al. 2012. 1,8-cineole, a TRPM8 agonist, is a novel natural antagonist of human TRPA1. *Mol Pain*, 8: 86.

Talavera, K. 2015. TRP channels as targets for modulation of taste transduction. In *TRP Channels in Sensory Transduction*, edited by R. Madrid and J. Bacigalupo. Cham: Springer International.

Talavera, K. et al. 2008. The taste transduction channel TRPM5 is a locus for bitter-sweet taste interactions. *FASEB J*, 22(5): 1343–1355.

Taruno, A. et al. 2013. CALHM1 ion channel mediates purinergic neurotransmission of sweet, bitter and umami tastes. *Nature*, 495(7440): 223–226.

Tominaga, M. et al. 1998. The cloned capsaicin receptor integrates multiple pain-producing stimuli. *Neuron*, 21(3): 531–543.

Treesukosol, Y. et al. 2007. A psychophysical and electrophysiological analysis of salt taste in Trpv1 null mice. *Am J Physiol Regul Integr Comp Physiol*, 292(5): R1799–R1809.

Viana, F. 2011. Chemosensory properties of the trigeminal system. *ACS Chem Neurosci*, 2(1): 38–50.

Vincis, R. and A. Fontanini. 2016. A gustocentric perspective to understanding primary sensory cortices. *Curr Opin Neurobiol*, 40: 118–124.

Vriens, J., B. Nilius, and R. Vennekens. 2008. Herbal compounds and toxins modulating TRP channels. *Curr Neuropharmacol*, 6(1): 79–96.

Wang, Y., R.P. Erickson, and S.A. Simon. 1995. Modulation of rat chorda tympani nerve activity by lingual nerve stimulation. *J Neurophysiol*, 73(4): 1468–1483.

Winnig, M. et al. 2008. Saccharin: Artificial sweetener, bitter tastant, and sweet taste inhibitor. In *Sweetness and Sweeteners*, edited by D.K. Weerasinghe and G. DuBois. Washington, DC: American Chemical Society.

Yamasaki, H., Y. Kubota, and M. Tohyama. 1985. Ontogeny of substance P-containing fibers in the taste buds and the surrounding epithelium. I. Light microscopic analysis. *Brain Res*, 350(1–2): 301–305.

Yarmolinsky, D.A., C.S. Zuker, and N. J. Ryba. 2009. Common sense about taste: From mammals to insects. *Cell*, 139(2): 234–244.

Ye, W. et al. 2016. The K+ channel KIR2.1 functions in tandem with proton influx to mediate sour taste transduction. *Proc Natl Acad Sci U S A*, 113(2): E229–E238.

Zhang, Z. et al. 2007. The transduction channel TRPM5 is gated by intracellular calcium in taste cells. *J Neurosci*, 27(21): 5777–5786.

Zygmunt, P.M. et al. 1999. Vanilloid receptors on sensory nerves mediate the vasodilator action of anandamide. *Nature*, 400(6743): 452–457.

8 TRP Channels and Pain

*Ricardo González-Ramírez, Yong Chen, Wolfgang B. Liedtke,
and Sara L. Morales-Lázaro*

CONTENTS

8.1 INTRODUCTION

Pain is one of the primary responses developed by our body to protect us from harm. However, there are numerous pathological conditions, such as diabetes, viral infections, nerve damage, and inflammation that produce persistent pain. Chronic pain has no apparent useful purpose and is, in most cases, refractory to current pharmacological treatments.

Following the onset of a painful peripheral stimulus, nociceptive neurons are activated to initiate a cascade of action potentials that propagate along the axons of the primary afferent fibers (C and Aδ fibers) to the nerve terminals found in laminae I and II of the dorsal horn in the spinal cord (Figure 8.1a). These nerve terminals release neurotransmitters such as glutamate, substance P, and calcitonin gene-related peptide (CGRP) to activate postsynaptic receptors located in spinothalamic

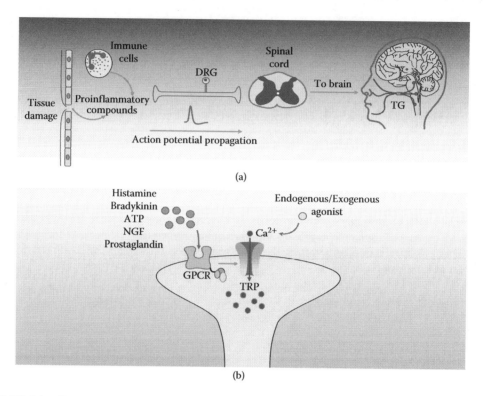

FIGURE 8.1 **(See color insert.)** Transduction of pain signal. (a) Nociceptor stimulation by tissue damage. Nociceptive signals and proinflammatory compounds trigger an action potential along the nociceptive fiber toward the central nervous system to be perceived as pain. (b) Nociceptive signals are detected by some members of the TRP family of ion channels, which are directly activated by their agonists and indirectly by proinflammatory mediators. (DRG, dorsal root ganglion; TG, trigeminal ganglion.)

tract neurons (Boadas-Vaello et al., 2016). The projections that reach the thalamus function in pain perception (Figure 8.1b). A variety of receptors and ion channels propagate and process pain signals.

Nociceptive neurons send signals from the periphery, through the afferent fibers, to the visceral, trigeminal, and somatic regions, and also connect the spinal cord to the brain, thus serving as mediators in painful stimulus transmission between the central and peripheral nervous systems (CNS and PNS) (Figure 8.1a). These neurons express a wide variety of receptors and ion channels that are distributed along the fibers and the somas. These are the molecules that detect noxious stimuli, transforming them into electrical signals and directing them to the CNS (Dubin and Patapoutian, 2010). The most important ion channel family that detects and transmits noxious stimuli is the transient receptor potential (TRP) channel family. This family contains proteins that are conserved nonselective calcium-permeable channels (Julius, 2013). In general, TRP channels act as molecular sensors of multiple stimuli, ranging from changes in pH, chemical agents, temperature, and osmolarity. The TRP family of ion channels is composed of 28 members divided into six subfamilies, classified as canonical (TRPC), vanilloid (TRPV), ankyrin (TRPA), melastatin (TRPM), polycystin (TRPP), and mucolipin (TRPML) (Wu et al., 2010).

The TRP channel structure varies considerably; however, there are certain shared domains that allow them to be grouped into the six subfamilies mentioned above. TRP channels consist of four subunits, each containing six transmembrane segments (S1–S6). A hydrophilic loop between the S5 and S6 forms the ion-conducting pore. The amino acids located before the pore confer channel selectivity. These channels are nonselective for cations but have preference for calcium (Owsianik et al., 2006b).

The most highly variable regions within the TRP channel sequences are the carboxyl and amino terminal ends. The ankyrin repeat is located at the amino terminus of the TRPC, TRPA, and TRPV subfamilies. The TRP box, which is a conserved six amino acid sequence found in the TRPC, TRPM, TRPA, and TRPV subfamilies, is located at the carboxyl end, and several studies have shown that the TRP box is important for channel gating (Valente et al., 2008). In addition to the ankyrin repeat and TRP box domains, TRP family members contain other domains, including the EF-hand, PDZ, or NUDIX domains. These domains are distributed among various TRP family members (Owsianik et al., 2006a). Because of their diversity in domain structures, TRP channels are able to respond to a wide variety of stimuli and form complexes with multiple proteins involved in different cellular processes. The ability to respond to different stimuli has positioned the TRP channels as the primary channels responsible for nociception in physiological and pathophysiological conditions such as chronic pain. In this chapter, we summarize the most relevant findings related to the TRPA, TRPM, and TRPV subfamilies in nociception and their importance in pain development and maintenance.

8.2   TRPA1 AS SENSOR OF NOXIOUS COLD TEMPERATURES

Nociceptive sensory neurons (nociceptors) present diversity in the type of molecular receptors that they express, as there are subpopulations of nociceptors with the capacity to detect thermal (heat, cold, or both), chemical, mechanical, and osmotic stimuli. The same nociceptor can respond to a combination of all these stimuli. This capacity of the nociceptive neurons is primarily associated with the expression of some TRP ion channel family receptors.

Among them is the TRPA1 channel, which owes its name to the fact that each monomer has a domain with several ankyrin repeats in its amino terminus (Figure 8.2). The cryo-electron microscopy structure of TRPA1 has recently been obtained, confirming the presence of six transmembrane segments, an extensive intracellular domain, and previously unknown structural domains (Paulsen et al., 2015). This channel is expressed in a subset of nociceptors that express the TRPV1 channel, but not the TRPM8 channel (Story et al., 2003).

The TRPA1 channel transduces noxious chemical signals because it is activated by the presence of exogenous irritant compounds and by endogenous compounds produced during tissue damage or neurogenic inflammation (Figure 8.2). Furthermore, it has been shown that this channel can function as a cold temperature thermal sensor, which can be activated by temperatures near 17°C (Story et al., 2003).

8.2.1   TRPA1 IN NOCICEPTIVE PAIN

Isothiocyanates are natural compounds that can activate the TRPA1 channel. These types of compounds are found in natural products such as wasabi, mustard, and horseradish, among others (Jordt et al., 2004). For example, the topical application of mustard oil is capable of generating pain and inflammation (Reeh et al., 1986) and is mediated by the detection of the irritant compound allyl isothiocyanate contained in mustard oil (Jordt et al., 2004). The transduction of noxious stimuli by isothiocyanates is closely related to its chemical structure (an isocyanate group in which oxygen is replaced by sulfur, rendering it with an electrophilic character) and its ability to cross the membrane barrier. These properties allow isothiocyanates to covalently interact with the thiol group of cysteines within TRPA1, causing TRPA1 activation (Hinman et al., 2006).

Another naturally occurring compound capable of producing a burning/stinging sensation is cinnamaldehyde (Bandell et al., 2004), which is found in cinnamon oil. The structure of this chemical compound consists of a phenolic group attached to an α-, β-unsaturated aldehyde, which is important for the interaction with the thiol group of cysteines within the TRPA1 channel, inducing channel activation (Macpherson et al., 2007). Cinnamaldehyde is a selective TRPA1 channel agonist because it does not activate any of the other nociceptive TRP (TRPV1, TRPM8, or TRPV4) channels (Bandell et al., 2004). Moreover, intraplantar injection of cinnamaldehyde into the paws of

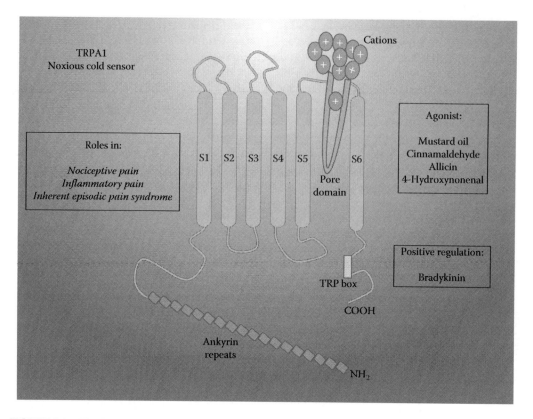

FIGURE 8.2   The TRPA1 ion channel is a noxious cold sensor. A subunit of the TRPA1 channel showing the transmembrane segments (S1–S6), a pore domain formed between S5 and S6, several ankyrin repeats located within the extensive amine end (NH_2), and the TRP box located in the proximal carboxyl end (COOH). The diagram depicts the most representative agonists of the TRPA1 channel, its role on pain perception, and positive regulators of TRPA1 activation.

mice causes a clear pain response, which persists in the TRPV1 null mice, indicating that this pain behavior is independent of TRPV1 channel activation and is associated with TRPA1 channel activation (Bandell et al., 2004). Furthermore, it has been demonstrated that cinnamaldehyde injection does not generate thermal hyperalgesia, a response known to be mediated by the TRPV1 channel, which confirms the specificity of cinnamaldehyde effects for causing acute pain through the TRPA1 channel (Bandell et al., 2004).

Within the repertoire of naturally occurring compounds that activate the TRPA1 channel is a compound found in fresh garlic, allicin (Macpherson et al., 2005; Bautista et al., 2005), the compound that gives the pungent sensation when eating raw garlic, which is caused by the stimulation of TRPA1- and TRPV1-expressing nociceptors (Bautista et al., 2005; Macpherson et al., 2005; Salazar et al., 2008). According to electrophysiological data, the TRPA1 channel is more sensitive to allicin activation compared to that observed in the TRPV1 channel (approximately 10 times more sensitive) (Macpherson et al., 2005), and no other TRP family member can be activated by this compound (Macpherson et al., 2005). In addition, intraplantar allicin injection into the paws of mice produces a clear acute pain response, which decreases by 40% in the TRPA1 and TRPV1 null animals (Salazar et al., 2008). This clearly indicates that one of the channels can compensate for the absence of the other to mediate a nociceptive response to allicin (Salazar et al., 2008). The molecular basis of this noxious signal is similar to that of allyl isothiocyanate because allicin is a thiosulfinate-type compound that also displays the ability to covalently modify cysteines within TRP channels to activate them (TRPA1 and TRPV1) (Salazar et al., 2008).

8.2.2 TRPA1 IN INFLAMMATORY PAIN AND MECHANICAL HYPERALGESIA

Bradykinin is released during tissue damage and inflammation; this inflammatory mediator indirectly activates TRPA1 channels to mediate the pain response. This effect requires bradykinin to interact with its receptor (B2R) activating the PLC-dependent signaling pathway, which promotes phosphatidylinositol 4,5, biphosphate (PIP2) hydrolysis to produce inositol triphosphate (IP_3) and diacylglycerol (DAG). Both metabolic intermediates can activate the TRPA1 channel (Bandell et al., 2004).

The nociceptive effects of the aforementioned stimuli have been confirmed through the use of the TRPA1 knockout mouse (Bautista et al., 2006; Kwan et al., 2006). For example, the TRPA1 knockout mouse fails to display a pain-like behavior upon application of mustard oil (Bautista et al., 2006; Kwan et al., 2006) or bradykinin (Kwan et al., 2006). These results confirm that TRPA1 is an important mediator of inflammatory pain (Bautista et al., 2006).

During tissue damage, ischemia, or cellular stress, a large number of proalgesic agents are produced, including reactive species, which can generate the oxidation of membrane lipids to produce compounds such as 4-hydroxynonenal, which is an endogenous aldehyde, capable of producing acute pain, releasing neuropeptide substances, and producing neurogenic inflammation through specific activation of the TRPA1 channel (Trevisani et al., 2007).

TRPA1 channel activation by all of the exogenous and endogenous compounds mentioned above, coupled with its activation by synthetic compounds such as formalin (McNamara et al., 2007), implicate that this channel is associated with inflammation processes and/or tissue damage.

Interestingly, the TRPA1 channel is also a specific mediator of the nociceptive response to the volatile compound acrolein (Bautista et al., 2006), which is present in tear gas and tobacco smoke and is the compound responsible for neurogenic inflammation that affects airways. Finally, it has been demonstrated that TRPA1 is involved in mechanical hyperalgesia and is an important signal transducer of noxious cold (Kwan et al., 2006).

8.2.3 TRPA1 IN AN INHERITED EPISODIC PAIN SYNDROME

In addition, the TRPA1 channel is associated with an inherited disorder known as familial episodic pain syndrome (Kremeyer et al., 2010), which is characterized by the presence of severe pain in the upper body accompanied by breathing difficulty, tachycardia, sweating, generalized pallor, and stiffness of the abdominal wall (Kremeyer et al., 2010). These episodes become present if an individual is in prolonged fasting, is fatigued and is exposed to cold and physical stress. This syndrome is caused by a point mutation in the TRPA1 channel, which generates a gain in its function (Kremeyer et al., 2010).

Such mutation is located at the transmembranal segment 4, where the asparagine 855 is substituted by a serine (N855S), leading to a shift in the voltage dependence to more negative potentials in a calcium-dependent manner, and increasing TRPA1 activity without modifying the affinity for its ligands (Kremeyer et al., 2010; Zima et al., 2015).

This evidence links the TRPA1 channel with a hereditary human channelopathy and highlights this channel as a therapeutic target for the generation of pharmaceutical alternatives that can counteract the pain produced in this hereditary syndrome and in inflammatory conditions associated with TRPA1 activation.

8.3 TRPM SUBFAMILY

The TRP melastatin subfamily (TRPM) consists of eight members, three of them associated with the transduction of noxious stimuli for producing a pain response (TRPM3, TRPM2, and TRPM8). These ion channels resemble the overall topology of the other TRP family members, although a major difference is the lack of ankyrin repeats in their amino end. We summarize some relevant findings related to TRPM ion channels in the transduction of noxious stimuli to produce pain responses.

8.3.1   TRPM2 in Inflammatory Pain

TRPM2 is a member of the melastatin-related transient receptor potential ion channel subfamily associated to inflammatory and neuropathic pain perception. It is expressed in brain, bone marrow, spleen, liver, lung, and some cells of the immune system (Perraud et al., 2001). The TRPM2 protein has a unique structural feature having a Nudix-like domain located at its carboxyl end; this domain allows TRPM2 to act as a pyrophosphatase to remove the ribose-5-phosphate moiety from ADP-ribose (Perraud et al., 2001). Furthermore, ADP-ribose is the main endogenous activator of TRPM2 channel and recently has been identified in TRPM2 as a sensor of warm temperatures (Tan and McNaughton, 2016), having a central role in preventing overheating when the temperature increases (Song et al., 2016).

Additionally, some reactive oxygen species (ROS) are positive regulators of TRPM2 activation; thus, this ion channel has a relevant role as sensor of the oxidative response in some cell systems (Kraft et al., 2004). Activation of TRPM2 has been linked to inflammatory processes such as ulcerative colitis. This channel is endogenously expressed in monocyte cells where it can be activated by H_2O_2 leading to calcium influx, and facilitating the stimulation of signaling pathways in order to produce and release important inflammation mediators such as the CXL2 chemokine, which is necessary for the recruitment of neutrophils. This, in turn, induces an exacerbation of the inflammation state, effects that are impaired on monocytes derived from the mice lacking TRPM2 expression (Yamamoto et al., 2008a). Moreover, TRPM2 knockout mice are resistant to the development of ulcerative colitis. The identification of TRPM2-positive expression rat dorsal root ganglion (DRG) neurons that innervate the distal colon confirms that TRPM2 contributes to visceral hypersensitivity in the induced rat colitis model (Matsumoto et al., 2016).

The inflammation process is an important state for inducing pain where TRPM2 activation plays a crucial role. By means of using the TRPM2 null mice, it has been demonstrated that the TRPM2 ion channel is necessary for the transduction of inflammatory pain, since TRPM2 null mice display an attenuated response to the second phase of pain induced by the formalin test, and they also show a decreased response to carrageenan-induced inflammatory pain (Haraguchi et al., 2012). In addition, TRPM2 expression is upregulated in the carrageenan-inflamed paws of wild-type (WT) mice, and the levels of H_2O_2, an endogenously produced stimulator of TRPM2 activity, significantly increase in the paws of the animals injected with carrageenan. These mice develop paw edema, mechanical allodynia, and thermal hyperalgesia, responses that are all decreased in the TRPM2 null mice (Haraguchi et al., 2012).

TRPM2 expression in nociceptive neurons from trigeminal ganglia (TG) has a role in peripheral inflammation, given that TRPM2 activation by H_2O_2 leads to the production of proinflammatory cytokines such as interleukin 6 (IL-6) and chemokine CXL2 (Chung et al., 2015). Furthermore, evidence from different experimental animal pain models, such as acetic acid–induced writhing behavior, induced osteoarthritis pain, and the diabetic neuropathic pain model suggest that TRPM2 plays a key role in pathological pain development, because in all of these, the pain response is decreased in mice lacking TRPM2 expression (So et al., 2015).

Thus, the TRPM2 ion channel plays an important role in nociceptive behavior mediated through the inflammatory response, rendering it a novel therapeutic target to reduce visceral and neuropathic pain produced in chronic inflammatory states.

8.3.2   TRPM3 in Neurogenic Pain Response

It has been determined that another member of the TRPM subfamily is an important mediator of noxious stimuli—the TRPM3 ion channel. This channel is expressed in neural (DRG and TG neurons) and nonneural tissue (kidney and pancreas). Initially, the TRPM3 ion channel was associated with insulin secretion because its activation by the neurosteroid pregnenolone sulfate causes insulin secretion from pancreatic cells (Wagner et al., 2008). Furthermore, TRPM3 ion channel expression in a subset of small-sensorial neurons has an important role for thermal (~40°C) and chemical sensation (Vriens et al., 2011). Pain behavior tests have demonstrated that pregnenolone sulfate evokes a pain response that is absent in mice lacking TRPM3 expression. Moreover, sensorial neurons isolated from these mice exhibit a reduced response to pregnenolone sulfate (Vriens et al., 2011). In addition, TRPM3 null mice show deficits in the detection of noxious heat stimuli (Vriens et al., 2011).

Until now, few compounds have been identified as ligands for the TRPM3 ion channel, and among them is the compound CIM0216, which is a potent small synthetic molecule that activates the TRPM3 ion channel (Held et al., 2015). Intradermal application of CIM0216 to the paws of mice evokes pain behavior in a TRPM3-dependent manner, and this pain response is more robust than that evoked by pregnenolone sulfate (Held et al., 2015). Interestingly, CIM0216 also induces the release of neuropeptides like CGRP, and this release is specifically through the activation of the TRPM3 ion channel (Held et al., 2015). This strongly suggests that TRPM3 activation has an important role in neurogenic inflammation.

8.3.3 TRPM8 in Cold Hypersensitivity and Neuropathic Pain

The TRPM8 ion channel is among the TRPM ion-channel members that has widely been related to pain responses in animal models. This ion channel is mainly expressed in sensory neurons from DRG and TG where TRPM8 expression is restricted to a subset of small-diameter sensory neurons (Peier, 2002). This channel is activated by warm and noxious cold temperatures in the range of 23°C–10°C, respectively, and by natural compounds such as menthol (Figure 8.3) (Peier, 2002; Bautista et al., 2007). TRPM8 activation produces the influx of calcium ions into neurons, changing the membrane potential and leading to transduction of noxious signals. The role of TRPM8 in cold thermosensation has been widely explored *in vivo* in mice lacking TRPM8 expression. These mice show a deficient response to cold thermal stimuli and chemical compounds that cause the

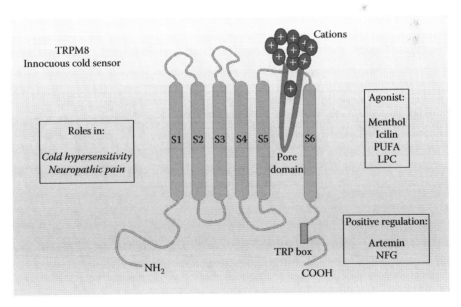

FIGURE 8.3 A TRPM8 protein—the transmembrane segments (S1–S6), a pore domain formed between S5 and S6, and the TRP box located in the proximal carboxyl end (COOH). The most representative agonists of the TRPM8 channel, its role on pain perception, and positive regulators of TRPM8 activation. (PUFA, polyunsaturated fatty acid; LPC, lysophosphatidylcholine; NGF, nerve growth factor.)

unpleasant feeling of coldness, such as icilin and acetone (Dhaka et al., 2007; Colburn et al., 2007; Bautista et al., 2007).

Tissue inflammation or nerve injury can result in extreme hypersensitivity to innocuous cooling stimuli or noxious cold, and regarding this it has been suggested that the TRPM8 channel is involved in cold allodynia (pain caused by a normally innocuous cold stimulus) (Allchorne et al., 2005). It has been demonstrated, by the use of a rat pain model where there is chronic constrictive nerve injury (Xing et al., 2007), that these animals display cold allodynia similar to that of patients who suffer from peripheral neuropathy. These animals show enhancement of withdrawal reflexes in response to cold agents such as acetone application, as well as an increase in the number of TRPM8 immunoreactive neurons and an increase in the number of neurons that are stimulated by menthol and cold temperatures (Xing et al., 2007).

Moreover, TRPM8 has been connected to cold hypersensitivity (enhanced sensitivity to a painful cold stimulus) (Allchorne et al., 2005). Unlike heat hyperalgesia, which is produced by several algogenic agents, TRPM8 cold sensitization is produced only by select neurotrophic factors such as artemin and nerve growth factor (NGF) that produce their effects through specific receptors coexpressed with the TRPM8 channel in a subset of sensory neurons (Lippoldt et al., 2013). Experimental data using the evaporative cooling assay, where the mentioned neurotrophic factors were intraplantarly injected and a drop of acetone was applied into the animal's paw, suggest that wild-type mice show an increase in their response to cold; however, in the TRPM8 knockout mice this response was abolished (Lippoldt et al., 2013).

The effects of TRPM8 ion channel on cold-induced pain can be modulated by some endogenously produced lipid compounds. For example, TRPM8 activation by cold temperatures and menthol requires the presence of PIP2 (a membrane phospholipid), which interacts directly with positively charged amino acids located in the TRP box of the channel (Rohacs et al., 2005). Furthermore, PIP2 alone is able to activate the channel (Rohacs et al., 2005). Therefore, it stems from this that cold allodynia and hypersensitivity produced by the activation of this ion channel could be attenuated by regulating PIP2 levels, where a regulatory pathway includes the activation of a calcium-dependent PLC which hydrolyzes PIP2 to produce DAG and IP3 (Rohacs et al., 2005). Then, DAG activates the protein kinase C (PKC) pathway that downregulates the activity of the TRPM8 ion channel, an effect associated with a decrease in the levels of TRPM8 protein with phosphorylated serine residues (Premkumar et al., 2005), resulting in ion channel desensitization.

In addition, TRPM8 activity is endogenously regulated by the iPLA2 pathway, a calcium-insensitive phospholipase that produces lysophospholipids and free polyunsaturated fatty acids (PUFAs) through the hydrolysis of sn-2-glycerophospholipids (Andersson et al., 2007). Interestingly, iPLA2 inhibition attenuates TRPM8 currents evoked by cold, menthol, and icilin (Andersson et al., 2007). Moreover, lysophosphocholine (LPC) released from the iPLA2 pathway is able to activate the TRPM8 ion channel even at 37°C (Andersson et al., 2007). Furthermore, it was found that LPC induces chemical cold hypersensitivity specifically through TRPM8 activation, because this behavior was absent in a TRPM8 null mouse (Gentry et al., 2010).

On the contrary, PUFAs, such as arachidonic, eicosapentaenoic, and docosahexaenoic acids, which also are released by iPLA2, are important endogenous inhibitors of the TRPM8 ion channel (Andersson et al., 2007).

Therefore, the iPLA2 pathway can produce contrasting effects on TRPM8 activation—a positive effect by lysophospholipids and negative effects through the PUFA. This would likely cancel the effects of both products; however, it has been shown that equimolar applications of LPC and arachidonic acid favor the activation of the TRPM8 channel (Andersson et al., 2007).

Under this scenario, the TRPM8 channel plays an important role in cold hypersensitivity produced by chronic pain conditions such as diabetes, trauma, or cancer. Thus, the development of the field study on TRPM8 ion channel regulation is important to find alternative ways to decrease cold hypersensitivity produced by the activation of this ion channel.

8.4 TRPV FAMILY

The TRP vanilloid subfamily of ion channels is named in accordance to its first characterized member, the TRPV1 channel, which is activated by vanilloid compounds such as capsaicin. Although TRPV1 is the only member of this subfamily sensitive to vanilloids compounds, other members share identity to TRPV1 in their sequence. This subfamily consists of six members (TRPV1 through TRPV6). All of the members are cation nonselective channels, where TRPV1–4 show high preference to Ca^{+2} and are broadly linked to nociception. We detail their roles in pain signal detection in the next section.

8.4.1 TRPV1 as a Widely Studied Portal to Pain

The TRPV1 channel is the most well-characterized TRP channel to date. TRPV1 can be activated by various stimuli, including temperature (~42°C), pH, and a wide range of both endogenous and exogenous compounds (Figure 8.4). Its main exogenous ligand is capsaicin, a compound from chili peppers. The beneficial effects of capsaicin on nociception were identified before the TRPV1 channel was discovered (Frias and Merighi, 2016). TRPV1 was cloned in 1997 from a cDNA library isolated from capsaicin- and temperature-stimulated nociceptor neurons (Caterina et al., 1997). Subsequent characterization of TRPV1 has revealed its key role in nociception. In addition, with the use of knockout mice, TRPV1 has been shown to be important in pain sensation (Caterina et al., 2000). Due to its importance in these physiological processes, several compounds have been developed to modulate the activity of TRPV1 in order to eliminate or reduce pain.

8.4.1.1 Expression of TRPV1 in the Nervous System

TRPV1 is preferentially expressed in the neurons of the PNS and CNS. In the PNS, TRPV1 is primarily expressed by the small and medium nociceptor neurons of the DRG, TG, nodal ganglion (NG), and sympathetic ganglion (SG), in the peptidergic and nonpeptidergic C fibers, and in some Aδ fibers (Cavanaugh et al., 2009; Huang et al., 2012; Hwang et al., 2005). In addition, it is expressed at lower levels in nerve fibers that innervate the bladder (Liu et al., 2014), lungs (Zhao et al., 2016; Dinh et al., 2004), and cochlea (Vass et al., 2004), as well as in the upper respiratory tract, where its function is to sense irritant compounds (De Logu et al., 2016; Lehmann et al., 2016).

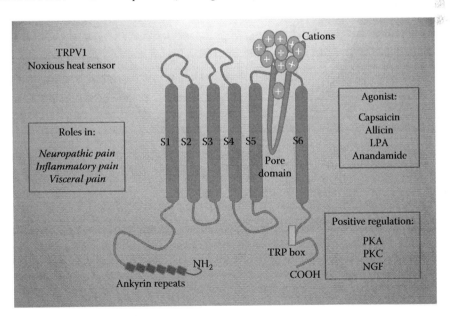

FIGURE 8.4 **(See color insert.)** A TRPV1 monomer showing the transmembrane segments (S1–S6), the most representative agonists known for the TRPV1 channel, its role on pain perception, and positive regulators of its activation.

TRPV1 is expressed in several regions of the CNS, specifically in laminae I and II of the dorsal horn of the spinal cord, where it modulates the synaptic transmission of nociceptive signals from the periphery (Spicarova and Palecek, 2009). TRPV1 is primarily expressed in the presynapse, although there have been studies that demonstrated its postsynaptic expression, albeit this has not been shown to be related to nociception (Spicarova and Palecek, 2008). The use of transgenic mice expressing reporter proteins driven by the TRPV1 promoter has revealed the specific areas in which TRPV1 is expressed (Cavanaugh et al., 2011). Besides DRG and TG neurons, some expression has been observed in the nerve fibers of small- and medium-diameter nociceptor neurons in the cornea, bladder, and skin, and in some regions of the brain, such as the brainstem, nucleus caudalis, nucleus ambiguous, olfactory bulb, and nucleus parabrachialis (Cavanaugh et al., 2011). Interestingly, TRPV1 is not expressed in the hypothalamus or hippocampus.

8.4.1.2 Regulation of TRPV1 Activity

Under normal conditions, only temperatures near 42°C activate TRPV1, but this thermal threshold decreases to 35°C–37°C after acidification of the medium (Tominaga et al., 1998). This phenomenon is very important during inflammation, because this condition drastically decreases the pH (up to a pH of 6.4) and rapidly activates TRPV1. Additionally, post-translational modifications, such as phosphorylation, can also regulate TRPV1 activity. TRPV1 sensitization is the phenomenon that facilitates the channel to be activated by low-intensity stimuli, and the process occurs after inflammation or tissue damage, which is triggered by various proinflammatory substances, such as NGF, substance P, bradykinin, and prostaglandins. These mediators primarily function by activating several kinases, such as PKC, mitogen-activated protein kinase (MAP kinase; MAPK), and protein kinase A (PKA). Once activated, these kinases phosphorylate various residues within the TRPV1 sequence, inducing TRPV1 sensitization (Cheng and Ji, 2008).

PKC and PKA phosphorylation of TRPV1 shifts its activation temperature from 43°C to 33°C–38°C, allowing it to open at body temperature, and also its activation by protons. In addition to direct regulation by phosphorylation by PKA and PKC, MAPK activation increases TRPV1 expression levels. The increased protein levels promote long-term nociceptive neuron sensitization (Ji et al., 2002). In addition, TRPV1 phosphorylation by tyrosine kinases increases the traffic of TRPV1-containing vesicles to the plasma membrane, causing an increase in the number of functional channels in the membrane (Jin et al., 2004; Zhang et al., 2005).

Another important mechanism in the regulation of TRPV1 activity is Ca^{2+}-induced channel desensitization. TRPV1 activation causes an increase in Ca^{2+} concentration, which binds to CaM (calmodulin, a calcium-modulated protein), and this complex activates PP2B (calcineurin), which dephosphorylates TRPV1, causing its desensitization. CaM interacts with the C-terminus end of the channel to regulate short-term desensitization (Mohapatra and Nau, 2005; Numazaki et al., 2003). Long-term desensitization causes endocytosis and channel degradation (Sanz-Salvador et al., 2012).

8.4.1.3 TRPV1 in Inflammatory Pain

When a skin lesion is produced, a wide variety of proinflammatory molecules are released, such as bradykinin, prostaglandins, leukotrienes, serotonin, histamine, substance P, thromboxanes, platelet-activating factor, adenosine and ATP, protons, and free radicals. During inflammation, cytokines, such as interleukins (IL), tumor necrosis factor, and neurotrophins, particularly NGF, are also generated. In general, these mediators sensitize TRPV1, increasing the probability of a stimulus to activate TRPV1. The first evidence of the role of TRPV1 in peripheral pain processing was generated by observations made in mice injected with the TRPV1 antagonist capsazepine. Capsazepine injection attenuated inflammation-induced thermal hyperalgesia (Caterina et al., 2000; Davis et al., 2000). TRPV1$^{-/-}$ mice do not respond to noxious temperatures and presented a dramatic attenuation of inflammation-induced thermal hyperalgesia. Additionally, TRPV1$^{-/-}$ mice did not experience thermal hyperalgesia after complete Freund's adjuvant (CFA) administration.

When bradykinin is released, it acts on β2 receptors, which in turn activate PKC to facilitate PIP2 hydrolysis by PLC to produce IP3 and DAG, two compounds that can affect TRPV1 function (Morales-Lazaro et al., 2013). DAG is a TRPV1 agonist that interacts with tyrosine 511 of its S3 domain (Woo et al., 2008). PIP2 binds to the C-terminal end of the channel to positively regulate TRPV1 (Brauchi et al., 2007). Prostaglandins increase capsaicin-induced currents in DRG neurons and reduce the temperature activation threshold of the channel, effects that have been corroborated in the TRPV1$^{-/-}$ mouse (Moriyama et al., 2005). TRPV1 integrates multiple proinflammatory stimuli, and for this reason, an extensive search for both exogenous and endogenous synthetic antagonists that inhibit these pain pathways is currently underway. For example, oleic acid, a compound that causes attenuation of the pain and itch responses, has recently been shown to inhibit TRPV1 activation (Morales-Lazaro et al., 2016).

8.4.1.4 TRPV1 in Neuropathic Pain

Neuropathic pain occurs when there is an injury or dysfunction of the CNS or PNS and is a condition that is difficult to treat. The TRPV1 channel is a polymodal nociceptor, and its principal function is to sense proinflammatory mediators. The process of sensitization increases excitability, causing hyperalgesia and allodynia. However, there is contradictory evidence regarding the role of TRPV1 in neuropathic pain. It has been found that in TRPV1$^{-/-}$ mice, there is no change in pain perception after nerve damage (Bolcskei et al., 2005; Caterina et al., 2000); however, pharmacological inhibition of TRPV1 decreases pain in several animal models of neuropathic pain (Kanai et al., 2005; Yamamoto et al., 2008b). Moreover, increased TRPV1 expression correlates with the development and maintenance of thermal hyperalgesia (Fukuoka et al., 2002). In diabetic neuropathy models, two phases have been shown to develop: there is hyperalgesia in the early stages of the disease and hypoalgesia in the later stages (Pabbidi et al., 2008). In the first stage, TRPV1 expression increases in DRG neurons and in the spinal cord (Hong and Wiley, 2005). This increase overlaps with the presence of large currents in DRG neurons isolated from these mice models. The role of TRPV1 in diabetic neuropathy has been demonstrated in TRPV1$^{-/-}$ mice that do not present thermal hyperalgesia after induction of diabetes (Cui et al., 2014). Finally, it has been found that TRPV1 plays an important role in cancer-induced chronic pain. In bone cancer, TRPV1 is overexpressed in DRG neurons, and this increased expression induces neuropathic pain. This overexpression enhances the response of TRPV1 to capsaicin (Pan et al., 2010). In addition, it has been shown that lysophosphatidic acid (LPA), which activates TRPV1 through a direct interaction of a lysine 710 located in the C-terminus of the channel, produces acute pain in animals injected in their paws with the lipid, and this response is attenuated in the TRPV1 null animal (Nieto-Posadas et al., 2011).

8.4.1.5 TRPV1 in Visceral Pain

TRPV1 is expressed in DRG neurons that innervate the colon, pancreas, stomach, duodenum, and bladder and is responsible for pain produced by gastrointestinal system inflammation (Beyak and Vanner, 2005). In tissue biopsies of patients with inflammatory bowel disease (IBD), an increase in TRPV1 expression has been observed, as well as in patients with rectal hypersensitivity. Notably, increased channel levels correlate with hypersensitivity levels (Hicks, 2006). The administration of acetic acid and capsaicin increases the activity of afferent fibers in the pelvis in control mice and in mice with dextran sulfate sodium (DSS)–generated colitis. However, the response to capsaicin in DSS-treated mice was greater than in controls (Makimura et al., 2012). Interestingly, TRPV1$^{-/-}$ mice were less sensitive to these treatments, highlighting the importance of TRPV1 in this type of pain (Jones et al., 2005).

8.4.2 TRPV2 IN INFLAMMATORY PAIN

TRPV2 is a vanilloid subfamily member that shares 49% sequence identity with TRPV1 (Caterina et al., 1999). This is an acid- and capsaicin-insensitive ion channel that is activated by high temperatures with a threshold of 52°C. Its expression is confined to medium and large primary afferent neurons, and it is also expressed in nonneuronal tissues such as spleen, lung, bladder, heart, and kidney (Caterina et al., 1999).

Chemical TRPV2 activation is produced by the natural compound cannabidiol (the major non-psychotropic cannabinoid produced by *Cannabis sativa*), stimulating the release of CGRP from DRG neurons in a fashion partially dependent of TRPV2 activation (Qin et al., 2008). Another specific TRPV2 agonist is the synthetic compound probenecid (an uricosuric agent); activation of the ion channel by this molecule evokes pain behavior in inflamed animal models (Bang et al., 2007).

Downregulation of TRPV2 activation is acquired through a conserved mechanism between some TRPV ion channels—PIP2 depletion from membranes which causes TRPV2 desensitization in a calcium-dependent manner, but independent of calmodulin's actions (Mercado et al., 2010). Desensitization produces a significant decrease on TRPV2 activation being that the channel enters into a refractory period.

TRPV2 expression in Aδ and Aβ fibers suggests a nociceptive role for transduction of high-thermal pain and for detection of chemical harmful stimuli through this protein. Moreover, TRPV2 expression is upregulated in peripheral inflammation induced after intraplantar injection of CFA (Shimosato et al., 2005). This upregulation is also observed after nerve injury (Frederick et al., 2007), suggesting the potential role of this ion channel in pain sensation. However, the nociceptive role of TRPV2 is controversial given that mice lacking TRPV2 expression display normal behavior responses to high temperatures and mechanical stimuli in basal and hyperalgesic conditions (Park et al., 2011). It is possible that the TRPV2 ion channel plays another important physiological role, given that TRPV2 KO mice exhibit certain degrees of perinatal lethality, and adult mice show body weight reduction (Park et al., 2011).

8.4.3 TRPV3 IN NOCICEPTIVE AND INFLAMMATION PAIN

A member of the TRP vanilloid subfamily important for warm and noxious temperature detection is the TRPV3 channel. This nonselective cation channel is abundantly expressed in skin (specifically in keratinocytes) (Peier et al., 2002) and has also been detected in neurons from DRG and TG (Xu et al., 2002). The TRPV3 protein shares 43% identity with the TRPV1 channel and is a warm temperature-sensitive channel but is insensitive to capsaicin and low pH (Peier et al., 2002; Xu et al., 2002).

Among the ligands for TRPV3 activation are compounds found in plants as camphor, carvacrol, and thymol (Vogt-Eisele et al., 2007); furthermore, TRPV3 activity is enhanced by some endogenously produced compounds such as arachidonic acid (AA), which is an unsaturated fatty acid released in inflammatory processes (Hu et al., 2006). Unlike TRPV1 activation that requires the metabolism of this fatty acid for currents to be activated, the enhancement of TRPV3 function is independent of AA oxidation (Hu et al., 2006). However, it is possible that in some skin pathologies, such as psoriasis, where AA levels are augmented (Brash, 2001), TRPV3 activation could be potentiated stimulating the inflammatory process and pain sensation.

Another TRPV3 agonist is farnesyl pyrophosphate (FPP), an endogenously produced intermediate metabolite from the mevalonate pathway (Holstein and Hohl, 2004), which produces isoprenoids such as cholesterol, a TRPV3 activator. FPP was identified as a novel pain-producing compound acting through specific activation of TRPV3 (Bang et al., 2010).

Although the physiological role of TRPV3 is still unclear, some studies suggest that this ion channel can contribute to nociception. For example, mice overexpressing TRPV3 in their keratinocytes release prostaglandin E2, and thermal nociception and hyperalgesia are produced (Huang et al., 2008). In contrast, mice lacking TRPV3 expression have impaired responses to warm and noxious temperatures (Moqrich et al., 2005).

Finally, some endogenous antagonists of TRPV3 activation have been also described. For example, some compounds synthetized from the omega-3 fatty acids, such as resolvins, are linked to anti-inflammatory and antinociceptive effects. One of them, the 17(R)-resolvin D1, is a specific inhibitor of TRPV3 channel (Bang et al., 2012). This resolvin affects the voltage dependence of TRPV3, decreasing channel activation (Bang et al., 2012). The inhibitory effects of this resolvin occur through a specific inhibition of TRPV3, because the pain behavior produced by capsaicin or cinnamaldehyde (activators of TRPV1 and TRPA1 channels, respectively) are unaffected. However, nociceptive behavior produced by intradermal injection of FPP (specific agonist to TRPV3) is attenuated by coinjection with resolvin D1 (Bang et al., 2012).

Taken together, these experimental observations suggest that TRPV3 channels are mediators of noxious thermal and chemical stimuli that produce nociceptive responses.

From all of these observations, it follows that it is necessary to intensify the search for knowledge on how this channel functions. This may allow us to understand all of its physiological roles and find alternatives to attenuate the transduction of harmful stimuli as a consequence of its activation.

8.4.4 TRPV4 AS A TRANSDUCTOR OF PAIN SIGNALS

As a member of the TRPV subfamily, TRPV4 is a multimodally activated, nonselective cation channel. Functional expression of TRPV4 has been detected in both neuronal and nonneuronal cell types (Liedtke et al., 2000). One of the initial discovery papers lays out a rationale on how TRPV4 can function in pain by demonstrating its expression in TG sensory neurons and commenting on a possible role for TRPV4 in pain signaling by examining the types of sensory neurons in which TRPV4 was expressed (Liedtke et al., 2000). This initial conjecture has now been confirmed in an impressive manner (e.g., a Medline search for "TRPV4 pain" generated 161 references, out of a total of 958 references for "TRPV4"). Primary sensory neurons with TRPV4 expression include pain-sensing neurons in DRG and TG (Liedtke and Friedman, 2003; Chen et al., 2013 ; Lietdke et al., 2000), most recently also satellite cells in sensory ganglia (Rajasekhar et al., 2015), in the CNS for pain transmission in astrocytes (Benfenati et al., 2007), microglial cells (Konno et al., 2012), and neurons (Shibasaki et al., 2007, 2015). Outside the nervous system, TRPV4 has been found in innervated cells such as chondrocytes (Phan et al., 2009), innervated epithelia such as skin keratinocytes (Moore et al., 2013; Chen et al., 2014), airway epithelial cells (Li et al., 2011), colonic epithelia (D'Aldebert et al., 2011), and odontoblasts (Kwon et al., 2014). Co-labeling experiments with neuronal markers and size determination revealed that TRPV4-expressing sensory neurons are also nociceptive (Chen et al., 2013). This is a finding in keeping with evidence that suggests that TRPV4 is involved in nociception, both physiologically and in sensitized states such as inflammation and nerve injury. Of note, TRPV4 expression in TG appears increased over that in DRGs, an observation, up to this day, not fully understood, and it has not been clarified how this difference correlates with sensory functions (Liedtke and Friedman, 2003; Vandewauw et al., 2013).

8.4.4.1 TRPV4 in Transduction of Mechanically Evoked Pain

Trpv4 null animals, soon generated after the initial description, were found to have defective mechanosensation in the physiologic, nonsensitized state (Suzuki et al., 2003; Liedtke and Friedman, 2003), bespeaking of a role for TRPV4 in nonsensitized mechanotransduction. This finding was supported and extended by assessing function of mammalian TRPV4 in ASH head nociceptor neurons of *Caenorhabditis elegans*, in a mutant line of animals that were lacking the proto-ancestral osmo-mechano-TRPV channel, OSM-9 (Lindy et al., 2014).

8.4.4.2 TRPV4 in Inflammatory Pain

TRPV4 in sensory neurons can be sensitized by proinflammatory mediators, such as prostaglandin E2, activator of proteinase-receptor 2 (PAR-2), an integrator of proteolytic signaling in inflammation, especially allergic inflammation, histamine, and/or serotonin, leading to increased nociception to hypotonic, mild hypertonic stimuli or mechanical stimuli (Alessandri-Haber et al., 2003, 2005, 2006). Grant et al. found that immunoreactive TRPV4 was coexpressed by rat DRG neurons with PAR2; intraplantar injection of PAR2 agonist caused mechanical hyperalgesia in mice and sensitized pain responses in a TRPV4-dependent manner (e.g., to the TRPV4 agonists 4α-PDD and hypotonic solutions). Deletion of Trpv4 prevented PAR2 agonist-induced mechanical hyperalgesia and sensitization (Grant et al., 2007). Further studies demonstrated that cathepsin S- (Zhao et al., 2014) or elastase-mediated activation of PAR-2 (Zhao et al., 2015) activate TRPV4 and sensitize nociceptors to function in a hypersensitive manner in inflammation. In the Liedtke-Lab, a key role for TRPV4 ion channels in the pain response to temporomandibular joint (TMJ) inflammation was recently established. In Trpv4$^{-/-}$ mice with TMJ inflammation, attenuation of bite force, a surrogate of TMJ injury–mediated pain in humans and functioning likewise in mice, was significantly and robustly reduced versus WT mice. Furthermore, TRPV4 protein expression in the TG was dramatically upregulated after TMJ inflammation, in synchrony with clinical severity after TMJ inflammatory injury, suggesting TRPV4 expression in the TG as a critical locale for behavioral sensitization (Chen et al., 2013). In favor of this concept, TRPV4 was expressed in TMJ-innervating sensory neurons, and TRPV4-expressing TG sensory neurons coexpressed phosphorylated ERK, a biochemical signaling activation marker in response to injury. pERK-TRPV4 coexpressing TG sensory neurons became more numerous in response to TMJ inflammatory injury, suggesting that TRPV4-mediated Ca^{2+} influx into TG sensory neurons evokes MAPK activation in these neurons, which functions as a molecular substrate of the sensitization response after inflammatory injury. At least as interesting was our finding that peripheral injury to the joint after inflammatory injury was independent of the genotype, whereas the pain response was strikingly dependent on Trpv4 or the ability of the animal to generate phosphorylated ERK. In another study by Denadai-Souza et al., TRPV4 expression in the TG was demonstrated (albeit without appropriate validating controls in Trpv4$^{-/-}$ animals), in addition to TMJ synovial cells (Denadai-Souza et al., 2012; Kochukov et al., 2006). Taken together, it emerges that TRPV4 plays a key role in inflammatory joint pain, perhaps rather at the level of joint-innervating sensory neurons than at peripheral TRPV4-expressing cells, and that therefore TRPV4 becomes an attractive therapeutic target to fight joint pain, a matter of increasing unmet medical need, given the increase in prevalence of age- and obesity-associated osteoarthritis and post-traumatic osteoarthritis (O'Conor et al., 2016).

The role of TRPV4 in facilitating and promoting inflammation and pain was also supported by observations in a mouse model of skin inflammation in which UVB radiation generated sunburn tissue damage (UV-burn) and associated pathological pain. Following UVB overexposure of their hindpaws, an area of mouse skin with increased resemblance to human skin, mice with induced Trpv4 deletions in keratinocytes and also mice treated with topical TRPV4 inhibitors to their hindpaws became virtually resistant to noxious thermal and mechanical stimuli versus control animals. Moreover, these animals also showed dramatically reduced skin inflammation. These findings indicate that TRPV4-expressing keratinocytes of the hindpaw can "moonlight" as nonneural sensing cells and pain-generating cells. In other words, activation of TRPV4 channels in skin keratinocytes, not in innervating sensory neurons, suffices to switch on neural pathological-pain circuits and response mechanisms. Importantly, we demonstrate a form of nonneural phototransduction in response to UVB radiation in skin keratinocytes, completely dependent on TRPV4. Exploring a possible underlying mechanism, we found that epidermal keratinocyte TRPV4 is essential for UVB-evoked skin tissue damage and increased expression of the proalgesic/algogenic mediator endothelin-1 (ET-1) (Moore et al., 2013). Of critical significance, we recorded a negative finding that will become the starting point for relevant future studies, namely, that keratinocyte-derived ET-1, dependent on TRPV4 function and UVB-mediated activation in these cells, was not the algogenic signal to innervating peripherals.

8.4.4.3 TRPV4 in Neuropathic Pain

A number of studies indicated that TRPV4 has been involved in a variety of rodent models of neuropathic pain, such as paclitaxel-induced neural injury leading to painful peripheral neuropathy (CIPN) and chronic compression of the DRG (Alessandri-Haber et al., 2004, 2009; Chen et al., 2011). Spinal administration of antisense oligodeoxynucleotides to TRPV4, which reduced the expression of TRPV4 in sensory neurons, abolished paclitaxel-induced mechanical hyperalgesia and attenuated hypotonic hyperalgesia (Alessandri-Haber et al., 2004). TRPV4 appears coexpressed with TRPC1 and TRPC6 in DRG neurons, and it has been proposed that the TRPC1 and TRPC6 may act in concert with TRPV4 to mediate mechanical hyperalgesia induced by paclitaxel and cis-platin (Alessandri-Haber et al., 2009), possibly representing a more general mechanism of pain in CIPN. In addition, paclitaxel can stimulate the release of mast cell tryptase, which activates PAR2 and, subsequently, PKA and PKC, resulting in mechanical and thermal hypersensitivity through TRPV4 sensitization (Chen et al., 2011). Following chronic compression of the DRG (CCD) in rat, intrathecal administration of NF-κB inhibitors pyrrolidine dithiocarbamate (PDTC) or BAY11-7082 induced significantly dose-dependent thermal hyperalgesia and a decrease in nitric oxide (NO) content in DRG when compared with control rats, which is suppressed by pretreatment with 4α-PDD (Wang et al., 2011). Further studies demonstrated that TRPV4-NO-cGMP-PKG and TRPV4-p38 MAPK pathways are involved in the development of thermal hyperalgesia following chronic compression of the DRG in rats (Ding et al., 2010; Qu et al., 2016). These results suggest that TRPV4 plays a crucial role in painful peripheral neuropathies caused by paclitaxel and related taxanes, affecting more than half of taxane-treated cancer patients. This is a large unmet medical need that not only translates to patient's quality of life but indeed to their ability to successfully undergo adjuvant chemotherapy and thus emerge with increased years of survival. We reach the same conclusion as for joint- or skin-mediated forms of pain and inflammation, as discussed above, namely, that TRPV4 is a promising target for anti-pain therapy. To that end, we recently developed novel TRPV4-inhibiting compounds in our group (Kanju et al., 2016). Two of these compounds also potently co-inhibited TRPA1 ion channels, which would be a highly beneficial combination for CIPN pain, as well as for other forms of pain reviewed in some detail below, namely, visceral pain of colon and pancreas and headaches (Kanju et al., 2016).

8.4.4.4   TRPV4 in Visceral Pain

Trpv4 mRNA is highly enriched in colonic sensory neurons as well, and its protein colocalized in a subset of fibers with the sensory neuropeptide CGRP in mice. Mechanosensory responses of colonic serosal and mesenteric afferents were enhanced by a TRPV4 agonist and dramatically reduced by targeted deletion of Trpv4. The behavioral responses to noxious colonic distention were also substantially reduced in mice lacking Trpv4 (Brierley et al., 2008). Similarly, another study found that the TRPV4 agonist 4α-PDD specifically activated a cationic current and calcium influx in colonic projections of DRG neurons and caused dose-dependent visceral hypersensitivity. TRPV4-targeted, but not mismatched siRNA, intervertebral treatments were effective at reducing basal visceral nociception, as well as 4α-PDD or PAR2 agonist-induced hypersensitivity (Cenac et al., 2008). PAR2 exacerbated visceromotor responses, indicative of mechanical hyperalgesia, are absent in $Trpv4^{-/-}$ mice, indicating TRPV4 is required for PAR2-induced mechanical hyperalgesia and excitation of colonic afferent neurons (Sipe et al., 2008). A more recent study showed that both TRPV4 and TRPA1 mediate colonic distension pain and CGRP release and appear to govern a wide and congruent dynamic range of distensions. Unlike TRPV4 and TRPA1, the role of TRPM8 seems to be confined to signaling extreme noxious distension (Mueller-Tribbensee et al., 2015). Interestingly, levels of the TRPV4 agonist 5,6-EET were increased in IBS colon biopsies compared with controls; increases correlated with pain and bloating scores. Small interfering RNA knockdown of TRPV4 in mouse primary afferent neurons inhibited the hypersensitivity caused by supernatants from IBS biopsies in mice. PUFA metabolites extracted from IBS biopsies or colons of mice with visceral hypersensitivity activated mouse sensory neurons *in vitro*, by activating TRPV4 (Cenac et al., 2015). These data indicate that TRPV4 contributes to visceral pain, with relevance to human disease.

Of note, TRPV4 also plays an important role in pancreatitis pain. Immunoreactive TRPV4 was detected in pancreatic nerve fibers and in DRG neurons innervating the pancreas, which were identified by retrograde tracing. Activation of TRPV4 with 4α-PDD increased intracellular Ca^{2+} in these neurons in culture. The secretagogue cerulein–induced c-Fos expression in spinal cord dorsal horn neurons and pain behavior were suppressed in TRPV4-deficient mice (Ceppa et al., 2010). Using a chronic pancreatitis model induced by a high-fat and alcohol (HFA) diet (Zhang et al. 2015) demonstrated that TRPV4 expression was increased in pancreatic stellate cells (PSCs). Calcium signals of PSCs from HFA-fed rats in response to 4α-PDD were dramatically higher than that of cells from control rats. Tumor necrosis factor-α (TNF-α) increased responses to 4α-PDD in control PSCs (Zhang et al., 2013). Furthermore, inhibition of TRPV4 with systemic injection of the selective blocker HC067047 effectively alleviated mechanical and thermal hypersensitivities of rats with AHF pancreatitis in a dose-dependent manner (Zhang et al., 2015). In the Liedtke-Lab, it was recently found that blockade of both TRPV4 and TRPA1 with the dual inhibitor 16-8 dramatically attenuated pancreas edema induced by caerulein. Furthermore, serum amylase, a marker of inflammatory injury of the pancreas, and pancreas inflammation were also significantly reduced by 16-8 treatment. Of note, pain behavior was virtually eliminated upon treatment with compound 16-8 (Kanju et al., 2016). Using the reliable preclinical models, these studies provided intriguing evidence that TRPV4 channel is a potential therapeutic target for treatment of pancreatitis pain.

8.4.4.5 TRPV4 in Headaches

Using *in vitro* patch-clamp electrophysiology of trigeminal neurons retrogradely labeled from the dura, around half of identified dural afferents generate currents in response to hypotonic solutions and 4α-PDD, indicating dural afferents do express TRPV4. Activation of meningeal TPRV4 using hypotonic solution or 4α-PDD *in vivo* resulted in facial allodynia that was blocked by the TRPV4 antagonist RN1734 (Wei et al., 2011). These data indicate that activation of TRPV4 within the meninges produces afferent nociceptive signaling from the head that may contribute to migraine headache. In the Liedtke-Lab, we recently found TRPV4 to be important for trigeminal nocifensive behavior evoked by formalin whiskerpad injections. This conclusion is supported by studies with Trpv4$^{-/-}$ mice and TRPV4-specific antagonists GSK205 and HC067047 (Chen et al., 2014). Our results suggest that TRPV4 acts as an important signaling molecule in irritation-evoked trigeminal pain. TRPV4-antagonistic therapies can therefore be envisioned as novel analgesics, possibly for specific targeting of trigeminal pain disorders, including headaches. TRPV4 therefore appears an appealing target molecule for treatment or prevention of headaches and other trigeminal pain disorders. Clearly, there is a mandate for future studies to generate more mechanistic insights.

Despite numerous interesting initial leads on the role of TRPV4 in pain, future studies are needed to address many important questions, a few of which are (1) how sensory-neuronal TRPV4 contributes to pain, particularly pathological pain, directly addressable via appropriate targeting strategies; and (2) the possible co-contributing role of glial cells expressing TRPV4 (Benfenati et al., 2007), co-signaling with other pain-TRP channels, in particular TRPA1 given the observed overlap in TRPV4–TRPA1 function in several important and relevant forms of pain, such as visceral pain of the colon and pancreas, chemotherapy-induced peripheral neuropathy pain (especially that mediated by taxane), headaches, formalin-evoked pain as an important model, and mechanically evoked pain in general (Kanju et al., 2016; Brierley et al., 2008; Cenac et al., 2008; Sipe et al., 2008; Mueller-Tribbensee et al., 2015; Cattaruzza et al., 2010; Cenac et al., 2015; Chen et al., 2011, 2014; McNamara et al., 2007; Nassini et al., 2012).

8.5 CONCLUDING REMARKS

Nociception is an important physiological process detecting harmful signals resulting in pain perception. In this chapter, we reviewed experimental evidence involving some TRP ion channels as molecular sensors of chemical, thermal, and mechanical noxious stimuli to produce pain sensation. Among them are the TRPA1 channel, members of the vanilloid subfamily (TRPV1, TRPV2, TRPV3, and TRPV4), and some members of the melastatin group (TRPM2, TRPM3, and TRPM8).

In recent years, remarkable advances have been made to elucidate how noxious signals from the periphery are transmitted to the central nervous system, and within these studies the importance of some TRP channels has been highlighted.

In view that pain perception can be a protective mechanism, these ion channels play an important role in the organism's survival. Furthermore, some pathologies are accompanied with chronic pain which severely impacts the quality of life of the people who suffer from it, and some TRP channels display an increased function suggesting that they are important portals to pain. Several studies have helped to understand the importance of these channels in producing pain and have promoted the search for novel compounds that downregulate the activity of these proteins to develop therapies more effective toward the relief of pain.

These channels are not only important in converting painful stimuli into an electrical signal, there is also evidence that each subtype of TRP channel fulfills a specific function in the nociceptor due to the activation of several signaling pathways. All of these events help to facilitate channel awareness and maintain pain.

Finally, the research and development of selective blockers of TRP channel activation being targeted by intensive and future research should be directed at dissecting the relationship between the structural function and the signaling pathways involved with each TRP channel subtype to generate the best pharmaceutical alternative to attenuate chronic pain sensation pathologies.

ACKNOWLEDGMENTS

This work was supported by a grant from Dirección General de Asuntos del Personal Académico (DGAPA)-Programa de Apoyo a Proyectos de Investigación e Innovación Tecnológica (PAPIIT) IA202717 to S.L.M.L. This work was supported, by grants from the U.S. National Institutes of Health DE018549, AR48182, AR48182-S1 to W.B.L., F33DE024668 to Y.C., mentor W.B.L, K12DE022793 to Y.C., U.S. Department of Defense W81XWH-13-1-0299 to W.B.L., and a Harrington Discovery Institute (Cleveland, OH) Scholar-Innovator Award to W.B.L. Helpful comments by Dr. Jon Levine (UCSF, CA) were appreciated for Figure 8.5.

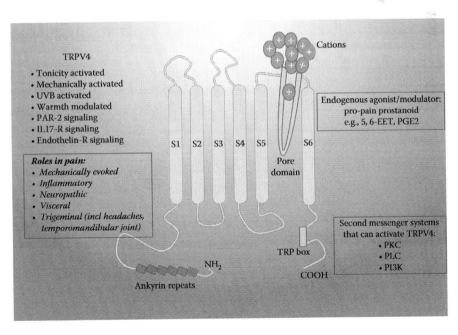

FIGURE 8.5 A TRPV4 monomer showing the S1–S6 transmembrane segments, roles in pain and stimuli, and second messengers that regulate the activity of this channel.

AUTHOR CONTRIBUTIONS

R.G.R. and S.L.M.L. Both authors contributed equally to this work. Y. C. and W. B. L. contributed to write the TRPV4 section.

REFERENCES

Alessandri-Haber, N. et al. 2003. Hypotonicity induces TRPV4-mediated nociception in rat. *Neuron*, 39(3): 497–511.

Alessandri-Haber, N. et al. 2004. Transient receptor potential vanilloid 4 is essential in chemotherapy-induced neuropathic pain in the rat. *J Neurosci*, 24(18): 4444–4452. doi:10.1523/JNEUROSCI.0242-04.2004.

Alessandri-Haber, N. et al. 2005. TRPV4 mediates pain-related behavior induced by mild hypertonic stimuli in the presence of inflammatory mediator. *Pain*, 118(1–2): 70–79. doi:10.1016/j.pain.2005.07.016.

Alessandri-Haber, N. et al. 2006. A transient receptor potential vanilloid 4-dependent mechanism of hyper-algesia is engaged by concerted action of inflammatory mediators. *J Neurosci*, 26(14): 3864–3874. doi:10.1523/JNEUROSCI.5385-05.2006.

Alessandri-Haber, N. et al. 2009. TRPC1 and TRPC6 channels cooperate with TRPV4 to mediate mechanical hyperalgesia and nociceptor sensitization. *J Neurosci*, 29(19): 6217–6228. doi:10.1523/JNEUROSCI.0893-09.2009.

Allchorne, A.J., D.C. Broom, and C.J. Woolf. 2005. Detection of cold pain, cold allodynia and cold hyperalge-sia in freely behaving rats. *Mol Pain*, 1: 36. doi:10.1186/1744-8069-1-36.

Andersson, D.A., M. Nash, and S. Bevan. 2007. Modulation of the cold-activated channel TRPM8 by lysophospholipids and polyunsaturated fatty acids. *J Neurosci*, 27(12): 3347–3355. doi:10.1523/JNEUROSCI.4846-06.2007.

Bandell, M. et al. 2004. Noxious cold ion channel TRPA1 is activated by pungent compounds and bradykinin. *Neuron*, 41(6): 849–857.

Bang, S. et al. 2007. Transient receptor potential V2 expressed in sensory neurons is activated by probenecid. *Neurosci Lett*, 425(2): 120–125. doi:10.1016/j.neulet.2007.08.035.

Bang, S. et al. 2010. Farnesyl pyrophosphate is a novel pain-producing molecule via specific activation of TRPV3. *J Biol Chem*, 285(25): 19362–19371. doi:10.1074/jbc.M109.087742.

Bang, S. et al. 2012. 17(R)-resolvin D1 specifically inhibits transient receptor potential ion channel vanil-loid 3 leading to peripheral antinociception. *Br J Pharmacol*, 165(3): 683–692. doi:10.1111/j.1476-5381.2011.01568.x.

Bautista, D.M. et al. 2005. Pungent products from garlic activate the sensory ion channel TRPA1. *Proc Natl Acad Sci U S A*, 102(34): 12248–12252. doi:10.1073/pnas.0505356102.

Bautista, D.M. et al. 2006. TRPA1 mediates the inflammatory actions of environmental irritants and proalgesic agents. *Cell*, 124(6): 1269–1282. doi:10.1016/j.cell.2006.02.023.

Bautista, D.M. et al. 2007. The menthol receptor TRPM8 is the principal detector of environmental cold. *Nature*, 448(7150): 204–208. doi:10.1038/nature05910.

Benfenati, V. et al. 2007. Expression and functional characterization of transient receptor potential vanilloid-related channel 4 (TRPV4) in rat cortical astrocytes. *Neuroscience*, 148(4): 876–892. doi:10.1016/j.neuro-science.2007.06.039.

Beyak, M.J. and S. Vanner. 2005. Inflammation-induced hyperexcitability of nociceptive gastrointestinal DRG neurones: The role of voltage-gated ion channels. *Neurogastroenterol Motil*, 17(2): 175–186.

Boadas-Vaello, P. et al. 2016. Neuroplasticity of ascending and descending pathways after somatosensory sys-tem injury: Reviewing knowledge to identify neuropathic pain therapeutic targets. *Spinal Cord*, 54(5): 330–340. doi:10.1038/sc.2015.225.

Bolcskei, K. et al. 2005. Investigation of the role of TRPV1 receptors in acute and chronic nociceptive pro-cesses using gene-deficient mice. *Pain*, 117(3): 368–376. doi:10.1016/j.pain.2005.06.024.

Brash, A.R. 2001. Arachidonic acid as a bioactive molecule. *J Clin Invest*, 107(11): 1339–1345. doi:10.1172/JCI13210.

Brauchi, S. et al. 2007. Dissection of the components for PIP2 activation and thermosensation in TRP channels. *Proc Natl Acad Sci U S A*, 104(24): 10246–10251. doi:10.1073/pnas.0703420104.

Brierley, S.M. et al. 2008. Selective role for TRPV4 ion channels in visceral sensory pathways. *Gastroenterology*, 134(7): 2059–2069. doi:10.1053/j.gastro.2008.01.074.

Caterina, M.J. et al. 1997. The capsaicin receptor: A heat-activated ion channel in the pain pathway. *Nature*, 389(6653): 816–824. doi:10.1038/39807.

Caterina, M.J. et al. 1999. A capsaicin-receptor homologue with a high threshold for noxious heat. *Nature*, 398(6726): 436–441. doi:10.1038/18906.

Caterina, M.J. et al. 2000. Impaired nociception and pain sensation in mice lacking the capsaicin receptor. *Science*, 288(5464): 306–313.

Cattaruzza, F. et al. 2010. Transient receptor potential ankyrin-1 has a major role in mediating visceral pain in mice. *Am J Physiol Gastrointest Liver Physiol*, 298(1): G81–G91. doi:10.1152/ajpgi.00221.2009.

Cavanaugh, D.J. et al. 2009. Distinct subsets of unmyelinated primary sensory fibers mediate behavioral responses to noxious thermal and mechanical stimuli. *Proc Natl Acad Sci U S A*, 106(22): 9075–9080. doi:10.1073/pnas.0901507106.

Cavanaugh, D.J. et al. 2011. Trpv1 reporter mice reveal highly restricted brain distribution and functional expression in arteriolar smooth muscle cells. *J Neurosci*, 31(13): 5067–5077. doi:10.1523/jneurosci.6451-10.2011.

Cenac, N. et al. 2008. Transient receptor potential vanilloid-4 has a major role in visceral hypersensitivity symptoms. *Gastroenterology*, 135(3): 937–946.e2. doi:10.1053/j.gastro.2008.05.024.

Cenac, N. et al. 2015. Quantification and potential functions of endogenous agonists of transient receptor potential channels in patients with irritable bowel syndrome. *Gastroenterology*, 149(2): 433–444.e7. doi:10.1053/j.gastro.2015.04.011.

Ceppa, E. et al. 2010. Transient receptor potential ion channels V4 and A1 contribute to pancreatitis pain in mice. *Am J Physiol Gastrointest Liver Physiol*, 299(3): G556–G571. doi:10.1152/ajpgi.00433.2009.

Chen, Y. et al. 2013. Temporomandibular joint pain: A critical role for Trpv4 in the trigeminal ganglion. *Pain*, 154(8): 1295–1304. doi:10.1016/j.pain.2013.04.004.

Chen, Y. et al. 2014. TRPV4 is necessary for trigeminal irritant pain and functions as a cellular formalin receptor. *Pain*, 155(12): 2662–2672. doi:10.1016/j.pain.2014.09.033.

Chen, Y., C. Yang, and Z.J. Wang. 2011. Proteinase-activated receptor 2 sensitizes transient receptor potential vanilloid 1, transient receptor potential vanilloid 4, and transient receptor potential ankyrin 1 in paclitaxel-induced neuropathic pain. *Neuroscience*, 193: 440–451. doi:10.1016/j.neuroscience.2011.06.085.

Cheng, J.K., and R.R. Ji. 2008. Intracellular signaling in primary sensory neurons and persistent pain. *Neurochem Res*, 33(10): 1970–1978. doi:10.1007/s11064-008-9711-z.

Chung, M.K. et al. 2015. The role of TRPM2 in hydrogen peroxide-induced expression of inflammatory cytokine and chemokine in rat trigeminal ganglia. *Neuroscience*, 297:160–169. doi:10.1016/j.neuroscience.2015.03.067.

Colburn, R.W. et al. 2007. Attenuated cold sensitivity in TRPM8 null mice. *Neuron*, 54(3): 379–386. doi:10.1016/j.neuron.2007.04.017.

Cui, Y.Y. et al. 2014. Spatio-temporal expression and functional involvement of transient receptor potential vanilloid 1 in diabetic mechanical allodynia in rats. *PLOS ONE*, 9(7): e102052. doi:10.1371/journal.pone.0102052.

D'Aldebert, E. et al. 2011. Transient receptor potential vanilloid 4 activated inflammatory signals by intestinal epithelial cells and colitis in mice. *Gastroenterology*, 140(1): 275–285. doi:10.1053/j.gastro.2010.09.045.

Davis, J.B. et al. 2000. Vanilloid receptor-1 is essential for inflammatory thermal hyperalgesia. *Nature*, 405(6783): 183–187. doi:10.1038/35012076.

De Logu, F. et al. 2016. TRP functions in the broncho-pulmonary system. *Semin Immunopathol*, 38(3): 321–329. doi:10.1007/s00281-016-0557-1.

Denadai-Souza, A. et al. 2012. Role of transient receptor potential vanilloid 4 in rat joint inflammation. *Arthritis Rheum*, 64(6): 1848–1858. doi:10.1002/art.34345.

Dhaka, A. et al. 2007. TRPM8 is required for cold sensation in mice. *Neuron*, 54(3): 371–378. doi:10.1016/j.neuron.2007.02.024.

Ding, X.L. et al. 2010. Involvement of TRPV4-NO-cGMP-PKG pathways in the development of thermal hyperalgesia following chronic compression of the dorsal root ganglion in rats. *Behav Brain Res*, 208(1): 194–201. doi:10.1016/j.bbr.2009.11.034.

Dinh, Q.T. et al. 2004. Substance P expression in TRPV1 and trkA-positive dorsal root ganglion neurons innervating the mouse lung. *Respir Physiol Neurobiol*, 144(1): 15–24. doi:10.1016/j.resp.2004.08.001.

Dubin, A.E., and A. Patapoutian. 2010. Nociceptors: The sensors of the pain pathway. *J Clin Invest*, 120(11): 3760–3772. doi:10.1172/jci42843.

Frederick, J. et al. 2007. Increased TRPA1, TRPM8, and TRPV2 expression in dorsal root ganglia by nerve injury. *Biochem Biophys Res Commun*, 358(4): 1058–1064. doi:10.1016/j.bbrc.2007.05.029.

Frias, B., and A. Merighi. 2016. Capsaicin, nociception and pain. *Molecules*, 21(6): 797. doi:10.3390/molecules21060797.

Fukuoka, T. et al. 2002. VR1, but not P2X(3), increases in the spared L4 DRG in rats with L5 spinal nerve ligation. *Pain*, 99(1–2): 111–120.

Gentry, C. et al. 2010. The roles of iPLA2, TRPM8 and TRPA1 in chemically induced cold hypersensitivity. *Mol Pain*, 6: 4. doi:10.1186/1744-8069-6-4.

Grant, A.D. et al. 2007. Protease-activated receptor 2 sensitizes the transient receptor potential vanilloid 4 ion channel to cause mechanical hyperalgesia in mice. *J Physiol*, 578(Pt 3): 715–733. doi:10.1113/jphysiol.2006.121111.

Haraguchi, K. et al. 2012. TRPM2 contributes to inflammatory and neuropathic pain through the aggravation of pronociceptive inflammatory responses in mice. *J Neurosci*, 32(11): 3931–3941. doi:10.1523/JNEUROSCI.4703-11.2012.

Held, K. et al. 2015. Activation of TRPM3 by a potent synthetic ligand reveals a role in peptide release. *Proc Natl Acad Sci U S A*, 112(11): E1363–E1372. doi:10.1073/pnas.1419845112.

Hicks, G.A. 2006. TRP channels as therapeutic targets: Hot property, or time to cool down? *Neurogastroenterol Motil*, 18(8): 590–594. doi:10.1111/j.1365-2982.2006.00823.x.

Hinman, A. et al. 2006. TRP channel activation by reversible covalent modification. *Proc Natl Acad Sci U S A*, 103(51): 19564–19568. doi:10.1073/pnas.0609598103.

Holstein, S.A. and R.J. Hohl. 2004. Isoprenoids: Remarkable diversity of form and function. *Lipids*, 39(4): 293–309.

Hong, S. and J.W. Wiley. 2005. Early painful diabetic neuropathy is associated with differential changes in the expression and function of vanilloid receptor 1. *J Biol Chem*, 280(1): 618–627. doi:10.1074/jbc.M408500200.

Hu, H.Z. et al. 2006. Potentiation of TRPV3 channel function by unsaturated fatty acids. *J Cell Physiol*, 208(1): 201–212. doi:10.1002/jcp.20648.

Huang, D. et al. 2012. Expression of the transient receptor potential channels TRPV1, TRPA1 and TRPM8 in mouse trigeminal primary afferent neurons innervating the dura. *Mol Pain*, 8: 66. doi:10.1186/1744-8069-8-66.

Huang, S.M. et al. 2008. Overexpressed transient receptor potential vanilloid 3 ion channels in skin keratinocytes modulate pain sensitivity via prostaglandin E2. *J Neurosci*, 28(51): 13727–13737. doi:10.1523/JNEUROSCI.5741-07.2008.

Hwang, S.J., J.M. Oh, and J.G. Valtschanoff. 2005. Expression of the vanilloid receptor TRPV1 in rat dorsal root ganglion neurons supports different roles of the receptor in visceral and cutaneous afferents. *Brain Res*, 1047(2): 261–266. doi:10.1016/j.brainres.2005.04.036.

Ji, R.R. et al. 2002. p38 MAPK activation by NGF in primary sensory neurons after inflammation increases TRPV1 levels and maintains heat hyperalgesia. *Neuron*, 36(1): 57–68.

Jin, X. et al. 2004. Modulation of TRPV1 by nonreceptor tyrosine kinase, c-Src kinase. *Am J Physiol Cell Physiol*, 287(2): C558–C563. doi:10.1152/ajpcell.00113.2004.

Jones, R.C., 3rd, L. Xu, and G.F. Gebhart. 2005. The mechanosensitivity of mouse colon afferent fibers and their sensitization by inflammatory mediators require transient receptor potential vanilloid 1 and acid-sensing ion channel 3. *J Neurosci*, 25(47): 10981–10989. doi:10.1523/jneurosci.0703-05.2005.

Jordt, S.E. et al. 2004. Mustard oils and cannabinoids excite sensory nerve fibres through the TRP channel ANKTM1. *Nature*, 427(6971): 260–265. doi:10.1038/nature02282.

Julius, D. 2013. TRP channels and pain. *Annu Rev Cell Dev Biol*, 29: 355–384. doi:10.1146/annurev-cellbio-101011-155833.

Kanai, Y. et al. 2005. Involvement of an increased spinal TRPV1 sensitization through its up-regulation in mechanical allodynia of CCI rats. *Neuropharmacology*, 49(7): 977–984. doi:10.1016/j.neuropharm.2005.05.003.

Kanju, P. et al. 2016. Small molecule dual-inhibitors of TRPV4 and TRPA1 for attenuation of inflammation and pain. *Sci Rep*, 6: 26894. doi:10.1038/srep26894.

Kochukov, M.Y. et al. 2006. Thermosensitive TRP ion channels mediate cytosolic calcium response in human synoviocytes. *Am J Physiol Cell Physiol*, 291(3): C424–C432. doi:10.1152/ajpcell.00553.2005.

Konno, M. et al. 2012. Stimulation of transient receptor potential vanilloid 4 channel suppresses abnormal activation of microglia induced by lipopolysaccharide. *Glia*, 60(5): 761–770. doi:10.1002/glia.22306.

Kraft, R. et al. 2004. Hydrogen peroxide and ADP-ribose induce TRPM2-mediated calcium influx and cation currents in microglia. *Am J Physiol Cell Physiol*, 286(1): C129–C137. doi:10.1152/ajpcell.00331.2003.

Kremeyer, B. et al. 2010. A gain-of-function mutation in TRPA1 causes familial episodic pain syndrome. *Neuron*, 66(5): 671–680. doi:10.1016/j.neuron.2010.04.030.

Kwan, K.Y. et al. 2006. TRPA1 contributes to cold, mechanical, and chemical nociception but is not essential for hair-cell transduction. *Neuron*, 50(2): 277–289. doi:10.1016/j.neuron.2006.03.042.

Kwon, M. et al. 2014. Single-cell RT-PCR and immunocytochemical detection of mechanosensitive transient receptor potential channels in acutely isolated rat odontoblasts. *Arch Oral Biol*, 59(12): 1266–1271. doi:10.1016/j.archoralbio.2014.07.016.

Lehmann, R. et al. 2016. The involvement of TRP channels in sensory irritation: A mechanistic approach toward a better understanding of the biological effects of local irritants. *Arch Toxicol*, 90(6): 1399–1413. doi:10.1007/s00204-016-1703-1.

Li, J. et al. 2011. TRPV4-mediated calcium influx into human bronchial epithelia upon exposure to diesel exhaust particles. *Environ Health Perspect*, 119(6): 784–793. doi:10.1289/ehp.1002807.

Liedtke, W. and J.M. Friedman. 2003. Abnormal osmotic regulation in trpv4$^{-/-}$ mice. *Proc Natl Acad Sci U S A*, 100(23): 13698–13703. doi:10.1073/pnas.1735416100.

Liedtke, W. et al. 2000. Vanilloid receptor-related osmotically activated channel (VR-OAC), a candidate vertebrate osmoreceptor. *Cell*, 103(3): 525–535.

Lindy, A.S. et al. 2014. TRPV channel-mediated calcium transients in nociceptor neurons are dispensable for avoidance behaviour. *Nat Commun*, 5: 4734. doi:10.1038/ncomms5734.

Lippoldt, E.K. et al. 2013. Artemin, a glial cell line-derived neurotrophic factor family member, induces TRPM8-dependent cold pain. *J Neurosci*, 33(30): 12543–12552. doi:10.1523/JNEUROSCI.5765-12.2013.

Liu, B.L. et al. 2014. Increased severity of inflammation correlates with elevated expression of TRPV1 nerve fibers and nerve growth factor on interstitial cystitis/bladder pain syndrome. *Urol Int*, 92(2): 202–208. doi:10.1159/000355175.

Macpherson, L.J. et al. 2005. The pungency of garlic: Activation of TRPA1 and TRPV1 in response to allicin. *Curr Biol*, 15(10): 929–934. doi:10.1016/j.cub.2005.04.018.

Macpherson, L.J. et al. 2007. Noxious compounds activate TRPA1 ion channels through covalent modification of cysteines. *Nature*, 445(7127): 541–545. doi:10.1038/nature05544.

Makimura, Y. et al. 2012. Augmented activity of the pelvic nerve afferent mediated by TRP channels in dextran sulfate sodium (DSS)-induced colitis of rats. *J Vet Med Sci*, 74(8): 1007–1013.

Matsumoto, K. et al. 2016. Role of transient receptor potential melastatin 2 (TRPM2) channels in visceral nociception and hypersensitivity. *Exp Neurol*, 285(Pt A): 41–50. doi:10.1016/j.expneurol.2016.09.001.

McNamara, C.R. et al. 2007. TRPA1 mediates formalin-induced pain. *Proc Natl Acad Sci U S A*, 104(33): 13525–13530. doi:10.1073/pnas.0705924104.

Mercado, J. et al. 2010. Ca^{2+}-dependent desensitization of TRPV2 channels is mediated by hydrolysis of phosphatidylinositol 4,5-bisphosphate. *J Neurosci*, 30(40): 13338–13347. doi:10.1523/JNEUROSCI.2108-10.2010.

Mohapatra, D.P. and C. Nau. 2005. Regulation of Ca^{2+}-dependent desensitization in the vanilloid receptor TRPV1 by calcineurin and cAMP-dependent protein kinase. *J Biol Chem*, 280(14): 13424–13432. doi:10.1074/jbc.M410917200.

Moore, C. et al. 2013. UVB radiation generates sunburn pain and affects skin by activating epidermal TRPV4 ion channels and triggering endothelin-1 signaling. *Proc Natl Acad Sci U S A*, 110(34): E3225–E3234. doi:10.1073/pnas.1312933110.

Moqrich, A. et al. 2005. Impaired thermosensation in mice lacking TRPV3, a heat and camphor sensor in the skin. *Science*, 307(5714): 1468–1472. doi:10.1126/science.1108609.

Morales-Lazaro, S.L., S.A. Simon, and T. Rosenbaum. 2013. The role of endogenous molecules in modulating pain through transient receptor potential vanilloid 1 (TRPV1). *J Physiol*, 591(13): 3109–3121. doi:10.1113/jphysiol.2013.251751.

Morales-Lazaro, S.L. et al. 2016. Inhibition of TRPV1 channels by a naturally occurring omega-9 fatty acid reduces pain and itch. *Nat Commun*, 7:13092. doi:10.1038/ncomms13092.

Moriyama, T. et al. 2005. Sensitization of TRPV1 by EP1 and IP reveals peripheral nociceptive mechanism of prostaglandins. *Mol Pain*, 1: 3. doi:10.1186/1744-8069-1-3.

Mueller-Tribbensee, S.M. et al. 2015. Differential contribution of TRPA1, TRPV4 and TRPM8 to colonic nociception in mice. *PLOS ONE*, 10(7): e0128242. doi:10.1371/journal.pone.0128242.

Nassini, R. et al. 2012. The "headache tree" via umbellulone and TRPA1 activates the trigeminovascular system. *Brain*, 135(Pt 2): 376–390. doi:10.1093/brain/awr272.

Nieto-Posadas, A. et al. 2011. Lysophosphatidic acid directly activates TRPV1 through a C-terminal binding site. *Nat Chem Biol*, 8(1): 78–85. doi:10.1038/nchembio.712.

Numazaki, M. et al. 2003. Structural determinant of TRPV1 desensitization interacts with calmodulin. *Proc Natl Acad Sci U S A*, 100(13): 8002–8006. doi:10.1073/pnas.1337252100.

O'Conor, C.J. et al. 2016. Cartilage-specific knockout of the mechanosensory ion channel TRPV4 decreases age-related osteoarthritis. *Sci Rep*, 6:29053. doi:10.1038/srep29053.

Owsianik, G. et al. 2006a. Structure-function relationship of the TRP channel superfamily. *Rev Physiol Biochem Pharmacol*, 156: 61–90.

Owsianik, G. et al. 2006b. Permeation and selectivity of TRP channels. *Annu Rev Physiol*, 68: 685–717. doi:10.1146/annurev.physiol.68.040204.101406.

Pabbidi, R.M. et al. 2008. Influence of TRPV1 on diabetes-induced alterations in thermal pain sensitivity. *Mol Pain*, 4: 9. doi:10.1186/1744-8069-4-9.

Pan, H.L., Y.Q. Zhang, and Z.Q. Zhao. 2010. Involvement of lysophosphatidic acid in bone cancer pain by potentiation of TRPV1 via PKCepsilon pathway in dorsal root ganglion neurons. *Mol Pain*, 6: 85. doi:10.1186/1744-8069-6-85.

Park, U. et al. 2011. TRP vanilloid 2 knock-out mice are susceptible to perinatal lethality but display normal thermal and mechanical nociception. *J Neurosci*, 31(32): 11425–11436.

Paulsen, C.E. et al. 2015. Structure of the TRPA1 ion channel suggests regulatory mechanisms. *Nature*, 520(7548): 511–517. doi:10.1038/nature14367.

Peier, A.M. 2002. A TRP channel that senses cold stimuli and menthol. *Cell*, 108: 705–715.

Peier, A.M. et al. 2002. A heat-sensitive TRP channel expressed in keratinocytes. *Science*, 296(5575): 2046–2049. doi:10.1126/science.1073140.

Perraud, A.L. et al. 2001. ADP-ribose gating of the calcium-permeable LTRPC2 channel revealed by Nudix motif homology. *Nature*, 411(6837): 595–599. doi:10.1038/35079100.

Phan, M.N. et al. 2009. Functional characterization of TRPV4 as an osmotically sensitive ion channel in porcine articular chondrocytes. *Arthritis Rheum*, 60(10): 3028–3037. doi:10.1002/art.24799.

Premkumar, L.S. et al. 2005. Downregulation of transient receptor potential melastatin 8 by protein kinase C-mediated dephosphorylation. *J Neurosci*, 25(49): 11322–11329. doi:10.1523/JNEUROSCI.3006-05.2005.

Qin, N. et al. 2008. TRPV2 is activated by cannabidiol and mediates CGRP release in cultured rat dorsal root ganglion neurons. *J Neurosci*, 28(24): 6231–6238. doi:10.1523/JNEUROSCI.0504-08.2008.

Qu, Y.J. et al. 2016. Effect of TRPV4-p38 MAPK pathway on neuropathic pain in rats with chronic compression of the dorsal root ganglion. *Biomed Res Int*, 2016: 6978923. doi:10.1155/2016/6978923.

Rajasekhar, P. et al. 2015. P2Y1 receptor activation of the TRPV4 ion channel enhances purinergic signaling in satellite glial cells. *J Biol Chem*, 290(48): 29051–29062. doi:10.1074/jbc.M115.689729.

Reeh, P.W., L. Kocher, and S. Jung. 1986. Does neurogenic inflammation alter the sensitivity of unmyelinated nociceptors in the rat? *Brain Res*, 384: 42–50.

Rohacs, T. et al. 2005. PI(4,5)P2 regulates the activation and desensitization of TRPM8 channels through the TRP domain. *Nat Neurosci*, 8(5): 626–634. doi:10.1038/nn1451.

Salazar, H. et al. 2008. A single N-terminal cysteine in TRPV1 determines activation by pungent compounds from onion and garlic. *Nat Neurosci*, 11(3): 255–261. doi:10.1038/nn2056.

Sanz-Salvador, L. et al. 2012. Agonist- and Ca^{2+}-dependent desensitization of TRPV1 channel targets the receptor to lysosomes for degradation. *J Biol Chem*, 287(23): 19462–19471. doi:10.1074/jbc.M111.289751.

Shibasaki, K. et al. 2007. Effects of body temperature on neural activity in the hippocampus: Regulation of resting membrane potentials by transient receptor potential vanilloid 4. *J Neurosci*, 27(7): 1566–1575. doi:10.1523/JNEUROSCI.4284-06.2007.

Shibasaki, K. et al. 2015. TRPV4 activation at the physiological temperature is a critical determinant of neuronal excitability and behavior. *Pflugers Arch*, 467(12): 2495–2507. doi:10.1007/s00424-015-1726-0.

Shimosato, G. et al. 2005. Peripheral inflammation induces up-regulation of TRPV2 expression in rat DRG. *Pain*, 119(1–3): 225–232. doi:10.1016/j.pain.2005.10.002.

Sipe, W.E. et al. 2008. Transient receptor potential vanilloid 4 mediates protease activated receptor 2-induced sensitization of colonic afferent nerves and visceral hyperalgesia. *Am J Physiol Gastrointest Liver Physiol*, 294(5): G1288–G1298. doi:10.1152/ajpgi.00002.2008.

So, K. et al. 2015. Involvement of TRPM2 in a wide range of inflammatory and neuropathic pain mouse models. *J Pharmacol Sci*, 127(3): 237–243. doi:10.1016/j.jphs.2014.10.003.

Song, K. et al. 2016. The TRPM2 channel is a hypothalamic heat sensor that limits fever and can drive hypothermia. *Science*, 353(6306): 1393–1398. doi:10.1126/science.aaf7537.

Spicarova, D. and J. Palecek. 2008. The role of spinal cord vanilloid (TRPV1) receptors in pain modulation. *Physiol Res*, 57(Suppl 3): S69–S77.

Spicarova, D. and J. Palecek. 2009. The role of the TRPV1 endogenous agonist N-oleoyldopamine in modulation of nociceptive signaling at the spinal cord level. *J Neurophysiol*, 102(1): 234–243. doi:10.1152/jn.00024.2009.

Story, G.M. et al. 2003. ANKTM1, a TRP-like channel expressed in nociceptive neurons, is activated by cold temperatures. *Cell*, 112(6): 819–829.

Suzuki, M. et al. 2003. Impaired pressure sensation in mice lacking TRPV4. *J Biol Chem*, 278(25): 22664–22668. doi:10.1074/jbc.M302561200.

Tan, C.H. and P.A. McNaughton. 2016. The TRPM2 ion channel is required for sensitivity to warmth. *Nature*, 536(7617): 460–463. doi:10.1038/nature19074.

Tominaga, M. et al. 1998. The cloned capsaicin receptor integrates multiple pain-producing stimuli. *Neuron*, 21(3): 531–543.

Trevisani, M. et al. 2007. 4-Hydroxynonenal, an endogenous aldehyde, causes pain and neurogenic inflammation through activation of the irritant receptor TRPA1. *Proc Natl Acad Sci U S A*, 104(33): 13519–13524. doi:10.1073/pnas.0705923104.

Valente, P. et al. 2008. Identification of molecular determinants of channel gating in the transient receptor potential box of vanilloid receptor I. *FASEB J*, 22(9): 3298–3309. doi:10.1096/fj.08-107425.

Vandewauw, I., G. Owsianik, and T. Voets. 2013. Systematic and quantitative mRNA expression analysis of TRP channel genes at the single trigeminal and dorsal root ganglion level in mouse. *BMC Neurosci*, 14: 21. doi:10.1186/1471-2202-14-21.

Vass, Z. et al. 2004. Co-localization of the vanilloid capsaicin receptor and substance P in sensory nerve fibers innervating cochlear and vertebro-basilar arteries. *Neuroscience*, 124(4): 919–927. doi:10.1016/j.neuroscience.2003.12.030.

Vogt-Eisele, A.K. et al. 2007. Monoterpenoid agonists of TRPV3. *Br J Pharmacol*, 151(4): 530–540. doi:10.1038/sj.bjp.0707245.

Vriens, J. et al. 2011. TRPM3 is a nociceptor channel involved in the detection of noxious heat. *Neuron*, 70(3): 482–494. doi:10.1016/j.neuron.2011.02.051.

Wagner, T.F. et al. 2008. Transient receptor potential M3 channels are ionotropic steroid receptors in pancreatic beta cells. *Nat Cell Biol*, 10(12): 1421–1430. doi:10.1038/ncb1801.

Wang, C. et al. 2011. Nuclear factor-kappa B mediates TRPV4-NO pathway involved in thermal hyperalgesia following chronic compression of the dorsal root ganglion in rats. *Behav Brain Res*, 221(1): 19–24. doi:10.1016/j.bbr.2011.02.028.

Wei, X. et al. 2011. Activation of TRPV4 on dural afferents produces headache-related behavior in a preclinical rat model. *Cephalalgia*, 31(16): 1595–1600. doi:10.1177/0333102411427600.

Woo, D.H. et al. 2008. Direct activation of transient receptor potential vanilloid 1(TRPV1) by diacylglycerol (DAG). *Mol Pain*, 4: 42. doi:10.1186/1744-8069-4-42.

Wu, L.J., T.B. Sweet, and D.E. Clapham. 2010. International Union of Basic and Clinical Pharmacology. LXXVI. Current progress in the mammalian TRP ion channel family. *Pharmacol Rev*, 62(3): 381–404. doi:10.1124/pr.110.002725.

Xing, H. et al. 2007. TRPM8 mechanism of cold allodynia after chronic nerve injury. *J Neurosci*, 27(50): 13680–13690. doi:10.1523/JNEUROSCI.2203-07.2007.

Xu, H. et al. 2002. TRPV3 is a calcium-permeable temperature-sensitive cation channel. *Nature*, 418(6894): 181–186. doi:10.1038/nature00882.

Yamamoto, S. et al. 2008a. TRPM2-mediated Ca^{2+} influx induces chemokine production in monocytes that aggravates inflammatory neutrophil infiltration. *Nat Med*, 14(7): 738–747. doi:10.1038/nm1758.

Yamamoto, W. et al. 2008b. Characterization of primary sensory neurons mediating static and dynamic allodynia in rat chronic constriction injury model. *J Pharm Pharmacol*, 60(6): 717–722. doi:10.1211/jpp.60.6.0006.

Zhang, L.P. et al. 2013. Prolonged high fat/alcohol exposure increases TRPV4 and its functional responses in pancreatic stellate cells. *Am J Physiol Regul Integr Comp Physiol*, 304(9): R702–R711. doi:10.1152/ajpregu.00296.2012.

Zhang, L.P. et al. 2015. Alcohol and high fat induced chronic pancreatitis: TRPV4 antagonist reduces hypersensitivity. *Neuroscience*, 311:166–179. doi:10.1016/j.neuroscience.2015.10.028.

Zhang, X., J. Huang, and P.A. McNaughton. 2005. NGF rapidly increases membrane expression of TRPV1 heat-gated ion channels. *Embo J*, 24(24): 4211–4423. doi:10.1038/sj.emboj.7600893.

Zhao, P. et al. 2014. Cathepsin S causes inflammatory pain via biased agonism of PAR2 and TRPV4. *J Biol Chem*, 289(39): 27215–27234. doi:10.1074/jbc.M114.599712.

Zhao, P. et al. 2015. Neutrophil elastase activates protease-activated receptor-2 (PAR2) and transient receptor potential vanilloid 4 (TRPV4) to cause inflammation and pain. *J Biol Chem*, 290(22): 13875–13887. doi:10.1074/jbc.M115.642736.

Zhao, Q. et al. 2016. TRPV1 and neuropeptide receptor immunoreactivity and expression in the rat lung and brainstem after lung ischemia-reperfusion injury. *J Surg Res*, 203(1): 183–192. doi:10.1016/j.jss.2016.03.050.

Zima, V. et al. 2015. Structural modeling and patch-clamp analysis of pain-related mutation TRPA1-N855S reveal inter-subunit salt bridges stabilizing the channel open state. *Neuropharmacology*, 93: 294–307. doi:10.1016/j.neuropharm.2015.02.018.

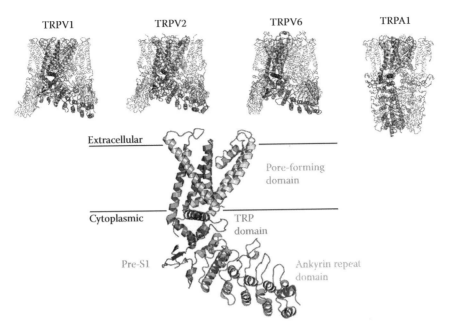

FIGURE 1.2 Structural arrangement of TRP channels whose structures have recently been elucidated. In the top row are structures of the indicated TRP channels, with one subunit shown in color. Below is a zoomed-in view of one subunit from TRPV1 to highlight the structure of the voltage sensor-like domain (blue), pore domain (yellow), the pre-S1 domain (pink), and the ankyrin repeat domain (green) for those channels.

rKv1.2 290 - LAILRVIRLVRVFRIFKLSRH - 310
rTRPV1 536 - EYVASMVFSLAMGWTNMLYYT - 556
rTRPM8 832 - FCLDYIIFTLRLIHIFTVSRN - 852

FIGURE 1.3 Sequence alignment of S4 membrane-spanning helices, with positive "gating charge" residues indicated in red. Histidine is colored pink to indicate its ability to be protonated and thus carry charge at near-physiological pH. The alignment shows that no charges are present in TRPV1, whereas TRPM8 retains at least some of the R-X-X-R motifs.

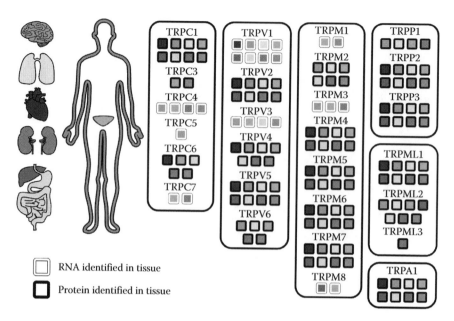

FIGURE 1.6 Expression profiles of TRP channels in humans. Colored boxes under each TRP channel name correspond to the colors of various organs on the left, indicating in which tissues each channel has been identified. Evidence for expression at the RNA level is indicated by a box with a white border, and evidence for expression at the protein level is indicated by a box with a black border. Tissue systems corresponding to each color are: orange (brain/nervous system), yellow (pulmonary system), red (cardiovascular system), purple (renal system), dark blue (liver), light blue (remainder of digestive system), pink (integumentary system), and green (reproductive system).

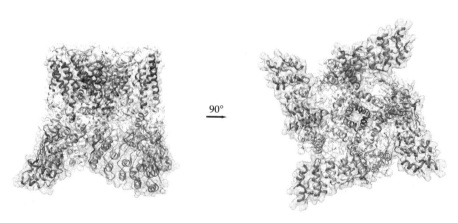

FIGURE 2.1 Structure of the TRPV1 ion channel as determined by cryo-electron microscopy. The channel is a homotetramer of subunits with six transmembrane domains and large intracellular regions formed by the N- and C-termini. The amino terminal ankyrin repeats interact between adjacent subunits at a specific region, which contains an unusually long alpha helix. This interaction also includes a beta strand from the C-terminus and part of the TRP box domain and might be determinant of heat activation. The coloring of regions in this figure was done according to b-factors, with blue hues being low values and red and orange high values. This coloring gives an indication of the regions with likely low and high mobility, respectively. The structure depicted 3J5R entry in the PDB database. The left panel is the side view and the right panel depicts a 90° rotation showing a view from the intracellular side.

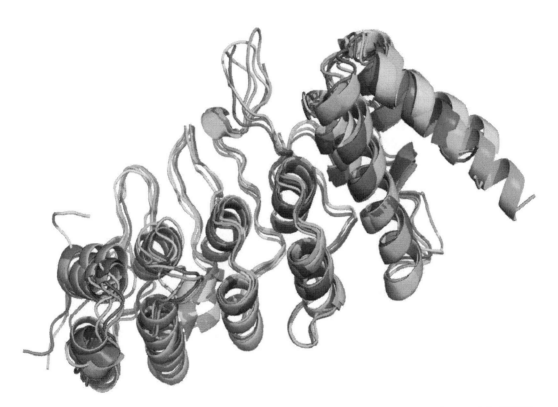

FIGURE 2.2 Comparison of the known structures of ankyrin repeat domains of the amino termini of TRV1, TRPV2, and TRPV6 channels. Structural alignment of the structures of the amino terminal ankyrin-repeat domains of some TRP channels. The coloring is as follows: pink, human TRPV2; green, rat TRPV2; cyan, rat TRPV1; yellow, TRPV6.

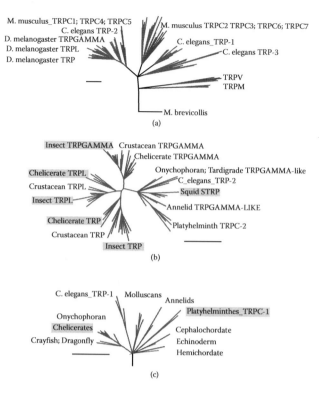

FIGURE 3.8 Evolution of TRPC channels: The complex phylogeny of TRPC channels. (a) Phylogram of TRPC, TRPV, and TRPM channel protein species mined from 43 animal genomes or transcriptomes released to GenBank. (b) Expanded phylogram of the clade, *trpγ-like*, containing *Drosophila* TRP and *C. elegans* TRP-2. (c) Expanded phylogram of the clade, *trp-1-like*, containing *C. elegans* TRP-1. Positions of proteins referred to in the text and of proteins belonging to major phyla of protostome (red) and deuterostome (blue) lineages are indicated. The protein sequences were aligned using MAFFT (Katoh et al., 2002; Katoh et al., 2005). Trees were calculated using the maximum likelihood method, GARLI (Bazinet et al., 2014; Zwickl, 2006), running on CIPRES (Miller et al., 2010), and plotted using Dendroscope (Huson and Scornavacca, 2012). Nodes with less than 90% bootstrap support were collapsed into polytomies. The scale bars indicate one substitution per site.

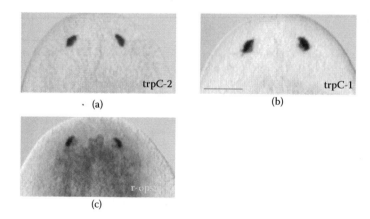

FIGURE 3.9 *In situ* hybridization patterns of probes directed against trpC-2 (a), trpC-1 (b), and rhabdomeral r-opsin mRNA sequences in the head of the platyhelminth, *S. mediterranea*. Densest areas of r-opsin localization indicate the photoreceptor cell bodies. (From Lapan, S.W. and P.W. Reddien, *Cell Rep.,* 2, 294–307, 2012. With permission.) Scale bar indicates 200 μm.

All	TRP-1-like	T	L	F/L	W	S	L	F	G/S	I/L/V	I/T	P/Q	I/P		
Basal	TRPGAMMA-like	S/T	L	Y	W	A/S	I	F/Y	G	L	I/V		L	/T	H/N
Arthropod	TRPGAMMA	T	L	F	W	A	A/S/V	F	G	L	I/V		L		N/S
Arthropod	TRP	S	L	F	W	A	S	F	G	L	V		L	T	S
Arthropod	TRPL	S	L	F	W	A	S	F	G	L/M	I/V	/G	I/L	/S	/S

D621

Pore helix ————— selectivity filter

FIGURE 3.10 Aligned consensus sequences of portions of the pore loop and selectivity filter of selected TRPC channel groups included in the trees of Figure 3.8. Inclusion in the consensus requires more than 20% presence of an amino acid at that position. Amino acid residue symbols are colored according to Lesk (2008); small (yellow); hydrophobic (green); polar (magenta); negatively charged (red); positively charged (blue).

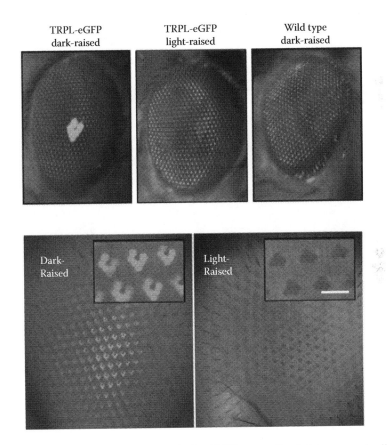

FIGURE 3.14 Light-dependent translocation of TRPL-eGFP between the rhabdomere and cell body *in vivo*. Upper panel: Deep pseudopupil (Franceschini and Kirschfeld, 1971) of TRPL-eGFP expressing flies (*trpl-eGFP*) raised in the dark (upper left panel). A fluorescence signal of the deep pseudopupil was observed in dark-raised flies and disappeared in light-raised flies (upper middle panel) and in wild-type eyes that do not express TRPL-eGFP (upper right panel), indicating that in the dark, TRPL-eGFP are predominantly localized in the rhabdomeres. Lower panel: Subcellular localization of TRPL-eGFP in dark- and light-raised transgenic flies (*trpl-eGFP*). Fluorescence was detected in intact eyes after optical neutralization of the cornea by water immersion. Flies were kept in the dark (lower left panel) or under continuous orange light for 16 hours (lower right panel). The insets show the central area of the eye at higher magnification showing localization of TRPL-eGFP in the rhabdomeres of dark-raised flies (green dots) and its absence in the rhabdomeres of light-raised flies (black dots). Scale bar, 15 μm. (From Meyer, N.E. et al., *Cell Sci.*, 119, 2592–2603, 2006. With permission.)

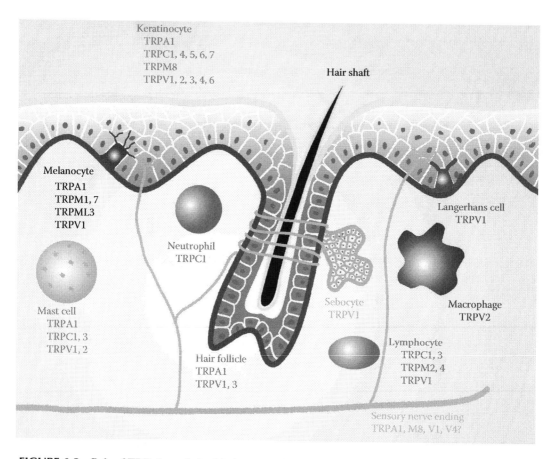

FIGURE 6.2 Role of TRP channels in skin homeostasis and skin diseases. The skin-expressed TRP channels can either contribute to maintaining normal skin physiology or play important roles in the pathogenesis of skin diseases.

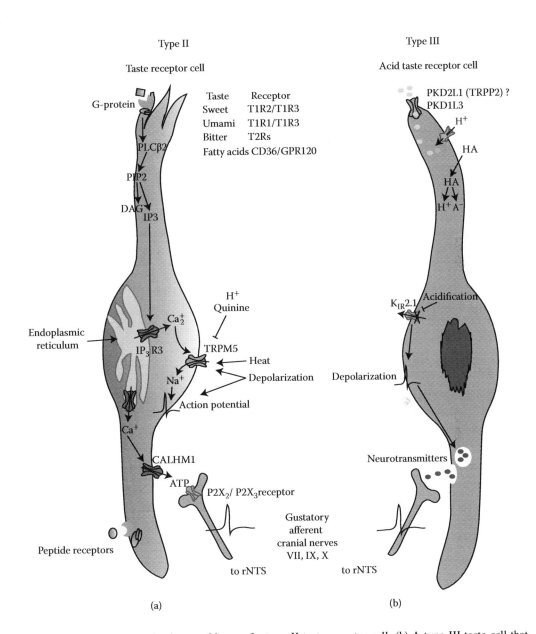

FIGURE 7.3 (a) The transduction machinery of a type II taste receptor cell. (b) A type III taste cell that responds to acidic stimuli. The subset of type III cells responsive to salt are not shown, because it is not known whether they contain TRP channels.

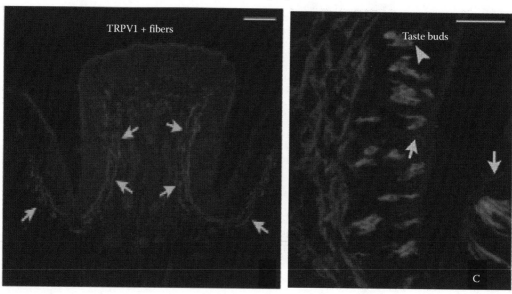

FIGURE 7.4 Left panel: Perigemmal expression of TRPV1 immunoreactive (TRPV1-IR) nerve fibers in rodent circumvallate papillae (in red). Right panel: The co-localization of TRPV1-IR immunoreactivity (red) and anti-gustducin alpha-subunit (green); the latter is present in type II taste receptor cells. Arrows indicate perigemmal TRPV1+ fibers. (From Ishida, Y. et al., *Mol Brain Res.,* 107, 17–22, 2002. With permission.)

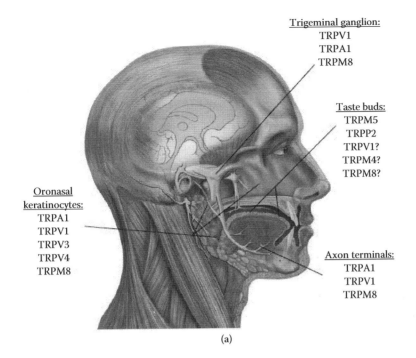

(a)

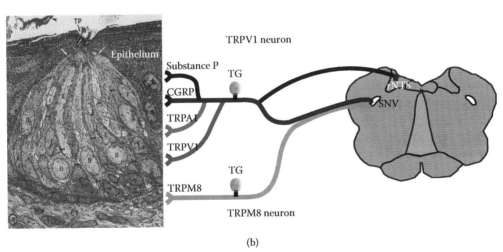

(b)

FIGURE 7.5 The oronasal cavity and its innervation, showing TRP channels involved in taste and chemesthesis. (a) TRP channels are expressed in sensory ganglion neurons, their axon terminals, and in epithelial keratinocytes throughout the oronasal cavity. The TRP channels illustrated here have known or implied functions in gustation and chemesthesis. Other TRP channels are expressed in these sites, but their functions in taste and chemesthesis are not well understood. (b) A cartoon of two types of trigeminal ganglion cells and their projection to medulla. The fiber containing TRPV1 projects both to the nucleus of the solitary tract (NST) where it can influence responses from taste fibers and the SNV where it can produce responses that will be perceived as pain. The activation of these fibers can also release peptides such as substance P and CGRP as well as other neuropeptides that can bind to receptors on taste cells as seen in Figure 7.3. Also shown is a perigemmal fiber containing TRPM8 (Abe et al., 2005) that projects to SNV where it provides information about cooling.

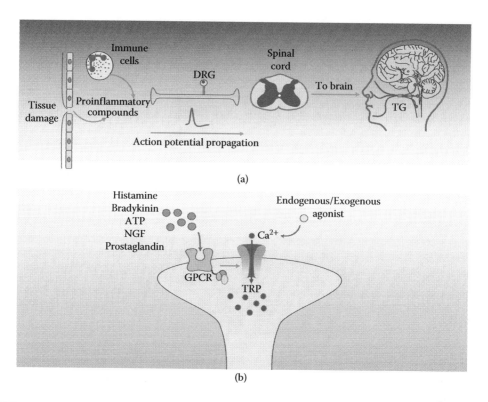

FIGURE 8.1 Transduction of pain signal. (a) Nociceptor stimulation by tissue damage. Nociceptive signals and proinflammatory compounds trigger an action potential along the nociceptive fiber toward the central nervous system to be perceived as pain. (b) Nociceptive signals are detected by some members of the TRP family of ion channels, which are directly activated by their agonists and indirectly by proinflammatory mediators. (DRG, dorsal root ganglion; TG, trigeminal ganglion.)

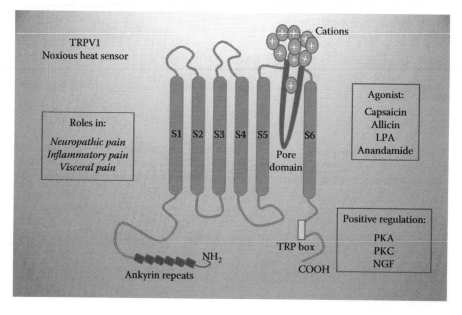

FIGURE 8.4 A TRPV1 monomer showing the transmembrane segments (S1–S6), the most representative agonists known for the TRPV1 channel, its role on pain perception, and positive regulators of its activation.

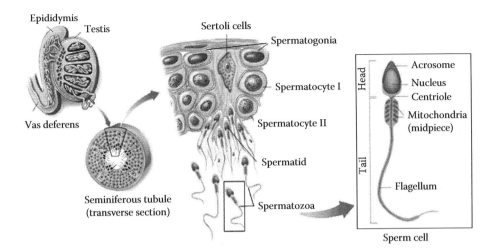

FIGURE 11.1 Human spermatogenesis. Sperm development takes place in the seminiferous tubules of the testis. When the male reaches puberty, spermatogonial stem cells near the basal lamina of the tubule start to divide mitotically to produce more spermatogonia and primary spermatocytes. The spermatocytes go through meiosis I to form haploid secondary spermatocytes that migrate further toward the tubular lumen where they divide by meiosis II to form spermatids. Throughout spermatogenesis, the germ cells are positioned in close proximity to Sertoli cells, which provide them with nutrients and growth factors needed for development. After forming the final structure with a head, containing the genomic material and the acrosome region, and the elongated tail, and becoming transcriptionally and translationally silent, spermatozoa are released into the testicular lumen. (Modified from Allais-Bonnet, A. and E. Pailhoux. *Front Cell Dev Biol.*, 2, 56, 2014.)

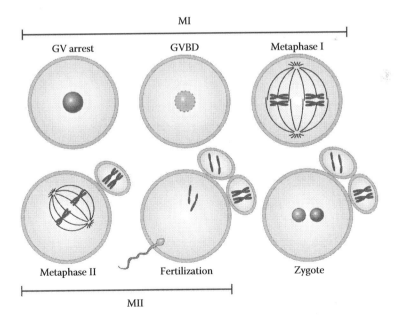

FIGURE 11.2 Oocyte maturation and fertilization. At birth, mammalian oocytes are arrested at the diplotene stage of the first meiosis (MI) prophase. Once the female reaches puberty, each estrus cycle triggers the breakdown of the germinal vesicle (GVBD) and allows the division of the oocyte to continue until metaphase II of the second meiosis (MII). Fertilization allows the oocyte to complete meiosis after which the haploid genomic material is combined with that of the male gamete (marked in blue).

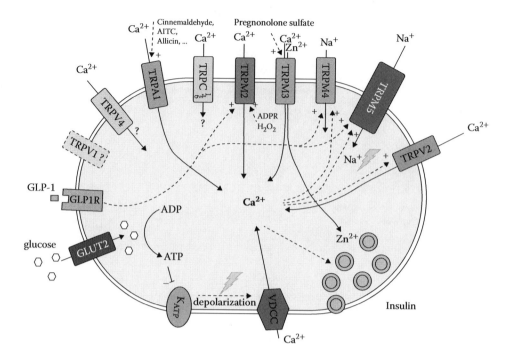

FIGURE 12.1 Expression of different TRP channels in the β-cell, important regulatory molecules and their action on the cell.

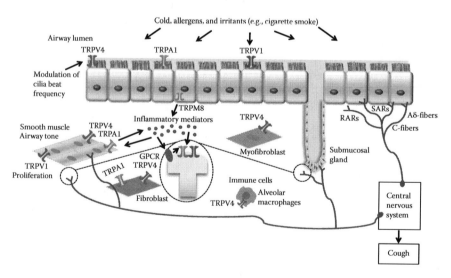

FIGURE 13.1 Localization and function of TRP channels in neuronal and nonneuronal cells of the respiratory tract. (RARs, rapidly adapting receptors; SARs, slowly adapting receptors.)

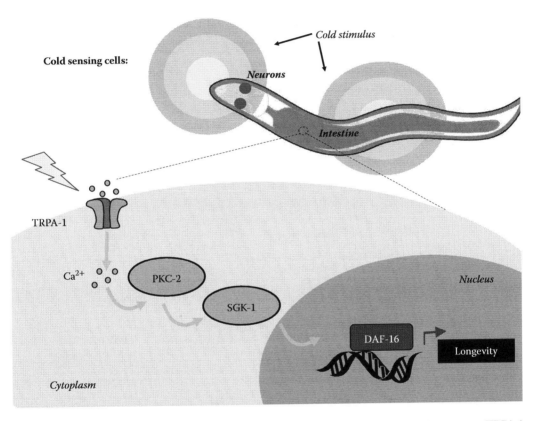

FIGURE 14.1 A genetic pathway that promotes longevity at cold temperatures in *C. elegans* upon TRPA-1 activation in cold sensing tissues (neurons and intestine). Calcium signaling triggers canonical Ca^{2+}-signaling cascade leading to the FOXO transcription factor DAF-16 to promote transcriptional programs that repress aging.

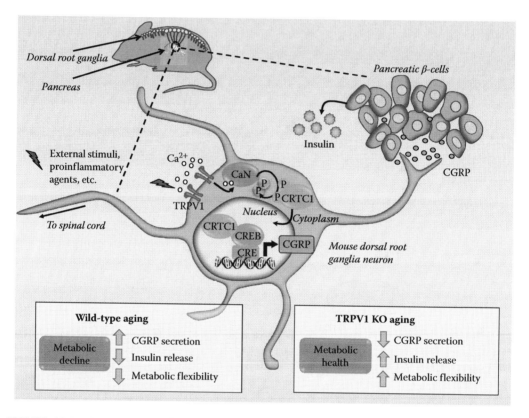

FIGURE 14.3 Model for the neuroendocrine regulation of metabolism by TRPV1-expressing neurons. Stimulation of TRPV1 by external stimuli promotes CGRP secretion from DRG neurons onto the pancreatic β-cells and inhibition of insulin release. TRPV1 activation results in Ca^{2+} influx and activation of calcineurin, allowing dephosphorylation of CRTC1 and release from 14-3-3 proteins, resulting in nuclear internalization of CRTC1 and transcription of its targets, such as CGRP. CGRP accumulation has detrimental effects on energy expenditure, glucose tolerance, and aging. In contrast, loss of TRPV1 promotes lifespan extension through increased insulin secretion, metabolic health by inactivation of the CRTC1/CREB signaling cascade.

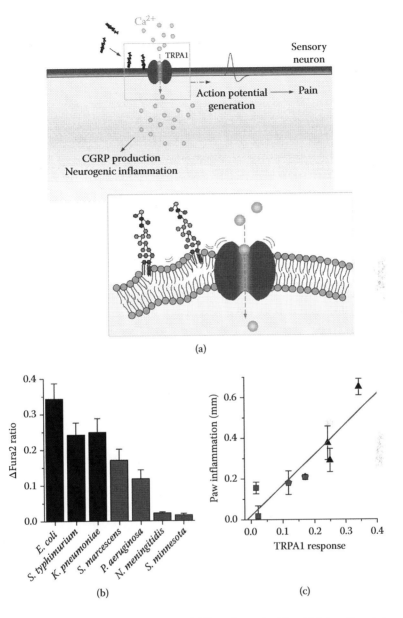

FIGURE 15.1 TRPA1 as a neuronal sensor of bacterial lipopolysaccharides. (a) Interactions between LPS and TRPA1 in the membrane of mammalian nociceptive neurons. Insertion of the lipid A moiety of LPS may induce mechanical perturbations in the plasma membrane, which are sensed by TRPA1. Activation of this channel leads to membrane depolarization and action potential firing, which results in pain, and to the release of CGRP and neurogenic inflammation. (b) Distinct effects of different LPS molecules on TRPA1. The blue bars represent cylindrical or lamellar LPS, the red bars semiconical LPS and the black bars conical LPS. (c) Correlation between the ability of different LPS to activate TRPA1 and to induce paw inflammation in the mouse. The blue squares represent cylindrical or lamellar LPS, the red pentagons semi-conical LPS, and the black triangles conical LPS. (Reproduced from Meseguer, V. et al., *Nat Commun.*, 5, 3125, 2014. With permission.)

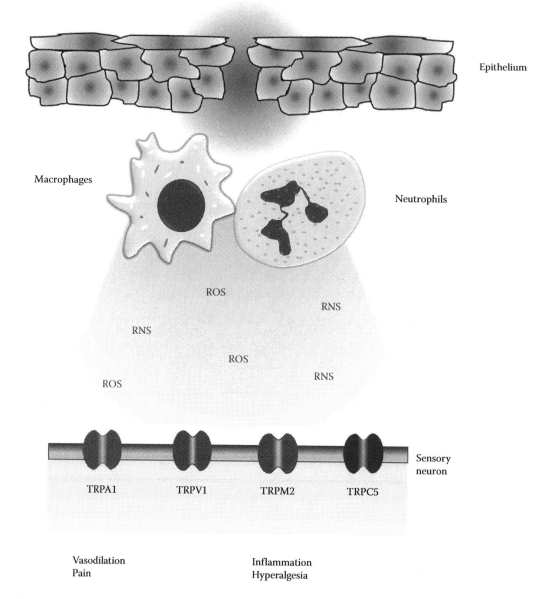

FIGURE 15.2 Neuronal TRP channels as sensors of immune effectors. Activation of macrophages and neutrophils by tissue injury results in the local release of reactive oxygen and nitrogen species that activate multiple TRP channels in sensory neurons. This leads to vasodilation, local inflammation, pain, and hyperalgesia.

9 TRP Channels in the Heart

Marc Freichel, Michael Berlin, Alexander Schürger,
Ilka Mathar, Lucas Bacmeister, Rebekka Medert,
Wiebke Frede, André Marx, Sebastian Segin,
and Juan E. Camacho Londoño

CONTENTS

9.1 INTRODUCTION

Calcium is an important second messenger in cardiac function. It is not only critical for the excitation-contraction coupling and relaxation of the heart (Bers, 2002), but it is also important for the activation of signal transduction pathways responsible for hypertrophic cardiac remodeling and heart failure, for example, by controlling gene transcription via Ca^{2+}-dependent signaling as well as for cardiac development, cardiac energy homeostasis, and eventually for cell death (Frey et al., 2000; Frey et al., 2004; Roderick et al., 2007). In beating cardiomyocytes, fast cycling changes in cytosolic Ca^{2+} concentration are the results of a timely coordinated interplay of voltage-gated Ca^{2+} channels, sodium-calcium-exchangers, ryanodine receptors, and the SERCA-ATPase (Bers, 2008). However, the channels and pumps mediating the fast Ca^{2+} cycling during beat-to-beat cardiac action are not only relevant for physiological cardiac functions but also for pathological processes such as development of pathological cardiac remodeling and development of heart failure. These pathological processes are essentially triggered by neuroendocrine stimuli such as noradrenaline, adrenaline, and angiotensin II, which subsequently lead to activation of G protein–dependent signaling pathways in cardiomyocytes that evoke Ca^{2+} entry and Ca^{2+}-dependent processes (e.g., activation of calcineurin/nuclear factor of activated T cells [NFAT], CaM-kinase, and protein kinase C inducing the development of myocyte growth and cardiac hypertrophy) (Heineke and Molkentin, 2006). Although the action of these sympathetic neurohormones represents an adaptive response that initially preserves cardiac function, the processes triggered by persistent activation during long-term cardiac stress leads to cardiac failure in many cardiovascular disease entities, including arterial hypertension and ischemic or valvular heart diseases. The sources of the Ca^{2+} elevation and the mechanisms whereby Ca^{2+} leads to calcineurin activation under repetitive Ca^{2+} concentration changes during the contraction cycle are still not entirely understood. Sustained elevation of diastolic Ca^{2+} levels has been identified as a mechanism (Dolmetsch et al., 1997) and can be achieved in cardiomyocytes (e.g., by an increase of Ca^{2+} transient frequency to trigger remodeling processes) (Colella et al., 2008; Tavi et al., 2004). On the molecular level this can be due to alterations in Ca^{2+} release from SR or Ca^{2+} transport mechanisms across the plasma membrane with changes in the expression or function in the SERCA2, RyR2, IP_3 receptor, sodium-calcium exchangers (NCX1), or Na^+/H^+ exchanger (NHE1) (Goonasekera and Molkentin, 2012). The Ca^{2+} entry pathways that are unrelated to those initiating contraction can evolve by targeting individual channels to subcellular microdomains such as caveolae, where a subset of L-type Ca^{2+} channels are functional (Makarewich et al., 2012) and may colocalize with beta-adrenergic receptors outside the junctional ryanodine receptor/T-tubular complex (Balijepalli et al., 2006). The complexity of Ca^{2+}-dependent regulation of cardiac hypertrophy by different molecular components *in vivo* becomes evident considering that even large increases of voltage-gated L-type Ca^{2+} channels (LTCC) result in only mild cardiac hypertrophy (Beetz et al., 2009) and that reduced LTCC activity can also stimulate hypertrophy most likely via compensatory neuroendocrine stress leading to sensitized and leaky SR Ca^{2+} release (Goonasekera and Molkentin, 2012). The concept of different Ca^{2+} pools regulating contractility ("contractile Ca^{2+}") and remodeling ("signaling Ca^{2+}") arising from distinct spatial localization of Ca^{2+} molecules is furthermore complicated by the developmental stage (neonatal/adult), the localization (atrial/ventricular), and the disease stage (nonfailing/failing) of the investigated cardiomyocytes.

In addition to the pathways directly associated with fast Ca^{2+} cycling, transient receptor potential (TRP) proteins have been uncovered in recent years as the molecular constituents of cation channels engaged by, for example, catecholamines or AngII in cardiac cells and as determinants of cardiac functions, although receptor- and store-operated Ca^{2+} entry pathways were previously described in cardiac cells (Freichel et al., 1999). TRP proteins form Na^+- and Ca^{2+}-conducting channels that can evoke changes in the Ca^{2+} homeostasis beyond the time scale of beat-to-beat Ca^{2+} transients and mediate longer-lasting modulation of Ca^{2+} levels. The mammalian 28 TRP proteins are classified according to structural homology into six subfamilies: TRPC (canonical), TRPV (vanilloid), TRPM (melastatin), TRPA (ankyrin), TRPML (mucolipin), and TRPP (polycystin). They are activated by numerous physical (e.g., mechanical stretch) and/or chemical stimuli (e.g., agonists including neurotransmitters) and can contribute to Ca^{2+} homeostasis by directly conducting Ca^{2+} or may contribute to Ca^{2+} entry indirectly via membrane depolarization and modulation of voltage-gated Ca^{2+} channels (Wu et al., 2010; Flockerzi and Nilius, 2014; Freichel et al., 2014). Thus, they have been proposed to be mediators of different physiological and pathophysiological cardiovascular processes (Inoue et al., 2006; Dietrich et al., 2007; Abramowitz and Birnbaumer, 2009; Watanabe et al., 2009; Dietrich et al., 2010; Vennekens, 2011). In the heart, initial attention has been placed to determine the role of TRP channels in the development of cardiac remodeling using *in vitro* and *in vivo* models (Guinamard and Bois, 2007; Nishida and Kurose, 2008; Eder and Molkentin, 2011). In this chapter we summarize the current knowledge regarding the expression and functional role of TRP channels for Ca^{2+} homeostasis in cardiomyocytes and cardiac fibroblasts, their contribution to cardiac contractility and conduction, as well as the development of arrhythmias and pathological remodeling processes as determined by overexpression studies in cardiac cells/tissues, by knockdown/knockout of the corresponding genes, or by the use of specific channel inhibitors. Based on the increasing experimental evidence for their role in cardiac (dys)function derived from animal models and disease-associated mutations, individual TRP channels are becoming promising therapeutic targets for cardiac diseases.

9.2 EXPRESSION OF TRPs IN CARDIAC CELLS

In this chapter we focus on the expression of the 28 TRPs that are found in mammals. It has been reported that all TRPs are expressed in the heart using different methods to detect TRP transcripts using reverse transcriptase polymerase chain reaction (RT-PCR), Northern blot analysis, in situ hybridizations, or microarray analysis, and TRP proteins by Western blots, immunocytochemistry, immunohistochemistry, or immunoelectron microscopy. These data are compiled in Table 9.1.

In addition, we summarize the large body of reports assigning TRP-channel expression in two cardiac cell types—cardiomyocytes (Table 9.2) and cardiac fibroblasts (Table 9.3). Knowledge about their occurrence in other cardiac nonmyocytes such as endothelial cells or leukocytes is still sparse. Regarding the protein expression analysis of TRPs reported so far, it is important to mention that antibody specificity was not always tested with cells or tissue from, for example, TRP-deficient mice, and the purity of cardiomyocyte or fibroblast preparations used has not always been reported.

In addition to expression in the healthy heart, TRP transcripts and proteins were found to be differentially expressed in cells and tissues from experimental models inducing cardiac remodeling or dysfunction as well as in myocardial biopsies from diseased human hearts. In different rodent models of cardiac hypertrophy an increased expression of TRPC1 (Ohba et al., 2006, 2007; Seth et al., 2009), TRPC2 (Ohba et al., 2006), TRPC3 (Bush et al., 2006; Ohba et al., 2006; Brenner and Dolmetsch, 2007; Seth et al., 2009; Koitabashi et al., 2010), and TRPC6 (Kuwahara et al., 2006; Ohba et al., 2006; Nishida et al., 2007; Seth et al., 2009; Koitabashi et al., 2010) has been reported. Additionally, increased expression of TRPC1 and TRPC5 in human failing hearts (Sucharov and Bristow, 2005; Bush et al., 2006), increased TRPC6 mRNA in human hearts with dilated cardiomyopathy (Kuwahara et al., 2006), and downregulation of TRPC4 in ventricular myocytes in biopsies from patients with ischemic cardiomyopathy were shown using a microarray screening

TABLE 9.1
Reported Expression of TRPs Using Cardiac Tissue and Analyzed by Different Methodologies

		Expression Information	
TRP Subfamily	**TRP**	**Expression Method/Tissue**	**Reference**
TRPC	hTRPC1	NB/heart	(Wes et al., 1995)
	rTRPC1	IsH/embryonic heart	(Wes et al., 1995)
	hTRPC1	NB/heart	(Zhu et al., 1995)
	bTRPC4	NB/heart	(Philipp et al., 1996)
	rTRPC1,3–6	RT-PCR/heart	(Garcia and Schilling, 1997)
	mTRPC1	NB/heart	(Sakura and Ashcroft, 1997)
	bTRPC4	NB/ventricle, atria	(Freichel et al., 1998)
	rbTRPC4	NB/heart	(Freichel et al., 1998)
	mTRPC1-4,6	RT-PCR/heart	(Freichel et al., 1999)
	mTRPC7	NB/heart	(Okada et al., 1999)
	hTRPC1,3–6	RT-PCR/heart	(Riccio et al., 2002)
	m-rTRPC3	WB/LV	(Bush et al., 2006)
	hTRPC5	WB/LV	(Bush et al., 2006)

(Continued)

TABLE 9.1 (*Continued*)
Reported Expression of TRPs Using Cardiac Tissue and Analyzed by Different Methodologies

TRP Subfamily	TRP	Expression Method/Tissue	Reference
		Expression Information	
	mTRPC1-3,6,7	RT-PCR/heart	(Kunert-Keil et al., 2006)
	mTRPC1,3,4,6	RT-PCR/heart	(Kuwahara et al., 2006)
	hTRPC6	RT-PCR/heart	(Kuwahara et al., 2006)
	mTRPC3	WB/heart	(Nakayama et al., 2006)
	mTRPC1-3,6	RT-PCR, WB, IHC/heart	(Ohba et al., 2006)
	rTRPC7	NB/heart	(Satoh et al., 2007)
	mTRPC1-4,6,7	RT-PCR/SAN	(Ju et al., 2007)
	mTRPC3	RT-PCR, WB/LV, SAN, AVN	(Ju et al., 2007)
	mTRPC1,3,4,6	WB/atria	(Ju et al., 2007)
	rTRPC1,3,5,6	RT-PCR, WB/heart	(Ohba et al., 2007)
	mTRPC1,3	WB/heart	(Shan et al., 2008)
	rTRPC1	RT-PCR, IHC/LV, RV, LA, RA	(Huang et al., 2009a)
	rTRPC4,5	WB/heart	(Liu et al., 2010a)
	mTRPC5	IHC, WB, MS/embryonic heart	(Nath et al., 2009)
	mTRPC1*,3,6	WB/heart	(Seth et al., 2009)
	mTRPC3,4,6	WB/heart	(Wu et al., 2010)
	mTRPC1,3,5–7	WB/heart	(Kitajima et al., 2011)
	rTRPC1,3,6	RT-PCR/atria, SAN	(Tellez et al., 2011)
	rTRPC1,3–6	WB/fetal, neonatal, and adult heart	(Jiang et al., 2014)
	rTRPC1	IHC/atria	(Shenton and Pyner, 2014)
	mTRPC1–7	RT-PCR/digested ventricles	(Camacho Londono et al., 2015)
	rTRPC7	WB/LV	(Cui et al., 2016)
TRPM	mTRPM7	NB/heart	(Runnels et al., 2001)
	hTRPM4b	RT-PCR/atria	(Guinamard et al., 2004)
	hTRPM1–7	RT-PCR/heart	(Fonfria et al., 2006)
	rTRPM4	RT-PCR/LV and atria	(Guinamard et al., 2006)
	mTRPM1–8	RT-PCR/heart	(Kunert-Keil et al., 2006)
	mTRPM4	RT-PCR, WB/SAN	(Demion et al., 2007)
	hTRPM4	RT-PCR/Purkinje fibers, SAN, R-atrium, LV, RV	(Kruse et al., 2009)
	mTRPM4*	WB/ventricle, atria	(Mathar et al., 2010)
	hTRPM4,7	RT-PCR, WB/RA	(Zhang et al., 2012b)
	mTRPM2	RT-PCR/heart	(Hiroi et al., 2013)
	hTRPM3	RT-PCR/interventricular septum	(Kuster et al., 2013)
	mTRPM2	RT-PCR/heart	(Miller et al., 2013)
	rTRPM7	RT-PCR/heart	(Tashiro et al., 2013)
	rTRPM4	RT-PCR/heart	(Wang et al., 2013)
	rTRPM2,4,6,7	RT-PCR/heart	(Demir et al., 2014)
	mTRPM6,7	RT-PCR, IsH/fetal and adult heart	(Cuffe et al., 2015)
	mTRPM4	RT-PCR/LV	(Jacobs et al., 2015)
	mTRPM4*	WB/LV-adult, neonatal heart	(Kecskes et al., 2015)
	hTRPM7	MA, WB, IHC/LV	(Parajuli et al., 2015)
	hTRPM6	RT-PCR, WB/RA	(Zhang et al., 2015)
	mTRPM7	WB/heart	(Antunes et al., 2016)

(Continued)

TABLE 9.1 (*Continued*)

Reported Expression of TRPs Using Cardiac Tissue and Analyzed by Different Methodologies

TRP Subfamily	TRP	Expression Information	
		Expression Method/Tissue	Reference
	rTRPV2	WB/heart	(Kobayashi et al., 2006)
TRPV	mTRPV2-4,6	RT-PCR/heart	(Kunert-Keil et al., 2006)
	rTRPV4	IHC/heart	(Willette et al., 2008)
	hTRPV1	IHC/atria (nerve fibers)	(Facer et al., 2011)
	hoTRPV5–6	RT-PCR, WB, IHC/heart	(Hwang et al., 2011)
	rTRPV1	WB/heart	(Ren et al., 2011)
	mTRPV1	WB/heart	(Guarini et al., 2012)
	mTRPV2	WB/heart	(Cohen et al., 2013)
	h, haTRPV2	WB, IHC/hearts, ventricles	(Iwata et al., 2013)
	mTRPV2	MA/LV	(Entin-Meer et al., 2014)
	rTRPV2	MA, RT-PCR, WB, IHC/LV	(Entin-Meer et al., 2014)
	mTRPV2*	WB/heart	(Katanosaka et al., 2014)
	rTRPV4	IHC/atria	(Shenton and Pyner, 2014)
	mTRPV1*	WB/heart	(Sun et al., 2014)
	mTRPV1	WB/LV	(Zheng et al., 2015)
TRPML	hTRPML1	NB/heart	(Bassi et al., 2000)
	mTRPML1	NB/fetal and adult heart	(Falardeau et al., 2002)
	mTRPML3	RT-PCR/heart	(Cuajungco and Samie, 2008)
	mTRPML2,3	RT-PCR/heart	(Samie et al., 2009)
	mTRPML1	RT-PCR/heart	(Grimm et al., 2010)
	hTRPML1,3	RT-PCR/heart	(Grimm et al., 2010)
TRPP	hTRPP3	NB, RNA-dot blot/heart, adult and fetal	(Nomura et al., 1998)
	hTRPP3	NB/heart	(Wu et al., 1998)
	mTRPP2	WB/heart	(Markowitz et al., 1999)
	hTRPP2	NB, IsH/embryonic and adult heart	(Chauvet et al., 2002)
	hTRPP5	RT-PCR/fetal heart	(Guo et al., 2000)
	mTRPP5	NB/heart	(Guo et al., 2000)
	mTRPP3	IHC/heart	(Basora et al., 2002)
	hTRPP2	WB/heart	(Li et al., 2003b)
	hTRPP3	WB/heart	(Li et al., 2003a)
	rTRPP2,3,5	RT-PCR/heart	(Volk et al., 2003)
	rTRPP2	WB/heart	(Li et al., 2005)
	hTRPP3	WB/heart	(Li et al., 2007)
	m, dTRPP2	WB/heart	(Anyatonwu et al., 2007)
	mTRPP2[a]	WB/LV	(Kuo et al., 2014, 2016)
TRPA	mTRPA1	RT-PCR/heart	(Kunert-Keil et al., 2006)
	mTRPA1	RT-PCR/heart	(Bodkin et al., 2014)
	mTRPA1	IsH/heart	(Pazienza et al., 2014)

AVN: atrioventricular node; b: bovine; d: dog; gp: guinea pig; h: human; ha: hamster; ho: horse; ICC: immunocytochemistry; IHC: immunohistochemistry; IsH: in situ hybridization; LA: left atrium; LV: left ventricle; m: mouse; MA: microarray; MS: mass spectrometry; NB: Northern blot; RA: right atrium; rb: rabbit; RT-PCR: reverse transcription polymerase chain reaction; r: rat; RV: right ventricle; SAN: sinoatrial node; WB: Western blot.

[a] In this case for antibody specificity, tissue samples from TRP-deficient mice were included or the antibody was previously validated using samples from the corresponding TRP-deficient mice.

TABLE 9.2

Reported Expression Pattern of TRPs in Cardiomyocytes Using Different Methodologies

Type of Expression Analysis — TRP Subfamily/Member	RT-PCR	Northern Blot	In Situ Hybridization	Western Blot	ICC	IHC	Other
TRPC							
TRPC1	(Ohba et al., 2006)-m (Brenner and Dolmetsch, 2007)-r (Ohba et al., 2007)-r (Ohba et al., 2009)-r (Vindis et al., 2010)-r (H9C2) (Chu et al., 2012)-r (Kiso et al., 2013)-r (Zhang et al., 2013)-h (Makarewich et al., 2014)-m (Seo et al., 2014b)-m (Sabourin et al., 2016)-r (Son et al. 2016)-r			(Nakayama et al., 2006)-r (C-nd) (Ohba et al., 2007)-r (C-nd) (Alvarez et al., 2008)-r (C-nd) (Ward et al., 2008)-m (C-ns) (Seth et al., 2009))-r (C-KO) (Vindis et al., 2010)-r(H9C2) (C-nd) (Feng et al., 2011)-r (C-nd) (Mohl et al., 2011)-m (C-pcp) (Chu et al., 2012)-r (C-nd) (Sabourin et al., 2012)-c (C-ns) (Sabourin et al., 2016)-r (C-nd)	(Ward et al., 2008)-m (C-pcp) (Huang et al., 2009a)-r (C-nd) (Seth et al., 2009)-r ·(C-KO) (Kojima et al., 2010)-m (C-nd) (Zhang et al., 2013)-h (C-nd) (Jiang et al., 2014)-r (C-pab)	(Ohba et al., 2006)-m (C-nd) (Ward et al., 2008)-m (C-pcp) (Huang et al., 2009a)-r (C-pcp) (Mohl et al., 2011)-m (C-nd)	
TRPC2	(Ohba et al., 2006)-m (Brenner and Dolmetsch, 2007)-r (Makarewich et al., 2014)-m						

(Continued)

TABLE 9.2 (Continued)

Reported Expression Pattern of TRPs in Cardiomyocytes Using Different Methodologies

Type of Expression Analysis	RT-PCR	Northern Blot	In Situ Hybridization	Western Blot	ICC	IHC	Other
TRPC3	(Bush et al., 2006)-r	(Kunert-Keil et al., 2006)-m		(Bush et al., 2006)-r-m (C-pab)	(Bush et al., 2006)-r -r (C-nd)	(Goel et al., 2007)-r (C-nd)	
	(Ohba et al., 2006)-m			(Nakayama et al., 2006)-r (C-nd)	(Brenner and Dolmetsch, 2007)-r (C-nd)		
	(Brenner and Dolmetsch, 2007)-r			(Onohara et al., 2006)-r (C-nd)	(Eder et al., 2007)-r (C-nd)		
	(Prasad and Inesi, 2009)-r			(Eder et al., 2007)-r (C-nd)	(Fauconnier et al., 2007)-m (C-pcp)		
	(Feng et al., 2011)-r			(Alvarez et al., 2008)-r (C-nd)	(Goel et al., 2007)-r (C-nd)		
	(Chu et al., 2012)-r			(Dyachenko et al., 2009)-r (C-nd)	(Kojima et al., 2010)-m (C-nd)		
	(Sowa et al., 2012)-r			(Prasad and Inesi, 2009)-r (C-nd)	(Jiang et al., 2014)-r (C-pab)		
	(Kiso et al., 2013)-r			(Feng et al., 2011)-r (C-nd)	(Doleschal et al., 2015))-m (C-nd)		
	(Zhang et al., 2013)-h			(Klaiber et al., 2011)-m (C-nd)	(Dominguez-Rodriguez et al., 2015)-r (C-nd)		
	(Makarewich et al., 2014)-m			(Chu et al., 2012)-r (C-nd)	(Qi et al., 2016)-m (C-nd) (derived from stem cells)		
	(Seo et al., 2014b)-m			(Sabourin et al., 2012)-c (C-ns)			
	(Dominguez-Rodriguez et al., 2015)-r			(Sowa et al., 2012)-r (C-nd)			
	(Hang et al., 2015)-r			(Dominguez-Rodriguez et al., 2015)-r (C-nd)			
	(Li et al., 2015)-r			(Li et al., 2015)-r (C-nd)			
	(Sabourin et al., 2016)-r						
TRPC4	(Seth et al., 2004)-r			(Seth et al., 2004)-r (C-nd)	(Kojima et al., 2010)-m (C-nd)		(Gronich et al., 2010)-h (MA)
	(Brenner and Dolmetsch, 2007)-r			(Nakayama et al., 2006)-r (C-nd)	(Wu et al., 2010)-m (C-nd)		
	(Kiso et al., 2013)-r			(Alvarez et al., 2008)-r (C-nd)	(Cooley et al., 2014)-r (C-nd)		
	(Zhang et al., 2013)-h			(Sabourin et al., 2012)-c (C-ns)	(Jiang et al., 2014)-r (C-pab)		
	(Cooley et al., 2014)-r			(Dominguez-Rodriguez et al., 2015)-r (C-nd) (Sabourin et al., 2016)-r (C-nd)	(Dominguez-Rodriguez et al., 2015)-r (C-nd)		
	(Makarewich et al., 2014)-m						
	(Dominguez-Rodriguez et al., 2015)-r						
	(Sabourin et al., 2016)-r						

(Continued)

TABLE 9.2 (Continued)

Reported Expression Pattern of TRPs in Cardiomyocytes Using Different Methodologies

Type of Expression Analysis	RT-PCR	Northern Blot	In Situ Hybridization	Western Blot	ICC	IHC	Other
TRPC5	(Seth et al., 2004)-r			(Seth et al., 2004)-r (C-nd)	(Jiang et al., 2014)-r (C-pab)		
	(Brenner and Dolmetsch, 2007)-r			(Bush et al., 2006)-h (C-pab)			
	(Kiso et al., 2013)-r			(Nakayama et al., 2006)-r (C-nd)			
	(Makarewich et al., 2014)-m			(Sowa et al., 2012)-r (C-nd)			
	(Sabourin et al., 2016)-r			(Sabourin et al., 2016)-r (C-nd)			
TRPC6	(Ohba et al., 2006)-m	(Kunert-Keil et al., 2006)-m		(Onohara et al., 2006) (C-nd)	(Kuwahara et al., 2006)-m (C-pcp)	(Ohba et al., 2006)-m (C-nd)	
	(Brenner and Dolmetsch, 2007)-r			(Nishida et al., 2007)-r (C-nd)	(Ward et al., 2008)-m (C-pcp) (Dyachenko	(Mohl et al., 2011)-m (C-nd)	
	(Koitabashi et al., 2010)-m-r			(Dyachenko et al., 2009)-m (C-nd)	et al., 2009)-m (C-nd)		
	(Vindis et al., 2010)-r(H9C2)			(Kinoshita et al., 2010)-r (C-nd)	(Mohl et al., 2011)-m (C-nd)		
	(Chu et al., 2012)-r			(Sun et al., 2010)-r (C-nd)	(Jiang et al., 2014)-r (C-pab)		
	(Kiso et al., 2013)-r			(Vindis et al., 2010)-r(H9C2) (C-nd)			
	(Zhang et al., 2013)-h			(Klaiber et al., 2011)-m (C-nd)			
	(Makarewich et al., 2014)-m			(Mohl et al., 2011)-m (C-pcp)			
	(Seo et al., 2014b)-m			(Chu et al., 2012)-r (C-nd)			
	(Hang et al., 2015)-r			(Sabourin et al., 2012)-c (C-ns)			
	(Sabourin et al., 2016)-r			(Qi et al., 2014)-r (C-nd)			
TRPC7	(Brenner and Dolmetsch, 2007)-r			(Onohara et al., 2006) (C-nd)	(Cui et al., 2016)-r (C-nd)	(Cui et al., 2016)-r (C-nd)	
	(Satoh et al., 2007)-r			(Alvarez et al., 2008)-r (C-nd)			
				(Sabourin et al., 2012)-c (C-ns)			
				(Cui et al., 2016)-r (C-nd)			
TRPM							
TRPM1							(Gronich et al., 2010)-h (MA)

(Continued)

TABLE 9.2 (Continued)
Reported Expression Pattern of TRPs in Cardiomyocytes Using Different Methodologies

Type of Expression Analysis	RT-PCR	Northern Blot	In Situ Hybridization	Western Blot	ICC	IHC	Other
TRPM2				(Roberge et al., 2014)-m (C-nd)	(Miller et al., 2013)-m (C-nd) (Hoffman et al., 2015)-m (C-nd)		
TRPM3							
TRPM4	(Zhang et al., 2012b)-h (Demion et al., 2014)-m (Mathar et al., 2014)-m (Kecskes et al., 2015)-m (Piao et al., 2015)-r (Son et al., 2016)-r			(Burt et al., 2013)-m(HL-1) (C-nd) (Kecskes et al., 2015)-m (C-KO) (Son et al., 2016)-r (Son et al., 2016)-m(HL-1)	(Zhang et al., 2012b)-h (C-nd) (Son et al., 2016)-r (C-nd)	(Liu et al., 2010b)-b (C-nd) (Burt et al., 2013)-m(HL-1) (C-nd) (Piao et al., 2015)-r (C-nd)	
TRPM5							
TRPM6							
TRPM7	(Zhang et al., 2015)-h (Gao et al., 2013)-h(hiPSC-CM) (Adapala et al., 2013)-r (Sah et al., 2013b)-m (SAN) (Sah et al., 2013a)-m (ventricle) (Tashiro et al., 2013)-r			(Zhang et al., 2015)-h (C-nd)	(Zhang et al., 2012b)-h (C-nd)		
TRPM8							
TRPV							
TRPV1	(Sun et al., 2014)-r(H9C2) (Qi et al., 2015b)-m			(Pei et al., 2014)-m-(C-nd) (Sun et al., 2014)-r(H9C2) (C-nd) (Wu et al., 2015)-m (C-nd) (Andrei et al., 2016)-m (C-KO)	(Sun et al., 2014)-r(H9C2) (C-nd) (Qi et al., 2015b)-m (C-pcp) (Andrei et al., 2016)-m (C-ns)	(Zhong and Wang, 2009)-m (C-KO) (Andrei et al., 2016)-m (C-KO)	

(Continued)

TABLE 9.2 (*Continued*)

Reported Expression Pattern of TRPs in Cardiomyocytes Using Different Methodologies

Type of Expression Analysis	RT-PCR	Northern Blot	In Situ Hybridization	Western Blot	ICC	IHC	Other
TRPV2	(Katanosaka et al., 2014)-m (Lorin et al., 2015)-m (Kang et al., 2016)-m			(Lorin et al., 2015)-m(C-nd) (Aguettaz et al., 2016)-m (C-nd) (Kang et al., 2016)-m (C-nd)	(Iwata et al., 2013)-ha (C-nd) (Lorin et al., 2015)-m(C-nd) (Aguettaz et al., 2016)-m (C-nd)	(Iwata et al., 2013)-m-ha (C-nd) (Katanosaka et al., 2014)-m (C-KO) (Aguettaz et al., 2016)-m (C-nd)	
TRPV3							
TRPV4	(Zhao et al., 2012)-r (Qi et al., 2015b)-h (Son et al., 2016)-r			(Zhao et al., 2012)-r (C-pab) (Qi et al., 2015b)-h-m-r (C-nd)	(Zhao et al., 2012)-r (C-blank control) (Qi et al., 2015a)-h (C-Co-transfected CHO-cells)	(Zhao et al., 2012)-r (C-nd)	
TRPV5							
TRPV6							
TRPML							
TRPML1							
TRPML2							
TRPML3							
TRPP							
TRPP2	(Volk et al., 2003)-r					(Chauvet et al., 2002)-h (C-nd)	
TRPP3							
TRPP5	(Volk et al., 2003)-r						
TRPA							
TRPA1	(Pazienza et al., 2014)-m		(Pazienza et al., 2014)-m	(Andrei et al., 2016)-m (C-KO)	(Andrei et al., 2016)-m (C-ns)	(Andrei et al., 2016)-m (C-KO)	

For those expression analyses using anti-TRPs antibodies the type of the corresponding control experiment(s) or a comment regarding the validation of the antibodies is included. ab: antibody; b: bovine; C-KD: control used siRNA-mediated knockdown; C-KO: control used corresponding knockout mouse specimen; C-nd: control not defined; C-ns: control not shown; C-pab: control used preabsorbed antibody; C-pcp: control preincubation with control peptide/antigen/preimmune serum; d: dog; gp: guinea pig; h: human; ha: hamster; ho: horse; ICC: immu-

TABLE 9.3

Reported Expression Pattern of TRPs in Cardiac Fibroblast Using Different Methodologies

Type of Expression Analysis	RT-PCR	Northern Blot	Western Blot	ICC	IHC
TRP Subfamily/ Member					
TRPC					
TRPC1	(Nishida et al., 2007)-m (Roderick et al., 2007)-r (Chen et al., 2010)-h (Du et al., 2010)-h,-m (C-nd) (Davis et al., 2012)-r (Ikeda et al., 2013)-h		(Harada et al., 2012) goat,-h,-d (C-nd) (Ikeda et al., 2013)-h (C-ns)	(Ikeda et al., 2013) (C-ns)	
TRPC2	(Rose et al., 2007)-r				
TRPC3	(Rose et al., 2007)-r (Chen et al., 2010)-h (Du et al., 2010)-m (C-nd) (Davis et al., 2012)-r (Ikeda et al., 2013)-h (Nishida et al., 2007)-m		(Harada et al., 2012) goat,-h,-d (C-nd) (Ikeda et al., 2013)-h (C-ns)	(Ikeda et al., 2013)-h (C-ns)	
TRPC4	(Rose et al., 2007)-r (Chen et al., 2010)-h (Du et al., 2010)-m (C-nd) (Davis et al., 2012)-r (Ikeda et al., 2013)-h		(Ikeda et al., 2013)-h (C-ns)	(Ikeda et al., 2013)-h (C-ns)	
TRPC5	(Rose et al., 2007)-r				
TRPC6	(Nishida et al., 2007)-m (Rose et al., 2007)-r (Chen et al., 2010)-h (Du et al., 2010)-h,-m (C-nd) (Davis et al., 2012)-r (Ikeda et al., 2013)-h (Kapur et al., 2014)-h+m		(Ikeda et al., 2013)-h (C-ns)	(Nishida et al., 2007)-m (C-nd) (Davis et al., 2012-r) (C-nd) (Ikeda et al., 2013) (C-ns) (Kapur et al., 2014)-h	
TRPC7	(Nishida et al., 2007)-m (Rose et al., 2007)-r				
TRPM					
TRPM1					
TRPM2	(Takahashi et al., 2012)-r		(Roberge et al., 2014)-m (C-nd)		
TRPM3	(Rose et al., 2007)-r (very law) (Du et al., 2010)-m (C-nd)				
TRPM4	(Rose et al., 2007)-r (Du et al., 2010)-h,-m (C-nd)		(Kecskes et al., 2015)-m (C-KO)		
TRPM5					
TRPM6	(Du et al., 2010)-m (C-nd)				

(Continued)

TABLE 9.3 (*Continued*)

Reported Expression Pattern of TRPs in Cardiac Fibroblast Using Different Methodologies

Type of Expression Analysis	RT-PCR	Northern Blot	Western Blot	ICC	IHC
TRPM7	(Rose et al., 2007)-r (Du et al., 2010)-h,-m (C-nd) (Adapala et al., 2013)-r (C-nd)	(Runnels et al., 2001)-m	(Guo et al., 2014)-r (C-nd) (Yu et al., 2014)-r (C-nd) (Zhou et al., 2015)-r (C-nd)	(Runnels et al., 2002)-r (C-nd)	
TRPM8					
TRPV					
TRPV1	(Rose et al., 2007)-r				
TRPV2	(Rose et al., 2007)-r (Hatano et al., 2009)-r (Du et al., 2010)-h,-m (C-nd)				
TRPV3					
TRPV4	(Rose et al., 2007)-r (Hatano et al., 2009)-r (Du et al., 2010)-h,-m (C-nd) (Adapala et al., 2013)-r (C-nd)		(Adapala et al., 2013)-r (C-nd)	(Hatano et al., 2009)-r (C-pcp)	
TRPV5	(Rose et al., 2007)-r (very low)				
TRPV6	(Rose et al., 2007)-r				
TRPML					
TRPML1					
TRPML2					
TRPML3					
TRPP					
TRPP2					(Markowitz et al., 1999)-m (C-pcp)
TRPP3					
TRPP5					
TRPA					
TRPA1	(Oguri et al., 2014)-h		(Oguri et al., 2014)-h (C-nd)	(Oguri et al., 2014)-h (C-nd)	

For those expression analyses using anti-TRPs antibodies, the type of the corresponding control experiment(s) or a comment regarding the validation of the antibodies is included. ab: antibody; b: bovine; C-KD: control used siRNA-mediated knockdown; C-KO: control used corresponding knockout mouse specimen; C-nd: control not defined; C-ns: control not shown; C-pab: control used preabsorbed antibody; C-pcp: control preincubation with control peptide/antigen/preimmune serum; d: dog; gp: guinea pig; h: human; ha: hamster; ho: horse; ICC: immunocytochemistry; IEM: immunoelectron microscopy; IHC: immunohistochemistry; m: mouse; MA: microarray; r: rat; rb: rabbit; SAN: sinoatrial node.

(Gronich et al., 2010). Additional examples in differential expression patterns of TRP channels include the following: TRPV2 expression is downregulated in rodent models of myocardial infarction (Entin-Meer et al., 2014) and upregulated together with TRPC1 in hearts from patients with dilated cardiomyopathy (Iwata et al., 2013). In patients with atrial fibrillation, TRPM6 mRNA and protein levels in the right atrium were significantly increased (Zhang et al., 2015). Analysis of mRNA and protein levels showed increased expression of TRPM7 in the left ventricle free wall from patients with ventricular tachycardia (Parajuli et al., 2015).

9.3 TRP PROTEINS AND THEIR CONTRIBUTION TO Ca^{2+} SIGNALING IN CARDIAC CELLS

9.3.1 TRP CHANNELS IN CARDIOMYOCYTES

9.3.1.1 TRPCs in Cardiomyocytes

In human atrial myocytes a nonselective cation current recorded using whole-cell patch voltage clamp under K^+-free conditions was enhanced by thapsigargin, endothelin-1 (ET-1) or angiotensin II (AngII). This enhancement was prevented by pipette inclusion of an antibody against TRPC1 (Zhang et al., 2013). Based on inhibition by an anti-TRPC1 antibody, it was also shown that the Ca^{2+} rise or paradox produced by restoration of extracellular Ca^{2+} after Ca^{2+}-free superfusion, was significantly reduced in adult ventricular cardiomyocytes (Kojima et al., 2010). In adult rat cardiomyocytes, a nonselective cation current (I_{ATP}) activated via P_2Y_2 receptors and the phospholipase C pathway, which exhibits several features of overexpressed TRPC3/TRPC7 channels, was inhibited upon addition of an anti-TRPC3 antibody to the patch pipette solution (Alvarez et al., 2008). Similarly, cation currents in mouse cardiomyocytes induced by OAG (1-oleoyl-2-acetyl-sn-glycerol; I_{OAG}), a diacylglycerol (DAG) analogue, or by OAG plus insulin were significantly reduced by inclusion of an anti-TRPC3 antibody in the patch pipette (Fauconnier et al., 2007). In mouse ventricular cardiomyocytes, a stretch-induced nonselective cation conductance is sensitive to antibodies against TRPC6 (Dyachenko et al., 2009). In ventricular cardiomyocytes isolated from a mouse overexpressing adrenergic α1A-receptors (α1-AR), a specific α1A-AR agonist, as well as AngII, resulted in increased intracellular Ca^{2+} concentration that was reduced by preincubation with an anti-TRPC6 but not with an anti-TRPC1 antibody (Mohl et al., 2011).

An AngII-induced TRPC-like cation current affecting membrane depolarization and Ca^{2+} oscillations was measured in neonatal rat ventricular myocytes (NRVM) (Onohara et al., 2006). It was proposed that stimulation of NRVM with AngII leads to the activation of TRPC3/TRPC6 channels via DAG, and its activation causes slow increases in the membrane potential, that is diminished upon TRPC6 siRNA treatment, leading to activation of L-type Ca^{2+}-channel activity and, finally, dephosphorylation of NFAT by calcineurin (Onohara et al., 2006). Ca^{2+} entry induced by depletion of intracellular Ca^{2+} stores (SOCE) together with AngII in adult mouse cardiomyocytes was enhanced in cardiomyocytes by TRPC3 channel overexpression in transgenic mice compared to wild-type controls (Nakayama et al., 2006). Similarly, double knockdown of TRPC3/TRPC6 in NRVM reduced SOCE evoked by ET-1 or AngII (Kinoshita et al., 2010). Overexpression of TRPC3 additionally led to increased Ca^{2+} entry triggered by combined store depletion via thapsigargin (TG) together with phenylephrine stimulation, and this was diminished in myocytes expressing a dominant-negative variant of TRPC4 (dnTRPC4) suggesting functional interaction of the TRPC1/TRPC4/TRPC5 and TRPC3/TRPC6/TRPC7 subgroups in cardiomyocytes (Wu et al., 2010). TRPC3 and dnTRPC4 could be co-immunoprecipitated from the heart of double transgenic mice, but since the specificities of most anti-TRPC antibodies including those that are commercially available in native tissues are not validated appropriately (Meissner et al., 2011), corresponding experiments using TRPC3- and/ or TRPC4-deficient mice could unequivocally demonstrate interaction of the two TRPC proteins in murine cardiomyocytes. SOCE (induced by cyclopiazonic acid, CPA) was found to be increased in ventricular cardiomyocytes from mice challenged with pressure overload via transverse aortic constriction (TAC), and this increase was largely diminished in myocytes expressing a dominant-negative variant of TRPC3 (dnTRPC3), TRPC6 (dnTRPC6), and TRPC4 (dnTRPC4), respectively (Wu et al., 2010). Additionally, adenovirus-mediated expression of dnTRPC4 or dnTRPC6 proteins in cultured adult feline myocytes reduced TRPC3-mediated Ca^{2+} entry evoked by a combined stimulation protocol using depletion of intracellular Ca^{2+} stores (by CPA) and simultaneous application of OAG (Makarewich et al., 2014). CPA/OAG stimulation also evoked Ca^{2+} elevation when TRPC4 or TRPC6 proteins were expressed in feline myocytes and TRPC6-mediated Ca^{2+} entry was also reduced by expression of dnTRPC4, again suggesting functional interaction of the TRPC1/TRPC4/TRPC5 and TRPC3/TRPC6/TRPC7 subgroups. In an independent study, dnTRPC3 also reduced Ca^{2+} entry evoked upon phospholipase C (PLC) stimulation and mediated by the sodium/calcium exchanger (NCX) in adult rat cardiomyocytes (Eder et al., 2007).

In ventricular cardiomyocytes from mice challenged with pressure overload (TAC), Seth et al. found a significant increase in nonselective cation currents compared to Sham-treated mice, and a similar increase in current density was observed upon AngII stimulation, but response to both chronic pressure overload and AngII was blunted in cardiomyocytes from TRPC1-deficient mice (Seth et al., 2009). However, the consequences of the blunted induction of this cation current for Ca^{2+} homeostasis were not investigated in the study. In a systematic analysis of multiple knockout mice using fluorescence imaging of electrically paced adult ventricular cardiomyocytes and Mn^{2+}-quench microfluorimetry, a background Ca^{2+} entry (BGCE) pathway that critically depends on TRPC1/TRPC4 proteins but not TRPC3/TRPC6 was recently identified. Despite numerous stimulation protocols used to assess the contribution of TRPC to cardiomyocytes Ca^{2+} signaling, in most studies comparably little attention was paid to the contribution of TRPC activity to Ca^{2+} transient amplitude and diastolic Ca^{2+} levels under steady-state pacing to reflect the condition of cardiomyocytes in the beating heart. Camacho Londoño et al. showed that reduction of BGCE in TRPC1/TRPC4-deficient murine cardiomyocytes lowers diastolic and systolic Ca^{2+} concentrations under steady-state pacing conditions without affecting cardiac contractility measured *in vitro* in isolated hearts and *in vivo* by echocardiography. In addition, it was shown that absolute and relative changes of Ca^{2+} transient amplitudes after AngII (500 nM) or isoproterenol stimulation (only absolute changes) were reduced in TRPC1/TRPC4-deficient cardiomyocytes but not in TRPC3/TRPC6-deficient cardiomyocytes (Camacho Londono et al., 2015). Another study described that in adult cardiomyocytes from TRPC3/TRPC6-deficient mice, the peak of AngII (10 nM)-induced Ca^{2+} transients and AngII-evoked L-type currents observed in cells from wild-type mice were abolished in cells from TRPC3/TRPC6-deficient mice (Klaiber et al., 2010). In cardiac hypertrophy, when transmembrane guanylyl cyclase (GC-A) is desensitized, atrial natriuretic peptide (ANP) evokes stimulatory effects on I_{Ca} and Ca^{2+} transients independent of cGMP/protein kinase G (PGK-I) signaling by direct GC-A/TRPC3/TRPC6 interaction. These stimulatory effects of ANP on Ca^{2+} transients and L-type Ca^{2+} currents were absent in TRPC3/TRPC6-deficient myocytes (Klaiber et al., 2011). In freshly isolated ventricular myocytes from wild-type mice that chronically received Isoproterenol, stimulation by endothelin-1 produced TRPC-like currents that were similar to currents observed in TRPC6-overexpressing cardiomyocytes. Importantly, these currents were not observed in cardiomyocytes from TRPC6-deficient mice and were inhibited by the addition of soluble Klotho proteins in myocytes isolated from TRPC6-overexpressing mice (Xie et al., 2012).

Using neonatal rat ventricular myocytes (NRVM), it was proposed that TRPC7 channels could mediate apoptosis of cardiomyocytes triggered by AngII when TRPC7 is overexpressed (Satoh et al., 2007). Recently, the involvement of TRPC channels in mineralocorticoid signaling was studied in NRVM. In these cells aldosterone treatment for 24 hours increased the expression levels of TRPC1, TRPC4, and TRPC5 and correlated with increased SOCE. These effects were prevented by TRPC1 and TRPC4 dominant negative mutants or by TRPC5 siRNA treatment (Sabourin et al., 2016).

With advances in cellular reprogramming, new cardiomyocyte cell models have become available to investigate the role of TRP proteins in cardiac cells. Recently TRPC3-like currents were evoked in stem cell–derived mouse cardiomyocytes (mESC-CMs) by OAG, which can be reduced by the TRPC3 blocker Pyr3 or by a TRPC3-dominant negative variant of TRPC3. The consequence of reducing those currents was a decrease in the pacemaker activity of the cells and decreased phosphorylation of RyR2 and phospholamban (Qi et al., 2016).

9.3.1.2 Knockdown and Knockout of TRPCs in Cardiomyocytes and Implications for Cardiac Hypertrophy Signaling

Based on the influence of Ca^{2+} signaling on cardiomyocyte physiology, a role of TRPC channels was proposed in signaling cascades mediating the development of cardiac hypertrophy and remodeling. Primary experimental evidence supporting this hypothesis was based on cellular experiments in NRVMs followed by experiments in cardiomyocytes from transgenic mice overexpressing TRPC proteins in the heart. A synopsis of the results that support a role of TRPCs in cellular hypertrophy is presented below.

Initial evidence pointing toward a participation of TRPC1 channels in cardiac hypertrophy came from a study reporting the fact that the murine TRPC1 gene has a neuron-restrictive silencer element (NRSE)-like sequence; this kind of sequence is a target of neuron-restrictive silencer factors (NRSFs), which are involved in the regulation of cardiac genes and cardiac remodeling. TRPC1 upregulation in transgenic α-MHC-dnNRSF mice was correlated with increased NFAT activity in NRVM overexpressing TRPC1 (Ohba et al., 2006). Furthermore, in this cell type it was shown that increases in cell size and in expression of ANP and brain natriuretic peptide (BNP) triggered by ET-1, AngII, or phenylephrine (PE), as well as the ET-1 induced Ca^{2+} entry were attenuated by TRPC1 knockdown (Ohba et al., 2007). Others reported that the hypertrophic response mediated by the serotonin receptor 5-HT_{2A} was suppressed by TRPC1 knockdown in H9C2 cells (Vindis et al., 2010).

In vitro models using cardiac cells indicated that TRPC3 expression induces cellular hypertrophy in NRVM and activates the calcineurin-NFAT signaling (Bush et al., 2006). By using siRNA in NRVM, it was shown that TRPC3 mediates the induction of ANP and BNP expression in response to numerous hypertrophic stimuli, but TRPC3 knockdown does not affect changes in cell size or beating frequency (Brenner and Dolmetsch, 2007). In a similar approach with NRVM, nicotine led to cardiomyocyte hypertrophy and increased SOCE, NFAT activity, and expression of cardiac hypertrophy genes in a dose-dependent manner that was diminished by TRPC3 siRNA treatment (Li et al., 2015). Additionally, NFAT consensus sequences in the promoter region of the TRPC6 gene have been detected (Kuwahara et al., 2006).

Likewise, in NRVM ET-1–induced hypertrophic responses, NFAT activation and Ca^{2+} oscillations were significantly reduced after treatment with siRNAs or dominant negative constructs directed against TRPC6 (Nishida et al., 2007; Nishida et al., 2010). Furthermore, based on experiments using a dnTRPC6, it was shown that PKG activated by phosphodiesterase (PDE) 5 inhibition phosphorylates TRPC6 proteins at Thr69 and prevents TRPC6-mediated Ca^{2+} influx. Substitution of Thr69 by Ala in TRPC6 abolished the antihypertrophic effects of PDE5 inhibition (Nishida et al., 2010). In line with this finding it was shown in NRVM that PDE5 inhibition blocks TRPC6 channel activation via cGMP- and protein kinase G (PKG)-dependent phosphorylation of TRPC6 channels and associated Cn/NFAT signaling that includes a further increase in TRPC6 expression (Koitabashi et al., 2010).

Additionally, experiments in cells expressing $PKG1α^{C42S}$ (redox-dead mutation) and a phosphosilenced mutant of TRPC6 ($TRPC6^{T70A,S322Q}$) that prevents TRPC6 modulation by PGK corroborate the concept that ET-1–evoked NFAT activation and BNP expression as signaling steps during the development of hypertrophy occur through TRPC6 activation and are balanced via PGK-mediated phosphorylation of TRPC6, which occurs most efficiently when PGK1α oxidation is prevented (Nakamura et al., 2015).

AngII-evoked NFAT activation was suppressed in rat neonatal cardiomyocytes by the TRPC3/Orai1 blocker Pyr3, which was initially identified as a TRPC3 blocker (Kiyonaka et al., 2009) but was later found to also inhibit Orai1 and SOCE currents (Schleifer et al., 2012). GSK2332255B, which is characterized as an antagonist of recombinant TRPC3 and TRPC6 channels with a low nanomolar IC_{50} (and an IC_{50} of 15 µM for I_{CRAC}), suppresses ET-1–induced upregulation of ANP, BNP, and the nuclear factor of activated T-cells (NFAT) reporter gene regulator of calcineurin 1 (RCAN1) in cardiomyocytes of wild-type animals in a similar way as TRPC3/TRPC6 gene inactivation does. Unfortunately, this antagonist has limited *in vivo* efficacy due to rapid metabolism and/or high protein binding (Seo et al., 2014b).

Two splice variants of TRPC4 (TRPC4α and TRPC4β) have been identified (Freichel et al., 2014). In neonatal rat cardiomyocytes, overexpression of TRPC4α, but not TRPC4β, increased cell size and the expression of hypertrophy-related gene signaling. In addition, a dominant-negative TRPC4 mutant reduced the PLCβ1b-induced cellular hypertrophy response (Cooley et al., 2014).

9.3.1.3   TRPMs in Cardiomyocytes

Stimulation of freshly isolated mouse ventricular myocytes with H_2O_2 evoked an elevation of the intracellular Ca^{2+} concentration, which was reduced in TRPM2-deficient myocytes (Miller et al., 2013). Adenosine diphosphate-ribose (ADPR) produced a current in myocytes isolated from the ventricle of wild-type mice that could not be recorded in cells from TRPM2 global or cardiomyocyte-specific deficient mice (Miller et al., 2014; Hoffman et al., 2015). Cardiomyocytes isolated from these mice after ischemia/reperfusion presented prolonged action potentials compared to wild-type cells (Miller et al., 2013). In TRPM2-deficient ventricular myocytes mitochondrial membrane potential, mitochondrial Ca^{2+} uptake, ATP levels, and O_2 consumption were lower, and mitochondrial superoxide levels were elevated, and these impairments became even more pronounced after ischemia/reperfusion (I/R) (Miller et al., 2014). Further investigation of the role of TRPM2 channels in the maintenance of bioenergetics in ventricular myocytes showed that Ca^{2+} entry via TRPM2 channels was crucial for cardiomyocyte ability to maintain mitochondrial function, reduce reactive oxygen species (ROS) levels, and increase O_2 consumption (Hoffman et al., 2015).

Cation currents with properties of those mediated by TRPM4b and TRPM5 have been described in freshly isolated human atrial cardiomyocytes (Guinamard et al., 2004; Zhang et al., 2012b). Currents with TRPM4 properties were reported to be present in ventricular myocytes from spontaneously hypertensive rats (SHRs) but were rarely detected in cells from Wistar Kyoto control rats (WKRs). The difference was attributed to increased TRPM4 expression in the hypertensive rats (Guinamard et al., 2006). TRPM4-like, Ca^{2+}-activated nonselective cation currents (NSC_{Ca}) were also characterized in sinoatrial node cells from the mouse (Demion et al., 2007). The TRPM4 blocker 9-phenanthrol decreased Ca^{2+} oscillations in HL-1 mouse cardiac myocytes where TRPM4 is expressed (Burt et al., 2013) and siRNA-mediated knockdown of TRPM4 prevented cell death induced by H_2O_2 in H9c2 rat cardiomyoblast-derived cells (Piao et al., 2015). Furthermore, TRPM4-like, 9-phenanthrol sensitive currents were also reported in 43% of freshly isolated Purkinje cells from rabbit heart but in none of the analyzed ventricular cells (Hof et al., 2016). In mice it was reported that the channel is functionally expressed in ventricular cardiomyocytes because TRPM4-like currents were characterized in wild-type cells and absent in TRPM4-deficient myocytes. Measurements of action potential duration showed a significantly decreased time for 50% and 90% repolarization in TRPM4-deficient ventricular myocytes leading to an increased driving force for the L-type Ca^{2+} current during the action potential. Voltage-gated L-type Ca^{2+} channels on their own were not affected just like the activity of other channels that are known to determine the duration of the action potential. Electrically evoked Ca^{2+} transients were comparable between wild-type and TRPM4-deficient myocytes under basal conditions, but upon β-adrenergic stimulation the peaks of the Ca^{2+} transients were enlarged in TRPM4-deficient myocytes, whereas the Ca^{2+} load of the SR was not different (Mathar et al., 2014). Adult ventricular cardiomyocytes from cardiomyocyte-specific TRPM4-deficient mice showed significantly increased SOCE upon AngII stimulation as well as enlarged area-under-the-curve of Ca^{2+} transients during constant pacing. However, the impact on basal diastolic Ca^{2+} levels produced by TRPM4 inactivation was not analyzed (Kecskes et al., 2015) . Experiments with neonatal cardiomyocytes from TRPM4-knockout mice demonstrated that the cellular hypertrophy after AngII treatment was exacerbated compared with wild-type mice (Kecskes et al., 2015). In rat and mouse atrial myocytes, as well as in HL-1 cells, shear stress activates a current that is mainly conducted by monovalent cations. This current was greatly reduced by 9-phenanthrol, intracellular dialysis of anti-TRPM4 antibodies, or siRNA against TRPM4; fascinatingly, the current was almost negligible in atrial myocytes from IP_3R_2-deficient mice (Son et al., 2016).

TRPM6/TRPM7-like currents have been recorded and characterized in freshly isolated ventricular cardiomyocytes from pig, rat, and guinea pig; their electrophysiological characteristic was suggested to correspond to the Mg^{2+}-inhibited nonselective cation current (I_{MIC}), and the currents showed to be modulated by ATP, PIP_2, and nonhydrolysable GTP analogs (Gwanyanya et al., 2004, 2006; Macianskiene et al., 2008). Later, similar currents were recorded in ventricular cardiomyocytes from adult mice and attributed to TRPM7 based on comparisons with cells from mice with cardiomyocyte-specific TRPM7 deletion. In TRPM7-deficient cardiomyocytes the action potential was prolonged, and transient outward currents were reduced (I_{to}), which was associated with transcriptional modifications including downregulation of Kcnd2 expression (Sah et al., 2013a). In addition, disruption of TRPM7 in cultured embryonic cardiomyocytes reduced spontaneous Ca^{2+} transient firing rates and was associated with robust downregulation of Hcn4, Cav3.1, and SERCA2a mRNA (Sah et al., 2013b). In a screening to identify inhibitors for TRPM7 channels, the compound NS8593 (IC_{50} 1.6 μM) was able to block TRPM7-like currents in cultured primary human cardiac myocytes (Chubanov et al., 2012) and reduced the Mg^{2+} entry in ventricular rat myocytes (Tashiro et al., 2014); however, the specificity of its action particularly in cardiomyocytes and a comparison with Trpm7 gene inactivation, remains to be demonstrated. TRPM7-like currents were also recorded in freshly isolated human right atrial cardiomyocytes from hearts in sinus rhythm and from both genders (Macianskiene et al., 2012; Zhang et al., 2012b); in myocytes from patients with atrial fibrillation the density of Mg_i^{2+}-sensitive TRPM7 current was enlarged compared to cells from patients with sinus rhythm (Zhang et al., 2012b) suggesting the involvement of TRPM7 in this pathology.

9.3.1.4 TRPVs in Cardiomyocytes

Pharmacological activation of TRPV1 with the agonist SA13353 prevented the increase in cardiomyocyte size evoked by ET-1, which increases in heart tissue upon prolonged cold exposure (Zhang et al., 2012a). H9C2 cells derived from rat cardiomyoblasts express TRPV1 proteins. Activation of TRPV1 with capsaicin induced loss of cell viability in a dose-dependent manner and was more prominent in ischemia/reperfusion conditions. Capsaicin induced mitochondrial superoxide production, mitochondrial depolarization, and elevation of intracellular Ca^{2+} concentration. These effects were blunted by the TRPV1 antagonist capsazepine or by siRNA against TRPV1 (Sun et al., 2014). Recently, mESC-CMs TRPV1 downregulation via shRNA blunted the rise in intracellular Ca^{2+} concentrations generated by capsaicin stimulation, reduced the percentage of cTnT-positive cardiomyocytes, reduced the diameter and percentage of beating embryonic bodies, and suppressed the expression of cardiomyocyte marker genes including cardiac actin, c-TnT, c-TnI, and α-MHC suggesting a role of TRPV1 in cardiomyocyte differentiation (Qi et al., 2015b).

Electrically stimulated cardiomyocytes from TRPV2-deficient mice have smaller (~50%) Ca^{2+} transients with prolonged time to 50% decay compared to wild-type cells. Analysis of caffeine-induced Ca^{2+} transients revealed that SR Ca^{2+} load was also reduced in TRPV2-deficient cells, whereas NCX activity was normal. No microscopic evidence of myocardial pathology such as inflammation, myofiber degeneration, or necrosis was identified in the TRPV2$^{-/-}$ mouse line (Rubinstein et al., 2014). In a second mouse model with cardiomyocyte-specific deletion of TRPV2 comparable defects in Ca^{2+} handling and cellular contractility were reported. However, neonatal cardiomyocytes of these mice form no intercalated discs, do not present stretch-induced increase in intracellular Ca^{2+} concentration, and lack secretion of insulin-like growth factor (IGF-1) in response to stretch stimulation (Katanosaka et al., 2014). In ventricular cardiomyocytes from the Duchenne muscular dystrophy mouse model, *mdx* mouse, the increase in intracellular Ca^{2+} concentration provoked by osmotic stress was largely reduced by pretreatment with anti-TRPV2 (but not anti-TRPC1) antibodies or by TRPV2 siRNA without affecting the SR Ca^{2+} content (Lorin et al., 2015). In a similar approach the stretch-induced Mn^{2+} entry in cardiomyocytes was abolished by anti-TRPV2 antibodies in cells from *mdx* mice but not from wild-type ones (Aguettaz et al., 2016). Human pluripotent stem cell–derived cardiomyocytes (hPSC-CMs) express TRPV4. Stimulation of these cells with the TRPV4 agonist 4α-PDD produced an increase in the intracellular Ca^{2+} concentration that was prevented by TRPV4 inhibitors and by transduction of a TRPV4-dominant negative construct (Qi et al., 2015a).

9.3.1.5   TRPPs in Cardiomyocytes

TRPP2-deficient mice (Pkd2$^{-/-}$) die *in utero* between embryonic day 12.5 and 13.5 and birth in part due to cardiac defects in septation and malformations (Wu et al., 2000; Pennekamp et al., 2002). TRPP2-deficient cardiomyocytes isolated at the embryonic day 17.5 exhibit a higher frequency of spontaneous Ca^{2+} oscillations, but the oscillations showed smaller peaks compared to those of wild-type cells. Furthermore, the Ca^{2+} release from the SR evoked by SERCA inhibition with thapsigargin, as well as the Ca^{2+} release evoked by caffeine were reduced, suggesting a reduced SR Ca^{2+} load (Anyatonwu et al., 2007). In line with this result, caffeine-induced Ca^{2+} transients in resting cardiomyocytes from 5-month-old TRPP2$^{+/-}$ mice are about 60% smaller compared to wild-type cells; interestingly, in electrically paced TRPP2$^{+/-}$ cells the magnitude of Ca^{2+} release was higher and presented a slower decay constant for Ca^{2+} reuptake compared to wild-type cells, but the contractility was similar. No differences in diastolic Ca^{2+} levels were detected (Kuo et al., 2014).

9.3.1.6   TRPA in Cardiomyocytes

In murine ventricular cardiomyocytes specific TRPA1 (allyl isothiocyanate, AITC) and TRPV1 (capsaicin) agonists, respectively, produced dose-dependent and transient rises in the intracellular Ca^{2+} concentration that were not observed in cells obtained from TRPA1- and TRPV1-deficient mice, or pretreated with selective TRPA1 or TRPV1 antagonists, AITC HC-030031 or antagonist SB366791, respectively (Andrei et al., 2016).

9.4   ROLE OF TRPs IN CARDIAC FIBROBLASTS

Within the past years increasing attention has been attributed to the role of cardiac fibroblasts that are involved in myocardial development as well as in pathologies characterized by changes in the extracellular matrix such as fibrosis. Fibroblasts regulate the environment in which cardiac myocytes are embedded through paracrine signaling by cytokines like IL-1, IL-6, TGFβ, and also by direct communication via cell-to-cell interaction. Thereby, they play a critical role in the development of cardiac remodeling and arrhythmias such as atrial fibrillation (Baudino et al., 2006; Souders et al., 2009; Kakkar and Lee, 2010). Expression of TRP channels in cardiac fibroblasts has been reported (Table 9.3), but the functional role of TRPs in cardiac fibroblasts is still poorly understood.

9.4.1   TRPCs in Cardiac Fibroblasts

In rat ventricular fibroblasts, nonselective cation currents triggered by stimulation of the C-type natriuretic peptide (CNP) receptors showed characteristics of TRPC channels (Rose et al., 2007). Also, spontaneous Ca^{2+} oscillations in human fibroblasts can be abolished by SOCE blocking approaches such as by La^{3+} or PLC inhibition (Chen et al., 2010). A nonselective cation current sensitive to a TRPC3 blocker (Pyr3) was described in cardiac fibroblast from adult rats. TRPC3 blockade with Pyr3 blunted AngII-induced Ca^{2+} entry, proliferation, and α-smooth muscle actin (α-SMA) protein expression in fibroblasts. Similarly, Pyr3 suppressed the increased proliferation and α-SMA expression observed in left atrial fibroblasts from an atrial fibrillation model in dogs

(Harada et al., 2012). Decreased proliferation was also observed in a shRNA-induced TRPC3 knockdown in canine atrial fibroblasts. TRPC6 has been postulated as a regulator of ET-1–induced myofibroblast differentiation in rat neonatal cardiac fibroblasts; fibroblasts expressing constitutively active $G\alpha_{12}$ or $G\alpha_{13}$ proteins or stimulated by ET-1 presented increased TRPC6 expression, enhanced Ca^{2+} influx, and increased NFAT activation; these responses were reduced by knockdown of TRPC6, but interestingly, myofibroblast formation was enhanced by TRPC6 inhibition suggesting that TRPC6 activity suppresses fibrotic responses in cardiac fibroblasts (Nishida et al., 2007). In another study, TRPC6 overexpression in neonatal rat cardiac fibroblasts led to the development of a myofibroblast phenotype with upregulated α-SMA expression associated with an increase in SOCE. SOCE was also increased upon transforming growth factor beta (TGF-β) expression in neonatal rat cardiac fibroblasts, and TGF-β–evoked increase in SOCE was also observed in dermal fibroblasts from wild-type but not from TRPC6-deficient mice (Davis et al., 2012). The increase in intracellular Ca^{2+} concentration elicited by OAG in human ventricular fibroblasts was inhibited by siRNA against TRPC6 (Ikeda et al., 2013). Silencing TRPC6 with a further siRNA approach, attenuated TGF-β1-mediated upregulation of α-SMA in human cardiac fibroblasts from the right ventricle (Kapur et al., 2014).

9.4.2 TRPMs in Cardiac Fibroblasts

A hypoxia-evoked cation current was investigated in primarily cultured fibroblasts from adult rat hearts. It exhibits a reversal potential of around −20 mV, was inhibited by clotrimazole, enhanced by ADP ribose, and correlated with increased expression of TRPM2. siRNA against TRPM2 prevented the development of the hypoxia-induced current in cardiac fibroblasts (Takahashi et al., 2012). H_2O_2 also induced a higher Ca^{2+} elevation in hypoxia-exposed adult rat cardiac fibroblasts compared to normoxia-exposed cells, which was attributed to TRPM2 (Takahashi et al., 2012). Based on electrophysiological recordings, Ca^{2+} imaging, and mRNA knockdown with shRNA, it was reported that right atrial human fibroblasts present TRPM7-like currents that mediate Ca^{2+} entry responsible for fibroblast differentiation and proliferation evoked by TGF-β1 stimulation (Du et al., 2010). Furthermore, the authors tried to detect currents from other TRP channels (TRPV2, TRPV4, and TRPC6) in atrial fibroblasts using conditions described for the corresponding channels in heterologous systems but did not succeed. Additionally, the treatment with siRNAs against TRPC1 did not alter Ca^{2+} influx that was evoked in these fibroblasts by application of a bath solution containing 20 mm Ca^{2+} (Du et al., 2010). A current with a current-voltage relationship identical to I_{TRPM7} was measured in fibroblasts obtained from ventricles of 1-day-old rats (Runnels et al., 2002). A TRPM7-like current as well as a time-dependent increase in current density upon AngII application has also been described in cardiac fibroblasts from adult rats (Zhou et al., 2015). Qin and coworkers described TRPM7-like currents in fibroblasts isolated from human atria or from adult murine hearts.

The currents were sensitive to TRPM7 antagonists characterized in TRPM7-overexpressing HEK-293 cells in the same study (Qin et al., 2013). Primary isolated cardiac fibroblasts from adult rats treated with H_2O_2 exhibit an elevated expression of fibrogenic growth factors such as collagen type I, fibronectin, smooth muscle α-actin, connective tissue growth factor (CTGF), and TGF-β1. Additionally, H_2O_2 induces a sustained increase in the intracellular Ca^{2+} concentration in these cells. All these responses were either blunted or reduced in fibroblasts where TRPM7 was silenced via RNA interference (Guo et al., 2014). With a similar approach the same group reported comparable results regarding the fibrogenic expression profile following AngII treatment. They also reported that AngII treatment increased intracellular Ca^{2+} (acutely within seconds) and Mg^{2+} (24 hours after stimulation) concentration, and both were reduced by TRPM7 shRNA treatment (Yu et al., 2014).

9.4.3 TRPVs in Cardiac Fibroblasts

It was shown that TRPV4 is required for TGF-β1-induced differentiation of cardiac fibroblasts into myofibroblasts. The TRPV4-specific antagonist AB159908 as well as the siRNA-mediated knockdown of TRPV4 inhibited TGF-β1-induced differentiation. TGF-β1 pretreatment augmented the increase in intracellular Ca^{2+} concentration produced by the TRPV4 agonist GSK1016790A: this response was reduced by pharmacological inhibition of TRPV4 (Adapala et al., 2013). *In vitro*, the AngII-induced proliferation of cardiac fibroblasts was reduced by capsaicin in cells from wild-type but not from TRPV1-deficient mice (Wang et al., 2014).

9.4.4 TRPA1 in Cardiac Fibroblasts

In primary human ventricular cardiac fibroblasts methylglyoxal provokes a sustained increase in the intracellular Ca^{2+} concentration that is largely reduced by treatment with HC030031, a selective TRPA1 antagonist, or by siRNA-induced knockdown of TRPA1 (Oguri et al., 2014). This is an interesting finding considering that methylglyoxal contribute to the development of pathologies associated with diabetes.

9.5 REGULATION OF CARDIAC CONTRACTILITY BY TRP CHANNELS

Blockade of TRPC6 using an anti-TRPC6 antibody inhibited the α_{1A}-AR-mediated increase in left ventricular pressure and dP/dtmax in Langendorff perfused mouse hearts from wild-type and α_{1A}-AR overexpressing mice (Mohl et al., 2011). Echocardiography revealed no difference in fractional shortening (FS), neither in TRPC6$^{-/-}$ mice nor in TRPC3$^{-/-}$ or TRPC3/TRPC6-deficient mice. Under chronic pressure overload, which evoked a reduction of FS in wild-type mice, TRPC3/TRPC6-deficient mice but not the single deficient mice were protected (Seo et al., 2014b). When isolated papillary muscles and cardiomyocytes from mice lacking either TRPC3 or TRPC6 were subjected to auxotonic load, the induced stress-stimulated contractility (SSC), also known as the Anrep effect, was diminished in TRPC6 but not TRPC3-deficient mice. Furthermore, the inhibition of the SSC response normally obtained by activation of the cGMP/PKG cascade was lacking in TRPC6-deficient mice. Using left ventricular pressure-volume loop (P-V loop) analysis, the effect was corroborated *in vivo* where an acute increase in cardiac afterload was induced by aortic constriction: whereas in control mice a reactive increase in cardiac inotropy was observed, this physiological response was impaired in TRPC6-deficient murine hearts. Additionally, TRPC6-deficient mice developed a delay in cardiac relaxation compared to wild-type mice under these conditions of increased afterload (Seo et al., 2014a). In comparison, the deletion of both TRPC1 and TRPC4 did not affect hemodynamically relevant cardiophysiological parameters measured under basal conditions, neither *in vivo* as evaluated by echocardiography, nor *in vitro* in working heart preparations (Camacho Londono et al., 2015).

Cardiac contractility analyzed by echocardiography in TRPM2-deficient mice (lacking exons 21 and 22) revealed no differences in comparison to wild-type hearts in left ventricular (LV) mass, heart rate, fractional shortening, and maximal change in dP/dt. However, FS and dP/dtmax were more severely reduced in TRPM2-deficient mice 2–3 days after ischemia/reperfusion injury (I/R, 30 minutes of ischemia), although no difference in infarct size was observed between genotypes (Miller et al., 2013). Similar results were described in cardiomyocyte-specific deficient mice in the I/R model (Hoffman et al., 2015).

On the contrary, in a second, global TRPM2-deficient mouse model no differences in the basal left ventricular contractility were observed in TRPM2-deficient mice, but in an I/R model using 45 minutes of ischemia and only 24 hours of reperfusion left ventricular contractility was less affected in TRPM2-deficient mice compared to wild-type mice, which the authors related to an attenuated exacerbation of the reperfused area probably due to less immigration of neutrophils (Hiroi et al., 2013).

Contractility measurements from isolated left ventricular papillary muscle *in vitro* and left ventricular pressure-volume loop analysis *in vivo* evidenced that hearts from TRPM4-deficient mice develop an increased β-adrenergic inotropic response (Mathar et al., 2014; Uhl et al., 2014). Interestingly, isolated papillary muscles from TRPM4-deficient mice showed improved contractile function in comparison to wild-type mice under ischemic conditions (Uhl et al., 2014).

Cardiac contractility was analyzed in TRPM7 kinase-deficient mice (heterozygous for TRPM7 kinase) and control mice under basal conditions and after chronic AngII treatment. Echocardiographic analysis revealed no differences under basal conditions; in contrast, ejection fraction (EF) was significantly reduced in the TRPM7 kinase-deficient mice compared to controls on AngII infusion (Antunes et al., 2016).

Cardiac contractility between TRPV1-deficient mice and wild-type littermates analyzed by pressure-volume analysis or echocardiography was not different (Pacher et al., 2004; Gao et al., 2014; Lang et al., 2015). Consistent with the previous reports no significant differences between wild-type and TRPV1-deficient mice that underwent a sham operation for myocardial infarction or for pressure overload–induced cardiac hypertrophy, were observed by echocardiography (Huang et al., 2009b, 2010; Buckley and Stokes, 2011; Wang et al., 2014). Cardiac function was studied in hearts isolated from TRPV1-deficient mice. Under conditions where confounding effects of neuro-peptides were excluded (blockage of calcitonin gene-related peptide [CGRP] and substance P [SP] receptors), TRPV1-deficient hearts exhibited improved cardiac function during ischemia/reperfusion (Sun et al., 2014). Furthermore, the TRPV1 inhibitor capsazepine reduced the positive inotropic effect produced by the $β_2$-adrenoreceptor agonist fenoterol or by H_2O_2 in isolated mouse left atria (Odnoshivkina et al., 2015). Changes in cardiac contractility produced by a high-salt diet were comparable between wild-type and TRPV1-deficient mice (Gao et al., 2014; Lang et al., 2015).

Probenecid is an agonist of TRPV2 with a positive inotropic effect *ex vivo* and *in vitro* (isolated hearts and cardiomyocytes). The specificity of this effect was evaluated taking advantage of TRPV2-deficient mice. An echocardiographic analysis revealed that a bolus of probenecid increased the contractile response in wild-type but not in TRPV2-deficient mice (Koch et al., 2012; Rubinstein et al., 2014). This effect was independent of changes in heart rate, which either was not affected (Koch et al., 2012) or was changed similarly in both genotypes (Rubinstein et al., 2014). Moreover, probenecid improved cardiac function in a murine model of I/R (Koch et al., 2013). In a more extended basal analysis of cardiac and vascular function in TRPV2-deficient mice, a decrease in stroke volume, ejection fraction, and cardiac output was observed in TRPV2-deficient mice by echocardiography analysis.

Differences in heart rate were observed in echocardiography but not in invasive measurements. Mean arterial pressure, LV systolic pressure, and LV end-diastolic pressure (LVEDP) were not different between wild-type and TRPV2-deficient mice under baseline conditions or during administration of probenecid. At the cellular level, cardiomyocytes from TRPV2-deficient mice have impaired contractile responses during electrical stimulation and present Ca^{2+} handling defects (Rubinstein et al., 2014). When TRPV2-deficient mice were challenged with exercise, the cardiac function was further deteriorated (Naticchioni et al., 2015). A cardiomyocyte-specific TRPV2-deficient mouse model with the option to inactivate TRPV2 in adulthood to circumvent developmental effects has been characterized regarding cardiac function. Already after 3 days of tamoxifen treatment, a ~95% reduction in TRPV2 protein could be observed in the hearts from TRPV2$^{fx/fx}$- αMHC-MerCreMer mice compared to TRPV2$^{fx/fx}$ mice. This reduction in TRPV2 expression was accompanied by reduced survival of the TRPV2-deficient mice and a rapid and severe decline in cardiac contractility (FS) combined with a massive dilation of the left ventricle evidencing the development of a dilated form of cardiomyopathy in a matter of days. Administration of IGF-1 to the cardiomyocyte-specific TRPV2-deficient mice prevented both the enlargement of the left ventricular diastolic dimensions and cardiac dysfunction. Additionally conduction defects were observed in the absence of TRPV2 in cardiomyocytes which were explained to be secondary to the disorganization of the intercalated discs (Katanosaka et al., 2014).

Severe acute cardiovascular effects were observed upon intravenous administration of the TRPV4 agonist GSK1016790A in dogs, where it reduced cardiac output associated with a dose-dependent reduction in blood pressure and a profound circulatory collapse most likely due to vascular leakage and tissue hemorrhage in the lung. However, this effect was not observed in isolated hearts from rats. The hemodynamic actions of GSK1016790A are most likely mediated by TRPV4, since the cardiovascular effects of the agonist produced in wild-type mice were not observed in TRPV4-deficient mice (Willette et al., 2008).

The defects in Ca^{2+} handling in cardiomyocytes from 5-month-old TRPP2$^{+/-}$ mice have an impact on cardiac contractility *in vivo*. Echocardiography in TRPP2$^{+/-}$ mice revealed increased EF after isoproterenol stimulation in comparison to wild-type mice. Remarkably, TRPP2$^{+/-}$ mice have thinner ventricular walls and longer cardiomyocytes, but no differences were observed in basal parameters of cardiac inotropy (Kuo et al., 2014). In a subsequent study cardiac contractility was compared between TRPP2$^{+/-}$ and wild-type mice at the ages of 1 and 9 months. The 9-month-old TRPP2$^{+/-}$ mice have signs of dilated cardiomyopathy denoted by a loss of EF and a thinner left ventricular posterior wall and septum, whereas in the younger animals only a thinner septum could be observed. In addition, the relative and also the absolute contractile response to isoproterenol was enhanced in 9-month-old, but not in 1-month-old TRPP2$^{+/-}$ mice. Interestingly, the increase in heart rate produced by isoproterenol was reduced by metoprolol in wild-type mice but not in TRPP2$^{+/-}$ (Kuo et al., 2016). In a previous study using another TRPP2-heterozygous mouse model (Pkd2$^{+/LacZ}$), Doppler echocardiography was performed in mice between 3.5 ±1 months and 16 ±2.3 months and from both genders, including young pregnant females. The measurements did not reveal main significant changes in cardiac contractility parameters between age-matched TRPP2-heterozygous mice and wild-type controls (Stypmann et al., 2007).

Echocardiographic analysis of 3-month-old TRPA1-deficient mice, and their corresponding wild-type controls did not reveal differences in cardiac function (Bodkin et al., 2014).

9.6 REGULATION OF CARDIAC REMODELING BY TRP CHANNELS *IN VIVO*

In 2006 two independent groups showed that overexpression either of TRPC3 or TRPC6 proteins under control of the cardiomyocyte-specific α-MHC promoter in mice leads to cardiac hypertrophy. Transgenic α-MHC-TRPC3 mice develop a progressive cardiac hypertrophy with aging. They present an increased AngII/PE- and transverse aortic banding (TAC)-induced hypertrophy, associated with increased NFAT activation and increased SOCE; in addition, 14 days after aortic banding TRPC3-transgenic mice developed cardiac failure (Nakayama et al., 2006). This hypertrophic phenotype was attenuated by the specific overexpression of the plasma membrane Ca^{2+} ATPase 4b (PMCA4b) in cardiomyocytes; the authors interpreted this finding as a result of the removal of Ca^{2+} excess provoked by TRPC3 overexpression (Wu et al., 2009). When isolated cardiomyocytes from TRPC3 overexpressing mice were challenged to a protocol of ischemia-reperfusion (I/R), the rate of apoptotic cardiomyocytes was increased compared to control cells (Shan et al., 2008). Additionally, it was reported that a pyrazole compound (Pyr3) that blocks overexpressed TRPC3 channels can attenuate TAC-induced hypertrophy in mice (Kiyonaka et al., 2009); however, recently it was shown that Pyr3 also blocks Orai1-mediated store-operated Ca^{2+} currents (Schleifer et al., 2012), which mediates hypertrophy in neonatal cardiomyocytes (Voelkers et al., 2010). Kuwahara and coworkers (2006) found that TRPC6 expression is upregulated upon activation of the calcineurin-NFAT pathway and analyzed the effect of TRPC6 overexpression in mouse cardiomyocytes. Transgenic α-MHC-TRPC6 mice with high expression levels die between 5 and 12 days after birth, possibly due to severe cardiomyopathy; transgenic mice with intermediate TRPC6 expression levels develop cardiomegaly and congestive heart failure around 30 weeks of age; founder mice with lower expression levels have no obvious defect, but they reveal a significant higher response to aortic banding measured by the heart weight–to–body weight ratio (Kuwahara et al., 2006). One of these TRPC6-transgenic mouse lines with decreased survival at 24 months of age and an increase in heart mass index and in cardiac fetal gene expression, was crossed with mice overexpressing Klotho under an ubiquitously active promotor, which resulted in improved survival, prevention of an increase in heart mass, and reduction in fetal gene expression (Xie et al., 2012).

Evidence for additional TRPC arose from other two *in vivo* approaches that showed that cardiomyocytes from hypertrophied hearts after aortic banding develop an increased Ca^{2+} influx or a non-selective current associated with TRPC channels that is not mediated by L-type Ca^{2+} channels or the Na^+/Ca^{2+} exchanger (Seth et al., 2009; Wu et al., 2010). The first approach attributed the nonselective current to TRPC1 channels, because this current was not detected in cardiomyocytes from TRPC1-deficient mice; furthermore, TRPC1-deficient mice had reduced TAC- and angiotensin II (AngII)-induced hypertrophy responses (Seth et al., 2009). The second approach reported that increase in Ca^{2+} entry in hypertrophied cardiomyocytes can occur via store depletion (SOCE).

Transgenic mice with specific cardiomyocyte expression of dominant-negative constructs either of TRPC3, TRPC4, or TRPC6 developed reduced pathological hypertrophy induced by TAC or AngII/PE infusion (Wu et al., 2010). The cardiac myocyte–specific expression of a dominant-negative TRPC4, which reduced the activity of TRPC3 and TRPC6 in feline myocytes upon CPA/OAG stimulation, was able to reduce hypertrophy and cardiac structural and functional remodeling after myocardial infarction while increasing survival in mice (Makarewich et al., 2014).

The recently described background Ca^{2+} entry (BGCE) mediated by TRPC1/TRPC4 mentioned above, has not only a relevance for the Ca^{2+} homeostasis in adult cardiomyocytes but also for development of pathological cardiac remodeling. Neurohumoral-induced cardiac hypertrophy as well as the expression of fetal genes (ANP, BNP) and genes regulated by Ca^{2+}-dependent signaling (RCAN1-4, myomaxin) was reduced in TRPC1/TRPC4-deficient (DKO) but not in TRPC1 or TRPC4 single deficient mice. Pressure overload–induced hypertrophy, interstitial fibrosis, as well as deterioration of cardiac function were all ameliorated in TRPC1/TRPC4-deficient mice, whereas they did not show alterations in other cardiovascular parameters contributing to systemic neurohumoral-induced hypertrophy such as renin secretion and blood pressure. Therefore, the TRPC1/TRPC4 gene inactivation protects against development of maladaptive cardiac remodeling without altering cardiac or extracardiac functions contributing to this pathogenesis (Camacho Londono et al., 2015). In the same study it was shown that TRPC3/TRPC6-deficient mice were not protected from the development of cardiac hypertrophy induced by AngII, isoproterenol, or pressure overload. Similar results were observed in an independent work showing no protection in TRPC3/TRPC6-deficient mice after AngII treatment, but a reduction of the hypertrophy response in TRPC3 but not in TRPC6 single deficient mice when compared to a control group that was composed of littermate mice containing at least one allele of the wild-type genes (Domes et al., 2015). In TRPC6-deficient mice a decrease in cardiac mass gain produced by isoproterenol was reported (Xie et al., 2012). Nevertheless, the relevance of TRPC3/TRPC6 was more prominent in a study using a model in the C57Bl/6J genetic background. Here, compound inactivation but not the single deletion of TRPC3 and TRPC6 was protective against pressure overload *in vivo* under this genetic background (Seo et al., 2014b). A possible explanation may be that the nicotinamide nucleotide transhydrogenase (Nnt) is inactivated in the C57Bl/6J mouse strain (Huang et al., 2006) and may lead to accumulation of ROS that enhance the activity of TRPC6-containing channels (Ding et al., 2011). A role of TRPC6 was also reported in another cardiac remodeling process (i.e., in scar formation after myocardial infarction), which was reduced in TRPC6-deficient mice associated with enhanced mortality in an LAD ligation model suggesting that TRPC6 activity is protective in that setting (Nishida et al., 2007; Davis et al., 2012). Echocardiographic measurements in these animals revealed a reduction in cardiac function and greater ventricular wall dilation compared with wild-type mice (Davis et al., 2012). In a canine model of atrial fibrillation administration of Pyr3, which inhibits TRPC3 and store-operated channels, suppressed atrial fibrillation while decreasing fibroblast proliferation and extracellular matrix gene expression (Harada et al., 2012).

A role of TRPM2 for cardiac remodeling was shown in a model of I/R (2–3 days after reperfusion). Although infarct size in TRPM2-global and cardiomyocyte-specific deficient mice was comparable to that produced in wild-type mice, cardiac performance was more strongly reduced in TRPM2-deficient mice, which was associated with differential expression of proteins determining EC coupling and was correlated with accumulation of ROS (Miller et al., 2013; Hoffman et al., 2015) by impairment of the mitochondrial energy homeostasis (Miller et al., 2014). Another group found a reduction of the infarct size in TRPM2-deficient mice after 1 day of I/R and less accumulation of neutrophils in hearts from these mice compared to control mice. However, when the ischemia was done without reperfusion, the infarct size between TRPM2-deficient mice and wild-type mice was similar (Hiroi et al., 2013). Also, doxorubicin-induced cardiomyopathy depends on the presence of TRPM2: doxorubicin administration impairs cardiac contractility and reduces survival in wild-type mice, and these malfunctions are exacerbated in mice with cardiomyocyte restricted deletion of TRPM2 (Hoffman et al., 2015).

Global TRPM4-deficient mice show mild cardiac hypertrophy at 6 months of age, which can develop secondary to the increase in systemic blood pressure that arises most likely due to an elevated tone of the sympathetic nervous system as reflected by the elevation of circulating catecholamine levels (Mathar et al., 2010). To determine the role of TRPM4 in cardiomyocytes for hypertrophy development in adult mice, a TRPM4 cardiomyocyte-specific deficient mouse was generated; after 1 year the cardiomyocyte deletion of Trpm4 did not provoke increase of cardiac mass (Kecskes et al., 2015). When the cardiomyocyte-specific TRPM4-deficient mice were challenged by chronic angiotensin II stimulation, cardiac hypertrophy indexes and expression of hypertrophy-related genes were enlarged compared to controls (Kecskes et al., 2015). TRPM4 global deficient mice were protected following induction of myocardial infarction and showed reduced mortality and better cardiac performance during β-adrenergic stimulation compared to wild-type mice. However, no differences were observed in basal contractility or infarct size (Jacobs et al., 2015). The analysis of a second TRPM4 global deficient mouse model showed that the development of cardiac hypertrophy during aging is associated with cardiomyocytes hyperplasia characterized by a higher density and smaller size of cardiomyocytes in TRPM4-deficient hearts that was evident from the neonatal stage on (Demion et al., 2014).

In a detailed and comprehensive analysis of the role of TRPM7 for ventricular function, several cardiac-targeted TRPM7-deficient mice were compared. Deletion of TRPM7 in heart before embryonic day 9 results in congestive heart failure and death by embryonic day 11.5 due to hypoproliferation of the compact myocardium. Deletion late during cardiogenesis (after E12.5, Trpm7$^{aMHC/flox}$) produces viable mice with normal adult ventricular size and function as well as no transcriptional changes in the heart suggesting that TRPM7 is critical for early events in embryonic cardiac development but is dispensable in late embryonic or adult myocardium. In contrast, upon TRPM7 deletion at an intermediate time point (between E9.5 and E12.5, Trpm7$^{aMHC/-}$) during cardiac development, 50% of mice develop cardiomyopathy associated with heart block, impaired repolarization, and ventricular arrhythmias, all together associated with extensive alterations in the myocardial transcriptional profile including elevations in transcripts of hypertrophy/remodeling genes and reductions in genes important for suppressing hypertrophy (Sah et al., 2013a). TRPM7 kinase-deficient mice (heterozygous for TRPM7 kinase) developed increased cardiac hypertrophy, fibrosis, and cardiac dysfunction after chronic AngII treatment (Antunes et al., 2016).

In a model of I/R in isolated Langendorff hearts, it was shown that TRPV1 gene deletion decreases injury-induced substance P release and impairs cardiac recovery function after I/R (Wang and Wang, 2005). Moreover, in perfused hearts a selective protease-activated receptor-2 (PAR2) agonist (SLIGRL) improved the recovery of hemodynamic parameters and reduced the infarct size in hearts from wild-type and TRPV1-deficient mice; the protective effect of SLIGRL was more prominent in wild-type than in TRPV1-deficient mice. Antagonists for CGRP, selective neurokinin-1 receptor, PKC-ε or PKA inhibitor, abolished SLIGRL protection in wild-type but not in TRPV1-deficient hearts (Zhong and Wang, 2009). Housing of mice for some weeks at low temperature (4°C) induced the development of cardiac hypertrophy, contractile dysfunction, fibrosis accumulation, and alterations in Ca^{2+} homeostasis in cardiomyocytes. Wild-type mice were subjected to cold stress in the absence or presence of the TRPV1 agonist SA13353. TRPV1 blockage attenuated or blunted cold exposure-induced hypertrophy and decreased fractional shortening among other functional cardiac and cellular parameters. Interestingly, cold stress-induced changes in fractional shortening and cardiomyocyte contraction were mimicked by short-term capsazepine treatment of wild-type mice kept at room temperature (Zhang et al., 2012a). The role of TRPV1 in myocardial infarction was evaluated, and 1 week after coronary ligation TRPV1-deficient mice showed an increased infarct size and mortality rate associated with enhanced myofibroblast infiltration, and excessive inflammation, capillary density, and collagen content. Echocardiography revealed deteriorated cardiac function in TRPV1-knockout mice compared with wild-type mice (Huang et al., 2009b, 2010). TRPV1-deficient mice are partially protected from pressure-overload induced cardiac hypertrophy. They exhibit a reduced increase in heart weight and extracellular matrix remodeling compared to wild-type mice; also the process of reduction in EF during hypertrophy development is slower in TRPV1-deficient mice (Buckley and Stokes, 2011). The effects of pressure overload–induced cardiac hypertrophy, such as an increase in heart weight index, enlargement of ventricular volume, decrease in cardiac function, and increase in cardiac fibrosis were attenuated in wild-type mice that received dietary capsaicin, compared to control wild-type mice. As proof of the specificity of the action of capsaicin, this attenuation was not observed in TRPV1-deficient mice receiving capsaicin (Wang et al., 2014). Capsaicin treatment was able to reduce the cardiac hypertrophy and changes in cardiac contractility induced by a high-salt diet in wild-type mice, but not in TRPV1-deficient mice which develop similar hypertrophy and fibrosis as wild-type mice (Gao et al., 2014; Lang et al., 2015). Reduction in the increase in cardiac mass after a high-fat diet in TRPV1-deficient mice was not observed in comparison to wild-type mice (Marshall et al., 2013). Multiple TRPV1-dependent signaling pathways may lead to protection from the development of cardiac remodeling as discussed in the studies cited above, but it should also be considered that long-term dietary administration of capsaicin improves endothelium-dependent vasorelaxation and prevents hypertension (Son et al., 2016).

A mouse model overexpressing TRPV2 under a cardiac promoter was also generated; these mice exhibit enlarged hearts, increased fibrosis, and myocardial structural defects, depending on expression levels (Iwata et al., 2003). TRPV2 accumulation in the plasma membrane could be blocked by overexpression of the amino-terminal (NT) domain of TRPV2. Targeted overexpression in cardiomyocytes of the NT domain prevented ventricular dilation and fibrosis, ameliorated contractile dysfunction in a mouse model of dilated cardiomyopathy, and improved survival (Iwata et al., 2013).

The cardiovascular effects of AngII infusion in TRPA1-deficient mice were similar to the ones observed in control mice. In both groups cardiac contractility was preserved, and the hypertension development was comparable. However, in TRPA1-deficient mice the increase in cardiac mass was more prominent (Bodkin et al., 2014).

9.7 TRP CHANNELS IN THE REGULATION OF CARDIAC CONDUCTION AND THEIR CONTRIBUTION TO ARRHYTHMIAS

Since the first cloning of cation channels of the TRP family, it was postulated that any of them that were identified in cardiac cells could influence the excitability of cardiomyocytes and cells of the conduction system, but a role in the development of cardiac arrhythmias by regulating the transcriptional control of cardiac fibroblast plasticity and pathological fibrosis has also emerged (Lighthouse and Small, 2016). Strong evidence in this direction has been reported for TRPM4, TRPC3, and TRPM7 and will be highlighted in the next paragraph.

TRPM4 functions as an essential subunit of Ca^{2+}-activated nonselective cation channels in many cell types including cardiomyocytes. The understanding of its contribution to cardiac conduction as well as the development of arrhythmias has been demonstrated by channel inhibitors such as 9-phenantrol, by the use of TRPM4-deficient mouse models and, third, by the identification of TRPM4 mutants that were found in a variety of inherited human cardiac arrhythmias. Two TRPM4-deficient mouse models have been described (Vennekens et al., 2007; Barbet et al., 2008). In the first TRPM4-deficient mouse model, with deletion of exons 15 and 16, there was no evidence for overt arrhythmias during long-term telemetric analysis of blood pressure (Mathar et al., 2010), and detailed analysis of surface ECG recordings revealed no differences in heart rate, QRS interval, QTC interval, and ST height compared to wild-type controls (Mathar et al., 2014; Jacobs et al., 2015). Also following LAD ligation, TRPM4-deficient mice showed infarct typical ECG patterns similar to wild-type controls, but mortality was significantly lower in TRPM4-deficient mice. Possibly TRPM4-deficient mice are more protected from the development of ischemia-induced arrhythmias evoked by Ca^{2+}-dependent triggered activity, but direct evidence for this mechanism is still lacking. In the second TRPM4-deficient mouse model, in which exons 3–6 were deleted (Barbet et al., 2008), multilevel cardiac conduction blocks were reported as attested by PR and QRS lengthening in surface ECGs and intercardiac exploration. In addition, this TRPM4-deficient mouse line exhibits Luciani-Wenckebach atrial ventricular blocks, which was explained by a parasympathetic overdrive (Demion et al., 2014). In cardiomyocytes from both TRPM4-deficient mouse models, a shortening in action potentials of either atrial cells (Simard et al., 2013; Demion et al., 2014) or ventricular cardiomyocytes (Mathar et al., 2014) was described. Interestingly, the reduction in ventricular action potential duration observed in microelectrode recordings from papillary muscles (Mathar et al., 2014) was not found in isolated ventricular cardiomyocytes (Demion et al., 2014). The discrepancy observed in the two mouse models remained obscure until now, and differences in experimental approaches and/or genetic background of the mice can be considered as the underlying reasons. In both mouse models, action potential shortening was unrelated to modifications of voltage-gated Ca^{2+} or K^+ currents that determine repolarization (Demion et al., 2014; Mathar et al., 2014).

The reduction of action potential duration evoked by TRPM4 deletion in atrial cells can be mimicked by 9-phenanthrol (IC_{50} 21 μM) at similar concentrations that were reported to inhibit heterologous TRPM4-mediated currents (Simard et al., 2013). At these concentrations, only a tiny reduction of action potential duration (APD) was observed in TRPM4-deficient mice. This finding indicates a specific TRPM4 inhibition of 9-phenanthrol in this preparation, although this cannot be extrapolated automatically to other cell types or organ preparations regarding the inhibitory effect of 9-phenanthrol on other cation channels including Ca^{2+}-activated chloride channels (Burris et al., 2015). Similarly, 9-phenanthrol inhibits dose dependently the APD and beating rate in freely beating mouse right atria, but the latter was not found in TRPM4-deficient mice (Hof et al., 2013; Hof et al., 2016), implicating a role for TRPM4 channels in sinoatrial node pacemaker activity. Interestingly, in both TRPM4-deficient mouse models, no modifications of basal heart rate were found (Mathar et al., 2010; Demion et al., 2014; Mathar et al., 2014).

Mutations in the human TRPM4 gene were shown to be associated with several entities of cardiac channelopathies comprising diverse forms of cardiac conduction block. At least 19 different mutations were described until now, most of them in the N-terminal cytosolic tail of the channel protein. All mutations were identified in diseases attributed to multiple levels of the conduction system, including progressive familial heart block type I (PFHB I) (Kruse et al., 2009; Daumy et al., 2016), isolated cardiac conduction disease (ICCD) (Liu et al., 2010b), Brugada syndrome (BrS) (Liu et al., 2013), progressive cardiac conduction defect (PCCD) (Daumy et al., 2016), as well as congenital or childhood atrioventricular block (Syam et al., 2016). All TRPM4 mutants have cardiac conduction block as a common feature in the clinical manifestation, but the mechanism of how the individual mutations may lead to this phenotype is still unresolved. The individual TRPM4 mutations found in the particular channelopathies as well as the consequences for TRPM4 function are summarized in recent reports (Guinamard et al., 2015; Stallmeyer et al., 2012). The mutated TRPM4 proteins were heterologously expressed in cell lines and were analyzed with respect to current density of whole cell currents, single channel conductance, cell surface protein expression, as well as Ca^{2+} sensitivity and voltage dependence, but a common mechanism underlying the pathogenesis of cardiac conduction blocks remains to be uncovered. For example, TRPM4 mutations associated with conduction blocks can lead to either an increased or decreased surface expression or current density, respectively.

For the initiation and maintenance of chronic arrhythmias, cardiac fibroblasts are also considered as a proarrhythmic substrate. Harada et al. (2012) showed that the expression of TRPC3 is upregulated in atria from patients with atrial fibrillation (AF) and that TRPC3 knockdown leads to a decrease of canine atrial fibroblast proliferation.

Further evidence for a role of TRPC3 channels for the maintenance of AF emerged from experiments in which inhibition of TRPC3 channels with Pyr3 reduced AF duration in a dog model of atrial tachypacing (Harada et al., 2012). A confinement regarding this interpretation comes from a study that shows that Pyr3 inhibits TRPC3- as well as Orai1-mediated Ca^{2+} entry and currents with similar potency (Schleifer et al., 2012). Additional evidence for TRP channels as mediators in the pathogenesis of AF was provided by a study of Du et al. (2010), in which a striking upregulation of TRPM7 currents and Ca^{2+} influx was reported. Moreover, TRPM7 knockdown reduced AF fibroblast differentiation evoked by TGF-β1 (Du et al., 2010). In a candidate gene screen in individuals with long Q-T syndrome, 13 variants leading to altered amino acid sequence were identified in the Trpm7 gene but not in more than 300 controls of the same continental ancestry (Arking et al., 2014). The analysis of time-dependent and cell type–specific TRPM7 inactivation in cardiac cells revealed that the timing of TRPM7 deletion variably perturbs cardiac development and function. Cre-mediated deletion of the TRPM7 gene early during cardiac development (by the use of Isl1-Cre or TnT-Cre) leads to thinning of the compact myocardium and to impaired septation, heart failure, and embryonic death (Sah et al., 2013a). Deletion of TRPM7 later during development (between E9.5 and E12.5, Trpm7$^{aMHC/-}$) results in contractile dysfunction, cardiac hypertrophy, and fibrosis as well as complex arrhythmias that were characterized by atrioventricular conduction block, impairment of ventricular repolarization, as well as a higher susceptibility for induction of ventricular arrhythmias. Mechanistically such arrhythmias do not depend only on the lack of TRPM7-mediated currents and the subsequent electrophysiological alterations in cardiomyocytes, but may be due to secondary multiple changes in myocardial transcription. In a comprehensive expression analysis in TRPM7-deficient hearts, a downregulation of Kcnd2, a K^+ voltage-gated channel subfamily D member, and the resulting reduction in transient outward currents as well as silencing of HCN4 expression, a hyperpolarization-activated cyclic-nucleotide-gated channel, leading to a slower diastolic membrane depolarization and automaticity in sinoatrial note cells was identified (Sah et al., 2013a, 2013b).

ACKNOWLEDGMENT/FUNDING

This work was supported by the *Klinische Forschergruppe* KFO-196, the DZHK (German Centre for Cardiovascular Research), the BMBF (German Ministry of Education and Research), and the TR-SFB 152.

Note: The reviewing procedure was based on an extensive search of all original articles found in PubMed until July 15, 2016. We used the following search words: TRP, heart, cardiomyocytes, cardiac fibroblasts, cardiac hypertrophy, cardiac remodeling, conduction, arrhythmia, and performed systematic combinations such as TRP and heart, TRPC and heart, TRPC1 and heart, and so on for every topic and every member of the mammalian TRP family (Flockerzi, 2007). Importantly, for TRPP2, TRPP3, and TRPP5, we included the synonyms PKD2, PKD2L1, and PKD2L2, respectively. For TRPMLs we also used the synonym MCOLN. Studies using nonselective cation channel blockers were not considered.

REFERENCES

Abramowitz, J. and L. Birnbaumer. 2009. Physiology and pathophysiology of canonical transient receptor potential channels. *FASEB J*, 23(2): 297–328. doi:10.1096/fj.08-119495.

Adapala, R.K. et al. 2013. TRPV4 channels mediate cardiac fibroblast differentiation by integrating mechanical and soluble signals. *J Mol Cell Cardiol*, 54: 45–52. doi:10.1016/j.yjmcc.2012.10.016.

Aguettaz, E. et al. 2016. Axial stretch-dependent cation entry in dystrophic cardiomyopathy: Involvement of several TRPs channels. *Cell Calcium*, 59(4): 145–155. doi:10.1016/j.ceca.2016.01.001.

Alvarez, J. et al. 2008. ATP/UTP activate cation-permeable channels with TRPC3/7 properties in rat cardiomyocytes. *Am J Physiol Heart Circ Physiol*, 295(1):H21–H28. doi:10.1152/ajpheart.00135.2008.

Andrei, S.R. et al. 2016. TRPA1 is functionally co-expressed with TRPV1 in cardiac muscle: Co-localization at z-discs, costameres and intercalated discs. *Channels (Austin)*, 10(5): 395–409. doi:10.1080/1933695 0.2016.1185579.

Antunes, T.T. et al. 2016. Transient receptor potential melastatin 7 cation channel kinase: New player in angiotensin II-Induced hypertension. *Hypertension*, 67(4): 763–773. doi:10.1161/HYPERTENSIONAHA.115.07021.

Anyatonwu, G.I. et al. 2007. Regulation of ryanodine receptor-dependent calcium signaling by polycystin-2. *Proc Natl Acad Sci U S A*, 104(15): 6454–6459. doi:10.1073/pnas.0610324104.

Arking, D.E. et al. 2014. Genetic association study of QT interval highlights role for calcium signaling pathways in myocardial repolarization. *Nat Genet*, 46(8): 826–836. doi:10.1038/ng.3014.

Balijepalli, R.C. et al. 2006. Localization of cardiac L-type Ca(2+) channels to a caveolar macromolecular signaling complex is required for beta(2)-adrenergic regulation. *Proc Natl Acad Sci U S A*, 103(19): 7500–7505. doi:10.1073/pnas.0503465103.

Barbet, G. et al. 2008. The calcium-activated nonselective cation channel TRPM4 is essential for the migration but not the maturation of dendritic cells. *Nat Immunol*, 9(10): 1148–1156. doi:10.1038/ni.1648.

Basora, N. et al. 2002. Tissue and cellular localization of a novel polycystic kidney disease-like gene product, polycystin-L. *J Am Soc Nephrol*, 13(2): 293–301.

Bassi, M.T. et al. 2000. Cloning of the gene encoding a novel integral membrane protein, mucolipidin-and identification of the two major founder mutations causing mucolipidosis type IV. *Am J Hum Genet*, 67(5): 1110–1120. doi:10.1016/S0002-9297(07)62941-3.

Baudino, T.A. et al. 2006. Cardiac fibroblasts: Friend or foe? *Am J Physiol Heart Circ Physiol*, 291(3): H1015–H1026. doi:10.1152/ajpheart.00023.2006.

Beetz, N. et al. 2009. Transgenic simulation of human heart failure-like L-type Ca^{2+}-channels: Implications for fibrosis and heart rate in mice. *Cardiovasc Res*, 84(3): 396–406. doi:10.1093/cvr/cvp251.

Bers, D.M. 2002. Cardiac excitation-contraction coupling. *Nature*, 415(6868): 198–205. doi:10.1038/415198a.

Bers, D.M. 2008. Calcium cycling and signaling in cardiac myocytes. *Annu Rev Physiol*, 70: 23–49. doi:10.1146/annurev.physiol.70.113006.100455.

Bodkin, J.V. et al. 2014. Investigating the potential role of TRPA1 in locomotion and cardiovascular control during hypertension. *Pharmacol Res Perspect*, 2(4): e00052. doi:10.1002/prp2.52.

Brenner, J.S. and R.E. Dolmetsch. 2007. TrpC3 regulates hypertrophy-associated gene expression without affecting myocyte beating or cell size. *PLoS One*, 2(8): e802. doi:10.1371/journal.pone.0000802.

Buckley, C.L. and A.J. Stokes. 2011. Mice lacking functional TRPV1 are protected from pressure overload cardiac hypertrophy. *Channels (Austin)*, 5(4): 367–374. doi:10.4161/chan.5.4.17083.

Burris, S.K. et al. 2015. 9-Phenanthrol inhibits recombinant and arterial myocyte TMEM16A channels. *Br J Pharmacol*, 172(10): 2459–2468. doi:10.1111/bph.13077.

Burt, R. et al. 2013. 9-Phenanthrol and flufenamic acid inhibit calcium oscillations in HL-1 mouse cardiomyocytes. *Cell Calcium*, 54(3): 193–201. doi:10.1016/j.ceca.2013.06.003.

Bush, E.W. et al. 2006. Canonical transient receptor potential channels promote cardiomyocyte hypertrophy through activation of calcineurin signaling. *J Biol Chem*, 281(44): 33487–33496. doi:10.1074/jbc.M605536200.

Camacho Londono, J.E. et al. 2015. A background Ca^{2+} entry pathway mediated by TRPC1/TRPC4 is critical for development of pathological cardiac remodelling. *Eur Heart J*, 36(33): 2257–2266. doi:10.1093/eurheartj/ehv250.

Chauvet, V. et al. 2002. Expression of PKD1 and PKD2 transcripts and proteins in human embryo and during normal kidney development. *Am J Pathol*, 160(3): 973–983. doi:10.1016/S0002-9440(10)64919-X.

Chen, J.B. et al. 2010. Multiple Ca^{2+} signaling pathways regulate intracellular Ca^{2+} activity in human cardiac fibroblasts. *J Cell Physiol*, 223(1): 68–75. doi:10.1002/jcp.22010.

Chu, W. et al. 2012. Mild hypoxia-induced cardiomyocyte hypertrophy via up-regulation of HIF-1alpha-mediated TRPC signalling. *J Cell Mol Med*, 16(9): 2022–2034. doi:10.1111/j.1582-4934.2011.01497.x.

Chubanov, V. et al. 2012. Natural and synthetic modulators of SK (K(ca)2) potassium channels inhibit magnesium-dependent activity of the kinase-coupled cation channel TRPM7. *Br J Pharmacol*, 166(4): 1357–1376. doi:10.1111/j.1476-5381.2012.01855.x.

Cohen, M.R. et al. 2013. Understanding the cellular function of TRPV2 channel through generation of specific monoclonal antibodies. *PLoS One*, 8(12): e85392. doi:10.1371/journal.pone.0085392.

Colella, M. et al. 2008. Ca^{2+} oscillation frequency decoding in cardiac cell hypertrophy: Role of calcineurin/NFAT as Ca^{2+} signal integrators. *Proc Natl Acad Sci U S A*, 105(8): 2859–2864. doi:10.1073/pnas.0712316105.

Cooley, N. et al. 2014. The phosphatidylinositol(4,5)bisphosphate-binding sequence of transient receptor potential channel canonical 4α is critical for its contribution to cardiomyocyte hypertrophy. *Mol Pharmacol*, 86(4): 399–405. doi:10.1124/mol.114.093690.

Cuajungco, M.P. and M.A. Samie. 2008. The varitint-waddler mouse phenotypes and the TRPML3 ion channel mutation: Cause and consequence. *Pflugers Arch*, 457(2): 463–473. doi:10.1007/s00424-008-0523-4.

Cuffe, J.S. 2015. Differential mRNA expression and glucocorticoid-mediated regulation of TRPM6 and TRPM7 in the heart and kidney throughout murine pregnancy and development. *PLoS One*, 10(2): e0117978. doi:10.1371/journal.pone.0117978.

Cui, L.B. 2016. Morphological identification of TRPC7 in cardiomyocytes from normal and renovascular hypertensive rats. *J Cardiovasc Pharmacol*, 67(2): 121–128. doi:10.1097/FJC.0000000000000321.

Daumy, X. et al. 2016. Targeted resequencing identifies TRPM4 as a major gene predisposing to progressive familial heart block type I. *Int J Cardiol*, 207: 349–358. doi:10.1016/j.ijcard.2016.01.052.

Davis, J. et al. 2012. A TRPC6-dependent pathway for myofibroblast transdifferentiation and wound healing in vivo. *Dev Cell*, 23(4): 705–715. doi:10.1016/j.devcel.2012.08.017.

Demion, M. et al. 2007. TRPM4, a Ca^{2+}-activated nonselective cation channel in mouse sino-atrial node cells. *Cardiovasc Res*, 73(3): 531–538. doi:10.1016/j.cardiores.2006.11.023.

Demion, M. et al. 2014. Trpm4 gene invalidation leads to cardiac hypertrophy and electrophysiological alterations. *PLoS One*, 9(12): e115256. doi:10.1371/journal.pone.0115256.

Demir, T. et al. 2014. Evaluation of TRPM (transient receptor potential melastatin) genes expressions in myocardial ischemia and reperfusion. *Mol Biol Rep*, 41(5): 2845–2849. doi:10.1007/s11033-014-3139-0.

Dietrich, A. et al. 2007. In vivo TRPC functions in the cardiopulmonary vasculature. *Cell Calcium*, 42(2): 233–244. doi:10.1016/j.ceca.2007.02.009.

Dietrich, A., H. Kalwa, and T. Gudermann. 2010. TRPC channels in vascular cell function. *Thromb Haemost*, 103(2): 262–270. doi:10.1160/TH09-08-0517.

Ding, Y. et al. 2011. Reactive oxygen species-mediated TRPC6 protein activation in vascular myocytes, a mechanism for vasoconstrictor-regulated vascular tone. *J Biol Chem*, 286(36): 31799–31809. doi:10.1074/jbc.M111.248344.

Doleschal, B. et al. 2015. TRPC3 contributes to regulation of cardiac contractility and arrhythmogenesis by dynamic interaction with NCX1. *Cardiovasc Res*, 106(1): 163–173. doi:10.1093/cvr/cvv022.

Dolmetsch, R.E. et al. 1997. Differential activation of transcription factors induced by Ca^{2+} response amplitude and duration. *Nature*, 386(6627): 855–858. doi:10.1038/386855a0.

Domes, K. et al. 2015. Murine cardiac growth, TRPC channels, and cGMP kinase I. *Pflugers Arch*, 467(10): 2229–2234. doi:10.1007/s00424-014-1682-0.

Dominguez-Rodriguez, A. et al. 2015. Proarrhythmic effect of sustained EPAC activation on TRPC3/4 in rat ventricular cardiomyocytes. *J Mol Cell Cardiol*, 87: 74–78. doi:10.1016/j.yjmcc.2015.07.002.

Du, J. et al. 2010. TRPM7-mediated Ca^{2+} signals confer fibrogenesis in human atrial fibrillation. *Circ Res*, 106(5): 992–1003. doi:10.1161/CIRCRESAHA.109.206771.

Dyachenko, V. et al. 2009. Mechanical deformation of ventricular myocytes modulates both TRPC6 and Kir2.3 channels. *Cell Calcium*, 45(1): 38–54. doi:10.1016/j.ceca.2008.06.003.

Eder, P. and J.D. Molkentin. 2011. TRPC channels as effectors of cardiac hypertrophy. *Circ Res*, 108(2): 265–272. doi:10.1161/CIRCRESAHA.110.225888.

Eder, P. et al. 2007. Phospholipase C-dependent control of cardiac calcium homeostasis involves a TRPC3-NCX1 signaling complex. *Cardiovasc Res*, 73(1): 111–119. doi:10.1016/j.cardiores.2006.10.016.

Entin-Meer, M. et al. 2014. The transient receptor potential vanilloid 2 cation channel is abundant in macrophages accumulating at the peri-infarct zone and may enhance their migration capacity towards injured cardiomyocytes following myocardial infarction. *PLoS One*, 9(8): e105055. doi:10.1371/journal.pone.0105055.

Facer, P. et al. 2011. Localisation of SCN10A gene product Na(v)1.8 and novel pain-related ion channels in human heart. *Int Heart J*, 52(3): 146–152.

Falardeau, J.L. et al. 2002. Cloning and characterization of the mouse Mcoln1 gene reveals an alternatively spliced transcript not seen in humans. *BMC Genomics*, 3: 3.

Fauconnier, J. et al. 2007. Insulin potentiates TRPC3-mediated cation currents in normal but not in insulin-resistant mouse cardiomyocytes. *Cardiovasc Res*, 73(2): 376–385. doi:10.1016/j.cardiores.2006.10.018.

Feng, S.L. et al. 2011. Activation of calcium-sensing receptor increases TRPC3 expression in rat cardiomyocytes. *Biochem Biophys Res Commun*, 406(2): 278–284. doi:10.1016/j.bbrc.2011.02.033.

Flockerzi, V. 2007. An introduction on TRP channels. *Handb Exp Pharmacol*, (179): 1–19. doi:10.1007/978-3-540-34891-7_1.

Flockerzi, V. and B. Nilius. 2014. TRPs: truly remarkable proteins. *Handb Exp Pharmacol*, 222: 1–12. doi:10.1007/978-3-642-54215-2_1.

Fonfria, E. et al. 2006. Tissue distribution profiles of the human TRPM cation channel family. *J Recept Signal Transduct Res*, 26(3): 159–178. doi:10.1080/10799890600637506.

Freichel, M. et al. 1999. Store-operated cation channels in the heart and cells of the cardiovascular system. *Cell Physiol Biochem*, 9(4–5): 270–283.

Freichel, M., V. Tsvilovskyy, and J.E. Camacho-Londono. 2014. TRPC4- and TRPC4-containing channels. *Handb Exp Pharmacol*, 222: 85–128. doi:10.1007/978-3-642-54215-2_5.

Freichel, M. et al. 1998. Alternative splicing and tissue specific expression of the 5' truncated bCCE 1 variant bCCE 1delta514. *FEBS Lett*, 422(3): 354–358.

Frey, N. et al. 2004. Hypertrophy of the heart: A new therapeutic target? *Circulation*, 109(13): 1580–1589. doi:10.1161/01.CIR.0000120390.68287.BB.

Frey, N. et al. 2000. Decoding calcium signals involved in cardiac growth and function. *Nat Med*, 6(11): 1221–1227. doi:10.1038/81321.

Gao, F. et al. 2014. TRPV1 Activation attenuates high-salt diet-induced cardiac hypertrophy and fibrosis through PPAR-delta upregulation. *PPAR Res*, 2014: 491963. doi:10.1155/2014/491963.

Gao, G. et al. 2013. Unfolded protein response regulates cardiac sodium current in systolic human heart failure. *Circ Arrhythm Electrophysiol*, 6(5): 1018–1024. doi:10.1161/CIRCEP.113.000274.

Garcia, R.L. and W.P. Schilling. 1997. Differential expression of mammalian TRP homologues across tissues and cell lines. *Biochem Biophys Res Commun*, 239(1): 279–283. doi:10.1006/bbrc.1997.7458.

Goel, M. et al. 2007. TRPC3 channels colocalize with Na^+/Ca^{2+} exchanger and Na^+ pump in axial component of transverse-axial tubular system of rat ventricle. *Am J Physiol Heart Circ Physiol*, 292(2): H874–H883. doi:10.1152/ajpheart.00785.2006.

Goonasekera, S.A. and J.D. Molkentin. 2012. Unraveling the secrets of a double life: Contractile versus signaling Ca^{2+} in a cardiac myocyte. *J Mol Cell Cardiol*, 52(2): 317–322. doi:10.1016/j.yjmcc.2011.05.001.

Grimm, C. et al. 2010. Small molecule activators of TRPML3. *Chem Biol*, 17(2): 135–148. doi:10.1016/j.chembiol.2009.12.016.

Gronich, N. et al. 2010. Molecular remodeling of ion channels, exchangers and pumps in atrial and ventricular myocytes in ischemic cardiomyopathy. *Channels (Austin)*, 4(2): 101–107.

Guarini, G. et al. 2012. Disruption of TRPV1-mediated coupling of coronary blood flow to cardiac metabolism in diabetic mice: Role of nitric oxide and BK channels. *Am J Physiol Heart Circ Physiol*, 303(2): H216–H223. doi:10.1152/ajpheart.00011.2012.

Guinamard, R. and P. Bois. 2007. Involvement of transient receptor potential proteins in cardiac hypertrophy. *Biochim Biophys Acta*, 1772(8): 885–894. doi:10.1016/j.bbadis.2007.02.007.

Guinamard, R. et al. 2004. Functional characterization of a Ca(2+)-activated non-selective cation channel in human atrial cardiomyocytes. *J Physiol*, 558(Pt 1): 75–83. doi:10.1113/jphysiol.2004.063974.

Guinamard, R. et al. 2006. Functional expression of the TRPM4 cationic current in ventricular cardiomyocytes from spontaneously hypertensive rats. *Hypertension*, 48(4): 587–594. doi:10.1161/01.HYP.0000237864.65019.a5.

Guinamard, R. et al. 2015. TRPM4 in cardiac electrical activity. *Cardiovasc Res*, 108(1): 21–30. doi:10.1093/cvr/cvv213.

Guo, J.L. et al. 2014. Transient receptor potential melastatin 7 (TRPM7) contributes to H_2O_2-induced cardiac fibrosis via mediating Ca(2+) influx and extracellular signal-regulated kinase 1/2 (ERK1/2) activation in cardiac fibroblasts. *J Pharmacol Sci*, 125(2): 184–192.

Guo, L. et al. 2000. Identification and characterization of a novel polycystin family member, polycystin-L2, in mouse and human: Sequence, expression, alternative splicing, and chromosomal localization. *Genomics*, 64(3): 241–251. doi:10.1006/geno.2000.6131.

Gwanyanya, A. et al. 2004. Magnesium-inhibited, TRPM6/7-like channel in cardiac myocytes: Permeation of divalent cations and pH-mediated regulation. *J Physiol*, 559(Pt3): 761–776. doi:10.1113/jphysiol.2004.067637.

Gwanyanya, A. et al. 2006. ATP and PIP2 dependence of the magnesium-inhibited, TRPM7-like cation channel in cardiac myocytes. *Am J Physiol Cell Physiol*, 291(4): C627–C635. doi:10.1152/ajpcell.00074.2006.

Hang, P. et al. 2015. Brain-derived neurotrophic factor regulates TRPC3/6 channels and protects against myocardial infarction in rodents. *Int J Biol Sci*, 11(5): 536–545. doi:10.7150/ijbs.10754.

Harada, M. et al. 2012. Transient receptor potential canonical-3 channel-dependent fibroblast regulation in atrial fibrillation. *Circulation*, 126(17): 2051–2064. doi:10.1161/CIRCULATIONAHA.112.121830.

Hatano, N., Y. Itoh, and K. Muraki. 2009. Cardiac fibroblasts have functional TRPV4 activated by 4α-phorbol 12, 13-didecanoate. *Life Sci*, 85(23–26): 808–814. doi:10.1016/j.lfs.2009.10.013.

Heineke, J. and J. D. Molkentin. 2006. Regulation of cardiac hypertrophy by intracellular signalling pathways. *Nat Rev Mol Cell Biol*, 7(8): 589–600. doi:10.1038/nrm1983.

Hiroi, T. et al. 2013. Neutrophil TRPM2 channels are implicated in the exacerbation of myocardial ischaemia/reperfusion injury. *Cardiovasc Res*, 97(2): 271–281. doi:10.1093/cvr/cvs332.

Hof, T. et al. 2013. Implication of the TRPM4 nonselective cation channel in mammalian sinus rhythm. *Heart Rhythm*, 10(11): 1683–1689. doi:10.1016/j.hrthm.2013.08.014.

Hof, T. et al. 2016. TRPM4 non-selective cation channels influence action potentials in rabbit Purkinje fibres. *J Physiol*, 594(2): 295–306. doi:10.1113/JP271347.

Hoffman, N.E. et al. 2015. Ca^{2+} entry via Trpm2 is essential for cardiac myocyte bioenergetics maintenance. *Am J Physiol Heart Circ Physiol*, 308(6): H637–H650. doi:10.1152/ajpheart.00720.2014.

Huang, H. et al. 2009a. TRPC1 expression and distribution in rat hearts. *Eur J Histochem*, 53(4): e26. doi:10.4081/ejh.2009.e26.

Huang, T.T. et al. 2006. Genetic modifiers of the phenotype of mice deficient in mitochondrial superoxide dismutase. *Hum Mol Genet*, 15(7): 1187–1194. doi:10.1093/hmg/ddl034.

Huang, W. et al. 2009b. Transient receptor potential vanilloid gene deletion exacerbates inflammation and atypical cardiac remodeling after myocardial infarction. *Hypertension*, 53(2): 243–250. doi:10.1161/HYPERTENSIONAHA.108.118349.

Huang, W. et al. 2010. Enhanced postmyocardial infarction fibrosis via stimulation of the transforming growth factor-beta-Smad2 signaling pathway: Role of transient receptor potential vanilloid type 1 channels. *J Hypertens*, 28(2): 367–376. doi:10.1097/HJH.0b013e328333af48.

Hwang, I. et al. 2011. Tissue-specific expression of the calcium transporter genes TRPV5, TRPV6, NCX1, and PMCA1b in the duodenum, kidney and heart of *Equus caballus*. *J Vet Med Sci*, 73(11): 1437–1444.

Ikeda, K. et al. 2013. Roles of transient receptor potential canonical (TRPC) channels and reverse-mode Na^+/Ca^{2+} exchanger on cell proliferation in human cardiac fibroblasts: Effects of transforming growth factor beta1. *Cell Calcium*, 54(3): 213–225. doi:10.1016/j.ceca.2013.06.005.

Inoue, R. et al. 2006. Transient receptor potential channels in cardiovascular function and disease. *Circ Res*, 99(2): 119–131. doi:10.1161/01.RES.0000233356.10630.8a.

Iwata, Y. et al. 2003. A novel mechanism of myocyte degeneration involving the Ca^{2+}-permeable growth factor-regulated channel. *J Cell Biol*, 161(5): 957–967. doi:10.1083/jcb.200301101.

Iwata, Y. et al. 2013. Blockade of sarcolemmal TRPV2 accumulation inhibits progression of dilated cardiomyopathy. *Cardiovasc Res*, 99(4): 760–768. doi:10.1093/cvr/cvt163.

Jacobs, G. et al. 2015. Enhanced beta-adrenergic cardiac reserve in Trpm4$^{-/-}$ mice with ischaemic heart failure. *Cardiovasc Res*, 105(3): 330–339. doi:10.1093/cvr/cvv009.

Jiang, Y. et al. 2014. Expression and localization of TRPC proteins in rat ventricular myocytes at various developmental stages. *Cell Tissue Res*, 355(1): 201–212. doi:10.1007/s00441-013-1733-1734.

Ju, Y.K. et al. 2007. Store-operated Ca^{2+} influx and expression of TRPC genes in mouse sinoatrial node. *Circ Res*, 100(11): 1605–1614. doi:10.1161/CIRCRESAHA.107.152181.

Kakkar, R. and R.T. Lee. 2010. Intramyocardial fibroblast myocyte communication. *Circ Res*, 106(1): 47–57. doi:10.1161/CIRCRESAHA.109.207456.

Kang, H.Y., Y.K. Choi, and E.B. Jeung. 2016. Inhibitory effect of progesterone during early embryonic development: Suppression of myocardial differentiation and calcium-related transcriptome by progesterone in mESCs: Progesterone disturb cardiac differentiation of mESCs through lower cytosolic Ca2. *Reprod Toxicol*, 64: 169–179. doi:10.1016/j.reprotox.2016.06.001.

Kapur, N. K. et al. 2014. Reducing endoglin activity limits calcineurin and TRPC-6 expression and improves survival in a mouse model of right ventricular pressure overload. *J Am Heart Assoc*, 3(4). doi:10.1161/JAHA.114.000965.

Katanosaka, Y. et al. 2014. TRPV2 is critical for the maintenance of cardiac structure and function in mice. *Nat Commun*, 5: 3932. doi:10.1038/ncomms4932.

Kecskes, M. et al. 2015. The Ca(2+)-activated cation channel TRPM4 is a negative regulator of angiotensin II-induced cardiac hypertrophy. *Basic Res Cardiol*, 110(4): 43. doi:10.1007/s00395-015-0501-x.

Kinoshita, H. et al. 2010. Inhibition of TRPC6 channel activity contributes to the antihypertrophic effects of natriuretic peptides-guanylyl cyclase-A signaling in the heart. *Circ Res*, 106(12): 1849–1860. doi:10.1161/CIRCRESAHA.109.208314.

Kiso, H. et al. 2013. Sildenafil prevents the up-regulation of transient receptor potential canonical channels in the development of cardiomyocyte hypertrophy. *Biochem Biophys Res Commun*, 436(3): 514–518. doi:10.1016/j.bbrc.2013.06.002.

Kitajima, N. et al. 2011. TRPC3-mediated Ca^{2+} influx contributes to Rac1-mediated production of reactive oxygen species in MLP-deficient mouse hearts. *Biochem Biophys Res Commun*, 409(1): 108–113. doi:10.1016/j.bbrc.2011.04.124.

Kiyonaka, S. et al. 2009. Selective and direct inhibition of TRPC3 channels underlies biological activities of a pyrazole compound. *Proc Natl Acad Sci U S A*, 106(13): 5400–5405. doi:10.1073/pnas.0808793106.

Klaiber, M. et al. 2010. Novel insights into the mechanisms mediating the local antihypertrophic effects of cardiac atrial natriuretic peptide: Role of cGMP-dependent protein kinase and RGS2. *Basic Res Cardiol*, 105(5): 583–595. doi:10.1007/s00395-010-0098-z.

Klaiber, M. et al. 2011. A cardiac pathway of cyclic GMP-independent signaling of guanylyl cyclase A, the receptor for atrial natriuretic peptide. *Proc Natl Acad Sci U S A*, 108(45): 18500–18505. doi:10.1073/pnas.1103300108.

Kobayashi, Y. et al. 2006. Identification and characterization of GSRP-56, a novel Golgi-localized spectrin repeat-containing protein. *Exp Cell Res*, 312(16): 3152–3164. doi:10.1016/j.yexcr.2006.06.026.

Koch, S.E. et al. 2012. Probenecid: Novel use as a non-injurious positive inotrope acting via cardiac TRPV2 stimulation. *J Mol Cell Cardiol*, 53(1): 134–144. doi:10.1016/j.yjmcc.2012.04.011.

Koch, S.E. et al. 2013. Probenecid as a noninjurious positive inotrope in an ischemic heart disease murine model. *J Cardiovasc Pharmacol Ther*, 18(3): 280–289. doi:10.1177/1074248412469299.

Koitabashi, N. et al. 2010. Cyclic GMP/PKG-dependent inhibition of TRPC6 channel activity and expression negatively regulates cardiomyocyte NFAT activation novel mechanism of cardiac stress modulation by PDE5 inhibition. *J Mol Cell Cardiol*, 48(4): 713–724. doi:10.1016/j.yjmcc.2009.11.015.

Kojima, A. et al. 2010. Ca^{2+} paradox injury mediated through TRPC channels in mouse ventricular myocytes. *Br J Pharmacol*, 161(8): 1734–1750. doi:10.1111/j.1476-5381.2010.00986.x.

Kruse, M. et al. 2009. Impaired endocytosis of the ion channel TRPM4 is associated with human progressive familial heart block type I. *J Clin Invest*, 119(9): 2737–2744. doi:10.1172/JCI38292.

Kunert-Keil, C. et al. 2006. Tissue-specific expression of TRP channel genes in the mouse and its variation in three different mouse strains. *BMC Genomics*, 7: 159. doi:10.1186/1471-2164-7-159.

Kuo, I.Y. et al. 2016. Decreased polycystin 2 levels result in non-renal cardiac dysfunction with aging. *PLoS One*, 11(4): e0153632. doi:10.1371/journal.pone.0153632.

Kuo, I.Y. et al. 2014. Decreased polycystin 2 expression alters calcium-contraction coupling and changes beta-adrenergic signaling pathways. *Proc Natl Acad Sci U S A*, 111(46): 16604–16609. doi:10.1073/pnas.1415933111.

Kuster, D.W. et al. 2013. MicroRNA transcriptome profiling in cardiac tissue of hypertrophic cardiomyopathy patients with MYBPC3 mutations. *J Mol Cell Cardiol*, 65: 59–66. doi:10.1016/j.yjmcc.2013.09.012.

Kuwahara, K. et al. 2006. TRPC6 fulfills a calcineurin signaling circuit during pathologic cardiac remodeling. *J Clin Invest*, 116(12): 3114–3126. doi:10.1172/JCI27702.

Lang, H. et al. 2015. Activation of TRPV1 attenuates high salt-induced cardiac hypertrophy through improvement of mitochondrial function. *Br J Pharmacol*, 172(23): 5548–5558. doi:10.1111/bph.12987.

Li, N. et al. 2015. Nicotine induces cardiomyocyte hypertrophy through TRPC3-mediated Ca^{2+}/NFAT signalling pathway. *Can J Cardiol*, 32: 1260.doi:10.1016/j.cjca.2015.12.015.

Li, Q. et al. 2003a. Troponin I binds polycystin-L and inhibits its calcium-induced channel activation. *Biochemistry*, 42(24): 7618–7625. doi:10.1021/bi034210a.

Li, Q. et al. 2003b. Polycystin-2 interacts with troponin I, an angiogenesis inhibitor. *Biochemistry*, 42(2): 450–457. doi:10.1021/bi0267792.

Li, Q. et al. 2005. Alpha-actinin associates with polycystin-2 and regulates its channel activity. *Hum Mol Genet*, 14(12): 1587–1603. doi:10.1093/hmg/ddi167.

Li, Q. et al. 2007. Direct binding of alpha-actinin enhances TRPP3 channel activity. *J Neurochem*, 103(6): 2391–2400. doi:10.1111/j.1471-4159.2007.04940.x.

Lighthouse, J.K. and E.M. Small. 2016. Transcriptional control of cardiac fibroblast plasticity. *J Mol Cell Cardiol*, 91: 52–60. doi:10.1016/j.yjmcc.2015.12.016.

Liu, F.F. et al. 2010a. Differential expression of TRPC channels in the left ventricle of spontaneously hypertensive rats. *Mol Biol Rep*, 37(6): 2645–2651. doi:10.1007/s11033-009-9792-z.

Liu, H. et al. 2010b. Gain-of-function mutations in TRPM4 cause autosomal dominant isolated cardiac conduction disease. *Circ Cardiovasc Genet*, 3(4): 374–385. doi:10.1161/CIRCGENETICS.109.930867.

Liu, H. et al. 2013. Molecular genetics and functional anomalies in a series of 248 Brugada cases with 11 mutations in the TRPM4 channel. *PLoS One*, 8(1): e54131. doi:10.1371/journal.pone.0054131.

Lorin, C., I. Vogeli, and E. Niggli. 2015. Dystrophic cardiomyopathy: Role of TRPV2 channels in stretch-induced cell damage. *Cardiovasc Res*, 106(1): 153–162. doi:10.1093/cvr/cvv021.

Macianskiene, R. et al. 2008. Inhibition of the magnesium-sensitive TRPM7-like channel in cardiac myocytes by nonhydrolysable GTP analogs: Involvement of phosphoinositide metabolism. *Cell Physiol Biochem*, 22(1–4): 109–118. doi:10.1159/000149788.

Macianskiene, R. et al. 2012. Characterization of Mg^{2+}-regulated TRPM7-like current in human atrial myocytes. *J Biomed Sci*, 19: 75. doi:10.1186/1423-0127-19-75.

Makarewich, C.A. et al. 2012. A caveolae-targeted L-type Ca^{2+} channel antagonist inhibits hypertrophic signaling without reducing cardiac contractility. *Circ Res*, 110(5): 669–674. doi:10.1161/CIRCRESAHA.111.264028.

Makarewich, C.A. et al. 2014. Transient receptor potential channels contribute to pathological structural and functional remodeling after myocardial infarction. *Circ Res*, 115(6): 567–580. doi:10.1161/CIRCRESAHA.115.303831.

Markowitz, G.S. et al. 1999. Polycystin-2 expression is developmentally regulated. *Am J Physiol*, 277(1 Pt 2): F17–F25.

Marshall, N.J. et al. 2013. A role for TRPV1 in influencing the onset of cardiovascular disease in obesity. *Hypertension*, 61(1): 246–252. doi:10.1161/HYPERTENSIONAHA.112.201434.

Mathar, I. et al. 2010. Increased catecholamine secretion contributes to hypertension in TRPM4-deficient mice. *J Clin Invest*, 120(9): 3267–3279. doi:10.1172/JCI41348.

Mathar, I. et al. 2014. Increased beta-adrenergic inotropy in ventricular myocardium from Trpm4$^{-/-}$ mice. *Circ Res*, 114(2): 283–294. doi:10.1161/CIRCRESAHA.114.302835.

Meissner, M. et al. 2011. Lessons of studying TRP channels with antibodies. In *TRP Channels*, edited by M.X. Zhu. Boca Raton, FL: CRC Press/Taylor & Francis.

Miller, B.A. et al. 2013. The second member of transient receptor potential-melastatin channel family protects hearts from ischemia-reperfusion injury. *Am J Physiol Heart Circ Physiol*, 304(7): H1010–H1022.

Miller, B.A. et al. 2014. TRPM2 channels protect against cardiac ischemia-reperfusion injury: Role of mitochondria. *J Biol Chem*, 289(11): 7615–7629. doi:10.1074/jbc.M113.533851.

Mohl, M.C. et al. 2011. Regulation of murine cardiac contractility by activation of alpha(1A)-adrenergic receptor-operated Ca^{2+} entry. *Cardiovasc Res*, 91(2): 310–319. doi:10.1093/cvr/cvr081.

Nakamura, T. et al. 2015. Prevention of PKG1alpha oxidation augments cardioprotection in the stressed heart. *J Clin Invest*, 125(6): 2468–2472. doi:10.1172/JCI80275.

Nakayama, H. et al. 2006. Calcineurin-dependent cardiomyopathy is activated by TRPC in the adult mouse heart. *FASEB J*, 20(10): 1660–1670. doi:10.1096/fj.05-5560com.

Nath, A.K. et al. 2009. Proteomic-based detection of a protein cluster dysregulated during cardiovascular development identifies biomarkers of congenital heart defects. *PLoS One*, 4(1): e4221. doi:10.1371/journal.pone.0004221.

Naticchioni, M. et al. 2015. Transient receptor potential vanilloid 2 regulates myocardial response to exercise. *PLoS One*, 10(9): e0136901. doi:10.1371/journal.pone.0136901.

Nishida, M. and H. Kurose. 2008. Roles of TRP channels in the development of cardiac hypertrophy. *Naunyn Schmiedebergs Arch Pharmacol*, 378(4): 395–406. doi:10.1007/s00210-008-0321-8.

Nishida, M. et al. 2007. Gα12/13-mediated up-regulation of TRPC6 negatively regulates endothelin-1-induced cardiac myofibroblast formation and collagen synthesis through nuclear factor of activated T cells activation. *J Biol Chem*, 282(32): 23117–23128. doi:10.1074/jbc.M611780200.

Nishida, M. et al. 2010. Phosphorylation of TRPC6 channels at Thr69 is required for anti-hypertrophic effects of phosphodiesterase 5 inhibition. *J Biol Chem*, 285 (17):13244-13253. doi: M109.074104 [pii]. 10.1074/jbc.M109.074104.

Nomura, H. et al. 1998. Identification of PKDL, a novel polycystic kidney disease 2-like gene whose murine homologue is deleted in mice with kidney and retinal defects. *J Biol Chem*, 273(40): 25967–25973.

Odnoshivkina, U.G. et al. 2015. β2-adrenoceptor agonist-evoked reactive oxygen species generation in mouse atria: Implication in delayed inotropic effect. *Eur J Pharmacol*, 765: 140–153. doi:10.1016/j.ejphar.2015.08.020.

Oguri, G. et al. 2014. Effects of methylglyoxal on human cardiac fibroblast: Roles of transient receptor potential ankyrin 1 (TRPA1) channels. *Am J Physiol Heart Circ Physiol*, 307(9): H1339–H1352. doi:10.1152/ajpheart.01021.2013.

Ohba, T. et al. 2006. Regulatory role of neuron-restrictive silencing factor in expression of TRPC1. *Biochem Biophys Res Commun*, 351(3): 764–770. doi:10.1016/j.bbrc.2006.10.107.

Ohba, T. et al. 2007. Upregulation of TRPC1 in the development of cardiac hypertrophy. *J Mol Cell Cardiol*, 42(3): 498–507. doi:10.1016/j.yjmcc.2006.10.020.

Ohba, T. et al. 2009. Essential role of STIM1 in the development of cardiomyocyte hypertrophy. *Biochem Biophys Res Commun*, 389(1): 172–176. doi:10.1016/j.bbrc.2009.08.117.

Okada, T. et al. 1999. Molecular and functional characterization of a novel mouse transient receptor potential protein homologue TRP7. Ca^{2+}-permeable cation channel that is constitutively activated and enhanced by stimulation of G protein-coupled receptor. *J Biol Chem*, 274(39): 27359–27370.

Onohara, N. et al. 2006. TRPC3 and TRPC6 are essential for angiotensin II-induced cardiac hypertrophy. *EMBO J*, 25(22): 5305–5316. doi:10.1038/sj.emboj.7601417.

Pacher, P., S. Batkai, and G. Kunos. 2004. Haemodynamic profile and responsiveness to anandamide of TRPV1 receptor knock-out mice. *J Physiol*, 558(Pt 2): 647–657. doi:10.1113/jphysiol.2004.064824.

Parajuli, N. et al. 2015. Determinants of ventricular arrhythmias in human explanted hearts with dilated cardiomyopathy. *Eur J Clin Invest*, 45(12): 1286–1296. doi:10.1111/eci.12549.

Pazienza, V. et al. 2014. The TRPA1 channel is a cardiac target of mIGF-1/SIRT1 signaling. *Am J Physiol Heart Circ Physiol*, 307(7): H939–H944. doi:10.1152/ajpheart.00150.2014.

Pei, Z. et al. 2014. α, β-Unsaturated aldehyde crotonaldehyde triggers cardiomyocyte contractile dysfunction: Role of TRPV1 and mitochondrial function. *Pharmacol Res*, 82: 40–50. doi:10.1016/j.phrs.2014.03.010.

Pennekamp, P. et al. 2002. The ion channel polycystin-2 is required for left-right axis determination in mice. *Curr Biol*, 12(11): 938–943.

Philipp, S. et al. 1996. A mammalian capacitative calcium entry channel homologous to *Drosophila* TRP and TRPL. *EMBO J*, 15(22): 6166–6171.

Piao, H. et al. 2015. Transient receptor potential melastatin-4 is involved in hypoxia-reoxygenation injury in the cardiomyocytes. *PLoS One*, 10(4): e0121703. doi:10.1371/journal.pone.0121703.

Prasad, A.M. and G. Inesi. 2009. Effects of thapsigargin and phenylephrine on calcineurin and protein kinase C signaling functions in cardiac myocytes. *Am J Physiol Cell Physiol*, 296(5): C992–C1002. doi:10.1152/ajpcell.00594.2008.

Qi, J. et al. 2014. Choline prevents cardiac hypertrophy by inhibiting protein kinase C-delta dependent transient receptor potential canonical 6 channel. *Int J Cardiol*, 172(3): e525–e526. doi:10.1016/j.ijcard.2014.01.072.

Qi, Y. et al. 2015a. Uniaxial cyclic stretch stimulates TRPV4 to induce realignment of human embryonic stem cell-derived cardiomyocytes. *J Mol Cell Cardiol*, 87: 65–73. doi:10.1016/j.yjmcc.2015.08.005.

Qi, Y. et al. 2015b. Role of TRPV1 in the differentiation of mouse embryonic stem cells into cardiomyocytes. *PLoS One*, 10(7): e0133211. doi:10.1371/journal.pone.0133211.

Qi, Z. et al. 2016. TRPC3 regulates the automaticity of embryonic stem cell-derived cardiomyocytes. *Int J Cardiol*, 203: 169–181. doi:10.1016/j.ijcard.2015.10.018.

Qin, X. et al. 2013. Sphingosine and FTY720 are potent inhibitors of the transient receptor potential melastatin 7 (TRPM7) channels. *Br J Pharmacol*, 168(6): 1294–1312. doi:10.1111/bph.12012.

Ren, J.Y. et al. 2011. Cardioprotection by ischemic postconditioning is lost in isolated perfused heart from diabetic rats: Involvement of transient receptor potential vanilloid 1, calcitonin gene-related peptide and substance P. *Regul Pept*, 169(1–3): 49–57. doi:10.1016/j.regpep.2011.04.004.

Riccio, A. et al. 2002. mRNA distribution analysis of human TRPC family in CNS and peripheral tissues. *Brain Res Mol Brain Res*, 109(1–2): 95–104.

Roberge, S. et al. 2014. TNF-α-mediated caspase-8 activation induces ROS production and TRPM2 activation in adult ventricular myocytes. *Cardiovasc Res*, 103(1): 90–99. doi:10.1093/cvr/cvu112.

Roderick, H.L. et al. 2007. Calcium in the heart: When it's good, it's very very good, but when it's bad, it's horrid. *Biochem Soc Trans*, 35(Pt 5): 957–961. doi:10.1042/BST0350957.

Rose, R.A. et al. 2007. C-type natriuretic peptide activates a non-selective cation current in acutely isolated rat cardiac fibroblasts via natriuretic peptide C receptor-mediated signalling. *J Physiol*, 580(Pt 1): 255–274. doi:10.1113/jphysiol.2006.120832.

Rubinstein, J. et al. 2014. Novel role of transient receptor potential vanilloid 2 in the regulation of cardiac performance. *Am J Physiol Heart Circ Physiol*, 306(4): H574–H584. doi:10.1152/ajpheart.00854.2013.

Runnels, L.W., L. Yue, and D.E. Clapham. 2001. TRP-PLIK, a bifunctional protein with kinase and ion channel activities. *Science*, 291(5506): 1043–1047. doi:10.1126/science.1058519.

Runnels, L.W., L. Yue, and D.E. Clapham. 2002. The TRPM7 channel is inactivated by PIP(2) hydrolysis. *Nat Cell Biol*, 4(5): 329–336. doi:10.1038/ncb781.

Sabourin, J. et al. 2012. Activation of transient receptor potential canonical 3 (TRPC3)-mediated Ca^{2+} entry by A1 adenosine receptor in cardiomyocytes disturbs atrioventricular conduction. *J Biol Chem*, 287(32): 26688–26701. doi:10.1074/jbc.M112.378588.

Sabourin, J. et al. 2016. Transient receptor potential canonical (TRPC)/Orai1-dependent store-operated Ca^{2+} channels: New targets of aldosterone in cardiomyocytes. *J Biol Chem*, 291(25): 13394–13409. doi:10.1074/jbc.M115.693911.

Sah, R. et al. 2013a. Timing of myocardial trpm7 deletion during cardiogenesis variably disrupts adult ventricular function, conduction, and repolarization. *Circulation*, 128(2): 101–114. doi:10.1161/CIRCULATIONAHA.112.000768.

Sah, R. et al. 2013b. Ion channel-kinase TRPM7 is required for maintaining cardiac automaticity. *Proc Natl Acad Sci U S A*, 110(32): E3037–E3046. doi:10.1073/pnas.1311865110.

Sakura, H. and F.M. Ashcroft. 1997. Identification of four trp1 gene variants murine pancreatic beta-cells. *Diabetologia*, 40(5): 528–532. doi:10.1007/s001250050711.

Samie, M.A. et al. 2009. The tissue-specific expression of TRPML2 (MCOLN-2) gene is influenced by the presence of TRPML1. *Pflugers Arch*, 459(1): 79–91. doi:10.1007/s00424-009-0716-5.

Satoh, S. et al. 2007. Transient receptor potential (TRP) protein 7 acts as a G protein-activated Ca^{2+} channel mediating angiotensin II-induced myocardial apoptosis. *Mol Cell Biochem*, 294(1–2): 205–215. doi:10.1007/s11010-006-9261-0.

Schleifer, H. et al. 2012. Novel pyrazole compounds for pharmacological discrimination between receptor-operated and store-operated Ca^{2+} entry pathways. *Br J Pharmacol*, 167(8): 1712–1722. doi:10.1111/j.1476-5381.2012.02126.x.

Seo, K. et al. 2014a. Hyperactive adverse mechanical stress responses in dystrophic heart are coupled to transient receptor potential canonical 6 and blocked by cGMP-protein kinase G modulation. *Circ Res*, 114(5): 823–832. doi:10.1161/CIRCRESAHA.114.302614.

Seo, K. et al. 2014b. Combined TRPC3 and TRPC6 blockade by selective small-molecule or genetic deletion inhibits pathological cardiac hypertrophy. *Proc Natl Acad Sci U S A*, 111(4): 1551–1556. doi:10.1073/pnas.1308963111.

Seth, M. et al. 2004. Sarco(endo)plasmic reticulum Ca^{2+} ATPase (SERCA) gene silencing and remodeling of the Ca^{2+} signaling mechanism in cardiac myocytes. *Proc Natl Acad Sci U S A*, 101(47): 16683–16688. doi:10.1073/pnas.0407537101.

Seth, M. et al. 2009. TRPC1 channels are critical for hypertrophic signaling in the heart. *Circ Res*, 105(10): 1023–1030. doi:10.1161/CIRCRESAHA.109.206581.

Shan, D., R.B. Marchase, and J.C. Chatham. 2008. Overexpression of TRPC3 increases apoptosis but not necrosis in response to ischemia-reperfusion in adult mouse cardiomyocytes. *Am J Physiol Cell Physiol*, 294(3): C833–C841. doi:10.1152/ajpcell.00313.2007.

Shenton, F.C. and S. Pyner. 2014. Expression of transient receptor potential channels TRPC1 and TRPV4 in veno-atrial endocardium of the rat heart. *Neuroscience*, 267: 195–204. doi:10.1016/j.neuroscience.2014.02.047.

Simard, C. et al. 2013. The TRPM4 non-selective cation channel contributes to the mammalian atrial action potential. *J Mol Cell Cardiol*, 59: 11–19. doi:10.1016/j.yjmcc.2013.01.019.

Son, M.J. et al. 2016. Shear stress activates monovalent cation channel transient receptor potential melastatin subfamily 4 in rat atrial myocytes via type 2 inositol 1,4,5-trisphosphate receptors and Ca^{2+} release. *J Physiol*, 594(11): 2985–3004. doi:10.1113/JP270887.

Souders, C.A., S.L. Bowers, and T.A. Baudino. 2009. Cardiac fibroblast: the renaissance cell. *Circ Res*, 105(12): 1164–1176. doi:10.1161/CIRCRESAHA.109.209809.

Sowa, N. et al. 2012. MicroRNA 26b encoded by the intron of small CTD phosphatase (SCP) 1 has an antagonistic effect on its host gene. *J Cell Biochem*, 113(11): 3455–3465. doi:10.1002/jcb.24222.

Stallmeyer, B. et al. 2012. Mutational spectrum in the Ca^{2+}-activated cation channel gene TRPM4 in patients with cardiac conductance disturbances. *Hum Mutat*, 33(1): 109–117. doi:10.1002/humu.21599.

Stypmann, J. et al. 2007. Cardiovascular characterization of Pkd2(+/LacZ) mice, an animal model for the autosomal dominant polycystic kidney disease type 2 (ADPKD2). *Int J Cardiol*, 120(2): 158–166. doi:10.1016/j.ijcard.2006.09.013.

Sucharov, C.C. and M.R. Bristow. 2005. Channels and β1-adrenergic-mediated activation of fetal gene program. *Circ Res*, 97: e9–e50.

Sun, Y.H. et al. 2010. Calcium-sensing receptor activation contributed to apoptosis stimulates TRPC6 channel in rat neonatal ventricular myocytes. *Biochem Biophys Res Commun*, 394(4): 955–961. doi:10.1016/j.bbrc.2010.03.096.

Sun, Z. et al. 2014. TRPV1 activation exacerbates hypoxia/reoxygenation-induced apoptosis in H9C2 cells via calcium overload and mitochondrial dysfunction. *Int J Mol Sci*, 15(10): 18362–18380. doi:10.3390/ijms151018362.

Syam, N. et al. 2016. Variants of transient receptor potential melastatin member 4 in childhood atrioventricular block. *J Am Heart Assoc*, 5(5). doi:10.1161/JAHA.114.001625.

Takahashi, K., K. Sakamoto, and J. Kimura. 2012. Hypoxic stress induces transient receptor potential melastatin 2 (TRPM2) channel expression in adult rat cardiac fibroblasts. *J Pharmacol Sci*, 118(2): 186–197.

Tashiro, M., H. Inoue, and M. Konishi. 2013. Magnesium homeostasis in cardiac myocytes of Mg-deficient rats. *PLoS One*, 8(9): e73171. doi:10.1371/journal.pone.0073171.

Tashiro, M., H. Inoue, and M. Konishi. 2014. Physiological pathway of magnesium influx in rat ventricular myocytes. *Biophys J*, 107(9): 2049–2058. doi:10.1016/j.bpj.2014.09.015.

Tavi, P., et al. 2004. Pacing-induced calcineurin activation controls cardiac Ca^{2+} signalling and gene expression. *J Physiol*, 554(Pt 2): 309–320. doi:10.1113/jphysiol.2003.053579.

Tellez, J.O. et al. 2011. Ageing-dependent remodelling of ion channel and Ca^{2+} clock genes underlying sino-atrial node pacemaking. *Exp Physiol*, 96(11): 1163–1178. doi:10.1113/expphysiol.2011.057752.

Uhl, S. et al. 2014. Adenylyl cyclase-mediated effects contribute to increased isoprenaline-induced cardiac contractility in TRPM4-deficient mice. *J Mol Cell Cardiol*, 74: 307–317. doi:10.1016/j.yjmcc.2014.06.007.

Vennekens, R. 2011. Emerging concepts for the role of TRP channels in the cardiovascular system. *J Physiol*, 589(Pt 7): 1527–1534. doi:10.1113/jphysiol.2010.202077.

Vennekens, R. et al. 2007. Increased IgE-dependent mast cell activation and anaphylactic responses in mice lacking the calcium-activated nonselective cation channel TRPM4. *Nat Immunol*, 8(3): 312–320. doi:10.1038/ni1441.

Vindis, C. et al. 2010. Essential role of TRPC1 channels in cardiomyoblasts hypertrophy mediated by 5-HT2A serotonin receptors. *Biochem Biophys Res Commun*, 391(1): 979–983. doi:10.1016/j.bbrc.2009.12.001.

Voelkers, M. et al. 2010. Orai1 and Stim1 regulate normal and hypertrophic growth in cardiomyocytes. *J Mol Cell Cardiol*, 48(6): 1329–1334. doi:10.1016/j.yjmcc.2010.01.020.

Volk, T. et al. 2003. A polycystin-2-like large conductance cation channel in rat left ventricular myocytes. *Cardiovasc Res*, 58(1): 76–88.

Wang, J. et al. 2013. 9-Phenanthrol, a TRPM4 inhibitor, protects isolated rat hearts from ischemia-reperfusion injury. *PLoS One*, 8(7): e70587. doi:10.1371/journal.pone.0070587.

Wang, L. and D.H. Wang. 2005. TRPV1 gene knockout impairs postischemic recovery in isolated perfused heart in mice. *Circulation*, 112(23): 3617–3623. doi:10.1161/CIRCULATIONAHA.105.556274.

Wang, Q. et al. 2014. Dietary capsaicin ameliorates pressure overload-induced cardiac hypertrophy and fibrosis through the transient receptor potential vanilloid type 1. *Am J Hypertens*, 27(12): 1521–1529. doi:10.1093/ajh/hpu068.

Ward, M.L. et al. 2008. Stretch-activated channels in the heart: Contributions to length-dependence and to cardiomyopathy. *Prog Biophys Mol Biol*, 97(2–3): 232–249. doi:10.1016/j.pbiomolbio.2008.02.009.

Watanabe, H. et al. 2009. The pathological role of transient receptor potential channels in heart disease. *Circ J*, 73(3): 419–427.

Wes, P.D. et al. 1995. TRPC1, a human homolog of a *Drosophila* store-operated channel. *Proc Natl Acad Sci U S A*, 92(21): 9652–9656.

Willette, R.N. et al. 2008. Systemic activation of the transient receptor potential vanilloid subtype 4 channel causes endothelial failure and circulatory collapse: Part 2. *J Pharmacol Exp Ther*, 326(2): 443–452. doi:10.1124/jpet.107.134551.

Wu, G. et al. 1998. Identification of PKD2L, a human PKD2-related gene: Tissue-specific expression and mapping to chromosome 10q25. *Genomics*, 54(3): 564–568. doi:10.1006/geno.1998.5618.

Wu, G. et al. 2000. Cardiac defects and renal failure in mice with targeted mutations in Pkd2. *Nat Genet*, 24(1): 75–78. doi:10.1038/71724.

Wu, L.J., T.B. Sweet, and D.E. Clapham. 2010. International Union of Basic and Clinical Pharmacology. LXXVI. Current progress in the mammalian TRP ion channel family. *Pharmacol Rev*, 62(3): 381–404. doi:10.1124/pr.110.002725.

Wu, X. et al. 2009. Plasma membrane Ca^{2+}-ATPase isoform 4 antagonizes cardiac hypertrophy in association with calcineurin inhibition in rodents. *J Clin Invest*, 119(4): 976–985. doi:10.1172/JCI36693.

Wu, X. et al. 2010. TRPC channels are necessary mediators of pathologic cardiac hypertrophy. *Proc Natl Acad Sci U S A*, 107(15): 7000–7005. doi:10.1073/pnas.1001825107.

Wu, Z. et al. 2015. α,β-Unsaturated aldehyde pollutant acrolein suppresses cardiomyocyte contractile function: Role of TRPV1 and oxidative stress. *Environ Toxicol*, 30(6): 638–647. doi:10.1002/tox.21941.

Xie, J. et al. 2012. Cardioprotection by Klotho through downregulation of TRPC6 channels in the mouse heart. *Nat Commun*, 3: 1238. doi:10.1038/ncomms2240.

Yu, Y. et al. 2014. TRPM7 is involved in angiotensin II induced cardiac fibrosis development by mediating calcium and magnesium influx. *Cell Calcium*, 55(5): 252–260. doi:10.1016/j.ceca.2014.02.019.

Zhang, Y. et al. 2012a. Cardiac-specific knockout of ET(A) receptor mitigates low ambient temperature-induced cardiac hypertrophy and contractile dysfunction. *J Mol Cell Biol*, 4(2): 97–107. doi:10.1093/jmcb/mjs002.

Zhang, Y.H. et al. 2012b. Evidence for functional expression of TRPM7 channels in human atrial myocytes. *Basic Res Cardiol*, 107(5): 282. doi:10.1007/s00395-012-0282-4.

Zhang, Y.H. et al. 2013. Functional transient receptor potential canonical type 1 channels in human atrial myocytes. *Pflugers Arch*, 465(10): 1439–1449. doi:10.1007/s00424-013-1291-3.

Zhang, Y.J. et al. 2015. Increased TRPM6 expression in atrial fibrillation patients contribute to atrial fibrosis. *Exp Mol Pathol*, 98(3): 486–490. doi:10.1016/j.yexmp.2015.03.025.

Zhao, Y. et al. 2012. Unusual localization and translocation of TRPV4 protein in cultured ventricular myocytes of the neonatal rat. *Eur J Histochem*, 56(3): e32. doi:10.4081/ejh.2012.e32.

Zheng, L.R. et al. 2015. Nerve growth factor rescues diabetic mice heart after ischemia/reperfusion injury via up-regulation of the TRPV1 receptor. *J Diabetes Complications*, 29(3): 323–328.

Zhong, B. and D.H. Wang. 2009. Protease-activated receptor 2-mediated protection of myocardial ischemia-reperfusion injury: Role of transient receptor potential vanilloid receptors. *Am J Physiol Regul Integr Comp Physiol*, 297(6): R1681–R1690. doi:10.1152/ajpregu.90746.2008.

Zhou, Y. et al. 2015. Effects of angiotensin II on transient receptor potential melastatin 7 channel function in cardiac fibroblasts. *Exp Ther Med*, 9(5): 2008–2012. doi:10.3892/etm.2015.2362.

Zhu, X. et al. 1995. Molecular cloning of a widely expressed human homologue for the *Drosophila* trp gene. *FEBS Lett*, 373(3): 193–198.

10 Renal Functions of TRP Channels in Health and Disease

Vladimir Chubanov, Sebastian Kubanek,
Susanne Fiedler, Lorenz Mittermeier,
Thomas Gudermann, and Alexander Dietrich

CONTENTS

10.1 KIDNEY FUNCTION AND EXPRESSION OF TRP CHANNELS

One key function of the kidneys is the ultrafiltration of plasma by the renal glomeruli in order to dispose of metabolic end products and excess electrolytes. However, in order to maintain body electrolyte homeostasis, the vast majority of filtrated salts and water are reabsorbed along the different segments of the renal tubular system. For instance, approximately 80% of total serum Mg^{2+} is filtered in the glomeruli with more than 95% being reabsorbed along the nephron. A vast array of ion channels and transporters is known to be involved in the filtration, secretion, and reabsorption of electrolytes.

In the last years, several members of the large transient receptor potential (TRP) family of ion channels have been shifted into the focus of renal research, because mutations in genes encoding members of three different TRP ion channel subfamilies have been linked to human kidney diseases (Gudermann, 2005). Mutations in *PKD1* (encoding TRPP1) and *PKD2* (coding for TRPP2) occur at high frequency in individuals with autosomal dominant polycystic kidney disease (Chang and Ong, 2008). The latter mutations and the pathophysiology of polycystic kidney disease will not be discussed further in this overview because the topic has been covered by numerous insightful review articles (Patel et al., 2009; Harris and Torres, 2009; Paul and Vanden Heuvel, 2014; Rangan et al., 2015; Trudel et al., 2016). Similarly, TRPV5 functioning as a key player in Ca^{2+} absorption in the distal tubule of the kidney will not be summarized in this review, because two recent articles are available (Hoorn and Zietse, 2013; Na and Peng, 2014).

Members of the classical TRP family (TRPC), its expression in different parts of the kidney, and its renal function are described in the first part of this review. TRPC1 the first cloned member of the mammalian TRPC family is differentially expressed in mesangial cells during diabetic nephropathy, and evidence for a role of TRPC3 in renal fibrosis was reported. Missense and nonsense mutations in the gene coding for TRPC6 have been found to segregate with autosomal dominant focal segmental sclerosis, a kidney disease that leads to progressive renal failure (Winn et al., 2005; Reiser et al., 2005).

Two members of the melastatin family of TRP channels (TRPM) and their function in mineral homeostasis in the kidney are the topic of the second part of this review. TRPM7 is ubiquitously expressed in the body, and together with TRPM6 are essential for kidney function. Loss-of-function mutations in TRPM6 are associated with hypomagnesemia with secondary hypocalcemia, a rare autosomal recessive disorder (Schlingmann et al., 2002; Chubanov et al., 2004; Konrad et al., 2004).

The involvement of TRP channel mutations in hereditary kidney disease has shed new light on the molecular pathogenesis of renal failure, and has significantly enhanced our appreciation of general cell physiological roles of TRP channels. In this short review article, which is also an update of our former manuscript (Dietrich et al., 2010), we focus on TRPC as well as TRPM6 and TRPM7 channels and discuss the role of these ion channels for renal physiology and pathophysiology.

10.2   TRPC CHANNELS IN KIDNEY FUNCTION AND DISEASE

Canonical or classical transient receptor potential (TRPC) channels were the first cloned mammalian homologues of the originally discovered *trp* channels in *Drosophila melanogaster*. The family of TRPC channels contains seven members and can be subdivided into subfamilies on the basis of amino acid similarity. While TRPC1 and TRPC2 are almost unique, TRPC4 and TRPC5 share ~64% amino acid identity. TRPC3, TRPC6, and TRPC7 form a structural and functional subfamily displaying 65%–78% identity at the amino acid level and share a common activator—that is, diacylglycerol (DAG) (Dietrich et al., 2005). DAG is produced by phospholipase C-isozymes activated after agonist binding to appropriate receptors (e.g., interaction of angiotensin II with AT1 receptors). All TRPC family members harbor an invariant sequence, the TRP box (containing the amino acid sequence: EWKFAR), in its C-terminal tail as well as three ankyrin repeats in the N-terminus (Figure 10.1a). The predicted transmembrane topology is similar to that of other TRP channels with intracellular N- and C-termini, six membrane-spanning helices (S1–S6), and a presumed pore forming loop (P) between S5 and S6 (Figure 10.1a). A functional TRPC ion channel complex is composed of four monomers of the same type in a homotetrameric complex or of four different TRPC monomers forming a heterotetrameric channel as illustrated in Figure 10.1b. A special feature of TRPC6 is two glycosylation sites that are responsible for its tight receptor-operated activation and its low basal activity compared to TRPC3 with high basal activity and glycosylated only at one site (Figure 10.1a) (Dietrich et al., 2003). This chapter focuses on TRPC1, TRPC3, TRPC5, and TRPC6 as they are prominently expressed in mesangial cells and podocytes in the kidney, and their renal function was at least initially characterized in recent years.

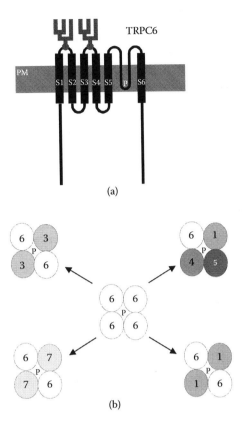

(a)

(b)

FIGURE 10.1 Structural features of TRPC6. (a) Topology of TRPC6 in the plasma membrane (PM). Transmembrane regions (S1–S6) and the predicted pore domain (P) are illustrated. Two glycosylated sites in TRPC6 are indicated by covalently bound carbohydrates (in gray). (b) Heteromultimerization potential of TRPC6. TRPC6 can directly interact with TRPC3 (3), TRPC7 (7), and with TRPC1 (1) to form functional tetramers but with TRPC4 (4) or TRPC5 (5) only in heteromeric complexes including TRPC1 (From Dietrich, A., *Handb. Exp. Pharmacol.* 222, 157-188, 2014.).

10.3 REGULATION OF TRPC1 EXPRESSION IN GLOMERULAR MESANGIAL CELLS

TRPC1 is almost ubiquitously expressed, but its exact function is still not known. Heterologously expressed TRPC1 is poorly translocated to the plasma membrane as a homotetramer and may not function as an ion channel on its own but may rather be a kind of β-subunit in heterotetrameric TRPC tetramers (summarized in Dietrich et al., 2014). Along these lines, TRPC1 was able to regulate TRPC1/TRPC4 and TRPC1/TRPC5 heteromeric channels (Strubing et al., 2001) and suppresses Ca^{2+} permeability of TRPC1/TRPC3, TRPC1/TRPC6 and TRPC1/TRPC7 complexes (Storch et al., 2012). An extended TRPC1 cDNA, which was discovered recently, may shed new light on TRPC1 characteristics when compared with short cDNA versions in functional assays (Ong et al., 2013).

Mesangial cells provide structural support for glomerular capillary loops and regulate capillary flow as well as the ultrafiltration surface (Schlondorff and Banas, 2009). TRPC1 is the most prominently expressed TRPC channel in cultured intraglomerular mesangial cells (see Figure 10.2d) and angiotensin II–induced contraction is inhibited by siRNA-mediated downregulation of TRPC1 (Du et al., 2007). However, these cells lack the expression of contractile α-smooth muscle actin *in vivo* that is only present in cell culture (summarized in Carlstrom et al., 2015). Therefore, the authors tested TRPC1 function on the glomerular filtration rate in rats *in vivo* using TRPC1-specific blocking antibodies and were able to reduce AngII-induced increases in blood pressure (Du et al., 2007).

We determined an important role of TRPC1 in cultured mesangial cells, because TRPC1-deficient cells show an increased rate of apoptotic cells compared to wild-type cells (Figure 10.2c). Moreover, mesangial cell proliferation and mesangial matrix expansion play an important role in a wide range of glomerular diseases, in particular in diabetic nephropathy (DN) (Dalla Vestra et al., 2001; Haneda et al., 2003), the most common form of kidney failure in the United States (Brownlee, 2001). TRPC1 mRNA expression is decreased in streptozotocin-treated and obese "Zucker" rats, which are animal models of diabetes types 1 and 2, respectively (Niehof and Borlak, 2008), and the TRPC1 gene is located in a chromosomal region harboring candidate genes for DN (McKnight et al., 2006). Downregulation of TRPC1 mRNA was also detected in diabetic *db/db* mice expressing a mutant leptin receptor (Zhang et al., 2009). We were able to confirm decreased TRPC1 mRNA expression in kidneys of mice homozygous but not heterozygous for the mutation (Figure 10.2a). Interestingly, micro-RNA miR-135a recently identified in DN patients downregulates the expression of TRPC1

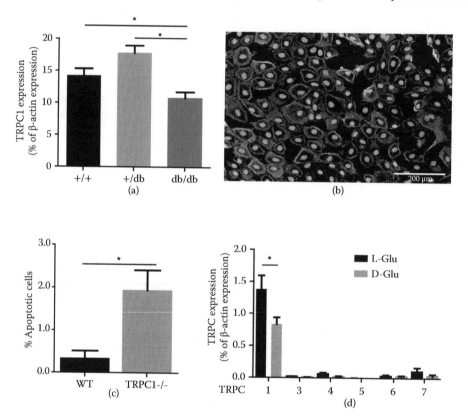

FIGURE 10.2 TRPC1 expression and function in murine kidneys and glomerular mesangial cells. (a) TRPC1 expression in kidneys from homozygous and heterozygous db/db (BKS.Cg-*Dock7^m Lepr^db*/+ +/J; Jackson Lab #000700) mice as well as wild-type (WT) controls. Quantitative reverse transcriptase polymerase chain reaction with murine TRPC1 primers was done as described. (b) Fluorescence image of mesangial cells stained with an α-SMA antibody coupled to a secondary FITC-coupled antibody and DAPI-staining of nuclei (From Kalwa, H., *J Cell Physiol*, 230, 2015.) Cells were freshly prepared from mouse kidneys and cultured according to a published protocol (Lichtnekert et al. 2009). (c) Quantification of apoptotic cells using the TUNEL method in cultures of mesangial cells from WT and TRPC1^−/− mice. TUNEL assays were done as described by the manufacturer (Click-iT TUNEL Alexa Fluor 488 Imaging Assay, Thermo Fisher Scientific #C10245). (d) TRPC mRNA expression levels after culturing mesangial cells in high D-glucose (25.6 nM) or low D-glucose (5.6 nM) medium. Equal osmolality was adjusted by adding L-glucose (20 nM) to the low D-glucose medium. After culturing 70%–80% confluent cells for 48 hours in low and high D-glucose medium mRNA was isolated and TRPC expression was quantified as described (From Kalwa, H., *J Cell Physiol,* 230, 2015.)

mRNA (He et al., 2014). Hepatocyte nuclear transcription factor 4α activates expression of TRPC1, and both proteins are downregulated in kidneys of diabetes patients (Niehof and Borlak, 2008). We mimicked DN conditions in cell culture by adding high concentrations of D-glucose to freshly isolated mesangial cells (see Figure 10.2b) and confirmed a downregulation of TRPC1 mRNA (see Figure 10.2d). Along these lines, TRPC1 polymorphisms are closely associated with type 2 diabetes and DN in a Han Chinese population (Chen et al., 2014), but not in an African American population (Zhang et al., 2009). Therefore, TRPC1 dysfunction may play an important role in the pathogenesis and development of DN, but its exact molecular function is rather elusive.

10.4 TRPC3 AND ITS ROLE IN RENAL FIBROSIS

TRPC3 is activated by DAG and is able to form heterotetrameric channels directly with TRPC1 (Storch et al., 2012), TRPC7, and TRPC6 (Hofmann et al., 2002). In a homomeric complex, TRPC3 displays higher basal activity than TRPC6 (Dietrich et al., 2003). Expression of TRPC3 was detected in renal fibroblasts, podocytes, and cells of the distal convoluted tubules as well as the cortical and medullary collecting ducts (Goel et al., 2006; Goel et al., 2007; Letavernier et al., 2012; Saliba et al., 2015). Fibroblast proliferation and extracellular signal-regulated protein kinases ERK1/2 phosphorylation were found to be regulated by increased Ca^{2+} entry through TRPC3 channels (Saliba et al., 2015), which might emerge as a pharmacological target in renal fibrosis. A physical interaction of TRPC3 with aquaporins was demonstrated in principal cells of the distal convoluted tubules where the channel is responsible for Ca^{2+} reabsorption (Goel et al., 2007).

10.5 TRPC5 FUNCTION DURING KIDNEY INJURY

TRPC5 forms heteromeric channels with TRPC1 and TRPC4 (Strubing et al., 2001, 2003), but its exact activation mechanism is still an important subject of current research. Genetic deletion of TRPC5 reduces damage to the glomerular filtration barrier in lipopolysaccharide (LPS)- and protamine sulfate (PS)-induced mouse models of kidney injury (Schaldecker et al., 2013). Glomeruli comprise mesangial cells and podocytes contributing to the slit diaphragm additionally formed by the fenestrated endothelium and the basal membrane. Until recently, podocytes were assumed not to be replaceable as these cells are not able to proliferate in adulthood, and podocyte dysfunction was believed to be the primary cause of glomerular damage. For this reason, TRPC5 function was extensively analyzed in immortalized podocyte cell lines, because freshly isolated podocytes do not proliferate *in vivo* (Tian et al., 2010). LPS activates TRPC5 (Beech et al., 2009), which stimulates Rac1-induced synaptopodin degradation leading to cytoskeletal collapse in cultured podocytes *in vitro* (Tian et al., 2010). The authors concluded that TRPC5 expression is induced in a disease state and initiates filter barrier damage, while TRPC6 keeps the filter barrier intact. However, functional analysis of TRPC5 currents in freshly isolated podocytes to lend credence to TRPC5 function *in vivo* is still lacking.

10.6 PHYSIOLOGICAL ROLE OF TRPC6 AND ALTERED FUNCTION BY INHERITED MUTATIONS IN THE KIDNEY

TRPC6 is a receptor-operated channel with low basal activity and forms heterotetramers with TRPC3, TRPC7 (Hofmann et al., 2002), and TRPC1 (Figure 10.1b) (Storch et al., 2012). As deduced from Northern blot analyses, TRPC6 is most prominently expressed in lung tissues (Boulay et al., 1997), and we were able to demonstrate an essential role of these channels in acute hypoxic pulmonary vasoconstriction (Weissmann et al., 2006) and lung edema formation (Weissmann et al., 2012).

TRPC6 in heteromeric complexes with TRPC3 is detected in the glomerulus and along the collecting duct (Hsu et al., 2007) and co-localizes with aquaporin 2 (Goel et al., 2006). Moreover, native TRPC3/TRPC6 heteromers have also been identified in Madin-Darby canine epithelial

(MDCK) cells (Bandyopadhyay et al., 2005). In polarized cultures of M1 and IMCD-3 collecting duct cells, however, TRPC3 localizes exclusively to the apical domain, whereas homotetrameric TRPC6 channels are found in basolateral and apical membranes (Goel et al., 2007).

In membranous nephropathy (MN), subepithelial deposition of autoantibodies leads to changes of the glomerular basement membrane with podocyte injury. Complement activation is made responsible for these morphological changes (Glassock, 2010). Most interestingly, very high levels of TRPC6 expression were identified in MN patients (Moller et al., 2007) and overexpression of TRPC6 in podocytes protected from complement-mediated injury, while TRPC6 deficiency increased the susceptibility of podocytes (Kistler et al., 2013).

Another important function of TRPC6 in the kidney was discovered by "reverse genetics." Several studies identified "gain-of-function" mutations in TRPC6 (see Figure 10.3a and Mottl et al., 2013, for a summary of all mutations) in patients with focal segmental glomerular sclerosis (FSGS) (reviewed in Dietrich et al., 2010). FSGS is morphologically characterized by sclerotic lesions in only a subfraction of renal glomeruli. FSGS typically entails massive proteinuria, hypertension, and nephrotic syndrome, eventually leading to end-stage renal disease both in children (7%–20% of cases) and in adults (approximately 35% of cases). In six families with hereditary forms of FSGS, two research groups independently identified gain-of-function mutations in the TRPC6 channel (see Figure 10.3a) leading to enhanced Ca^{2+} influx (P112Q) (Winn et al., 2005) and increased current amplitudes in electrophysiological recordings of heterologously expressed ion channels (R895C, E897K) (Reiser et al., 2005). TRPC6 expression in the kidney was localized to the glomeruli and tubuli (Winn et al., 2005) as well as podocytes (Reiser et al., 2005) as analyzed by immunohistochemistry with a TRPC6-specific antibody. Three mutations identified by Reiser et al. (N143S, S270T, and K874*) did not affect current amplitudes but were recently reported to result in channels with a higher open probability than wild-type TRPC6 channels (Moller et al., 2009). Moreover, channel density at the plasma membrane was significantly higher in HEK293 cells expressing the P112Q mutant. A seventh mutation (Q889K) was found in 31 Chinese pedigrees with late-onset FSGS (Zhu et al., 2009). HEK293 cells expressing the mutant channel protein exhibited significantly higher Ca^{2+} influx after stimulation by 1-oleoyl-2-acetyl-glycerol (OAG), a membrane-permeable analogue of DAG, than cells expressing the wild-type form of TRPC6—an observation that might be attributable to a significantly higher expression level of the mutant protein. *TRPC6* mutation analysis in 130 Spanish patients with FSGS led to the identification of three additional, as yet functionally uncharacterized *TRPC6* missense mutations, two in the N-terminus close to the ankyrin repeats (G109S, N125S) and a third one adjacent to the TRP domain (L780P) (Santin et al., 2009). Notably, the discovery of an 11th *TRPC6* mutation (M132T), which showed the highest mean inward Ca^{2+} current amplitude—10-fold higher than wild-type TRPC6—in a child argues against the prevailing hypothesis that FSGS induced by *TRPC6* mutations manifests only in adults (Du et al., 2007). Confusingly, five of the identified mutations (N125S, L395A, G757D, L870P, and R895L) were reported to be loss-of-function mutations after heterologous expression in HEK293 cells (Figure 10.3a) (Riehle et al., 2016). Therefore, only a careful analysis of TRPC6 mutants in their natural environment (e.g., in freshly isolated podocytes from knock-in mouse lines carrying the mutations) justifies conclusions as to a gain- or loss-of-function phenotype *in vivo*. In aggregate, the mutations described so far are indicative of a frequency of 6% of *TRPC6* mutations in familial FSGS and approximately 2% in sporadic cases (Santin et al., 2009).

Data from gene-deficient and transgenic mouse lines published so far support a gain-of-function concept of TRPC6-induced FSGS. TRPC6-deficient mice are protected from angiotensin II (AngII)-induced albuminuria (Eckel et al., 2011), while *in vivo* delivery of cDNA encoding TRPC6 to mice induces proteinuria (Moller et al., 2007). Therefore overactive TRPC6 together with other proteins, for example, nephrin and podocin (reviewed in (Eckel et al., 2011 and presented in Figure 10.3b), are responsible for podocyte dysfunction and finally kidney failure in FSGS. Overexpression of wild-type or mutant TRPC6 in transgenic mice is sufficient to induce a late onset of proteinuria and glomerular lesions; two cardinal symptoms also found in FSGS patients (Krall et al., 2010).

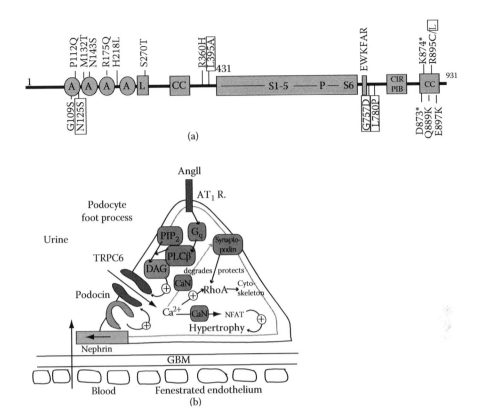

FIGURE 10.3 (a) Domain structure of TRPC6 with a detailed localization of mutations in the human TRPC6 channel identified in patients with focal segmental glomerulosclerosis. Gain-of-function mutations are indicated in black, while loss-of-function mutations are illustrated in *A*, ankyrin repeat; *cc*, coiled coil domain; *CIRPIB*, Ca^{2+}/CaM PI binding site; *EWKFAR*, conserved TRP-box motif; *L*, DAG sensitive lipid trafficking domain and mutants in patients with proteinuria (e.g., *P112Q*) (From Dietrich, A., *Handb. Exp. Pharmacol.*, 222, 157-188, 2014.). (b) Simplified summary of TRPC6/Ca^{2+}–dependent signal transduction pathways in foot processes of glomerular podocytes. Angiotensin II (AngII) activates angiotensin 1 (AT$_1$) receptors, opening TRPC6 channels via phospholipase C (PLC) produced diacylglycerol (DAG) from phosphatidylinositol-bisphosphate (PIP$_2$). A physiological cytosolic Ca^{2+} concentration induced by normal TRPC6 activity adapts the actin cytoskeleton to the renal flow by RhoA closing the slit diaphragm by nephrin. Pathophysiological Ca^{2+} concentrations, however, may activate the phosphatase calcineurin (CaN), which dephosphorylates nuclear factor of activated T cells (NFAT) inducing podocyte hypertrophy or dephosphorylates synaptopodin resulting in its degradation. After a loss of synaptopodin RhoA is no longer protected from degradation, and proper function of the actin cytoskeleton is not ensured.

Ultrafiltration of plasma to dispose of metabolic end products, excess electrolytes, and water is one of the key functions of the kidney glomerulus that is destructed in FSGS patients. Glomeruli comprise mesangial cells and podocytes contributing to the slit diaphragm additionally formed by the fenestrated endothelium and the basement membrane (see Figure 10.3b). Until recently podocytes were assumed not to be replaceable as these cells are not able to proliferate in adulthood, and podocyte dysfunction was believed to be the primary cause of FSGS. For this reason TRPC6 function was extensively analyzed in immortalized podocyte cell lines. Channel mutants characterized by increased Ca^{2+} influx (P112Q) (Winn et al., 2005) or larger current amplitudes (R895C and E897K) (Reiser et al., 2005) in a heterologous expression system were expressed in this podocyte cell line and resulted in basal NFAT-mediated transcription (Schlondorff et al., 2009). Moreover, TRPC6 activation of Erk1/2 phosphorylation is increased by FSGS gain-of-function mutations (Chiluiza et al., 2013). Association of TRPC6 with PLCγ and nephrin is altered by the mutations and may increase TRPC6 levels and activity at the plasma membrane (Kanda et al., 2011). These changes result in a disruption of the actin cytoskeleton in cultured podocytes (Moller et al., 2007) responsible for the defective filtration process. TRPC6-induced Ca^{2+} influx activates the Ca^{2+}-sensitive phosphatase calcineurin (CaN) leading to synaptopodin dephosphorylation and decreasing protection of RhoA from proteasomal degradation (Figure 10.3b) (Faul et al., 2008). Moreover, the actin-binding protein synaptopodin is important for podocyte function and is downregulated in various glomerular diseases (Barisoni et al., 1999). Its expression is inversely correlated with TRPC6 expression in cultured podocytes (Yu et al., 2016). Overexpression of TRPC6 and its FSGS mutants in synaptopodin-depleted podocytes induced apoptosis (Yu et al., 2016). Interestingly, cyclosporine A, an unspecific blocker of autoimmune disorders, is used in primary FSGS, stabilizes synaptopodin, and reverses the LPS-induced increase in TRPC6 surface expression *in vivo* (Yu et al., 2016).

Klotho is a membrane protein predominantly produced in the kidney, which forms coreceptors with fibroblast growth factor (FGF) receptor for the ligand FGF23 (Urakawa et al., 2006). Its large extracellular domain can be cleaved and released as soluble Klotho into the systemic circulation, urine and cerebrospinal fluid (Imura et al., 2004). As it was demonstrated that soluble Klotho acting on insulin growth factor (IGF) receptor is able to inhibit phosphoinositide 3-kinase-dependent translocation of TRPC6 to the plasma membrane of ventricular myocytes and is protective against cardiac hypertrophy (Xie et al., 2012), an important role of this protein in the kidney was likely. Indeed, Klotho was able to inhibit translocation of TRPC6 to the plasma membrane in cultured podocytes and administration of Klotho to mice overexpressing TRPC6 reduced proteinuria (Kim et al., 2017). Along this line, heterozygous Klotho deficiency induces proteinuria and damage to the glomerular filter increased urinary Klotho excretion (Kim et al., 2017). Therefore, the authors proposed a protective role of soluble Klotho for the glomerular filter by reducing TRPC6 plasma membrane levels in podocytes (Kim et al., 2017).

TRPC6 might also interact with podocin detecting mechanical forces exerted by the glomerular filtration process, resulting in TRPC6 activation (Figure 10.3b) (Anderson et al., 2013; Kim et al., 2013). Accordingly, heterologously expressed podocin regulates TRPC6 activity in a cholesterol-dependent manner (Huber et al., 2007). Another TRPC6 interacting protein, nephrin, is an essential component of the slit diaphragm. Nephrin deficiency leads to overexpression and mislocalization of TRPC6 in podocytes, supporting the concept of a signalling complex with nephrin, podocin, and probably AT_1-receptors (reviewed in Gudermann, 2005). However, all these data were obtained in a cultured podocyte cell line that might not reliably reflect the *in vivo* situation. Therefore, we isolated podocytes from wild-type and TRPC6-deficient mice to express TRPC6 gain-of-function mutants in their natural environment by means of lentiviral integration of cDNAs. Transfection of WT and TRPC6$^{-/-}$ podocytes with recombinant lentiviruses coding for TRPC6 and its FSGS mutants, however, was lethal, while HEK293 cells survived the transfection procedure and expressed wild-type and mutant channels at the plasma membrane (Kalwa et al., 2015). We were able to identify a close interaction of phospholipase C-ε with TRPC6 (Kalwa et al., 2015). PLCε-deficient podocytes were smaller but showed similar rates of DNA synthesis, comparable angiotensin II (AngII)–induced formation of actin stress fibers, and almost identical GTPγS-induced currents as compared to WT cells. In stark contrast to these results, GTPγS-induced currents in TRPC6-deficient podocytes did not differ from basal levels, highlighting the importance of TRPC6 for G protein-induced signal transduction cascades in native podocytes (Kalwa et al., 2015). In contrast to earlier publications about a podocyte cell line (Kim et al., 2013; Roshanravan and Dryer, 2014; Anderson et al., 2013), there were no differences in stretch-activated currents in TRPC6$^{-/-}$ and TRPC1/TRPC3/TRPC6$^{-/-}$ podocytes compared to wild-type podocytes, because stretch-induced currents and cytoskeletal disorganization were solely dependent on ATP-induced activation of P2X purinoceptor4 (P2X$_4$) (Forst et al., 2016).

A sophisticated technique to study TRPC6 activity in situ in podocytes of freshly isolated glomeruli was published in 2013 (Ilatovskaya and Staruschenko, 2013) and might be used in further studies of TRPC6 activity in native podocytes, because the exact physiological and pathophysiological functions of TRPC6 in these cells are still elusive.

However, restricting TRPC6 function in kidney to podocytes only may not reflect the whole truth. Parietal epithelial cells (PECs) were discovered in the wall of the glomerulus and could be differentiated to podocytes by incubation with all-trans retinoic acid (ATRA) *in vitro* (Zhang et al., 2012). In light of this discovery, a comparative analysis of TRPC6 function in PECs and freshly isolated podocytes as well as in other cell types of the glomerulus-like endothelial and mesangial cells is highly desirable.

Other forms of glomerulosclerosis are induced by type 2 diabetes and may result in diabetic nephropathy (DN), the most frequent form of end-stage renal disease (Brownlee, 2001). Apart from podocytes and renal endothelial cells, mesangial cells are held responsible for DN. Rats with streptozotocin-induced diabetes display reduced TRPC6 expression in mesangial cells (Graham et al., 2007). Moreover, chronic application of high glucose to cultured mesangial cells induced ROS production and PKC activation followed by decreased TRPC6 expression (Graham et al., 2011) mediated by binding of NF-κB transcription factors to the TRPC6 promoter (Wang et al., 2013). Decreased TRPC6 function may be responsible for mesangial hypocontractility and increased glomerular filtration rates that initiate proteinuria like in FSGS patients. However, we were not able to monitor fluctuations of TRPC6 mRNA levels and detected only marginal levels of TRPC6 message in cultured mesangial cells (Figure 10.2d). Therefore, TRPC6 currents need to be analyzed in freshly isolated mesangial cells better representing the *in vivo* situation.

10.7 EXPRESSION AND FUNCTION OF TRPM7 AND TRPM6 CHANNELS IN THE KIDNEY

10.7.1 CHANNEL PROPERTIES OF TRPM7

TRPM7 is a large multifunctional membrane protein. TRPM7 contains a TRP-like ion channel segment linked to a cytosolic α-type serine/threonine protein kinase domain. TRPM channels are highly conserved throughout the animal kingdom (Mederos et al., 2008; Hofmann et al., 2010). However, only TRPM7 and its close relative protein TRPM6 are covalently linked to α-kinase domains. α-Kinases represent a class of atypical serine/threonine protein kinases characterized by a low amino acid sequence homology to conventional protein kinases (Ryazanov et al., 1997, 1988). In addition to TRPM7 and TRPM6, mammals have four other proteins containing α-kinase domains such as elongation factor-2 kinase, heart α-kinase, muscle α-kinase and lymphocyte α-kinase (Ryazanov et al., 1997, 1988). A more detailed description of the domain topology of TRPM7 is discussed next using TRPM6 as a template (Figure 10.4).

Mechanistically, the functional relationship between TRPM7 kinase and channel moieties remains poorly understood. *In vitro*, TRPM7 kinase is able to phosphorylate serine/threonine residues of TRPM6 (Brandao et al., 2014), annexin A1 (Dorovkov and Ryazanov, 2004), myosin II isoforms (Clark et al., 2008b), eEF2-k (Perraud et al., 2011), and PLCγ2 (Deason-Towne et al., 2012). Furthermore, TRPM7 kinase is able to phosphorylate its own residues in a "substrate" segment located upstream of the kinase domain (Clark et al., 2008a; Matsushita et al., 2005). A structural analysis suggested that TRPM7 kinase domains exist as a dimer. In addition, the kinase domain of TRPM7 can be cleaved off, and the released cytosolic kinase domain may subsequently translocate to the cell nucleus (Krapivinsky et al., 2014; Desai et al., 2012). In the nucleus, the cleaved fragment phosphorylates histones to modulate the chromatin modification landscape (Krapivinsky et al., 2014).

The channel segment of TRPM7 forms a constitutively active ion channel that is selective for divalent cations such as Zn^{2+}, Ca^{2+}, and Mg^{2+} (Krapivinsky et al., 2014; Nadler et al., 2001; Runnels et al., 2001; Monteilh-Zoller et al., 2003). It has been proposed that TRPM7-mediated influx of all these cations is physiologically relevant (Nadler et al., 2001; Runnels et al., 2001; Monteilh-Zoller et al., 2003). Mutagenesis of the pore-forming sequence of TRPM7 defined specific amino acid residues that contribute to the "selectivity filter" of the channel pore (Mederos et al., 2008;

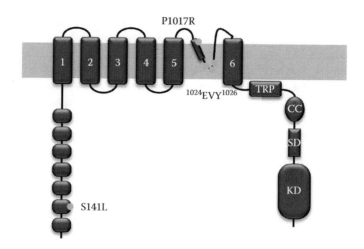

FIGURE 10.4 Domain topology of TRPM6. The plasma membrane channel segment of TRPM7 comprises six transmembrane helices (*1–6*). A short stretch between the 5 and 6 helices contains a predicted pore helix and pore-forming loop. A large cytosolic N-terminus of TRPM7 contains a set of domains that are highly conserved among the TRPM gene family and resemble ankyrin repeats as revealed by three-dimensional modeling. (From Chubanov, V. and T. Gudermann., *Handb Exp Pharmacol.*, 222, 2014.) A C-terminus of TRPM7 contains a highly conserved transient receptor potential (*TRP*) domain, a coiled-coil (*CC*) domain, a kinase substrate domain (*SD*), and a kinase domain (*KD*). Missense mutations discovered in HSH patients are indicated by black dots. Point mutations affecting the function of the ion selectivity filter (^{1024}EVY^{1026}) or the catalytic activity of the kinase (K1804R) are illustrated by gray dots.

Li et al., 2007). In contrast, processes underlying channel gating of TRPM7 remain unclear. The prevailing models are mainly resting upon three observations. First, perfusion of cells with a Mg^{2+} free cytosolic solution activates TRPM7 currents. Conversely, internally applied MgATP inhibits TRPM7 currents. Therefore, it was proposed that intracellular Mg^{2+} and MgATP are physiological negative regulators of the channel (Nadler et al., 2001; Demeuse et al., 2006; Schmitz et al., 2007). Experiments with a "kinase-dead" knock-in point mutation (K1646R in mouse TRPM7 corresponding to K1804R in human TRPM6) (Figure 10.4) or a channel variant lacking the whole kinase domain led to the concept that the kinase domain modulates the sensitivity of the TRPM7 channel to Mg^{2+} and MgATP (Demeuse et al., 2006; Schmitz et al., 2003). More recently, it was shown that a point mutation of a conserved serine residue in the TRP domain (S1107E in mouse TRPM7) is sufficient to create a constitutively active TRPM7 channel variant insensitive to intracellular regulation by Mg^{2+} and suggesting that the TRP domain plays a key role in channel gating of TRPM7 (Hofmann et al., 2014).

The second model relies on the finding that the TRPM7 channel is tightly regulated by the plasma membrane phospholipid phosphatidylinositol-4,5-bisphosphate (PIP$_2$) (Runnels et al., 2002). Stimulation of phospholipase C (PLC)–coupled G protein-coupled receptors (GPCRs) causes depletion of membrane PIP$_2$ and, subsequently, inactivation of TRPM7 currents even in the absence of Mg^{2+} (Runnels et al., 2002). Furthermore, it was suggested (Kozak et al., 2005) that cytosolic Mg^{2+} can interact with negatively charged PIP$_2$ to interfere with channel gating of TRPM7. Recently, Xie et al. (2011) found that neutralization of basic residues in the TRP domain leads to nonfunctional or dysfunctional TRPM7 variant with impaired regulation by PIP$_2$, suggesting that the TRP domain may interact with PIP$_2$.

The third model has emerged recently during a screen for small organic compounds acting as inhibitors or activators of the TRPM7 channel (Hofmann et al., 2014). Among the identified compounds, naltriben was characterized in greater detail (Hofmann et al., 2014). Naltriben reversibly and potently activates recombinant and native TRPM7 channels without prior depletion of intracellular Mg^{2+} and even after PIP_2 depletion. Consequently, it was suggested that the identified activators act on the TRPM7 channel via a ligand-binding site, and that a putative endogenous agonist can activate the TRPM7 channel in a similar manner as naltriben.

TRPM7 is a ubiquitously expressed protein, and endogenous TRPM7 currents were detected in all cells investigated so far (Penner and Fleig, 2007; Paravicini et al., 2012; Runnels, 2010; Bates-Withers et al., 2011; Fleig and Chubanov, 2014). Genetic or pharmacological targeting of TRPM7 in cultured cells revealed that TRPM7 regulates cellular Mg^{2+} levels (Chubanov et al., 2004; Schmitz et al., 2003; Ryazanova et al., 2010; Sahni and Scharenberg, 2008), cell motility (Su et al., 2006; Wei et al., 2009; Clark et al., 2006; Meng et al., 2013; Siddiqui et al., 2012; Kuras et al., 2012; Su et al., 2011; Chen et al., 2010), proliferation/cell survival (Desai et al., 2012; Nadler et al., 2001; Schmitz et al., 2003; Sahni and Scharenberg, 2008; Chen et al., 2013; Zierler et al., 2011), differentiation (Zhang et al., 2012; Abed et al., 2011), mechanosensitivity (Wei et al., 2009; Numata et al., 2007; Oancea et al., 2006), and exocytosis (Brauchi et al., 2008; Zierler et al., 2016). At the organismal level, it was suggested that TRPM7 plays a role in anoxic neuronal death (Aarts et al., 2003), hypertension (Touyz, 2008), neurodegenerative disorders (Hermosura et al., 2005; Tseveleki et al., 2010), cardiac fibrosis (Du et al., 2010), and tumor growth/progression (Guilbert et al., 2009; Kim et al., 2008; Jiang et al., 2007; Hanano et al., 2004; Middelbeek et al., 2012; Rybarczyk et al., 2012; Chen et al., 2013; Gao et al., 2011; Zhang et al., 2014a; Norenberg et al., 2016).

The *in vivo* role of TRPM7 has been extensively studied in genetically tractable animal models including mouse, zebrafish, and frog. Three independent *Trpm7* mutant mice were generated and were referred to as *Trpm7$^{\beta geo}$*, *Trpm7$^{\Delta 17}$*, and *Trpm7$^{\Delta kinase}$* strains (Ryazanova et al., 2010; Jin et al., 2008). Despite the fact that different targeting strategies were used, all three lines are now regarded as true null mutations of *Trpm7*, including the *Trpm7$^{\Delta kinase}$* allele (*Trpm7$^{\Delta kinase/\Delta kinase}$* embryonic stem cells devoid fully TRPM7 currents) (Ryazanova et al., 2010). Consistently, mice homozygous for the latter alleles die at embryonic day 6.5–7.5 (e6.5–e7.5) (Jin et al., 2008; Ryazanova et al., 2010). However, *Trpm7$^{\Delta kinase/+}$* heterozygotes show reduced Mg^{2+} levels in blood, bone, and urine (Ryazanova et al., 2010). In contrast to control mice, a substantial fraction of *Trpm7$^{\Delta kinase/+}$* die shortly after placing them on a Mg^{2+}-deficient diet, suggesting that a key aspect of TRPM7 function is the regulation of organismal Mg homeostasis. A follow-up study with *Trpm7$^{\Delta kinase/+}$* mice suggested that TRPM7 plays a role in Ang II–induced hypertension and associated vascular and target organ damage (Yogi et al., 2011).

Conditional mutagenesis of mice carrying a floxed *Trpm7* allele (*Trpm7^{fl}*) was employed to study the spatiotemporal requirement for *Trpm7* during embryonic development. An epiblast-restricted ablation of *Trpm7* resulted in lethality, indicating that *Trpm7* is required within the proper embryo (Jin et al., 2008). Global inactivation of *Trpm7* at e7–9 also caused embryonic lethality. In contrast, mutagenesis of *Trpm7* at e14.5 produced healthy *Trpm7* null pups, suggesting that *Trpm7* is required only before and during organogenesis (Jin et al., 2008). Consequently, several *Cre* transgenic lines were used to perform organ-restricted ablations of *Trpm7*. Deletion of *Trpm7* in the T-cell lineage disrupts thymopoiesis (Jin et al., 2008). Inactivation of *Trpm7* in the embryonic ureteric bud (metanephrogenic diverticulum) causes ablation of TRPM7 in collecting ducts of the postnatal kidney without obvious morphological alterations (Jin et al., 2012). In contrast, deletion of *Trpm7* in the embryonic metanephric mesenchyme resulted in inactivation of TRPM7 in renal tubules of adult mice. The latter mutants show a reduction in the number of glomeruli, renal tubular dilation, and formation of cysts in the proximal tubules, suggesting that *Trpm7* is required for nephrogenesis. Inactivation of *Trpm7* in the neural crest cells at e10.5 results in loss of dorsal root ganglion neurons and skin melanocytes (Jin et al., 2012). Cardiac deletion of *Trpm7* at e9.0 results in heart failure and death (Sah et al., 2013). In contrast, inactivation of cardiac *Trpm7* at e13.0 produces viable mice with normal heart function. Deletion of *Trpm7* at an intermediate time-point reduces the viability of the mutants to 50%. The surviving mutant mice display cardiomyopathy associated with heart block, impaired repolarization, and ventricular arrhythmias (Sah et al., 2013). Finally, inactivation of *Trpm7* specifically in the embryonic myocardium or the sinoatrial node revealed a key role of

TRPM7 in cardiac automaticity (Sah et al., 2013). Hence, conditional inactivation of *Trpm7* in mice indicates that TRPM7 plays a critical role in early embryonic development and morphogenesis of internal organs including the kidney. However, the specific molecular mechanisms underlining the prenatal requirements of *Trpm7* remain to be established.

In contrast to *Trpm7* null mutations, specific genetic ablation of the kinase activity is not harmful for embryonic development and survival of mice. Kaitsuka et al. (2014) found out that a mouse strain carrying a constitutive "kinase-dead" knock-in (KI) point mutation in the catalytic site of the TRPM7 kinase domain (K1646R) displays normal prenatal development and postnatal survival, unchanged Ca^{2+} and Mg^{2+} serum levels, and is devoid of obvious pathological phenotypes indicating that the *Trpm7* null phenotypes are caused due to impaired channel activity of TRPM7 (Kaitsuka et al., 2014). In an alternative study, it was shown that adult *Trpm7* KI mice are more resistant to dietary Mg^{2+} deprivation in terms of survival and development of Mg^{2+} deficiency (Ryazanova et al., 2014). These findings were interpreted to mean that TRPM7 kinase is required for senesing of the organismal Mg^{2+} status and the regulation of systemic responses to Mg^{2+} deprivation (Ryazanova et al., 2014).

Several loss-of-function mutations in zebrafish *Trpm7* (*drTrpm7*) have been reported. *drTrpm7*-deficient fish display normal early morphogenesis. However, mutant larvae develop multiple defects such as impaired touch responsiveness, defective melanin synthesis and apoptotic death of melanophores, defective proliferation of epithelial cells in the exocrine pancreas, and lethality in late larval life (Yee et al., 2011; McNeill et al., 2007; Low et al., 2011; Elizondo et al., 2005, 2010). *Trpm7* mutant larvae show reduced organismal levels of Mg^{2+} and Ca^{2+} (Yee et al., 2011; Elizondo et al., 2010). Mg^{2+}, but not Ca^{2+}, supplementation partially rescues melanophore survival and proliferation of cells in the exocrine pancreas (Yee et al., 2011; Elizondo et al., 2010). *Trpm7* mutants develop kidney stones, express higher levels of stanniocalcin 1 (stc1) and the anti-hyperphosphatemic factor, fibroblast growth factor 23 (fgf23) (Elizondo et al., 2010). Stc1 modulates total Mg^{2+} and Ca^{2+} levels both in mutant and wild-type larvae. The concentrations of Mg^{2+} and Ca^{2+} can be normalized in *drTrpm7* mutants by a block of stc1 activity, whereas the formation of kidney stones can be prevented by knockdown of fgf23 (Elizondo et al., 2010).

The role of TRPM7 in embryonic development has also been studied by genetic manipulation of *Xenopus leavis* embryos. Knock-down of *Xenopus Trpm7* (*xTrpm7*) using morpholino oligonucleotides reveals that *xTrpm7,* in conjunction with the noncanonical Wnt, pathway regulates cell polarity and migration during gastrulation (Liu et al., 2011). The gastrulation defect can be rescued by either Mg^{2+} supplementation or by overexpression of the SLC41A2 Mg^{2+} transporter indicating that TRPM7-mediated entry of Mg^{2+} plays an important role in *X. leavis* gastrulation (Liu et al., 2011).

In summary, experiments with genetically tractable animal models support the concept that the channel activity of TRPM7 is essential for early development, organogenesis, and regulation of mineral homeostasis.

10.7.2 ROLE OF TRPM7 IN HUMAN DISEASES ASSOCIATED WITH ALTERED MINERAL HOMEOSTASIS

Recent genetic studies indicate that TRPM7 plays a causal pathophysiological role in humans. Thus, the T1482I polymorphism in TRPM7 is associated with familial amyotrophic lateral sclerosis and parkinsonism dementia in Guam (Hermosura et al., 2005) but not in Kii, Japan (Hara et al., 2010). T1482I is also associated with an elevated risk of colorectal adenoma, especially in individuals with a high ratio of Ca:Mg intake (Dai et al., 2007).

More recently, a screen for inherited forms of macrothrombocytopenia in humans led to the identification of two pedigrees associated with point mutations in the human *TRPM7* gene (Stritt et al., 2016). The affected patients are heterozygous for C721G or R902C mutations in TRPM7, displaying impaired thrombopoiesis due to altered cellular Mg^{2+} homeostasis and cytoskeletal architecture (Stritt et al., 2016). C721G and R902C variants are located in the channel segment of TRPM7. Introduction of C721G and R902C mutations into the recombinant TRPM7 protein entails suppression of TRPM7 currents, suggesting that impaired channel TRPM7 activity underlies macrothrombocytopenia in the affected individuals (Stritt et al., 2016). In line with this idea, megakaryocyte-restricted inactivation of *Trpm7* in mice recapitulates the phenotype of patients carrying C721G and R902C variants, whereas the kinase-dead *Trpm7* KI strain does not display any obvious alterations of thrombopoiesis (Stritt et al., 2016). It remains to be established whether C721G and R902C mutations in TRPM7 result in abnormal systemic Mg^{2+} homeostasis.

10.7.3 Channel Properties of TRPM6

The human *TRPM6* gene is located on chromosome 9 (9q21.13) and comprises 39 exons. Its orthologs were found in all vertebrate species. However, most studies on TRPM6 were performed with human and mouse TRPM6. The human TRPM6 gene expresses multiple mRNA isoforms. For instance, alternative 5′ exons of the gene can be spliced in-frame to a common second exon resulting in three full-length mRNA variants, *TRPM6a*, *TRPM6b*, and *TRPM6c* (Chubanov et al., 2004). In addition, several alternatively spliced transcripts lacking exons coding for the channel segment have been identified. These splice variants were named M6-kinases 1, 2, and 3 (Chubanov et al., 2004). *TRPM6a* was commonly used in studies with recombinant expression systems, whereas functional properties of the other TRPM6 gene products remain unknown.

It is commonly accepted that TRPM6 expression is mainly limited to the intestine, kidney, and testis (Schlingmann et al., 2002; Walder et al., 2002). Further, experiments on microdissected rat nephrons revealed high expression levels of TRPM6 in the distal convoluted tubule (DCT) and low levels in the proximal tubule and collecting duct. Immunolocalization of mouse TRPM6 using a polyclonal TRPM6-specific antibody demonstrated that the protein is localized at the apical surface of DCT cells in the kidney, and in the brush-border of epithelial cells in the duodenum (Voets et al., 2004). It has been suggested that TRPM6 protein levels on the apical surface of DCT cells are tightly regulated by serum Mg^{2+} and EGF (Groenestege et al., 2007; Thebault et al., 2009).

The human TRPM6a isoform is a large protein (2022 amino acids), which, like TRPM7, comprises transmembrane and cytosolic domains (Figure 10.4). The overall architecture of the pore-forming segment of TRPM6 (as well as TRPM7) is similar to that of voltage-gated potassium channels (Ramsey et al., 2006; Nilius et al., 2007; Venkatachalam and Montell, 2007). The channel segment of TRPM6/M7 comprises six transmembrane helixes (S1–S6). A short stretch of amino acids between S5 and S6 contains a predicted hydrophobic pore helix followed by a pore loop. TRPM6/TRPM7 channel subunits assemble into homo- and heterotetramers (Hofmann et al., 2002). Accordingly, the loops of four channel subunits contribute to a common structure serving as an "ion selectivity filter." Functional characterization of TRPM6 and TRPM7 variants with mutations in the predicted pore loop revealed that a short sequence motif (^{1024}EVY^{1026} in human TRPM6a and ^{1047}EVY^{1049} in mouse TRPM7) determines the permeability of divalent cations (Figure 10.4) (Mederos et al., 2008; Li et al., 2007).

The channel segment of TRPM6 is linked to large cytosolic N- and C-terminal segments (Figure 10.4). Similar to most TRP channels, a highly conserved TRP domain (including the EWKFAR "TRP box" amino acid sequence) (Ramsey et al., 2006; Nilius et al., 2007; Venkatachalam and Montell, 2007) is located immediately downstream of the TRPM6 S6 helix. C-terminal to the TRP domain, TRPM6 contains a coiled-coil domain (Fujiwara and Minor, 2008) followed by a poorly conserved linker sequence known as the "Ser/Thr-rich" or "kinase substrate domain" (Clark et al., 2008a; Matsushita et al., 2005), which is then followed by the kinase domain (Figure 10.4). The TRPM6 kinase domain shares a high amino acid similarity to the TRPM7 kinase. The cytosolic N-terminal segment of TRPM6 comprises ~700 residues and is highly conserved in vertebrate and nonvertebrate TRPM channels but does not exhibit any noteworthy primary sequence homology to other domains (Figure 10.4). Nevertheless, our efforts to model the three-dimensional topology of N-terminal segments of TRPM6 and TRPM7 invariably predict the presence of multiple "ankyrin-like" repeats (Chubanov and Gudermann, 2014). This prediction does not come as a surprise, because many other TRP channels contain multiple ankyrin repeats in similar positions (Phelps et al., 2007).

As mentioned above, TRPM6 forms functional heterotetrameric channel complexes together with TRPM7. Five domains may be directly involved in the assembly of TRPM6/TRPM7 complexes. Thus, coexpression of wild-type TRPM7 with a TRPM6 variant carrying a dominant negative point mutation in its pore-forming region (P1017R) resulted in a suppression of TRPM7 currents, indicating that the transmembrane segments of TRPM6 and TRPM7 associate to form a common channel pore (Figure 10.4) (Chubanov et al., 2007). Another mutation, S141L in the intracellular N-terminus of human TRPM6 (S138L in mouse TRPM7), disrupts the channel assembly, underscoring the critical role of the N-terminus for the formation of functional channel complexes (Figure 10.4) (Chubanov et al., 2004). In addition, it was suggested that the coiled-coil domains of TRPM7 and TRPM6 are directly involved in the assembly of the channel-kinase complexes (Figure 10.4) (Fujiwara and Minor, 2008). Structural and biochemical experiments revealed that the kinase domains of TRPM6 and TRPM7 form functional homo- and heterodimers, and that such dimerization is required for kinase activity (Fujiwara and Minor, 2008; Crawley and Cote, 2009). Finally, a dimerization motif (amino acids 1700–1730 in TRPM6a) was identified as the determinant of TRPM6 kinase dimer formation (van der Wijst et al., 2014). Mutagenesis of critical residues in this domain blocks kinase activity and negatively regulates channel activity (van der Wijst et al., 2014).

Physiological substrates of the TRPM6 kinase remain unknown. *In vitro* studies established that, similar to TRPM7, TRPM6 kinases phosphorylate myosin IIA, IIB, and IIC on identical residues (Clark et al., 2008b; Clark et al., 2006). Therefore, it is possible that TRPM7 and TRPM6 kinases phosphorylate the same proteins. Analogous to TRPM7, TRPM6 kinase can autophosphorylate its own Ser/Thr residues (Clark et al., 2008a). Furthermore, upon coexpression with TRPM7, TRPM6 is able to cross-phosphorylate residues in the TRPM7 protein (Schmitz et al., 2005). Physiological consequences of TRPM6 kinase-mediated phosphorylation processes are unclear. The kinase domain of TRPM6 may interact with the repressor of estrogen receptor activity (REA), the receptor of activated protein kinase C 1 (RACK1), and methionine sulfoxide reductase B1 (MsrB1) (Cao et al., 2008, 2009, 2010). Coexpression of recombinant REA, RACK1, and MsrB1 with human TRPM6 resulted in moderate inhibition of TRPM6 channel activity (Cao et al., 2008, 2009, 2010). It remains unknown whether TRPM6 kinase phosphorylates the latter proteins.

Functional studies on TRPM6 channel properties have so far been restricted to the recombinant human TRPM6a isoform in heterologous expression systems. These experiments resulted in two models for the physiological role of endogenous TRPM6. Two groups used multiple expression systems to show that TRPM6 displays distinct features as compared to TRPM7 (Chubanov et al., 2004, 2007; Brandao et al., 2014; Schmitz et al., 2005). Specifically, the researchers noted that TRPM6 does not efficiently form homomultimeric channel complexes in the plasma membrane, but requires TRPM7 to be co-targeted to the cell surface (Chubanov et al., 2004, 2007; Brandao et al., 2014; Schmitz et al., 2005). Within the heterooligomeric channels formed, TRPM6 alters the biophysical properties of the heteromultimer as compared to homomeric TRPM7 channel properties such as increased channel activity and reduced sensitivity to intracellular MgATP (Chubanov et al., 2004, 2007; Schmitz et al., 2005; Li et al., 2006). Since TRPM6 is invariantly coexpressed with the ubiquitous TRPM7, it was suggested that native TRPM6 functions mainly in heteromeric TRPM6/TRPM7 complexes, allowing amplification of the cellular uptake of Mg^{2+} on the apical surface of transporting epithelial cells (Chubanov and Gudermann, 2014). Consequently, such increased Mg^{2+} influx would enhance basolateral Mg^{2+} extrusion resulting in vectorial transcellular Mg^{2+} transport by renal or intestinal epithelial cells.

The second model assumes that TRPM6 represents a functional copy of the TRPM7 channel and that in epithelial cells TRPM6 operates independently from TRPM7 (van der Wijst et al., 2014). This model relies on the finding that transient expression of the human TRPM6a variant using the pCI-Neo-IRES-GFP vector (Xie et al., 2011; Voets et al., 2004; Li et al., 2006), but not any other expression vectors (Chubanov et al., 2004; Zhang et al., 2014 b), allows detection of functional TRPM6 homo-oligomers with biophysical characteristics resembling TRPM7. Like TRPM7, TRPM6 was found to be highly permeable to Mg^{2+} or Ca^{2+}, and sensitive to intracellular Mg^{2+}, MgATP, and PI(4,5) P_2 levels (Xie et al., 2011; Voets et al., 2004; Li et al., 2006). It was also found that TRPM6 currents are negatively regulated by hydrogen peroxide, 17β-estradiol, and sphingosine (Xie et al., 2011; Voets et al., 2004; Cao et al., 2010; Li et al., 2006; Qin et al., 2012). Furthermore, insulin and FGF acting via Rac1 may increase the cell surface expression of recombinant TRPM6 (Thebault et al., 2009; Nair et al., 2012). It remains elusive why different expression vectors yield disparate results. More importantly, both models still await stringent testing by investigating the endogenous TRPM6 protein: Any attempts to measure native TRPM6 currents either in wild-type or TRPM7-deficient cells were not successful so far (Ryazanova et al., 2010; Zhang et al., 2014a; Li et al., 2006).

Two independent studies attempted to explore *Trpm6* gene-deficient mice. Walder et al. (2009) found that *Trpm6* null mice die at e12.5. Mutants that could survive beyond this stage develop neural tube closure defects (NTDs). A high Mg^{2+} diet of parents did not rescue the mortality of *Trpm6* null embryos (Walder et al., 2009). The authors suggested that TRPM6 plays a direct role in neural tube closure resulting in embryonic mortality of mutant mice (Walder et al., 2009). The exact mechanisms triggering NTD in *Trpm6* null mice remain elusive. A second study (Woudenberg-Vrenken et al., 2010) also showed that *Trpm6*-deficient mutants are not viable. However, *Trpm6*$^{+/-}$ mice are characterized by moderately reduced serum Mg^{2+} levels. Unexpectedly, urinary Mg^{2+} excretion was unaffected in *Trpm6*$^{+/-}$ animals (Woudenberg-Vrenken et al., 2010).

10.7.4 ROLE OF TRPM6 IN INHERITED HYPOMAGNESEMIA

Two laboratories independently showed that mutations in the human *TRPM6* gene give rise to autosomal recessive hypomagnesemia with secondary hypocalcemia (HSH) (Schlingmann et al., 2002, 2005; Walder et al., 2002). Affected infants present with generalized convulsions and muscle spasms. Clinical assessment at the time of manifestation reveals low serum levels of Mg^{2+} (0.1–0.3 mM) and Ca^{2+} (1–1.6 mM) as compared to healthy individuals (0.7–1.1 mM Mg^{2+} and 2.2–2.9 mM Ca^{2+}). Hypomagnesemia and other symptoms were relieved after administration of high doses of Mg^{2+}. Despite this treatment, serum Mg^{2+} levels remain in the subnormal range (Schlingmann et al., 2002, 2005, 2007; Konrad et al., 2004; Walder et al., 2002).

Mechanistically, many aspects of HSH are not well understood. It was suggested that hypomagnesemia is likely caused by a defect in Mg^{2+} reabsorption in the kidney (Schlingmann et al., 2002, 2005; Walder et al., 2002). Currently this model represents the common view on HSH, but, to our best knowledge, direct functional evidence for such a mechanism is still lacking. Yet, the etiology of hypocalcaemia remains unclear. Several explanations have been offered, including end-organ unresponsiveness to parathyroid hormone (PTH), altered release of PTH, and impaired formation of 1,25-dihydroxy vitamin D3 (Schlingmann et al., 2005; Konrad and Weber, 2003; Woodard et al., 1972).

In addition, a very low frequency of HSH and a high degree of consanguinity in the majority of HSH pedigrees complicate the identification of clear-cut genotype-phenotype relationships in affected individuals. For instance, several individuals homozygous for *TRPM6* mutations are asymptomatic, while other HSH patients present with additional symptoms such as mental retardation, osteoporosis, cardiac arrhythmia, severe failure to thrive, and bilateral basal ganglia calcification (Schlingmann et al., 2002, 2005; Chubanov et al., 2004; Walder et al., 2002; Chubanov et al., 2007; Guran et al., 2011; Habeb et al., 2012; Esteban-Oliva et al., 2009; Apa et al., 2008; Zhao et al., 2013; Altincik et al., 2016). In addition, a substantial predominance of male patients has been observed in several HSH families (Chery et al., 1994) interpreted as an indication for genetic heterogeneity of the disease (Chery et al., 1994). Unfortunately, studies with *Trpm6*-deficient mice

(Walder et al., 2009) failed to confirm the monogenic basis and the pathomechanisms of HSH. Nevertheless, experiments with *Trpm6*-deficient mice (Walder et al., 2009) suggest that mutations in *TRPM6* may impact on the development of the human fetus, thus offering an explanation for the fact that only few HSH families have been identified worldwide so far.

It is commonly accepted that HSH is an autosomal recessive disorder associated with loss-of-function (null) mutations in *TRPM6*. The majority of HSH mutations introduce stop and frame-shift mutations, or impair exon splicing (Schlingmann et al., 2002, 2005; Chubanov et al., 2004; Walder et al., 2002; Chubanov et al., 2007; Guran et al., 2011; Habeb et al., 2012; Esteban-Oliva et al., 2009; Apa et al., 2008; Zhao et al., 2013; Altincik et al., 2016; Katayama et al., 2015; Coulter et al., 2015; Astor et al., 2015; Lainez et al., 2014). In addition, a few point mutations have been identified (Figure 10.4). Some of the missense mutations are located within the cytosolic N-terminus of TRPM6 suggestive of a critical role of this domain for channel function. A remarkable example is S141L variant: TRPM6^{S141L} is unable to associate with TRPM7 (Chubanov et al., 2004). Consistently, introduction of the homologous mutation into TRPM7 (S138L) also affects ion channel complex assembly (Figure 10.4) (Chubanov et al., 2004). Several missense HSH mutations are located in the channel segment of TRPM6 and invariably ablate the TRPM6 function. A notable example is the P1017R variant, which is located in the pore-forming region of TRPM6 upstream to ^{1024}EVY^{1026} residues forming a cation selectivity filter (Figure 10.4) (Chubanov et al., 2007) . TRPM6^{P1017R} impairs channel activity of TRPM6/TRPM7 heteromers by dominant-negative suppression, suggesting that a block of cation fluxes via TRPM6/TRPM7 complexes is sufficient for the development of HSH.

The currently available information about TRPM6 allows suggestion of the following mechanistic model leading to a negative Mg^{2+} balance resulting in hypomagnesemia in HSH (Figure 10.5). Two different Mg^{2+} transport pathways operate in the kidney: an active transcellular transport route and a passive paracellular transport route. Tight junctions containing

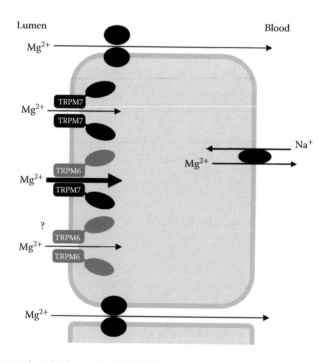

FIGURE 10.5 Suggested model for a role of TRPM6 in renal reabsorption. The kidneys comprise an active transcellular transport route and a passive paracellular Mg^{2+} transport route. The transcellular pathways consist of apical Mg^{2+} entry mediated by Mg^{2+} channels such as TRPM6 and TRPM7 into epithelial cells and a basolateral extrusion step maintained by Mg^{2+} exchangers.

claudin-16 and claudin-19 are involved in the paracellular transport (Kausalya et al., 2006; Simon et al., 1999; Will et al., 2010; Muller et al., 2006; Almeida et al., 2014). The transcellular pathways consist of apical Mg^{2+} entry mediated by Mg^{2+} channels into epithelial cells and a basolateral extrusion step maintained by Mg^{2+} exchangers such as CNNM4 and SLC41A1 (Konrad et al., 2004; Schlingmann et al., 2007; Dai et al., 2001; Quamme and de Rouffignac, 2000; Yamazaki et al., 2016; Funato et al., 2014a, 2014b; Hirata et al., 2014; Yamazaki et al., 2013; Hurd et al., 2013). In the kidney, the transcellular and paracellular routes are arranged sequentially along the nephron. Most of the filtered Mg^{2+} is reabsorbed in the thick ascending limb via passive paracellular mechanisms driven by positive transepithelial voltage. Only 5%–10% of the filtered Mg^{2+} is reabsorbed in the distal convoluted tubule (DCT) solely via the transcellular pathway. There is no Mg^{2+} reabsorption in more distal nephron segments, suggesting that the DCT determines the final urinary Mg^{2+} content (Konrad et al., 2004; Schlingmann et al., 2007; Dai et al., 2001; Quamme and de Rouffignac, 2000). As pointed out earlier, both TRPM6 and TRPM7 are expressed in the DCT (Chubanov et al., 2004). Accordingly, both proteins may be involved in the apical Mg^{2+} uptake in the DCT and two distinct pathomechanisms of HSH can be suggested (Figure 10.5). According to the first model, TRPM6/TRPM7 heterooligomers mainly contribute to Mg^{2+} uptake in DCT cells, while activity of TRPM7 homomultimers is not sufficient to fulfill this function. Alternatively, a still unidentified unique characteristic of homomeric TRPM6 channel complexes may be essential for DCT function, implying that TRPM7 cannot qualitatively compensate for the lack of TRPM6 (Figure 10.5). Nevertheless, both scenarios assume that a defect in TRPM6-dependent Mg^{2+} reabsorption in the DCT is sufficient to induce a negative organismal Mg^{2+} balance leading to organismal Mg^{2+} deficiency.

10.8 GENOME-WIDE ASSOCIATIONS WITH *TRPM6*

It has been suggested that Mg^{2+} deficiency is associated with an increased risk of insulin resistance and type 2 diabetes (Sales and Pedrosa Lde, 2006; Takayanagi et al., 2015). A recent screen for genome-wide associations reported that two single nucleotide polymorphisms (SNPs) in the TRPM6 coding region, rs3750425 and rs2274924, might confer susceptibility for type 2 diabetes upon women with low Mg^{2+} intake (Song et al., 2009). Furthermore, these SNPs may be responsible for a higher likelihood of developing gestational diabetes mellitus (Nair et al., 2012). Genome-wide association studies with more than 15,000 individuals showed that rs11144134 in TRPM6 is associated with lower serum Mg^{2+} levels (Meyer et al., 2010). In another study, genome-wide association studies to link Mg^{2+} intake and fasting glucose and insulin levels were carried out in more than 50.000 healthy Europeans (Hruby et al., 2013). The authors reported that higher dietary Mg^{2+} intake is associated with lowered glucose and insulin concentrations, whereas rs2274924 in TRPM6 is associated with higher glucose levels (Hruby et al., 2013). Finally, Shuen et al. (2009) identified an association between estrogen receptor alpha (ESR1) polymorphisms and serum Mg^{2+} levels. Taken together, TRPM6 variants emerge as a diagnostic tool for insulin resistance and metabolic disorders.

10.9 CONCLUSION

In the past few years, TRP channels have emerged as central players in human physiology, and mutations in these proteins are associated with frequently occurring kidney diseases as well as other pathologies (Xie et al., 2011). As more information on the *in vivo* role of TRP channels in suitable animal models and additional clinical data from patients carrying mutations in TRP channel genes become available, our knowledge of the role of TRP channels in renal physiology and pathophysiology will expand considerably. Further progress in this field will improve our mechanistic understanding of TRP channel function in general and may help identify novel therapeutic targets.

ACKNOWLEDGMENTS

We would like to thank Bettina Braun for excellent technical assistance. This work was supported by the Deutsche Forschungsgemeinschaft (TRR 152, project 15 and 16).

REFERENCES

Aarts, M. et al. 2003. A key role for TRPM7 channels in anoxic neuronal death. *Cell*, 115: 863–877.

Abed, E., C. Martineau, and R. Moreau. 2011. Role of melastatin transient receptor potential 7 channels in the osteoblastic differentiation of murine MC3T3 cells. *Calcif Tissue Int*, 88: 246–253. doi:10.1007/s00223-010-9455-z.

Almeida, J.R. et al. 2014. Five years results after intrafamilial kidney post-transplant in a case of familial hypomagnesemia due to a claudin-19 mutation. *J Bras Nefrol*, 36: 401–405.

Altincik, A., K.P. Schlingmann, and M.S. Tosun. 2016. A novel homozygous mutation in the transient receptor potential melastatin 6 gene: A case report. *J Clin Res Pediatr Endocrinol*, 8: 101–104. doi:10.4274/jcrpe.2254.

Anderson, M. et al. 2013. Opposing effects of podocin on the gating of podocyte TRPC6 channels evoked by membrane stretch or diacylglycerol. *Am J Physiol Cell Physiol*, 305: C276–C289. doi:10.1152/ajpcell.00095.2013.

Apa, H. et al. 2008. A case of hypomagnesemia with secondary hypocalcemia caused by Trpm6 gene mutation. *Indian J Pediatr*, 75: 632–634. doi:10.1007/s12098-008-0121-7.

Astor, M.C. et al. 2015. Hypomagnesemia and functional hypoparathyroidism due to novel mutations in the Mg-channel TRPM6. *Endocr Connect*, 4: 215–222. doi:10.1530/EC-15-0066.

Bandyopadhyay, B.C. et al. 2005. Apical localization of a functional TRPC3/TRPC6-Ca^{2+}-signaling complex in polarized epithelial cells. Role in apical Ca^{2+} influx. *J Biol Chem*, 280: 12908–12916. doi:10.1074/jbc.M410013200.

Barisoni, L. et al. 1999. The dysregulated podocyte phenotype: A novel concept in the pathogenesis of collapsing idiopathic focal segmental glomerulosclerosis and HIV-associated nephropathy. *J Am Soc Nephrol*, 10: 51–61.

Bates-Withers, C., R. Sah, and D.E. Clapham. 2011. TRPM7, the Mg^{2+} inhibited channel and kinase. *Adv Exp Med Biol*, 704: 173–183. doi:10.1007/978-94-007-0265-3_9.

Beech, D.J. et al. 2009. TRPC channel lipid specificity and mechanisms of lipid regulation. *Cell Calcium*, 45: 583–588. doi:10.1016/j.ceca.2009.02.006.

Boulay, G. et al. 1997. Cloning and expression of a novel mammalian homolog of *Drosophila* transient receptor potential (Trp) involved in calcium entry secondary to activation of receptors coupled by the Gq class of G protein. *J Biol Chem*, 272: 29672–29680.

Brandao, K. et al. 2014. TRPM6 kinase activity regulates TRPM7 trafficking and inhibits cellular growth under hypomagnesic conditions. *Cell Mol Life Sci*, 71: 4853–4867. doi:10.1007/s00018-014-1647-7.

Brauchi, S. et al. 2008. TRPM7 facilitates cholinergic vesicle fusion with the plasma membrane. *Proc Natl Acad Sci U S A*, 105: 8304–8308. doi:0800881105 [pii]10.1073/pnas.0800881105.

Brownlee, M. 2001. Biochemistry and molecular cell biology of diabetic complications. *Nature*, 414: 813–820. doi:10.1038/414813a.

Cao, G. et al. 2008. RACK1 inhibits TRPM6 activity via phosphorylation of the fused alpha-kinase domain. *Curr Biol*, 18: 168–176. doi:10.1016/j.cub.2007.12.058.

Cao, G. et al. 2009. Regulation of the epithelial Mg^{2+} channel TRPM6 by estrogen and the associated repressor protein of estrogen receptor activity (REA). *J Biol Chem*, 284: 14788–14795. doi:10.1074/jbc.M808752200.

Cao, G. et al. 2010. Methionine sulfoxide reductase B1 (MsrB1) recovers TRPM6 channel activity during oxidative stress. *J Biol Chem*, 285: 26081–26087. doi:10.1074/jbc.M110.103655.

Carlstrom, M., C.S. Wilcox, and W.J. Arendshorst. 2015. Renal autoregulation in health and disease. *Physiol Rev*, 95: 405–511. doi:10.1152/physrev.00042.2012.

Chang, M.Y. and A.C. Ong. 2008. Autosomal dominant polycystic kidney disease: Recent advances in pathogenesis and treatment. *Nephron Physiol*, 108: p1–p7. doi:10.1159/000112495.

Chen, J.P. et al. 2010. TRPM7 regulates the migration of human nasopharyngeal carcinoma cell by mediating Ca^{2+} influx. *Cell Calcium*, 47: 425–432. doi:10.1016/j.ceca.2010.03.003.

Chen, K. et al. 2014. Genome-wide association study identifies new susceptibility loci for epithelial ovarian cancer in Han Chinese women. *Nat Commun*, 5: 4682. doi:10.1038/ncomms5682.

Chen, K.H. et al. 2013. TRPM7 channels regulate proliferation and adipogenesis in 3T3-L1 preadipocytes. *J Cell Physiol*, 229: 60–67. doi:10.1002/jcp.24417.

Chen, Y.F. et al. 2013. Remodeling of calcium signaling in tumor progression. *J Biomed Sci*, 20: 23. doi:10.1186/1423-0127-20-23.

Chery, M. et al. 1994. Hypomagnesemia with secondary hypocalcemia in a female with balanced X;9 translocation: Mapping of the Xp22 chromosome breakpoint. *Hum Genet*, 93: 587–591.

Chiluiza, D. et al. 2013. Gain-of-function mutations in transient receptor potential C6 (TRPC6) activate extracellular-signal-regulated kinases Erk1/2. *J Biol Chem*, 288(25): 18407–18420. doi:10.1074/jbc.M113.463059.

Chubanov, V. and T. Gudermann. 2014. Trpm6. *Handb Exp Pharmacol*, 222: 503–520. doi:10.1007/978-3-642-54215-2_20.

Chubanov, V. et al. 2004. Disruption of TRPM6/TRPM7 complex formation by a mutation in the TRPM6 gene causes hypomagnesemia with secondary hypocalcemia. *Proc Natl Acad Sci U S A*, 101: 2894–2899. doi:10.1073/pnas.0305252101010305252101.

Chubanov, V. et al. 2007. Hypomagnesemia with secondary hypocalcemia due to a missense mutation in the putative pore-forming region of TRPM6. *J Biol Chem*, 282: 7656–7667. doi:10.1074/jbc.M611117200.

Clark, K. et al. 2006. TRPM7, a novel regulator of actomyosin contractility and cell adhesion. *EMBO J*, 25: 290–301. doi:10.1038/sj.emboj.7600931.

Clark, K. et al. 2008a. Massive autophosphorylation of the Ser/Thr-rich domain controls protein kinase activity of TRPM6 and TRPM7. *PLOS ONE*, 3: e1876. doi:10.1371/journal.pone.0001876.

Clark, K. et al. 2008b. TRPM7 regulates myosin IIA filament stability and protein localization by heavy chain phosphorylation. *J Mol Biol*, 378: 790–803. doi:10.1016/j.jmb.2008.02.057.

Coulter, M. et al. 2015. Hypomagnesemia due to two novel TRPM6 mutations. *J Pediatr Endocrinol Metab*, 28: 1373–1378. doi:10.1515/jpem-2014-0394.

Crawley, S.W. and G.P. Cote. 2009. Identification of dimer interactions required for the catalytic activity of the TRPM7 alpha-kinase domain. *Biochem J*, 420: 115–122. doi:10.1042/BJ20081405.

Dai, L.J. et al. 2001. Magnesium transport in the renal distal convoluted tubule. *Physiol Rev*, 81: 51–84.

Dai, Q. et al. 2007. The relation of magnesium and calcium intakes and a genetic polymorphism in the magnesium transporter to colorectal neoplasia risk. *Am J Clin Nutr*, 86: 743–751.

Dalla Vestra, M. et al. 2001. Role of mesangial expansion in the pathogenesis of diabetic nephropathy. *J Nephrol*, 14(Suppl 4): S51–S57.

Deason-Towne, F., A.L. Perraud, and C. Schmitz. 2012. Identification of Ser/Thr phosphorylation sites in the C2-domain of phospholipase Cγ2 (PLCγ2) using TRPM7-kinase. *Cell Signal*, 24: 2070–2075. doi:10.1016/j.cellsig.2012.06.015.

Demeuse, P., R. Penner, and A. Fleig. 2006. TRPM7 channel is regulated by magnesium nucleotides via its kinase domain. *J Gen Physiol*, 127: 421–434. doi:10.1085/jgp.200509410.

Desai, B.N. et al. 2012. Cleavage of TRPM7 releases the kinase domain from the ion channel and regulates its participation in Fas-induced apoptosis. *Dev Cell*, 22: 1149–1162. doi:10.1016/j.devcel.2012.04.006.

Dietrich, A. et al. 2003. N-linked protein glycosylation is a major determinant for basal TRPC3 and TRPC6 channel activity. *J Biol Chem*, 278: 47842–47852.

Dietrich, A. et al. 2005. Functional characterization and physiological relevance of the TRPC3/6/7 subfamily of cation channels. *Naunyn Schmiedebergs Arch Pharmacol*, 371: 257–265. doi:10.1007/s00210-005-1052-8.

Dietrich, A., V. Chubanov, and T. Gudermann. 2010. Renal TRPathies. *J Am Soc Nephrol*, 21: 736–744.

Dietrich, A., M. Fahlbusch, and T. Gudermann. 2014. Classical transient receptor potential 1 (TRPC1): Channel or channel regulator? *Cells*, 3: 939–962. doi:10.3390/cells3040939.

Dorovkov, M.V. and A.G. Ryazanov. 2004. Phosphorylation of annexin I by TRPM7 channel-kinase. *J Biol Chem*, 279: 50643–50646. doi:10.1074/jbc.C400441200.

Du, J. et al. 2007. Canonical transient receptor potential 1 channel is involved in contractile function of glomerular mesangial cells. *J Am Soc Nephrol*, 18: 1437–1445. doi:10.1681/ASN.2006091067.

Du, J. et al. 2010. TRPM7-mediated Ca^{2+} signals confer fibrogenesis in human atrial fibrillation. *Circ Res*, 106: 992–1003. doi:10.1161/CIRCRESAHA.109.206771.

Eckel, J. et al. 2011. TRPC6 enhances angiotensin II-induced albuminuria. *J Am Soc Nephrol*, 22: 526–535. doi:10.1681/ASN.2010050522.

Elizondo, M.R. et al. 2005. Defective skeletogenesis with kidney stone formation in dwarf zebrafish mutant for trpm7. *Curr Biol*, 15: 667–671. doi:10.1016/j.cub.2005.02.050.

Elizondo, M.R., E.H. Budi, and D.M. Parichy. 2010. Trpm7 regulation of in vivo cation homeostasis and kidney function involves stanniocalcin 1 and fgf23. *Endocrinology*, 151: 5700–5709. doi:10.1210/en.2010-0853.

Esteban-Oliva, D., G. Pintos-Morell, and M. Konrad. 2009. Long-term follow-up of a patient with primary hypomagnesaemia and secondary hypocalcaemia due to a novel TRPM6 mutation. *Eur J Pediatr*, 168: 439–442. doi:10.1007/s00431-008-0767-1.

Faul, C. et al. 2008. The actin cytoskeleton of kidney podocytes is a direct target of the antiproteinuric effect of cyclosporine A. *Nat Med*, 14: 931–938. doi:10.1038/nm.1857.

Fleig, A. and V. Chubanov. 2014. Trpm7. *Handb Exp Pharmacol*, 222: 521–546. doi:10.1007/978-3-642-54215-2_21.

Forst, A.L. et al. 2016. Podocyte purinergic P2X4 channels are mechanotransducers that mediate cytoskeletal disorganization. *J Am Soc Nephrol*, 27: 848–862. doi:10.1681/ASN.2014111144.

Fujiwara, Y. and D.L. Jr. Minor. 2008. X-ray crystal structure of a TRPM assembly domain reveals an antiparallel four-stranded coiled-coil. *J Mol Biol*, 383: 854–870. doi:10.1016/j.jmb.2008.08.059.

Funato, M. et al. 2014a. A complement factor B mutation in a large kindred with atypical hemolytic uremic syndrome. *J Clin Immunol*, 34: 691–695. doi:10.1007/s10875-014-0058-8.

Funato, Y. et al. 2014b. Membrane protein CNNM4-dependent Mg^{2+} efflux suppresses tumor progression. *J Clin Invest*, 124: 5398–5410. doi:10.1172/JCI76614.

Gao, H. et al. 2011. EGF enhances the migration of cancer cells by up-regulation of TRPM7. *Cell Calcium*, 50: 559–568. doi:10.1016/j.ceca.2011.09.003.

Glassock, R.J. 2010. The pathogenesis of idiopathic membranous nephropathy: A 50-year odyssey. *Am J Kidney Dis*, 56: 157–167. doi:10.1053/j.ajkd.2010.01.008.

Goel, M. et al. 2006. Identification and localization of TRPC channels in the rat kidney. *Am J Physiol Renal Physiol*, 290: F1241–F1252. doi:10.1152/ajprenal.00376.2005.

Goel, M. et al. 2007. TRPC3 channels colocalize with Na^+/Ca^{2+} exchanger and Na^+ pump in axial component of transverse-axial tubular system of rat ventricle. *Am J Physiol Heart Circ Physiol*, 292: H874–H883. doi:10.1152/ajpheart.00785.2006.

Graham, S. et al. 2007. Downregulation of TRPC6 protein expression by high glucose, a possible mechanism for the impaired Ca^{2+} signaling in glomerular mesangial cells in diabetes. *Am J Physiol Renal Physiol*, 293: F1381–1390. doi:10.1152/ajprenal.00185.2007.

Graham, S. et al. 2011. Abundance of TRPC6 protein in glomerular mesangial cells is decreased by ROS and PKC in diabetes. *Am J Physiol Cell Physiol*, 301: C304–C315. doi:10.1152/ajpcell.00014.2011.

Groenestege, W.M. et al. 2007. Impaired basolateral sorting of pro-EGF causes isolated recessive renal hypomagnesemia. *J Clin Invest*, 117: 2260–2267. doi:10.1172/JCI31680.

Gudermann, T.A. 2005. New TRP to kidney disease. *Nat Genet*, 37: 663–664. doi:10.1038/ng0705-663.

Guilbert, A. et al. 2009. Evidence that TRPM7 is required for breast cancer cell proliferation. *Am J Physiol Cell Physiol*, 297: C493–C502. doi:10.1152/ajpcell.00624.2008.

Guran, T. et al. 2011. Clinical and molecular characterization of Turkish patients with familial hypomagnesaemia: Novel mutations in TRPM6 and CLDN16 genes. *Nephrol Dial Transplant*, 27: 667–673. doi:10.1093/ndt/gfr300.

Habeb, A.M., H. Al-Harbi, and K.P. Schlingmann. 2012. Resolving basal ganglia calcification in hereditary hypomagnesemia with secondary hypocalcemia due to a novel TRPM6 gene mutation. *Saudi J Kidney Dis Transpl*, 23: 1038–1042. doi:10.4103/1319-2442.100945.

Hanano, T. et al. 2004. Involvement of TRPM7 in cell growth as a spontaneously activated Ca^{2+} entry pathway in human retinoblastoma cells. *J Pharmacol Sci*, 95: 403–419.

Haneda, M. et al. 2003. Overview of glucose signaling in mesangial cells in diabetic nephropathy. *J Am Soc Nephrol*, 14: 1374–1382.

Hara, K. et al. 2010. TRPM7 is not associated with amyotrophic lateral sclerosis-parkinsonism dementia complex in the Kii peninsula of Japan. *Am J Med Genet B Neuropsychiatr Genet*, 153B: 310–313. doi:10.1002/ajmg.b.30966.

Harris, P.C. and V.E. Torres. 2009. Polycystic kidney disease. *Annu Rev Med*, 60: 321–337. doi:10.1146/annurev.med.60.101707.125712.

He, F. et al. 2014. MiR-135a promotes renal fibrosis in diabetic nephropathy by regulating TRPC1. *Diabetologia*, 57: 1726–1736. doi:10.1007/s00125-014-3282-0.

Hermosura, M. C. et al. 2005. A TRPM7 variant shows altered sensitivity to magnesium that may contribute to the pathogenesis of two Guamanian neurodegenerative disorders. *Proc Natl Acad Sci U S A*, 102: 11510–11515.

Hermosura, M.C. et al. 2005. A TRPM7 variant shows altered sensitivity to magnesium that may contribute to the pathogenesis of two Guamanian neurodegenerative disorders. *Proc Natl Acad Sci U S A*, 102: 11510–11515. doi:10.1073/pnas.0505149102.

Hirata, Y. et al. 2014. Mg^{2+}-dependent interactions of ATP with the cystathionine-β-synthase (CBS) domains of a magnesium transporter. *J Biol Chem*, 289: 14731–14739. doi:10.1074/jbc.M114.551176.

Hofmann, T. et al. 2002. Subunit composition of mammalian transient receptor potential channels in living cells. *Proc Natl Acad Sci U S A*, 99: 7461–7466.

Hofmann, T. et al. 2010. *Drosophila* TRPM channel is essential for the control of extracellular magnesium levels. *PLOS ONE*, 5: e10519. doi:10.1371/journal.pone.0010519.

Hofmann, T. et al. 2014. Activation of TRPM7 channels by small molecules under physiological conditions. *Pflugers Archiv*, 466(12): 2177–2189. doi:10.1007/s00424-014-1488-0.

Hoorn, E.J. and R. Zietse. 2013. Disorders of calcium and magnesium balance: A physiology-based approach. *Pediatr Nephrol*, 28: 1195–1206. doi:10.1007/s00467-012-2350-2.

Hruby, A. et al. 2013. Higher magnesium intake is associated with lower fasting glucose and insulin, with no evidence of interaction with select genetic loci, in a meta-analysis of 15 CHARGE Consortium Studies. *J Nutr*, 143: 345–353. doi:10.3945/jn.112.172049.

Hsu, Y.J., J.G. Hoenderop, and R.J. Bindels. 2007. TRP channels in kidney disease. *Biochim Biophys Acta*, 1772: 928–936. doi:10.1016/j.bbadis.2007.02.001.

Huber, T.B., B. Schermer, and T. Benzing. 2007. Podocin organizes ion channel-lipid supercomplexes: Implications for mechanosensation at the slit diaphragm. *Nephron Exp Nephrol*, 106: e27–e31. doi:10.1159/000101789.

Hurd, T.W. et al. 2013. Mutation of the Mg^{2+} transporter SLC41A1 results in a nephronophthisis-like phenotype. *J Am Soc Nephrol*, 24: 967–977. doi:10.1681/ASN.2012101034.

Ilatovskaya, D.V. and A. Staruschenko. 2013. Single-channel analysis of TRPC channels in the podocytes of freshly isolated Glomeruli. *Methods Mol Biol*, 998: 355–369. doi:10.1007/978-1-62703-351-0_28.

Imura, A. et al. 2004. Secreted Klotho protein in sera and CSF: Implication for post-translational cleavage in release of Klotho protein from cell membrane. *FEBS Lett*, 565: 143–147. doi:10.1016/j.febslet.2004.03.090.

Jiang, J. et al. 2007. Transient receptor potential melastatin 7-like current in human head and neck carcinoma cells: Role in cell proliferation. *Cancer Res*, 67: 10929–10938. doi:10.1158/0008-5472.CAN-07-1121.

Jin, J. et al. 2008. Deletion of Trpm7 disrupts embryonic development and thymopoiesis without altering Mg^{2+} homeostasis. *Science*, 322: 756–760. doi:10.1126/science.1163493.

Jin, J. et al. 2012. The channel kinase, TRPM7, is required for early embryonic development. *Proc Natl Acad Sci U S A*, 109: E225–E233. doi:10.1073/pnas.1120033109.

Kaitsuka, T. et al. 2014. Inactivation of TRPM7 kinase activity does not impair its channel function in mice. *Sci Rep*, 4: 5718. doi:10.1038/srep05718.

Kalwa, H. et al. 2015. Phospholipase C epsilon (PLCepsilon) induced TRPC6 activation: A common but redundant mechanism in primary podocytes. *J Cell Physiol*, 230: 1389–1399. doi:10.1002/jcp.24883.

Kanda, S. et al. 2011. Tyrosine phosphorylation-dependent activation of TRPC6 regulated by PLC-γ1 and nephrin: Effect of mutations associated with focal segmental glomerulosclerosis. *Mol Biol Cell*, 22: 1824–1835. doi:10.1091/mbc.E10-12-0929.

Katayama, K. et al. 2015. New TRPM6 mutation and management of hypomagnesaemia with secondary hypocalcaemia. *Brain Dev*, 37: 292–298. doi:10.1016/j.braindev.2014.06.006.

Kausalya, P.J. et al. 2006. Disease-associated mutations affect intracellular traffic and paracellular Mg^{2+} transport function of Claudin-16. *J Clin Invest*, 116: 878–891. doi:10.1172/JCI26323.

Kim, B.J. et al. 2008. Suppression of transient receptor potential melastatin 7 channel induces cell death in gastric cancer. *Cancer Sci*, 99: 2502–2509. doi:10.1111/j.1349-7006.2008.00982.x.

Kim, E.Y. et al. 2013. NOX2 interacts with podocyte TRPC6 channels and contributes to their activation by diacylglycerol: Essential role of podocin in formation of this complex. *Am J Physiol Cell Physiol*. 305: C960–C971. doi:10.1152/ajpcell.00191.2013.

Kim, J.H. et al. 2017. Klotho may ameliorate proteinuria by targeting TRPC6 channels in podocytes. *J Am Soc Nephrol*, 28(1): 140–151. doi:10.1681/ASN.2015080888.

Kistler, A.D. et al. 2013. Transient receptor potential channel 6 (TRPC6) protects podocytes during complement-mediated glomerular disease. *J Biol Chem*, 288: 36598–36609. doi:10.1074/jbc.M113.488122.

Konrad, M. and S. Weber. 2003. Recent advances in molecular genetics of hereditary magnesium-losing disorders. *J Am Soc Nephrol*, 14: 249–260.

Konrad, M., K.P. Schlingmann, and T. Gudermann. 2004. Insights into the molecular nature of magnesium homeostasis. *Am J Physiol Renal Physiol*, 286: F599–F605.

Kozak, J.A. et al. 2005. Charge screening by internal pH and polyvalent cations as a mechanism for activation, inhibition, and rundown of TRPM7/MIC channels. *J Gen Physiol*, 126: 499–514.

Krall, P. et al. 2010. Podocyte-specific overexpression of wild type or mutant trpc6 in mice is sufficient to cause glomerular disease. *PLOS ONE*, 5: e12859. doi:10.1371/journal.pone.0012859.

Krapivinsky, G. et al. 2014. The TRPM7 chanzyme is cleaved to release a chromatin-modifying kinase. *Cell*, 157: 1061–1072. doi:10.1016/j.cell.2014.03.046.

Kriz, W. 2005. TRPC6—A new podocyte gene involved in focal segmental glomerulosclerosis. *Trends Mol Med*, 11: 527–530. doi:10.1016/j.molmed.2005.10.001.

Kuras, Z. et al. 2012. KCa3.1 and TRPM7 channels at the uropod regulate migration of activated human T cells. *PLOS ONE*, 7: e43859. doi:10.1371/journal.pone.0043859.

Lainez, S. et al. 2014. New TRPM6 missense mutations linked to hypomagnesemia with secondary hypocalcemia. *Eur J Hum Genet*, 22: 497–504. doi:10.1038/ejhg.2013.178.

Letavernier, E. et al. 2012. Williams-Beuren syndrome hypercalcemia: Is TRPC3 a novel mediator in calcium homeostasis? *Pediatrics*, 129: E1626–E1630. doi:10.1542/peds.2011-2507.

Li, M., J. Jiang, and L. Yue. 2006. Functional characterization of homo- and heteromeric channel kinases TRPM6 and TRPM7. *J Gen Physiol*, 127: 525–537. doi:10.1085/jgp.200609502.

Li, M. et al. 2007. Molecular determinants of Mg^{2+} and Ca^{2+} permeability and pH sensitivity in TRPM6 and TRPM7. *J Biol Chem*, 282: 25817–25830. doi:10.1074/jbc.M608972200.

Lichtnekert, J. et al. 2009. Trif is not required for immune complex glomerulonephritis: Dying cells activate mesangial cells via Tlr2/Myd88 rather than Tlr3/Trif. *Am J Physiol Renal Physiol*, 296: F867–874. doi:10.1152/ajprenal.90213.2008.

Liu, W. et al. 2011. TRPM7 regulates gastrulation during vertebrate embryogenesis. *Dev Biol*, 350: 348–357. doi:10.1016/j.ydbio.2010.11.034.

Low, S.E. et al. 2011. TRPM7 is required within zebrafish sensory neurons for the activation of touch-evoked escape behaviors. *J Neurosci*, 31: 11633–11644. doi:10.1523/JNEUROSCI.4950-10.2011.

Matsushita, M. et al. 2005. Channel function is dissociated from the intrinsic kinase activity and autophosphorylation of TRPM7/ChaK1. *J Biol Chem*, 280: 20793–20803.

McKnight, A.J. et al. 2006. A genome-wide DNA microsatellite association screen to identify chromosomal regions harboring candidate genes in diabetic nephropathy. *J Am Soc Nephrol*, 17: 831–836. doi:10.1681/Asn.2005050493.

McNeill, M.S. et al. 2007. Cell death of melanophores in zebrafish trpm7 mutant embryos depends on melanin synthesis. *J Invest Dermatol*, 127: 2020–2030. doi:10.1038/sj.jid.5700710.

Mederos y Schnitzler, M. et al. 2008. Evolutionary determinants of divergent calcium selectivity of TRPM channels. *FASEB J*, 22: 1540–1551. doi:10.1096/fj.07-9694com.

Meng, X. et al. 2013. TRPM7 mediates breast cancer cell migration and invasion through the MAPK pathway. *Cancer Lett*, 333: 96–102. doi:10.1016/j.canlet.2013.01.031.

Meyer, T.E. et al. 2010. Genome-wide association studies of serum magnesium, potassium, and sodium concentrations identify six loci influencing serum magnesium levels. *PLOS GENETICS*, 6(8): e1001045. doi:10.1371/journal.pgen.1001045.

Middelbeek, J. et al. 2012. TRPM7 is required for breast tumor cell metastasis. *Cancer Res*, 72: 4250–4261. doi:10.1158/0008-5472.CAN-11-3863.

Moller, C.C., J. Flesche, and J. Reiser. 2009. Sensitizing the slit diaphragm with TRPC6 ion channels. *J Am Soc Nephrol*, 20: 950–953. doi:10.1681/ASN.2008030329.

Moller, C.C. et al. 2007. Induction of TRPC6 channel in acquired forms of proteinuric kidney disease. *J Am Soc Nephrol*, 18: 29–36. doi:10.1681/ASN.2006091010.

Monteilh-Zoller, M.K. et al. 2003. TRPM7 provides an ion channel mechanism for cellular entry of trace metal ions. *J Gen Physiol*, 121: 49–60.

Mottl, A.K. et al. 2013. A novel TRPC6 mutation in a family with podocytopathy and clinical variability. *BMC Nephrol*, 14: 104. doi:10.1186/1471-2369-14-104.

Muller, D. et al. 2006. Familial hypomagnesemia with hypercalciuria and nephrocalcinosis: Blocking endocytosis restores surface expression of a novel Claudin-16 mutant that lacks the entire C-terminal cytosolic tail. *Hum Mol Genet*, 15: 1049–1058. doi:10.1093/hmg/ddl020.

Na, T. and J.B. Peng. 2014. TRPV5: A Ca^{2+} channel for the fine-tuning of Ca^{2+} reabsorption. *Handb Exp Pharmacol*, 222: 321–357. doi:10.1007/978-3-642-54215-2_13.

Nadler, M.J. et al. 2001. LTRPC7 is a Mg.ATP-regulated divalent cation channel required for cell viability. *Nature*, 411: 590–595.

Nair, A.V. et al. 2012. Loss of insulin-induced activation of TRPM6 magnesium channels results in impaired glucose tolerance during pregnancy. *Proc Natl Acad Sci U S A*, 109: 11324–11329. doi:10.1073/pnas.1113811109.

Niehof, M. and J. Borlak. 2008. HNF4α and the Ca-channel TRPC1 are novel disease candidate genes in diabetic nephropathy. *Diabetes*, 57: 1069–1077. doi:10.2337/db07-1065.

Nilius, B. et al. 2007. Transient receptor potential cation channels in disease. *Physiol Rev*, 87: 165–217. doi:10.1152/physrev.00021.2006.

Norenberg, W. et al. 2016. TRPM7 is a molecular substrate of ATP-evoked P2X7-like currents in tumor cells. *J Gen Physiol*, 147: 467–483. doi:10.1085/jgp.201611595.

Numata, T., T. Shimizu, and Y. Okada. 2007. Direct mechano-stress sensitivity of TRPM7 channel. *Cell Physiol Biochem*, 19: 1–8. doi:10.1159/000099187.

Oancea, E., J.T. Wolfe, and D.E. Clapham. 2006. Functional TRPM7 channels accumulate at the plasma membrane in response to fluid flow. *Circ Res*, 98: 245–253. doi:10.1161/01.RES.0000200179.29375.cc.

Ong, E.C. et al. 2013. A TRPC1 protein-dependent pathway regulates osteoclast formation and function. *J Biol Chem*, 288: 22219–22232. doi:10.1074/jbc.M113.459826.

Paravicini, T.M., V. Chubanov, and T. Gudermann. 2012. TRPM7: A unique channel involved in magnesium homeostasis. *Int J Biochem Cell Biol*, 44: 1381–1384. doi:10.1016/j.biocel.2012.05.010.

Patel, V., R. Chowdhury, and P. Igarashi. 2009. Advances in the pathogenesis and treatment of polycystic kidney disease. *Curr Opin Nephrol Hypertens*, 18: 99–106.

Paul, B.M. and G.B. Vanden Heuvel. 2014. Kidney: Polycystic kidney disease. *Wiley Interdiscip Rev Dev Biol*, 3: 465–487. doi:10.1002/wdev.152.

Penner, R. and A. Fleig. 2007. The Mg^{2+} and Mg^{2+}-nucleotide-regulated channel-kinase TRPM7. *Handb Exp Pharmacol*, 179: 313–328. doi:10.1007/978-3-540-34891-7_19.

Perraud, A.L. et al. 2011. The channel-kinase TRPM7 regulates phosphorylation of the translational factor eEF2 via eEF2-k. *Cell Signal*, 23: 586–593. doi:10.1016/j.cellsig.2010.11.011.

Phelps, C.B. et al. 2007. Insights into the roles of conserved and divergent residues in the ankyrin repeats of TRPV ion channels. *Channels (Austin)*, 1: 148–151.

Qin, X. et al. 2012. Sphingosine and FTY720 are potent inhibitors of the transient receptor potential melastatin 7 (TRPM7) channels. *Br J Pharmacol*, 168: 1294–1312. doi:10.1111/bph.12012.

Quamme, G.A. and C. de Rouffignac. 2000. Epithelial magnesium transport and regulation by the kidney. *Front Biosci*, 5: D694–D711.

Ramsey, I.S., M. Delling, and D.E. Clapham. 2006. An introduction to TRP channels. *Annu Rev Physiol*, 68: 619–647. doi:10.1146/annurev.physiol.68.040204.100431.

Rangan, G.K. et al. 2015. Autosomal dominant polycystic kidney disease: A path forward. *Semin Nephrol*, 35: 524–537. doi:10.1016/j.semnephrol.2015.10.002.

Reiser, J. et al. 2005. TRPC6 is a glomerular slit diaphragm-associated channel required for normal renal function. *Nat Genet*, 37: 739–744. doi:10.1038/ng1592.

Riehle, M. et al. 2016. TRPC6 G757D Loss-of-function mutation associates with FSGS. *J Am Soc Nephrol*, 27(9): 2771–2783. doi:10.1681/ASN.2015030318.

Roshanravan, H. and S.E. Dryer. 2014. ATP acting through P2Y receptors causes activation of podocyte TRPC6 channels: Role of podocin and reactive oxygen species. *Am J Physiol Renal Physiol*, 306: F1088–1097. doi:10.1152/ajprenal.00661.2013.

Runnels, L.W. 2010. TRPM6 and TRPM7: A Mul-TRP-PLIK-cation of channel functions. *Curr Pharm Biotechnol*, 12: 42–53.

Runnels, L.W., L. Yue, and D.E. Clapham. 2001. TRP-PLIK, a bifunctional protein with kinase and ion channel activities. *Science*, 291: 1043–1047.

Runnels, L.W., L. Yue, and D.E. Clapham. 2002. The TRPM7 channel is inactivated by PIP(2) hydrolysis. *Nat Cell Biol*, 4: 329–336.

Ryazanov, A.G. et al. 1997. Identification of a new class of protein kinases represented by eukaryotic elongation factor-2 kinase. *Proc Natl Acad Sci U S A*, 94: 4884–4889.

Ryazanov, A.G., E.A. Shestakova, and P.G. Natapov. 1988. Phosphorylation of elongation factor 2 by EF-2 kinase affects rate of translation. *Nature*, 334: 170–173. doi:10.1038/334170a0.

Ryazanova, L.V. et al. 2010. TRPM7 is essential for Mg^{2+} homeostasis in mammals. *Nat Commun*, 1: 109. doi:10.1038/ncomms1108.

Ryazanova, L.V. et al. 2014. Elucidating the role of the TRPM7 α-kinase: TRPM7 kinase inactivation leads to magnesium deprivation resistance phenotype in mice. *Sci Rep*, 4: 7599. doi:10.1038/srep07599.

Rybarczyk, P. et al. 2012. Transient receptor potential melastatin-related 7 channel is overexpressed in human pancreatic ductal adenocarcinomas and regulates human pancreatic cancer cell migration. *Int J Cancer*, 131: E851–E861. doi:10.1002/ijc.27487.

Sah, R. et al. 2013. Ion channel-kinase TRPM7 is required for maintaining cardiac automaticity. *Proc Natl Acad Sci U S A*, 110: E3037–E3046. doi:10.1073/pnas.1311865110.

Sah, R. et al. 2013. The timing of myocardial trpm7 deletion during cardiogenesis variably disrupts adult ventricular function, conduction and repolarization. *Circulation*, 128(2): 101–114. doi:10.1161/CIRCU-LATIONAHA.112.000768.

Sahni, J. and A.M. Scharenberg. 2008. TRPM7 ion channels are required for sustained phosphoinositide 3-kinase signaling in lymphocytes. *Cell Metab*, 8: 84–93. doi:10.1016/j.cmet.2008.06.002.

Sales, C.H. and F. Pedrosa Lde. 2006. Magnesium and diabetes mellitus: Their relation. *Clin Nutr*, 25: 554–562. doi:10.1016/j.clnu.2006.03.003.

Saliba, Y. et al. 2015. Evidence of a role for fibroblast transient receptor potential canonical 3 Ca^{2+} channel in renal fibrosis. *J Am Soc Nephrol*, 26: 1855–1876. doi:10.1681/Asn.2014010065.

Santin, S. et al. 2009. TRPC6 mutational analysis in a large cohort of patients with focal segmental glomerulosclerosis. *Nephrol Dial Transplant*, 24(10): 3089–3096. doi:10.1093/ndt/gfp229.

Schaldecker, T. et al. 2013. Inhibition of the TRPC5 ion channel protects the kidney filter. *J Clin Invest*, 123: 5298–5309. doi:10.1172/JCI71165.

Schlingmann, K.P. et al. 2002. Hypomagnesemia with secondary hypocalcemia is caused by mutations in TRPM6, a new member of the TRPM gene family. *Nat Genet*, 31: 166–170.

Schlingmann, K.P. et al. 2005. Novel TRPM6 mutations in 21 families with primary hypomagnesemia and secondary hypocalcemia. *J Am Soc Nephrol*, 16: 3061–3069.

Schlingmann, K.P. et al. 2007. TRPM6 and TRPM7—Gatekeepers of human magnesium metabolism. *Biochim Biophys Acta*, 1772: 813–821. doi:10.1016/j.bbadis.2007.03.009.

Schlondorff, D. and B. Banas. 2009. The mesangial cell revisited: No cell is an island. *J Am Soc Nephrol*, 20: 1179–1187. doi:10.1681/ASN.2008050549.

Schlondorff, J. et al. 2009. TRPC6 mutations associated with focal segmental glomerulosclerosis cause constitutive activation of NFAT-dependent transcription. *Am J Physiol Cell Physiol*, 296: C558–C569. doi:10.1152/ajpcell.00077.2008.

Schmitz, C. et al. 2003. Regulation of vertebrate cellular Mg^{2+} homeostasis by TRPM7. *Cell*, 114: 191–200.

Schmitz, C. et al. 2005. The channel kinases TRPM6 and TRPM7 are functionally nonredundant. *J Biol Chem*, 280: 37763–37771.

Schmitz, C., F. Deason, and A.L. Perraud. 2007. Molecular components of vertebrate Mg^{2+}-homeostasis regulation. *Magnes Res*, 20: 6–18.

Shuen, A.Y. et al. 2009. Genetic determinants of extracellular magnesium concentration: Analysis of multiple candidate genes, and evidence for association with the estrogen receptor alpha (ESR1) locus. *Clin Chim Acta*, 409: 28–32. doi:10.1016/j.cca.2009.08.007.

Siddiqui, T.A. et al. 2012. Regulation of podosome formation, microglial migration and invasion by Ca^{2+}-signaling molecules expressed in podosomes. *J Neuroinflammation*, 9: 250. doi:10.1186/1742-2094-9-250.

Simon, D.B. et al. 1999. Paracellin-1, a renal tight junction protein required for paracellular Mg^{2+} resorption. *Science*, 285: 103–106.

Song, Y. et al. 2009. Common genetic variants of the ion channel transient receptor potential membrane melastatin 6 and 7 (TRPM6 and TRPM7), magnesium intake, and risk of type 2 diabetes in women. *BMC Med Genet*, 10: 4. doi:10.1186/1471-2350-10-4.

Storch, U. et al. 2012. Transient receptor potential channel 1 (TRPC1) reduces calcium permeability in heteromeric channel complexes. *J Biol Chem*, 287: 3530–3540. doi:10.1074/jbc.M111.283218.

Stritt, S. et al. 2016. Defects in TRPM7 channel function deregulate thrombopoiesis through altered cellular Mg^{2+} homeostasis and cytoskeletal architecture. *Nat Commun*, 7: 11097. doi:10.1038/ncomms11097.

Strubing, C. et al. 2001. TRPC1 and TRPC5 form a novel cation channel in mammalian brain. *Neuron*, 29: 645–655.

Strubing, C. et al. 2003. Formation of novel TRPC channels by complex subunit interactions in embryonic brain. *J Biol Chem*, 278: 39014–39019.

Su, L. T. et al. 2006. TRPM7 regulates cell adhesion by controlling the calcium-dependent protease calpain. *J Biol Chem*, 281: 11260–11270. doi:10.1074/jbc.M512885200.

Su, L.T. et al. 2011. TRPM7 regulates polarized cell movements. *Biochem J*, 434: 513–521. doi:10.1042/BJ20101678.

Takayanagi, K. et al. 2015. Downregulation of transient receptor potential M6 channels as a cause of hypermagnesiuric hypomagnesemia in obese type 2 diabetic rats. *Am J Physiol Renal Physiol*, 308: F1386–F1397. doi:10.1152/ajprenal.00593.2013.

Thebault, S. et al. 2009. EGF increases TRPM6 activity and surface expression. *J Am Soc Nephrol*, 20: 78–85. doi:10.1681/ASN.2008030327.

Tian, D. et al. 2010. Antagonistic regulation of actin dynamics and cell motility by TRPC5 and TRPC6 channels. *Sci Signal*, 3(145): 77. doi:10.1126/scisignal.2001200.

Touyz, R. M. 2008. Transient receptor potential melastatin 6 and 7 channels, magnesium transport, and vascular biology: Implications in hypertension. *Am J Physiol Heart Circ Physiol*, 294: H1103–H1118. doi:10.1152/ajpheart.00903.2007.

Trudel, M., Q. Yao, and F. Qian. 2016. The role of G-protein-coupled receptor proteolysis site cleavage of polycystin-1 in renal physiology and polycystic kidney disease. *Cells*, 5(1) : E3. doi:10.3390/cells5010003.

Tseveleki, V. et al. 2010. Comparative gene expression analysis in mouse models for multiple sclerosis, Alzheimer's disease and stroke for identifying commonly regulated and disease-specific gene changes. *Genomics*, 96: 82–91. doi:10.1016/j.ygeno.2010.04.004.

Urakawa, I. et al. 2006. Klotho converts canonical FGF receptor into a specific receptor for FGF23. *Nature*, 444: 770–774. doi:10.1038/nature05315.

van der Wijst, J. et al. 2014. Kinase and channel activity of TRPM6 are co-ordinated by a dimerization motif and pocket interaction. *Biochem J*, 460: 165–175. doi:10.1042/BJ20131639.

van der Wijst, J., R.J. Bindels, and J.G. Hoenderop. 2014. Mg^{2+} homeostasis: The balancing act of TRPM6. *Curr Opin Nephrol Hypertens*, 23: 361–369. doi:10.1097/01.mnh.0000447023.59346.ab.

Venkatachalam, K. and C. Montell. 2007. TRP channels. *Annu Rev Biochem*, 76: 387–417.

Voets, T. et al. 2004. TRPM6 forms the Mg^{2+} influx channel involved in intestinal and renal Mg^{2+} absorption. *J Biol Chem*, 279: 19–25. doi:10.1074/jbc.M311201200.

Walder, R.Y. et al. 2002. Mutation of TRPM6 causes familial hypomagnesemia with secondary hypocalcemia. *Nat Genet*, 31: 171–174.

Walder, R.Y. et al. 2009. Mice defective in Trpm6 show embryonic mortality and neural tube defects. *Hum Mol Genet*, 18: 4367–4375. doi:10.1093/hmg/ddp392.

Wang, Y. et al. 2013. Nuclear factor kappaB mediates suppression of canonical transient receptor potential 6 expression by reactive oxygen species and protein kinase C in kidney cells. *J Biol Chem*, 288: 12852–12865. doi:10.1074/jbc.M112.410357.

Wei, C. et al. 2009. Calcium flickers steer cell migration. *Nature* 457: 901–905. doi:10.1038/nature07577.

Weissmann, N. et al. 2006. Classical transient receptor potential channel 6 (TRPC6) is essential for hypoxic pulmonary vasoconstriction and alveolar gas exchange. *Proc Natl Acad Sci U S A*, 103: 19093–19098. doi:10.1073/pnas.0606728103.

Weissmann, N. et al. 2012. Activation of TRPC6 channels is essential for lung ischaemia-reperfusion induced oedema in mice. *Nat Commun*, 3: 649. doi:10.1038/ncomms1660.

Will, C. et al. 2010. Targeted deletion of murine Cldn16 identifies extra- and intrarenal compensatory mechanisms of Ca^{2+} and Mg^{2+} wasting. *Am J Physiol Renal Physiol*, 298: F1152–F1161. doi:10.1152/ajprenal.00499.2009.

Winn, M.P. et al. 2005. A mutation in the TRPC6 cation channel causes familial focal segmental glomerulosclerosis. *Science*, 308: 1801–1804.

Woodard, J.C., P.D. Webster, and A.A. Carr. 1972. Primary hypomagnesemia with secondary hypocalcemia, diarrhea and insensitivity to parathyroid hormone. *Am J Dig Dis*, 17: 612–618.

Woudenberg-Vrenken, T.E. et al. 2010. Transient receptor potential melastatin 6 knockout mice are lethal whereas heterozygous deletion results in mild hypomagnesemia. *Nephron Physiol*, 117: p11–p19. doi:10.1159/000320580.

Xie, J. et al. 2011. Phosphatidylinositol 4,5-bisphosphate (PIP(2)) controls magnesium gatekeeper TRPM6 activity. *Sci Rep*, 1: 146. doi:10.1038/srep00146.

Xie, J. et al. 2012. Cardioprotection by Klotho through downregulation of TRPC6 channels in the mouse heart. *Nat Commun*, 3: 1238. doi:10.1038/ncomms2240.

Yamazaki, D. et al. 2013. Basolateral Mg^{2+} extrusion via CNNM4 mediates transcellular Mg^{2+} transport across epithelia: A mouse model. *PLOS GENETICS*, 9: e1003983. doi:10.1371/journal.pgen.1003983.

Yamazaki, D. et al. 2016. The Mg^{2+} transporter CNNM4 regulates sperm Ca^{2+} homeostasis and is essential for reproduction. *J Cell Sci*, 129: 1940–1949. doi:10.1242/jcs.182220.

Yee, N.S., W. Zhou, and I.C. Liang. 2011. Transient receptor potential ion channel Trpm7 regulates exocrine pancreatic epithelial proliferation by Mg^{2+}-sensitive Socs3a signaling in development and cancer. *Dis Model Mech*, 4: 240–254. doi:10.1242/dmm.004564.

Yogi, A. et al. 2011. Transient receptor potential melastatin 7 (TRPM7) cation channels, magnesium and the vascular system in hypertension. *Circ J*, 75: 237–245.

Yu, H. et al. 2016. Synaptopodin limits TRPC6 podocyte surface expression and attenuates proteinuria. *J Am Soc Nephrol*, 27(11): 3308–3319. doi:10.1681/ASN.2015080896.

Zhang, D. et al. 2009. Evaluation of genetic association and expression reduction of TRPC1 in the development of diabetic nephropathy. *Am J Nephrol*, 29: 244–251. doi:10.1159/000157627.

Zhang, J. et al. 2012. Retinoids augment the expression of podocyte proteins by glomerular parietal epithelial cells in experimental glomerular disease. *Nephron Exp Nephrol*, 121: e23–e37. doi:10.1159/000342808.

Zhang, Z. et al. 2012. Upregulation of TRPM7 channels by angiotensin II triggers phenotypic switching of vascular smooth muscle cells of ascending aorta. *Circ Res*, 111: 1137–1146. doi:10.1161/CIRCRESAHA.112.273755.

Zhang, Z. et al. 2014a. N-Myc-induced up-regulation of TRPM6/TRPM7 channels promotes neuroblastoma cell proliferation. *Oncotarget*, 5: 7625–7634. doi:10.18632/oncotarget.2283.

Zhang, Z. et al. 2014b. The TRPM6 kinase domain determines the Mg.ATP sensitivity of TRPM7/M6 heteromeric ion channels. *J Biol Chem*, 289: 5217–5227. doi:10.1074/jbc.M113.512285.

Zhao, Z. et al. 2013. Novel TRPM6 mutations in familial hypomagnesemia with secondary hypocalcemia. *Am J Nephrol*, 37: 541–548. doi:10.1159/000350886.

Zhu, B. et al. 2009. Identification and functional analysis of a novel TRPC6 mutation associated with late onset familial focal segmental glomerulosclerosis in Chinese patients. *Mutat Res*, 664: 84–90.

Zierler, S. et al. 2011. Waixenicin A inhibits cell proliferation through magnesium-dependent block of transient receptor potential melastatin 7 (TRPM7) channels. *J Biol Chem*, 286: 39328–39335. doi:10.1074/jbc.M111.264341.

Zierler, S. et al. 2016. TRPM7 kinase activity regulates murine mast cell degranulation. *J Physiol*, 594: 2957–2970. doi:10.1113/JP271564.

11 Fertility and TRP Channels

Ida Björkgren and Polina V. Lishko

CONTENTS

11.1 INTRODUCTION: TRP CHANNELS IN REPRODUCTIVE TISSUES

Since their discovery in late 1970, transient receptor potential (TRP) channels have been implicated in a variety of cellular and physiological functions (Minke, 2010). The superfamily of TRP channels consists of nearly 30 members that are organized into seven major subgroups based on their specific function and sequence similarities (Owsianik et al., 2006; Ramsey et al., 2006). With the exception of TRPN channels that are only found in invertebrates and fish, mammalian genomes contain representatives of all six subfamilies: (1) TRPV (vanilloid); (2) TRPC (canonical); (3) TRPM (melastatin); (4) TRPA (ankyrin); (5) TRPML (mucolipin); and (6) TRPP (polycystin). TRP channels play crucial regulatory roles in many physiological processes, including those associated with reproductive tissues. As calcium-permeable cation channels that respond to a variety of signals (Clapham et al., 2003; Wu et al., 2010), TRP channels exert their role as sensory detectors in both male and female gametes, and play regulatory functions in germ cell development and maturation. Recent evidence obtained from *Caenorhabditis elegans* studies point to the importance of these proteins during fertilization where certain sperm TRP channels could migrate from a spermatozoon into an egg to ensure successful fertilization and embryo development. In this chapter we discuss how TRP channels can regulate both female and male fertility in different species and their specific roles.

11.2 TRP CHANNELS AND MALE FERTILITY

Fertility is defined by a capability of an organism to produce an offspring; male fertility is defined by the ability of male organisms to produce a sufficient number of healthy and motile male germ cells, spermatozoa, that can find and fertilize an oocyte, resulting in healthy embryo development. Sperm development begins in the male reproductive tract, which consists of the testes, epididymis, and ejaculatory duct: vas deferens and urethra. As the sperm cell differentiates from its precursor and migrates down to the male reproductive tract, it experiences many stages of cellular remodeling, in which the sperm intracellular calcium levels play an important part.

11.2.1 ROLE OF TRP CHANNELS DURING SPERMATOGENESIS

The process of male germ cell differentiation in mammals (i.e., spermatogenesis) takes place in the seminiferous tubules of the testis. Closest to the basal membrane lay the spermatogonial stem cells that, during puberty, start to proliferate and form spermatocytes (Figure 11.1) (de Kretser et al., 1998; Neto et al., 2016; Paniagua and Nistal, 1984). The decision of spermatogonia to self-renew, differentiate, or go through apoptosis is mainly dependent on an input from the surrounding Sertoli cells. The close association between Sertoli and germ cells is kept throughout sperm development as a means to provide the germ cells with nutrients and growth factors for cell survival and differentiation (reviewed in Hai et al., 2014). After the initial mitotic divisions of spermatogonia, the resulting round-shaped primary spermatocytes go through meiosis to form haploid round spermatids that further differentiate into elongated spermatids upon growing flagella (Figure 11.1). Elongated spermatids eventually discard a significant portion of their cytoplasm, becoming transcriptionally and translationally silent and eventually acquiring the terminal shape of spermatozoa. As the germ cells differentiate, they move out toward the lumen of the seminiferous tubule. Once the cells have gained their final structure, with a head containing the densely packed DNA and a long tail, the spermatozoa detach from the Sertoli cells and travel from the lumen of the seminiferous tubules into the first segments of the epididymis for further maturation (de Kretser et al., 1998; Neto et al., 2016).

This high rate of cell division and differentiation in the testis requires a vast network of regulatory proteins such as the transient receptor potential vanilloid 1 (TRPV1) (Tominaga et al., 1998) ion channel. TRPV1 shows a species-specific expression pattern in differentiating sperm cells. In rat, TRPV1 is found in premeiotic spermatogonia, while early spermatocytes and spermatids show from weak to no expression, respectively (Mizrak et al., 2008). In mice, the expression pattern is the opposite, with increased levels of TRPV1 during meiosis (Grimaldi et al., 2009). However, both rodent species display a regulatory role for TRPV1 in germ cell survival. *In vivo*, TRPV1 is activated by high temperatures and bioactive lipids, such as the endocannabinoids anandamide (AEA) (Caterina

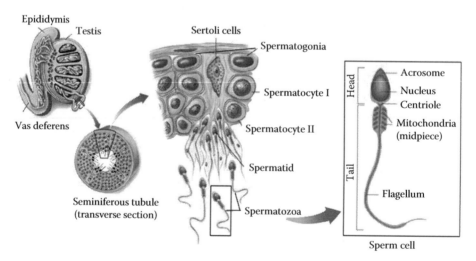

FIGURE 11.1 **(See color insert.)** Human spermatogenesis. Sperm development takes place in the seminiferous tubules of the testis. When the male reaches puberty, spermatogonial stem cells near the basal lamina of the tubule start to divide mitotically to produce more spermatogonia and primary spermatocytes. The spermatocytes go through meiosis I to form haploid secondary spermatocytes that migrate further toward the tubular lumen where they divide by meiosis II to form spermatids. Throughout spermatogenesis, the germ cells are positioned in close proximity to Sertoli cells, which provide them with nutrients and growth factors needed for development. After forming the final structure with a head, containing the genomic material and the acrosome region, and the elongated tail, and becoming transcriptionally and translationally silent, spermatozoa are released into the testicular lumen. (Modified from Allais-Bonnet, A. and E. Pailhoux. *Front Cell Dev Biol.*, 2, 56, 2014.)

et al., 1997; Zygmunt et al., 1999), 2-arachydonoyl glycerol (Zygmunt et al., 2013), and lysophosphatidic acid (Nieto-Posadas et al., 2012). Mice lacking *Trpv1* showed increased apoptosis of germ cells after incubation of testis in a water bath of 42°C (Mizrak and van Dissel-Emiliani, 2008). The location of testis in the scrotum gives the tissue a lower core temperature than the rest of the body. A common condition among infant boys is cryptorchidism where the testis does not descend into the scrotum but remains in the abdomen even after birth. This gives rise to elevated temperatures that could have a detrimental effect on fertility even though the testis in most cases descends into the scrotum during the first year of life (Virtanen and Toppari, 2015). The activation of TRPV1 during these conditions could thus play a beneficial role in cell survival. Contradictory to the study with knockout mice (Mizrak and van Dissel-Emiliani, 2008), *in vitro* cultures of rat spermatogonial cell lines showed increased apoptosis after activation of TRPV1 by the agonist capsaicin (Mizrak et al., 2008). In addition, intraperitoneal injection of capsaicin into male mice gave rise to increased apoptosis of differentiating germ cells (Nagabhushan and Bhide, 1985). However, a study using similar dosage of the agonist reported no changes in testicular weight or histology after injection (Muralidhara and Narasimhamurthy, 1988). These conflicting results were hypothesized to be caused by the use of different experimental methods, but they could also be due to differences in the physiological context of which TRPV1 is activated. It is possible that heat stress causes additional changes in the testis that, in concert with the function of TRPV1, lead to germ cell survival. If activated under normal physiological conditions, TRPV1 could instead cause apoptosis of mouse meiotic germ cells. Interestingly, the mouse testis produces a steady state of AEA throughout spermatogenesis, which could activate TRPV1 (Grimaldi et al., 2009). As a large part of the developing germ cells are discarded before becoming fully differentiated (Neto et al., 2016), TRPV1s' activation by AEA could play a selective role in this process and thereby regulate access to nutrients and growth factors for the remaining sperm cells. However, considering the fact that Trpv1$^{-/-}$ male mice are fertile, the absence of TRPV1 could be compensated by some other regulatory ion channel. Further studies are needed to clarify the activity and function of TRPV1 in mammalian germ cell differentiation.

Sertoli cells contain a variety of ion channels including calcium- and voltage-dependent Cl$^-$ channels (Auzanneau et al., 2003; Lalevee and Joffre, 1999), voltage-dependent Ca^{2+} channels (Lalevee et al., 1997; Taranta et al., 1997), and TRPV1 channels (Rossi et al., 2007). An *in vitro* study of rat Sertoli cells has shown a role for TRPV1 in the regulation of an acid-sensing Cl$^-$ channel (ASCC) (Auzanneau et al., 2008). Acidic conditions activate ASCC and produce an anionic outward current (Auzanneau et al., 2003). In the presence of 1 µM capsaicin, complete inhibition of the current was observed while 10 µM of the TRPV1 antagonist capsazepine caused potentiation of ASCC activity (Auzanneau et al., 2008). Interestingly, conditions under an acidic pH (pH < 5.9) lead to activation of TRPV1 and decrease the temperature threshold of this ion channel (Tominaga et al., 1998). Thus, the low pH needed to activate ASCC could also allow for activation of TRPV1 further regulating Cl$^-$ secretion by Sertoli cells. Increased extracellular Cl$^-$ levels affect Sertoli cell pH through inhibition of HCO$_3$$^-$/Cl$^-$ exchangers in the cell membrane (Oliveira et al., 2009). In addition, the acidic microenvironment of Sertoli cells is required for synthesis of membrane-bound stem cell factor, which promotes survival and proliferation of spermatogonia, and the acidic environment of the testes also keeps spermatozoa in a quiescent state, preventing early activation of motility and inhibiting acrosomal enzymes (Acott and Carr, 1984; Darszon et al., 2006). The role of TRPV1 as a regulator of ASCC could thus have a great impact on spermatogenesis.

11.2.2 Function of TRP Channels during Sperm Maturation

11.2.2.1 TRP Channels in the Epididymis

Upon leaving the testis, sperm cells have gained their final structure with a head where the haploid DNA is stored, minimal cytoplasm, and a long flagellum required for motility. They also become transcriptionally and translationally silent, with no additional protein synthesis observed from the time of differentiation into spermatids. However, further maturation is needed for spermatozoa to become fully motile and able to fertilize the oocyte. The first post-testicular maturation step takes place in the epididymis where the lipid and protein structure of the sperm membrane is further modified to enable the cell to react to signals in the female reproductive tract (Jones, 2002). Although the epididymis consists of one long convoluted duct, it is divided into three to four segments: initial segment, caput, corpus, and cauda, each with its own pattern of gene expression. The highly regulated protein and ion secretion of the segments gives rise to the luminal environment required for sperm maturation, while concurrently keeping the germ cells in a quiescent state (reviewed in Dacheux and Dacheux, 2013; Sipilä and Björkgren, 2016). As sperm cells descend from the seminiferous tubules and enter the caput epididymis, they experience gradually declining levels of extracellular Ca^{2+}, Na^+, Cl^-, bicarbonate, and pH. Both bicarbonate and Ca^{2+} are required for sperm activation, and the low ion concentrations thus contribute to keeping the sperm in a quiescent state throughout epididymal transit (Dacheux and Dacheux, 2013; Liu et al., 2012; Wandernoth et al., 2010; Wennemuth et al., 2003). In rat, Na^+ and Cl^- drop from 150 to 75 mM and 10 mM, respectively. Bicarbonate drops from 20 to 1 mM, and pH acidifies from 7.4 to 6.5. Interestingly, K^+ concentration initially decreases from 50 mM in testes to 5 mM in the caput, only to rise back to the same 50 mM in distal parts of the epididymis and vas deferens (Levine and Kelly, 1978; Levine and Marsh, 1971). The Ca^{2+} content of the epididymal lumen decreases from the proximal to the distal segments. In caput of the rat epididymis, the free luminal ionic Ca^{2+} concentration is 0.8 mM, while distal cauda, where sperm are stored, show a concentration of 0.25 mM (Jenkins et al., 1980). To reduce luminal Ca^{2+} concentrations the epididymis utilizes the highly Ca^{2+} selective TRPV6 channel (Peng et al., 1999), expressed in epididymal epithelial cells (Weissgerber et al., 2011). In addition to *Trpv6*, the epididymis shows expression of *Trpv5*. However, ablation of *Trpv5* does not affect Ca^{2+} uptake in the epididymis (Hoenderop et al., 2003). *Trpv6* knockout mice, either displaying a full knockout of the protein (Weissgerber et al., 2012) or with a point mutation in the pore region of the channel rendering it nonfunctional (Weissgerber et al., 2011), show a 10-fold increase of the Ca^{2+} concentration in cauda epididymis. However, the high luminal Ca^{2+} levels did not lead to an increased intracellular level of this cation in sperm (Weissgerber et al., 2011), at least based on sperm population measurements. This could in part be explained by the active export of Ca^{2+} through the Ca^{2+}-ATPases in the sperm membrane, which increase both in activity and number during sperm epididymal transit (Brandenburger et al., 2011; Dacheux and Dacheux, 2013; Wennemuth et al., 2003). Although sperm Ca^{2+} levels appeared normal, the change in Ca^{2+} concentration of the epididymis led to hypofertility of the knockout mice due to increased apoptosis and reduced progressive motility of spermatozoa (Weissgerber et al., 2012; Weissgerber et al., 2011). A similar phenotype has also been detected in a rat model with overexpression of the Ca^{2+}-binding protein regucalcin (RGN), where increased levels of RGN led to a reduced uptake of Ca^{2+} from the epididymal lumen negatively affecting the motility of sperm (Correia et al., 2013). Sperm motility requires influx of Ca^{2+} into mature spermatozoa, which, in turn, activates an atypical sperm-soluble adenylyl cyclase (sAC) and leads to an increased production of cyclic AMP (cAMP) and flagellar activity (Esposito et al., 2004; Hess et al., 2005). Although the higher Ca^{2+} concentration of caput should give rise to increased cAMP levels in spermatozoa, this does not take place, and the *Trpv6* knockout mouse sperm show normal cAMP levels, both prior to and after induction of sperm capacitation, although the motility of spermatozoa is significantly reduced (Weissgerber et al., 2011). It could also be possible that increased extracellular Ca^{2+} concentration would negatively impact spermatozoal Ca^{2+} pumps, such as Ca-ATPse. These exchangers work by exporting Ca^{2+} outside and importing protons with a 1–2 ratio. Less effective Ca-ATPses would result in alkalinization of sperm cytoplasm and lead to premature sperm activation, which can be detrimental for fertility (Babcock and Pfeiffer, 1987). Although this would lead to increased Ca^{2+} levels in sperm, the change is localized to the tail and might not be noticeable by the methods used in the current study (Weissgerber et al., 2011). Thus, more studies are needed to explain the connection between the Ca^{2+} concentration gradient observed in the epididymis and the role of TRPV6 in sperm motility regulation. Furthermore, the effect of TRPV6 ablation could be partially compensated by TRPV5 overexpression in the epididymis. Studies utilizing a double knockout of the channels may reveal a more severe reduction in fertility.

11.2.2.2 Role of TRP Channels in Prostate

Upon ejaculation, sperm cells are mixed with fluids from prostate and seminal vesicles, which, combined with fluids from the testis and epididymis, create seminal plasma. It has been shown that the first fractions of an ejaculate contain the highest sperm number, but also the highest percentage of prostate secretions. The alkaline and ionic composition of these fractions promotes the motility and viability of sperm in the female reproductive tract (Mann and Lutwak-Mann, 1981). The involvement of prostate TRP channels in ion secretion has also been suggested to play a role in male fertility. For example, the previously mentioned *Trpv6* knockout mouse line displayed defects in Ca^{2+} homeostasis resulting in an increased diameter of the prostate lobes together with Ca^{2+} precipitates in the ductal lumen (Weissgerber et al., 2011). Although increased seminal Ca^{2+} concentrations could have a detrimental effect on sperm motility, the phenotype of epididymal sperm was too severe to draw any conclusions on the impact of the observed prostate dysfunction. Another candidate for regulation of the ion homeostasis of semen is the cold-sensitive Ca^{2+}-channel TRP melastatin member 8 (TRPM8) (McKemy et al., 2002), which was first identified in the human prostate (Tsavaler et al., 2001). The gene showed low expression levels in normal prostate tissue while increased expression was detected in several malignancies, including prostate, breast, and colon cancers (Tsavaler et al., 2001). Further reverse transcriptase polymerase chain reaction (RT-PCR) and immunohistochemical staining showed *Trpm8* expression throughout both the rat and human genitourinary tract (Stein et al., 2004) as well as in sensory neurons (McKemy et al., 2002). Although prostate cancer cell lines require TRPM8 for cell cycle progression, proliferation of normal prostate tissue is not dependent on *Trpm8* expression (Valero et al., 2012). Of the three cell types that make up the prostate—basal cells, transit amplifying/intermediate (TA/I) cells, and apical cells—the TA/I cells are the most proliferative (Garraway et al., 2003). However, only the highly differentiated apical cells contain TRPM8 in the plasma membrane (Bidaux et al., 2007). Expression of full-length TRPM8 requires androgen receptor (AR) activity and TA/I cells, which lack the receptor, only express a truncated form of the protein (Bidaux et al., 2007). Thus, instead of regulating cell proliferation, TRPM8 was suggested to play a role in Ca^{2+} uptake by the apical cells and to indirectly take part in sperm cell fertilizing capacity. Interestingly, TRPM8 has also been shown to directly bind testosterone (Asuthkar et al., 2015a). Incubation of prostate cells with testosterone elicited an increased uptake of Ca^{2+} through activation of TRPM8 (Asuthkar et al., 2015a, 2015b). A testosterone-induced increase in single-channel currents was also detected when TRPM8 was incorporated into artificial lipid membranes, thereby excluding other cellular processes and metabolic pathways that could affect the observed response (Asuthkar et al., 2015b). As the prostate is located inside the body where fluctuations in temperature are unlikely to occur, testosterone could serve as the main regulator of TRPM8 activity and Ca^{2+} uptake by the apical cells. However, further studies are needed to verify this interesting hypothesis.

11.2.2.3 Sperm TRP Channels: Role in Motility and Acrosome Reaction

Once released from the ejaculatory duct, spermatozoa rely on their motility and specific navigation skills to find and fertilize an egg. Change in motility, ability to undergo an acrosome reaction, and sperm response to female reproductive tract cues, as well as the process of fusion with an egg, rely on fast intracellular ion changes within spermatozoa. Ion channels are indispensable for sperm physiology, as they control several essential milestones on the sperm's route toward the egg: activation of sperm motility, sperm terminal maturation in the female reproductive tract, and hyperactivation of sperm motility in the vicinity of the egg (Visconti et al., 2011). Ion channels control sperm activity by regulating membrane potential and intracellular levels of Ca^{2+} and H^+:

intracellular Ca^{2+} stimulates sperm motility and fertility, whereas intracellular H^+ inhibits these processes (Inaba, 2011). Sperm swimming behavior is controlled by rises in intracellular Ca^{2+} that changes flagellar beat pattern through Ca^{2+}-binding proteins calaxins (Mizuno et al., 2012). Ca^{2+}-bound calaxins inhibit the activity of dynein motors within the sperm tail cytoskeleton known as the axoneme, resulting in high-amplitude asymmetric flagellar bending, or hyperactivation. The latter is essential for the ability of mammalian sperm to overcome the protective vestments of the egg. Additionally, rises in intracellular Ca^{2+} plays a pivotal role in acrosome exocytosis (Florman et al., 2008; Florman et al., 1989; Publicover et al., 2007; Yanagimachi and Usui, 1974). For many species, such as human, rodent, and sea urchin, a principal Ca^{2+} channel of sperm was determined to be CatSper (cationic channel of sperm) (Kirichok et al., 2006; Lishko et al., 2012; Ren et al., 2001); however, there are species where functional CatSper is not detected. For example, worms, birds, and teleost fish genomes seem to be devoid of CatSper, yet their sperm cells still rely on Ca^{2+} changes (Cai et al., 2014). Therefore, a possibility of TRP channels to be involved in sperm physiology has certain merits. Several TRP channels have recently been proposed to function in mature sperm cells. These include TRPM8, TRPV1, TRPC2, and others (De Blas et al., 2009; Jungnickel et al., 2001; Maccarrone et al., 2005; Wissenbach et al., 1998). However, the precise contribution of transient receptor potential ion channels in sperm vital physiological processes, such as sperm motility and acrosome reaction, is yet to be established, as most of the evidence indicating TRP-channel expression in mature sperm is based on biochemical evidence, antibody binding, and Ca^{2+} imaging experiments. In addition, mice deficient in TRPC1–7, TRPV1–4, TRPA1, TRPM1–4, and TRPM8 have no obvious defects in sperm physiology or male fertility (Lishko et al., 2012). However, one has to bear in mind that mammalian spermatozoa are quite diverse in their choice of molecular regulators, and the molecular effectors that regulate the physiology of murine sperm cells may not necessarily work for other species. Therefore, the absence of certain TRP channels may not impact rodent fertility potential but could be detrimental for human fertility. Direct application of electrophysiology to mature sperm cells should shed more light on the molecular identities and physiological regulation of sperm TRP ion channels, resulting in new tools to control the behavior of spermatozoa and to increase or decrease male fertility. For example, the Ca^{2+} rise in response to menthol in human sperm cells, which was initially attributed to TRPM8 presence in human spermatozoa (De Blas et al., 2009), has been shown to be caused by CatSper channel activity rather than TRPM8 (Brenker et al., 2012). Recently, TRPV4 has been detected in human spermatozoa (Kumar et al., 2016); however, direct recording of this channel activity, as well as human genetics are required to prove TRPV4-channel functional importance for human fertility. Interestingly, sperm cells of invertebrates can donate their TRP channels to the egg during fertilization to initiate and maintain Ca^{2+} oscillations that are followed by a fertilization event. Work from Onami group has identified sperm TRP-3 channel as the one required for such Ca^{2+} oscillations (Takayama and Onami, 2016). The role of this channel in egg activation will be discussed later in this chapter.

Several members of the TRPC family were shown to be expressed in murine sperm cells, such as TRPC1, TRPC2, TRPC3, and TRPC5 (Jungnickel et al., 2001; Trevino et al., 2001). Specifically, mouse TRPC2 was shown to be expressed in the subacrosomal region of the anterior sperm head (Jungnickel et al., 2001). It was suggested to play a role in the initiation of the acrosome reaction, via a ligand binding to zona pellucida (ZP) proteins (Arnoult et al., 1999; Jungnickel et al., 2001). As sperm acrosome reaction (AR) begins upon the sperm cell entering the cumulus oophorus and binding to ZP proteins, it coincides with a Ca^{2+} rise in the head, which together with intracellular alkalization drives acrosomal exocytosis. However, later research has shown that sperm binding to ZP proteins is not sufficient to trigger AR (Baibakov et al., 2007), and rather mechanical stimulus is needed. As sperm cells push their way through the cumulus oophorus, such actions open sperm mechanosensitive ion channels that give rise to the increased intracellular Ca^{2+} levels required for AR. Several TRP channels were shown to be mechanosensitive, among which are polycystin-2 or TRPP2, and are implicated in the activation of AR (Barritt and Rychkov, 2005; Neill et al., 2004). However, TRPC2-deficient mice are fertile, while the human TRPC2 homolog is a pseudogene

(Leypold et al., 2002; Stowers et al., 2002; Wes et al., 1995). Thus, it remains to be seen whether the other mechanosensory ion channels are functionally expressed in mammalian sperm cells and whether they can play a key role in sperm fertility, including the initiation of AR.

11.3 TRP CHANNELS AND FEMALE FERTILITY

11.3.1 Role in Oocyte Maturation and Activation

Ca^{2+} signaling plays a vital role in oocyte maturation by maintaining cell homeostasis and possibly contributing to oocyte metabolism. The immature oocyte is arrested at the prophase stage (germinal vesicle [GV]) of meiosis I (MI) until puberty when a surge of luteinizing hormone triggers the cell to progress to the metaphase stage of meiosis II (MII) where it undergoes a second arrest (Figure 11.2). During oocyte maturation Ca^{2+} is taken up and stored in the endoplasmic reticulum (ER) to be utilized during fertilization (reviewed in Wakai and Fissore, 2013; Wang and Machaty, 2013). When the spermatozoon fuses with the egg, an increase in the intracellular levels of Ca^{2+} can be detected as a wave that progresses through the fertilized egg. This initial burst of Ca^{2+} signaling launches embryonic cell divisions and activates protein production, ATP synthesis, and cell metabolism (reviewed in Whitaker, 2006).

The requirement of Ca^{2+} for egg activation is evolutionarily conserved and found throughout the animal kingdom, from arthropods to higher primates. However, unlike vertebrates, the increase in intracellular Ca^{2+} levels in *Drosophila* eggs is observed as the gamete travels through the female reproductive tract and is independent of fertilization (Heifetz et al., 2001). Ovulation activates mechanosensitive ion channels in the egg membrane and causes an influx of Ca^{2+} from the external environment, which in turn initiates a release of Ca^{2+} from the ER stores (Horner and Wolfner, 2008; Kaneuchi et al., 2015). The resulting Ca^{2+} wave begins in one or both poles of the egg and then spreads throughout the cytoplasm (Kaneuchi et al., 2015). Three members of the TRP family of proteins—*trpm* (TRPM3), *trpml* (TRPP1/Pkd2), and *painless* (TRPA1)—are found in the ovaries

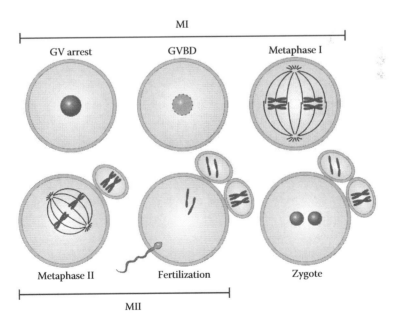

FIGURE 11.2 (See color insert.) Oocyte maturation and fertilization. At birth, mammalian oocytes are arrested at the diplotene stage of the first meiosis (MI) prophase. Once the female reaches puberty, each estrus cycle triggers the breakdown of the germinal vesicle (GVBD) and allows the division of the oocyte to continue until metaphase II of the second meiosis (MII). Fertilization allows the oocyte to complete meiosis after which the haploid genomic material is combined with that of the male gamete (marked in blue).

of fruit flies, and it is likely that one or several of these are responsible for the Ca^{2+} wave (Sartain and Wolfner, 2013). Although Ca^{2+} influx is blocked by incubation of the egg with the TRP channel inhibitor N-(p-amylcinnamoyl) anthranilic acid (ACA), further studies are required to determine the specific channel(s) responsible for initiating the Ca^{2+} wave (Kaneuchi et al., 2015).

In most species the intracellular Ca^{2+} wave stems from opening of intracellular Ca^{2+} channels, the inositol trisphosphate receptors (IP_3R), in the ER. In sea urchins, fish, and frogs, fertilization triggers a single Ca^{2+} rise in the form of a Ca^{2+} wave that initiates at the site of gamete fusion and spreads throughout the egg. In mammals, the Ca^{2+} signal adopts the form of repetitive rises, oscillations, which can last for several hours (Clapham, 2007). In mice, the Ca^{2+} waves occur every 10 minutes, while in mammals with larger eggs, such as those of human, cow, and pig, the rises occur at significantly longer intervals (Figure 11.3) (reviewed in (Kashir et al., 2013; Miyazaki, 2006; Swann and Lai, 2013). These continuous oscillations also require Ca^{2+} influx from the extracellular environment to replenish the ER stores (Igusa and Miyazaki, 1983; Miao et al., 2012; Winston et al., 1995).

Recently, the temperature-dependent TRPV3 channel was suggested to play a role in Ca^{2+} uptake of mammalian egg cells (Carvacho et al., 2013). In mice, the expression of *TrpV3* begins when oocyte maturation resumes and increases until MII. In accordance with the expression data, activation of TRPV3 by 2-aminoethoxydiphenyl borate (2-APB) led to an increased Ca^{2+} current from the MI to MII stage, while immature oocyte at the GV stage did not display TRPV3 activation (Figure 11.4) (Carvacho et al., 2013). In mature, MII eggs, Ca^{2+} influx after TRPV3 activation by 2-APB or the

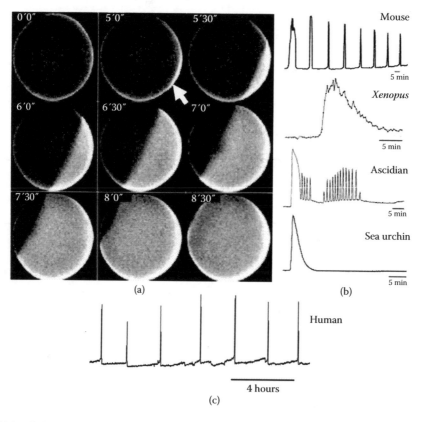

FIGURE 11.3 Calcium oscillations in fertilized oocytes. (a) The single Ca^{2+} wave of a *Xenopus* egg detected by calcium-green-1-dextran after gamete fusion (the arrow indicates the site of sperm entry). (b) The Ca^{2+} transient and subsequent oscillations in different species. (c) Ca^{2+} oscillations in the fertilized human egg. (Modified from Kaneuchi, T. et al., *Proc Natl Acad Sci U S A*, 112, 3, 2015; Ramadan, W.M. et al., *Cell Commun Signal*, 10, 1, 2012.)

agonist carvacrol induced egg activation (Carvacho et al., 2013; Lee et al., 2016). Furthermore, it was shown that strontium (Sr^{2+})-mediated egg activation requires TRPV3 (Carvacho et al., 2013). Sr^{2+} causes Ca^{2+} oscillations in rodent oocytes and is widely used to induce parthenogenesis (Kono et al., 1996; Li et al., 2009; Tomashov-Matar et al., 2005). Although it is evident that TRPV3 is able to cause Ca^{2+} influx in oocytes, TRPV3 knockout mice did not display significant reduction in fertility or alteration in Ca^{2+} oscillations after fertilization (Carvacho et al., 2013). Other molecules implicated in Ca^{2+} influx in oocyte maturation and fertilization are the stromal interacting molecules (STIM1 and STIM2) and Orai1, which are part of the store-operated Ca^{2+} entry (SOCE) mechanism. After Ca^{2+} depletion from the ER stores, STIM1, which is located in the organelle membrane and acts as a Ca^{2+} sensor, activates Ca^{2+} influx through the Orai1 channels located in the plasma membrane (Prakriya and Lewis, 2015). In pigs, downregulation of *Orai1* by siRNAs blocks Ca^{2+} entry after depletion of the ER stores and prevents Ca^{2+} oscillations induced by fertilization (Wang et al., 2012). However, Ca^{2+} oscillations in mice were not dependent on Orai1, as shown by several studies using Orai1-specific inhibitors, which failed to block oscillation induced by IVF (Miao et al., 2012; Takahashi et al., 2013). Another candidate Ca^{2+} channel is $Ca_v3.2$, which is expressed in oocytes, and null female mice for this voltage-gated Ca^{2+}-channel showed a mild impairment of fertility (Bernhardt et al., 2015). These results were somewhat surprising since the membrane potential of mouse eggs does not change significantly after fertilization (Igusa et al., 1983; Jaffe et al., 1983). However, Bernhardt et al. hypothesized that a small number of $Ca_v3.2$ channels would be active even when the voltage lies in between values rendering maximum activation or total inactivation. The mild effects of $Ca_v3.2$ knockout could be caused by ablation of these small, so called, window currents (Bernhardt et al., 2015). $Ca_v3.2^{-/-}$ mice showed significantly reduced concentrations of Ca^{2+} in the ER of MII eggs, which in turn affected the amplitude of the first transient after IVF; a number of these $Ca_v3.2$ knockout eggs also showed reduced persistent oscillations after IVF. However, these female mice produced litters with only a slightly reduced number of pups, suggesting that other Ca^{2+} channels, such as TRPV3 or other unknown channels, are required to support influx during fertilization (Bernhardt et al., 2015). It is also clear from studies in pig and mouse oocytes that there are species-specific differences regarding the use of Ca^{2+}-channels during fertilization, and further studies are required, especially to identify the ion channels that underlie influx in human oocytes.

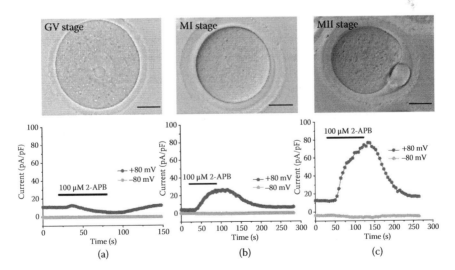

FIGURE 11.4 2-APB activation of TRPV3 current during oocyte maturation. Mouse (a) GV, germinal vesicle oocyte; (b) MI, first meiosis egg; (c) MII, second meiosis egg. Scale bar 50 mm. Lower panel displays whole-cell patch-clamp recordings from oocytes at different maturation stages, in response to 100 μM 2-APB (2-aminoethoxydiphenyl borate, black bar). (From Carvacho, I. et al., *Cell Rep*, 5, 5, 2013. With permission.)

11.3.2 TRP Channels' Role in Fertilization

There are several theories on how the spermatozoon stimulates Ca^{2+} signaling/waves in eggs. Spermatozoa could, for example, contain factors that, either by directly binding to receptors on the egg membrane or by diffusing into the cytoplasm, activate Ca^{2+} channels and/or Ca^{2+} release. The latter has also been shown in several species, including mammals, where introduction of sperm-specific phospholipase C (PLC) zeta into the eggs triggers IP3 production from phosphatidylinositol 4,5-bisphosphate (PI[4,5]P2) (Saunders et al., 2002; Kashir et al., 2013; Escoffier et al., 2015). Another formal possibility, which is supported by a recent study in *C. elegans*, shows that sperm-specific Ca^{2+} channels could be incorporated into the egg's membrane during the fusion of the gametes (Takayama and Onami, 2016). This would give rise to a localized Ca^{2+} uptake from the extracellular environment, which, in turn, could initiate further signaling events in the zygote. By utilizing high-speed confocal imaging, Takayama and Onami were able to observe a rapid local Ca^{2+} rise immediately after fusion of *C. elegans* gametes, followed by the global Ca^{2+} wave that initiates embryogenesis (Takayama and Onami, 2016). *C. elegans* spermatozoa contain a Ca^{2+}-permeable TRP channel, TRP-3 (also known as SPE-41), which is needed for sperm-egg fusion (Xu and Sternberg, 2003). Unlike mammalian oocytes, the egg cell of *C. elegans* does not contain a TRP channel. Instead, the sperm TRP-3 is incorporated into the egg membrane during fertilization and was shown to induce the local increase in Ca^{2+} levels. However, ablation of the channel in spermatozoa did not prevent activation of the zygote, but only delayed the onset of the global Ca^{2+} wave (Takayama and Onami, 2016). Thus, TRP-3 might be one of several factors in *C. elegans* sperm that are required for the initiation of embryogenesis. Considering that mammalian spermatozoa express many ion channels, it would be interesting to study the possibility of a similar mechanism in other species. However, the size of the sperm cell in comparison to the egg is considerably larger in *C. elegans* than in mammals, and the number of membrane channels incorporated during gamete fusion could therefore have a greater impact on the activation of *C. elegans* eggs than in other species, especially mammals.

11.4 TRP CHANNELS' ROLE IN SEX DISCRIMINATION

Pheromones are important triggers for sexual behavior, such as mating, male dominance, and offspring protection by females. Most vertebrates convey pheromone signals through an accessory olfactory organ, the vomeronasal organ (VNO). During embryonic development the human VNO also begins to develop. However, the organ and the surrounding cartilage dissolve before birth and cannot be found in adult individuals (Tirindelli et al., 2009). Although humans lack functional vomeronasal receptors for pheromone signaling, sexual behavior can still be influenced by pheromones. For example, women living together may experience synchronized menstrual cycles because of a yet unknown pheromone(s) (McClintock, 1971). A similar phenomenon can be seen in other mammals where the main olfactory epithelium, in addition to the VNO, is able to respond to certain pheromones (Tirindelli et al., 2009).

One of the main channels to convey pheromone signaling in VNO neurons is the cation channel TRPC2. Humans lack functioning TRPC2 due to addition of several stop codons within the coding sequence (Wes et al., 1995). In mice, ablation of *Trpc2* gave rise to behavioral changes. Male *Trpc2*−/− mice displayed normal mating behavior when placed with female mice but did not show any aggression toward male mice. Instead they tried to mount the male intruders placed in the cage with them (Leypold et al., 2002; Stowers et al., 2002). This behavior took place whether or not the castrated male intruder had been swabbed with urine containing typical male pheromones, which would trigger aggression in wild-type mice (Stowers et al., 2002). *Trpc2*−/− females showed a similar lack of aggression. *Trpc2* ablation caused lactating female mice, which normally display aggression toward male intruders, to behave in a sexually receptive manner (Kimchi et al., 2007; Leypold et al., 2002). Unlike mice with surgically removed VNO, *Trpc2*−/− male mice did not show a reduction in mounting of partners but instead displayed a lack in gender discrimination, where they would mount female and male partners equally (Leypold et al., 2002; Stowers et al., 2002). Interestingly, $Trpc2^{-/-}$ females displayed male sexual behavior with solicitation, mounting, and pelvic thrust toward both males and females (Kimchi et al., 2007). There are several theories as to how $Trpc2$ ablation elicits such a response in mice. In the first study by Stowers et al., they claim that $Trpc2^{-/-}$ mice completely lack pheromone response and therefore a functional VNO (Stowers et al., 2002). However, it is now known that the VNO contains three other ion channels that, together with TRPC2, are activated by pheromone signaling. Electrophysiological measurements in $Trpc2^{-/-}$ mice evidenced both Cl⁻ and K⁺ currents that were not entirely dependent on TRPC2 activation (Kim et al., 2012). Instead of a purely genetic reason as to why the $Trpc2^{-/-}$ mice display behavioral changes, it has been suggested that the changes are neomorphic in character (Yu, 2015). There is a marked loss of basal layer neurons in the $Trpc2^{-/-}$ VNO (Kim et al., 2012; Stowers et al., 2002). As a result, the posterior accessory olfactory bulb, which receives projections from the basal layer, is also reduced (Hasen and Gammie, 2009). In addition, the lower activation level of the $Trpc2^{-/-}$ VNO could lead to an altered input to neurons in the ventral medial hypothalamic nucleus. Depending on the activation level, these neurons evoke either mating or aggressive behavior and alter signaling from the VNO that could thus lead to a behavioral change (Lee et al., 2014; Yu, 2015). Not only may $Trpc2$ ablation cause problems in the activity of the neuronal circuit, but it could also lead to developmental issues. The migration of gonadotropin-releasing hormone (GnRH) neurons into the brain is guided by vomeronasal projections (Schwanzel-Fukuda, 1999). Thus, loss of the VNO basal layer could lead to aberrant GnRH circuit formation and further affect animal behavior. In a review, Yu also hypothesizes that deficient pheromone signaling during postnatal development could cause altered hormone levels and masculinization of the female brain. In support of this theory, he points out the increased levels of free testosterone detected in $Trpc2^{-/-}$ females (Yu, 2015). However, Kimchi et al., who performed the original study, did not support this idea but claimed that the testosterone levels are not high enough to cause a change in female behavior. Instead, they presented a theory in which the female brain would contain both female and male circuitry, where TRPC2 is part of a signaling pathway of gender-specific modulators that decide which of the neuronal circuits is activated (Kimchi et al., 2007). Further studies are needed to shed light on which hypothesis is accurate, especially since expression of $Trpc2$ has also been detected in tissues other than the VNO, for example, in the main olfactory epithelium, embryonic brain tissues, and murine erythroid precursors (Boisseau et al., 2009; Elg et al., 2007; Hirschler-Laszkiewicz et al., 2012; Omura and Mombaerts, 2014). As deletion of other signaling factors in the VNO does not give rise to masculine behavior in females or to male-male mounting, it has been suggested that TRPC2 in other organs might contribute to the observed phenotype (Yu, 2015). Although the VNO has an important function in most mammals, higher primates, including humans, do not possess this organ. However, as previously mentioned, humans are able to detect pheromones in a yet unknown manner, and further studies are needed to determine if any TRP family members are involved in this process.

11.5 CONCLUSIONS

One of the main functions of the TRP ion channels is serving as cellular sensors and conducting Ca^{2+} in response to different polymodal stimuli. TRP channels are modulated by a wide variety of effector molecules, and many cells express more than one type or even several splice isoforms of the same channel. Many TRP channels have been shown to be expressed in reproductive tissues of invertebrates, fish, birds, and mammals. Among these are TRPV1, TRPV5, TRPV6, TRPM8, TRPC1, TRPC2, and others. However, the ablation of TRP channels rarely leads to infertility, rather resulting in a subfertility phenotype. This could be explained by the fact that rather than having a regulatory role, many TRP channels exert mainly sensory functions. It is also possible that the absence of one type of TRP could be compensated by another, and in order to see a clear phenotypical effect on fertility, several TRP channels in a selective tissue must be ablated. Additional research that combines various techniques, including selective ablation of TRP channels in reproductive tissues, is pivotal to clearly elucidate the regulatory function of these proteins in male and female fertility, especially in mammals.

ACKNOWLEDGMENTS

The authors thank Rafael A. Fissore (University of Massachusetts, Amherst) for the insightful comments on the manuscript. This work was supported by Packer Wentz Endowment Will, Pew Scholars and Alfred P. Sloan awards, NIH R01GM111802, and R21HD081403 to P.V.L.

REFERENCES

Acott, T.S. and D.W. Carr. 1984. Inhibition of bovine spermatozoa by caudal epididymal fluid: II. Interaction of pH and a quiescence factor. *Biol Reprod*, 30(4): 926–935.

Allais-Bonnet, A. and E. Pailhoux. 2014. Role of the prion protein family in the gonads. *Front Cell Dev Biol*, 2: 56. doi:10.3389/fcell.2014.00056.

Arnoult, C. et al. 1999. Control of the low voltage-activated calcium channel of mouse sperm by egg ZP3 and by membrane hyperpolarization during capacitation. *Proc Natl Acad Sci U S A*, 96(12): 6757–6762.

Asuthkar, S. et al. 2015a. The TRPM8 protein is a testosterone receptor: II. Functional evidence for an ionotropic effect of testosterone on TRPM8. J Biol Chem, 290(5): 2670–2688. doi:10.1074/jbc.M114.610873.

Asuthkar, S. et al. 2015b. The TRPM8 protein is a testosterone receptor: I. Biochemical evidence for direct TRPM8-testosterone interactions. J Biol Chem, 290(5): 2659–2669. doi:10.1074/jbc.M114.610824.

Auzanneau, C. et al. 2003. A novel voltage-dependent chloride current activated by extracellular acidic pH in cultured rat Sertoli cells. *J Biol Chem*, 278(21): 19230–19236. doi:10.1074/jbc.M301096200.

Auzanneau, C. et al. 2008. Transient receptor potential vanilloid 1 (TRPV1) channels in cultured rat Sertoli cells regulate an acid sensing chloride channel. *Biochem Pharmacol*, 75(2): 476–483. doi:S0006-2952(07)00610-7.

Babcock, D.F. and D.R. Pfeiffer. 1987. Independent elevation of cytosolic [Ca^{2+}] and pH of mammalian sperm by voltage-dependent and pH-sensitive mechanisms. *J Biol Chem*, 262(31): 15041–15047.

Baibakov, B. et al. 2007. Sperm binding to the zona pellucida is not sufficient to induce acrosome exocytosis. *Development*, 134(5): 933–943. doi:10.1242/dev.02752.

Barritt, G. and G. Rychkov. 2005. TRPs as mechanosensitive channels. *Nat Cell Biol*, 7(2): 105–107. doi:10.1038/ncb0205-105.

Bernhardt, M.L. et al. 2015. CaV3.2 T-type channels mediate Ca^{2+} entry during oocyte maturation and following fertilization. *J Cell Sci*, 128(23): 4442–4452. doi:10.1242/jcs.180026.

Bidaux, G. et al. 2007. Prostate cell differentiation status determines transient receptor potential melastatin member 8 channel subcellular localization and function. *J Clin Invest*, 117(6): 1647–1657. doi:10.1172/JCI30168.

Boisseau, S. et al. 2009. Heterogeneous distribution of TRPC proteins in the embryonic cortex. *Histochem Cell Biol*, 131(3): 355–363. doi:10.1007/s00418-008-0532-6.

Brandenburger, T. et al. 2011. Switch of PMCA4 splice variants in bovine epididymis results in altered isoform expression during functional sperm maturation. *J Biol Chem*, 286(10): 7938–7946. doi:10.1074/jbc.M110.142836.

Brenker, C. et al. 2012. The CatSper channel: A polymodal chemosensor in human sperm. *EMBO J*, 31(7): 1654–1665. doi:10.1038/emboj.2012.30.

Cai, X., X. Wang, and D.E. Clapham. 2014. Early evolution of the eukaryotic Ca^{2+} signaling machinery: Conservation of the CatSper channel complex. *Mol Biol Evol*, 31(10):2735–2740. doi:10.1093/molbev/msu218.

Carvacho, I. et al. 2013. TRPV3 channels mediate strontium-induced mouse-egg activation. *Cell Rep*, 5(5): 1375–1386. doi:10.1016/j.celrep.2013.11.007.

Caterina, M.J. et al. 1997. The capsaicin receptor: A heat-activated ion channel in the pain pathway. *Nature*, 389(6653): 816–824. doi:10.1038/39807.

Clapham, D.E. 2007. Calcium signaling. *Cell*, 131(6): 1047–1058.

Clapham, D.E. et al. 2003. International Union of Pharmacology. XLIII. Compendium of voltage-gated ion channels: Transient receptor potential channels. *Pharmacol Rev*, 55(4): 591–596. doi:10.1124/pr.55.4.6.

Correia, S. et al. 2013. Sperm parameters and epididymis function in transgenic rats overexpressing the Ca^{2+}-binding protein regucalcin: A hidden role for Ca^{2+} in sperm maturation? *Mol Hum Reprod*, 19(9): 581–589. doi:10.1093/molehr/gat030.

Dacheux, J.L. and F. Dacheux. 2013. New insights into epididymal function in relation to sperm maturation. *Reproduction*, 147(2): R27–R42. doi:10.1530/REP-13-0420.

Darszon, A. et al. 2006. Sperm channel diversity and functional multiplicity. *Reproduction*, 131(6): 977–988. doi:10.1530/rep.1.00612.

De Blas, G.A. et al. 2009. TRPM8, a versatile channel in human sperm. *PLOS ONE*, 4(6): e6095. doi:10.1371/journal.pone.0006095.

de Kretser, D.M. et al. 1998. Spermatogenesis. *Hum Reprod*, 13(Suppl 1): 1–8.

Elg, S. et al. 2007. Cellular subtype distribution and developmental regulation of TRPC channel members in the mouse dorsal root ganglion. *J Comp Neurol*, 503(1): 35–46. doi:10.1002/cne.21351.

Escoffier, J. et al. 2015. Subcellular localization of phospholipase Czeta in human sperm and its absence in DPY19L2-deficient sperm are consistent with its role in oocyte activation. *Mol Hum Reprod* 21 (2):157-168. doi: 10.1093/molehr/gau098.

Esposito, G. et al. 2004. Mice deficient for soluble adenylyl cyclase are infertile because of a severe sperm-motility defect. *Proc Natl Acad Sci U S A*, 101(9): 2993–2998. doi:10.1073/pnas.0400050101.

Florman, H.M., M.K. Jungnickel, and K.A. Sutton. 2008. Regulating the acrosome reaction. *Int J Dev Biol*, 52(5–6): 503–510. doi:10.1387/ijdb.082696hf.

Florman, H.M. et al. 1989. An adhesion-associated agonist from the zona pellucida activates G protein-promoted elevations of internal Ca^{2+} and pH that mediate mammalian sperm acrosomal exocytosis. *Dev Biol*, 135(1): 133–146.

Garraway, L.A. et al. 2003. Intermediate basal cells of the prostate: In vitro and in vivo characterization. *Prostate*, 55(3): 206–218. doi:10.1002/pros.10244.

Grimaldi, P. et al. 2009. The endocannabinoid system and pivotal role of the CB2 receptor in mouse spermato-genesis. *Proc Natl Acad Sci U S A*, 106(27): 11131–11136. doi:10.1073/pnas.0812789106.

Hai, Y. et al. 2014. The roles and regulation of Sertoli cells in fate determinations of spermatogonial stem cells and spermatogenesis. *Semin Cell Dev Biol*, 29: 66–75. doi:10.1016/j.semcdb.2014.04.007.

Hasen, N.S. and S.C. Gammie. 2009. *Trpc2* gene impacts on maternal aggression, accessory olfactory bulb anat-omy and brain activity. *Genes, Brain, and Behav*, 8(7): 639–649. doi:10.1111/j.1601-183X.2009.00511.x.

Heifetz, Y., J. Yu, and M.F. Wolfner. 2001. Ovulation triggers activation of *Drosophila* oocytes. *Dev Biol*, 234(2): 416–424. doi:10.1006/dbio.2001.0246.

Hess, K.C. et al. 2005. The "soluble" adenylyl cyclase in sperm mediates multiple signaling events required for fertilization. *Dev Cell*, 9(2): 249–259.

Hirschler-Laszkiewicz, I. et al. 2012. Trpc2 depletion protects red blood cells from oxidative stress-induced hemolysis. *Exp Hematol*, 40(1): 71–83. doi:10.1016/j.exphem.2011.09.006.

Hoenderop, J.G. et al. 2003. Renal Ca^{2+} wasting, hyperabsorption, and reduced bone thickness in mice lacking TRPV5. *J Clin Invest*, 112(12): 1906–1914. doi:10.1172/JCI19826.

Horner, V.L. and M.F. Wolfner. 2008. Mechanical stimulation by osmotic and hydrostatic pressure activates *Drosophila* oocytes in vitro in a calcium-dependent manner. *Dev Biol*, 316(1): 100–109. doi:10.1016/j.ydbio.2008.01.014.

Igusa, Y. and S. Miyazaki. 1983. Effects of altered extracellular and intracellular calcium concentration on hyperpolarizing responses of the hamster egg. *J Physiol*, 340: 611–632.

Igusa, Y., S. Miyazaki, and N. Yamashita. 1983. Periodic hyperpolarizing responses in hamster and mouse eggs fertilized with mouse sperm. *J Physiol*, 340: 633–647.

Inaba, K. 2011. Sperm flagella: Comparative and phylogenetic perspectives of protein components. *Mol Hum Reprod*, 17(8): 524–538. doi:10.1093/molehr/gar034.

Jaffe, L.A., A.P. Sharp, and D.P. Wolf. 1983. Absence of an electrical polyspermy block in the mouse. *Dev Biol*, 96(2): 317–323.

Jenkins, A.D., C.P. Lechene, and S.S. Howards. 1980. Concentrations of seven elements in the intraluminal fluids of the rat seminiferous tubules, rate testis, and epididymis. *Biol Reprod*, 23(5): 981–987.

Jones, R. 2002. Plasma membrane composition and organization during maturation of spermatozoa in the epididymis. In *The Epididymis: From Molecules to Clinical Practice: A Comprehensive Survey of the Efferent Ducts, the Epididymis and the Vas Deferens*, edited by B. Robaire and B.T. Hinton, 405–416. Boston, MA: Springer

Jungnickel, M.K. et al. 2001. Trp2 regulates entry of Ca^{2+} into mouse sperm triggered by egg ZP3. *Nat Cell Biol*, 3(5): 499–502.

Kaneuchi, T. et al. 2015. Calcium waves occur as *Drosophila* oocytes activate. *Proc Natl Acad Sci U S A*, 112(3): 791–796. doi:10.1073/pnas.1420589112.

Kashir, J. et al. 2013. Comparative biology of sperm factors and fertilization-induced calcium signals across the animal kingdom. *Mol Reprod Dev*, 80(10): 787–815. doi:10.1002/mrd.22222.

Kim, S. et al. 2012. Paradoxical contribution of SK3 and GIRK channels to the activation of mouse vomeronasal organ. *Nat Neurosci*, 15(9): 1236–1244. doi:10.1038/nn.3173.

Kimchi, T., J. Xu, and C. Dulac. 2007. A functional circuit underlying male sexual behaviour in the female mouse brain. *Nature*, 448(7157): 1009–1014.

Kirichok, Y., B. Navarro, and D.E. Clapham. 2006. Whole-cell patch-clamp measurements of spermatozoa reveal an alkaline-activated Ca^{2+} channel. *Nature*, 439(7077): 737–740. doi:10.1038/nature04417.

Kono, T. et al. 1996. A cell cycle-associated change in Ca^{2+} releasing activity leads to the generation of Ca^{2+} transients in mouse embryos during the first mitotic division. *J Cell Biol*, 132(5): 915–923.

Kumar, A. et al. 2016. TRPV4 is endogenously expressed in vertebrate spermatozoa and regulates intracellular calcium in human sperm. *Biochem Biophys Res Commun*, 473(4): 781–788. doi:10.1016/j.bbrc.2016.03.071.

Lalevee, N. and M. Joffre. 1999. Inhibition by cAMP of calcium-activated chloride currents in cultured Sertoli cells from immature testis. *J Membr Biol*, 169(3): 167–174.

Lalevee, N., F. Pluciennik, and M. Joffre. 1997. Voltage-dependent calcium current with properties of T-type current in Sertoli cells from immature rat testis in primary cultures. *Biol Reprod*, 56(3): 680–687.

Lee, H. et al. 2014. Scalable control of mounting and attack by Esr1+ neurons in the ventromedial hypothalamus. *Nature*, 509(7502): 627–632. doi:10.1038/nature13169.

Lee, H.C. et al. 2016. TRPV3 channels mediate Ca^{2+} influx induced by 2-APB in mouse eggs. *Cell Calcium*, 59(1): 21–31. doi:10.1016/j.ceca.2015.12.001.

Levine, N. and H. Kelly. 1978. Measurement of pH in the rat epididymis in vivo. *J Reprod Fertil*, 52(2): 333–335.

Levine, N. and D.J. Marsh. 1971. Micropuncture studies of the electrochemical aspects of fluid and electrolyte transport in individual seminiferous tubules, the epididymis and the vas deferens in rats. *J Physiol*, 213(3): 557–570.

Leypold, B.G. et al. 2002. Altered sexual and social behaviors in trp2 mutant mice. *Proc Natl Acad Sci U S A*, 99(9): 6376–6381. doi:10.1073/pnas.082127599.

Li, C. et al. 2009. Production of normal mice from spermatozoa denatured with high alkali treatment before ICSI. *Reproduction*, 137(5): 779–792. doi:10.1530/REP-08-0476.

Lishko, P.V. et al. 2012. The control of male fertility by spermatozoan ion channels. *Annu Rev Physiol*, 74: 453–475. doi:10.1146/annurev-physiol-020911-153258.

Liu, Y., D.K. Wang, and L.M. Chen. 2012. The physiology of bicarbonate transporters in mammalian reproduction. *Biol Reprod*, 86(4): 99. doi:10.1095/biolreprod.111.096826.

Maccarrone, M. et al. 2005. Characterization of the endocannabinoid system in boar spermatozoa and implications for sperm capacitation and acrosome reaction. *J Cell Sci*, 118(Pt19): 4393–4404. doi:10.1242/jcs.02536.

Mann, T. and C. Lutwak-Mann. 1981. *Male Reproductive Function and Semen*. Vol. 1. New York, NY: Springer-Verlag.

McClintock, M.K. 1971. Menstrual synchorony and suppression. *Nature*, 229(5282): 244–245.

McKemy, D.D., W.M. Neuhausser, and D. Julius. 2002. Identification of a cold receptor reveals a general role for TRP channels in thermosensation. *Nature*, 416(6876): 52–58. doi:10.1038/nature719.

Miao, Y.L. et al. 2012. Calcium influx-mediated signaling is required for complete mouse egg activation. *Proc Natl Acad Sci U S A*, 109(11): 4169–4174. doi:10.1073/pnas.1112333109.

Minke, B. 2010. The history of the *Drosophila* TRP channel: The birth of a new channel superfamily. *J Neurogenet*, 24(4): 216–233. doi:10.3109/01677063.2010.514369.

Miyazaki, S. 2006. Thirty years of calcium signals at fertilization. *Semin Cell Dev Biol*, 17(2): 233–243.

Mizrak, S.C. and F.M. van Dissel-Emiliani. 2008. Transient receptor potential vanilloid receptor-1 confers heat resistance to male germ cells. *Fertil Steril*, 90(4): 1290–1293. doi:10.1016/j.fertnstert.2007.10.081.

Mizrak, S.C. et al. 2008. Spermatogonial stem cell sensitivity to capsaicin: An in vitro study. *Reprod Biol Endocrinol: RB&E*, 6: 52. doi:10.1186/1477-7827-6-52.

Mizuno, K. et al. 2012. Calaxin drives sperm chemotaxis by Ca^{2+}-mediated direct modulation of a dynein motor. *Proc Natl Acad Sci U S A*, 109(50): 20497–20502. doi:10.1073/pnas.1217018109.

Muralidhara and K. Narasimhamurthy. 1988. Non-mutagenicity of capsaicin in albino mice. *Food Chem Toxicol*, 26(11–12): 955–958.

Nagabhushan, M. and S.V. Bhide. 1985. Mutagenicity of chili extract and capsaicin in short-term tests. *Environ Mutagen*, 7(6): 881–888.

Neill, A.T., G.W. Moy, and V.D. Vacquier. 2004. Polycystin-2 associates with the polycystin-1 homolog, suREJ3, and localizes to the acrosomal region of sea urchin spermatozoa. *Mol Reprod Dev*, 67(4): 472–477. doi:10.1002/mrd.20033.

Neto, F.T. et al. 2016. Spermatogenesis in humans and its affecting factors. *Semin Cell Dev Biol*, 59: 10–26.

Nieto-Posadas, A. et al. 2012. Lysophosphatidic acid directly activates TRPV1 through a C-terminal binding site. *Nat Chem Biol*, 8(1): 78–85. doi:10.1038/nchembio.712.

Oliveira, P.F. et al. 2009. Intracellular pH regulation in human Sertoli cells: Role of membrane transporters. *Reproduction*, 137(2): 353–359. doi:10.1530/REP-08-0363.

Omura, M. and P. Mombaerts. 2014. Trpc2-expressing sensory neurons in the main olfactory epithelium of the mouse. *Cell Rep*, 8(2): 583–595. doi:10.1016/j.celrep.2014.06.010.

Owsianik, G. et al. 2006. Structure-function relationship of the TRP channel superfamily. *Rev Physiol Biochem Pharmacol*, 156: 61–90.

Paniagua, R. and M. Nistal. 1984. Morphological and histometric study of human spermatogonia from birth to the onset of puberty. *J Anat*, 139(Pt 3): 535–552.

Peng, J.B. et al. 1999. Molecular cloning and characterization of a channel-like transporter mediating intestinal calcium absorption. *J Biol Chem*, 274(32): 22739–22746.

Prakriya, M. and R.S. Lewis. 2015. Store-operated calcium channels. *Physiol Rev*, 95(4): 1383–1436. doi:10.1152/physrev.00020.2014.

Publicover, S., C.V. Harper, and C. Barratt. 2007. [Ca^{2+}]i signalling in sperm—Making the most of what you've got. *Nat Cell Biol*, 9(3): 235–242. doi:10.1038/ncb0307-235.

Ramadan, W.M. et al. 2012. Oocyte activation and phospholipase C zeta (PLCzeta): Diagnostic and therapeutic implications for assisted reproductive technology. *Cell Commun Signal*, 10(1): 12-811X-10–12. doi:10.1186/1478-811X-10-12.

Ramsey, I.S., M. Delling, and D.E. Clapham. 2006. An introduction to TRP channels. *Annu Rev Physiol*, 68: 619–647. doi:10.1146/annurev.physiol.68.040204.100431.

Ren, D. et al. 2001. A sperm ion channel required for sperm motility and male fertility. *Nature*, 413(6856): 603–609. doi:10.1038/35098027.

Rossi, G. et al. 2007. Follicle-stimulating hormone activates fatty acid amide hydrolase by protein kinase A and aromatase-dependent pathways in mouse primary Sertoli cells. *Endocrinology*, 148(3): 1431–1439.

Sartain, C.V. and M.F. Wolfner. 2013. Calcium and egg activation in *Drosophila*. *Cell Calcium*, 53(1): 10–15. doi:10.1016/j.ceca.2012.11.008.

Saunders, C. M. et al. 2002. PLC zeta: a sperm-specific trigger of Ca(2+) oscillations in eggs and embryo development. *Development (Cambridge, England)*, 129(15): 3533-3544.

Schwanzel-Fukuda, M. 1999. Origin and migration of luteinizing hormone-releaing hormone neurons in mammals. *Microsc Res Tech*, 44(1): 2–10. doi:10.1002/(SICI)1097-0029(19990101)44:1<2::AID-JEMT2>3.0.CO;2-4.

Sipilä, P. and I. Björkgren. 2016. Segment-specific regulation of epididymal gene expression. *Reproduction*, 152: R91–R99.

Stein, R.J. et al. 2004. Cool (TRPM8) and hot (TRPV1) receptors in the bladder and male genital tract. *J Urol*, 172(3): 1175–1178.

Stowers, L. et al. 2002. Loss of sex discrimination and male-male aggression in mice deficient for TRP2. *Science*, 295(5559): 1493–1500. doi:10.1126/science.1069259.

Swann, K. and F.A. Lai. 2013. PLCzeta and the initiation of Ca^{2+} oscillations in fertilizing mammalian eggs. *Cell Calcium*, 53(1): 55–62. doi:10.1016/j.ceca.2012.11.001.

Takahashi, T. et al. 2013. Ca^{2+} influx-dependent refilling of intracellular Ca^{2+} stores determines the frequency of Ca^{2+} oscillations in fertilized mouse eggs. *Biochem Biophys Res Commun*, 430(1): 60–65. doi:10.1016/j.bbrc.2012.11.024.

Takayama, J. and S. Onami. 2016. The sperm TRP-3 channel mediates the onset of a Ca^{2+} wave in the fertilized *C. elegans* oocyte. *Cell Rep*, 15(3): 625–637. doi:10.1016/j.celrep.2016.03.040.

Taranta, A. et al. 1997. ω-Conotoxin-sensitive Ca^{2+} voltage-gated channels modulate protein secretion in cultured rat Sertoli cells. *Mol Cell Endocrinol*, 126(2): 117–123.

Tirindelli, R. et al. 2009. From pheromones to behavior. *Physiol Rev*, 89(3): 921–956. doi:10.1152/physrev.00037.2008.

Tomashov-Matar, R. et al. 2005. Strontium-induced rat egg activation. *Reproduction*, 130(4): 467–474.

Tominaga, M. et al. 1998. The cloned capsaicin receptor integrates multiple pain-producing stimuli. *Neuron*, 21(3): 531–543.

Trevino, C.L. et al. 2001. Identification of mouse trp homologs and lipid rafts from spermatogenic cells and sperm. *FEBS Lett*, 509(1): 119–125.

Tsavaler, L. et al. 2001. Trp-p8, a novel prostate-specific gene, is up-regulated in prostate cancer and other malignancies and shares high homology with transient receptor potential calcium channel proteins. *Cancer Res*, 61(9): 3760–3769.

Valero, M.L. et al. 2012. TRPM8 ion channels differentially modulate proliferation and cell cycle distribution of normal and cancer prostate cells. *PLOS ONE*, 7(12): e51825. doi:10.1371/journal.pone.0051825.

Virtanen, H.E. and J. Toppari. 2015. Cryptorchidism and fertility. *Endocrinol Metab Clin North Am*, 44(4): 751–760. doi:10.1016/j.ecl.2015.07.013.

Visconti, P.E. et al. 2011. Ion channels, phosphorylation and mammalian sperm capacitation. *Asian J Androl*, 13(3): 395–405. doi:10.1038/aja.2010.69.

Wakai, T. and R.A. Fissore. 2013. Ca^{2+} homeostasis and regulation of ER Ca^{2+} in mammalian oocytes/eggs. *Cell Calcium*, 53(1): 63–67. doi:10.1016/j.ceca.2012.11.010.

Wandernoth, P.M. et al. 2010. Role of carbonic anhydrase IV in the bicarbonate-mediated activation of murine and human sperm. *PLOS ONE*, 5(11): e15061.

Wang, C. and Z. Machaty. 2013. Calcium influx in mammalian eggs. *Reproduction*, 145(4): R97–R105. doi:10.1530/REP-12-0496.

Wang, C. et al. 2012. Orai1 mediates store-operated Ca^{2+} entry during fertilization in mammalian oocytes. *Dev Biol*, 365(2): 414–423. doi:10.1016/j.ydbio.2012.03.007.

Weissgerber, P. et al. 2011. Male fertility depends on Ca^{2+} absorption by TRPV6 in epididymal epithelia. *Sci Signal*, 4(171): ra27. doi:10.1126/scisignal.2001791.

Weissgerber, P. et al. 2012. Excision of *Trpv6* gene leads to severe defects in epididymal Ca^{2+} absorption and male fertility much like single D541A pore mutation. *J Biol Chem*, 287(22): 17930–17941. doi:10.1074/jbc.M111.328286.

Wennemuth, G., D.F. Babcock, and B. Hille. 2003. Calcium clearance mechanisms of mouse sperm. *J Gen Physiol*, 122(1): 115–128. doi:10.1085/jgp.200308839.

Wennemuth, G. et al. 2003. Bicarbonate actions on flagellar and Ca^{2+}-channel responses: Initial events in sperm activation. *Development*, 130(7): 1317–1326.

Wes, P.D. et al. 1995. TRPC1, a human homolog of a *Drosophila* store-operated channel. *Proc Natl Acad Sci U S A*, 92(21): 9652–9656.

Whitaker, M. 2006. Calcium at fertilization and in early development. *Physiol Rev*, 86(1): 25–88.

Winston, N.J. et al. 1995. The exit of mouse oocytes from meiotic M-phase requires an intact spindle during intracellular calcium release. *J Cell Sci*, 108(Pt 1): 143–151.

Wissenbach, U. et al. 1998. Structure and mRNA expression of a bovine trp homologue related to mammalian trp2 transcripts. *FEBS Lett*, 429(1): 61–66.

Wu, L.J., T.B. Sweet, and D.E. Clapham. 2010. International Union of Basic and Clinical Pharmacology. LXXVI. Current progress in the mammalian TRP ion channel family. *Pharmacol Rev*, 62(3): 381–404. doi:10.1124/pr.110.002725.

Xu, X. Z. and P. W. Sternberg. 2003. A C. elegans sperm TRP protein required for sperm-egg interactions during fertilization. *Cell*, 114(3): 285-297. doi: S0092867403005658.

Yanagimachi, R. and N. Usui. 1974. Calcium dependence of the acrosome reaction and activation of guinea pig spermatozoa. *Exp Cell Res*, 89(1): 161–174.

Yu, C.R. 2015. TRICK or TRP? What $Trpc2^{-/-}$ mice tell us about vomeronasal organ mediated innate behaviors. *Front Neurosci*, 9: 221. doi:10.3389/fnins.2015.00221.

Zygmunt, P.M. et al. 1999. Vanilloid receptors on sensory nerves mediate the vasodilator action of anandamide. *Nature*, 400(6743): 452–457. doi:10.1038/22761.

Zygmunt, P.M. et al. 2013. Monoacylglycerols activate TRPV1—A link between phospholipase C and TRPV1. *PLOS ONE*, 8(12): e81618. doi:10.1371/journal.pone.0081618.

12 The Role of TRP Channels in the Pancreatic Beta-Cell

Koenraad Philippaert and Rudi Vennekens

CONTENTS

12.1 TRP CHANNELS IN DIABETES

Diabetes is the most common metabolic disease in humans. The hallmark of diabetes is an elevated blood glucose concentration, but this abnormality is just one of many biochemical and physiological alterations that occur in the disease. Both genetic and environmental factors contribute to its pathogenesis, which involves insufficient insulin secretion, reduced responsiveness to endogenous or exogenous insulin, increased glucose production, and/or abnormalities in fat and protein metabolism. Diabetes is not one distinct disorder, but it can arise as a result of numerous defects in regulation of the synthesis, secretion, and action of insulin. In type I diabetes, idiopathic or immune-mediated β-cell destruction lies at the basis of complete insulin deficiency (Thompson Coon, 2010). In type II diabetes, the cause of hyperglycemia is more complex. It ranges from predominantly insulin resistance with relative insulin deficiency to a predominantly insulin secretory defect with insulin resistance. There are several genetic defects that are risk factors toward the development of type II diabetes such as functional alterations of transient receptor potential (TRP) channels. Furthermore, there is a connection between old age, a high-calorie diet, and diabetes type II (Herder and Roden, 2011). Mice studies revealed that feeding a high-fat diet leads to gradual development of obesity (Hariri and Thibault, 2010), glucose intolerance through insulin resistance, and ultimately diabetes (Winzell and Ahrén, 2004). A subset of type II diabetic patients develops hyperglycemia secondary to hypoinsulinemia instead of peripheral insulin resistance (Ashcroft and Rorsman, 2013; Pende et al., 2000; Phillips, 1996). This can be induced by either β-cell dysfunction or loss of functional β-cell mass (Peiris et al., 2012). Patients with type II diabetes are usually treated with pharmacologic agents in combination with lifestyle modifications (Lebovitz, 2011). The used pharmaceuticals fall mainly into three categories: those that stimulate insulin secretion, those that reduce peripheral insulin resistance, and those that mimic insulin action.

Sulfonylureas, which block K_{ATP} channels independent of glucose metabolism, are commonly used insulin secretagogues but have the associated risk of inducing hypoglycemia and weight gain if incorrectly dosed. As obesity is an important risk factor for type II diabetes, weight neutrality of insulin secretagogue medication is an important factor. More recently, incretin mimetics became available for the treatment of type II diabetes. These enhance the secretion of insulin in response to nutrient intake but have the adverse effects of nausea, headache, and increased risks of hypoglycemia, and they have a poor patient compliance (Joy, Rodgers, and Scates, 2005). Clearly, there is still a need for improvement of diabetes therapies, with new pharmaceuticals that are safe to use and have limited adverse effects.

12.2 TRPS IN PANCREATIC BETA-CELLS

Insulin secretion is regulated by postprandial glucose levels, and many modulating signaling pathways are present in the pancreatic β-cell. An increasing amount of data point to a role for TRP channels in pancreatic β-cells, and the regulation of insulin secretion. These data suggest that TRP channels might be interesting drug targets for the treatment of diabetes and that they could be involved in the development of this important disease (Figure 12.1).

Modulation of TRP channel activity can lead to increased production and/or secretion of insulin. Apart from the pancreatic pathology, diabetes is also associated with the development of neuropathy, vasculopathy, and nephropathy. Considering the wide expression pattern of TRP channels, it is not surprising that TRP channels have also been associated with these aspects of diabetes, and increasing data indicate that their potential as drug targets for the treatment of diabetic comorbidities is gaining in importance.

TRP channels have been described both in primary β-cells and insulin-secreting cell lines (Marigo et al., 2009; Colsoul, Vennekens, and Nilius, 2011). TRPV2, TRPV4, TRPC1, TRPC4,

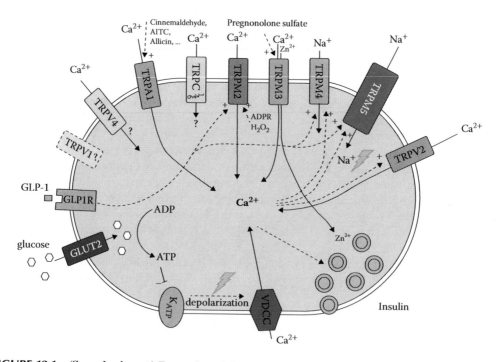

FIGURE 12.1 **(See color insert.)** Expression of different TRP channels in the β-cell, important regulatory molecules and their action on the cell.

TRPC6, TRPM2, TRPM3, TRPM4, and TRPM5 are identified in insulinoma cell lines such as MIN6 (mouse) or INS-1 (rat). Expression of TRPV2, TRPV4, TRPC1, TRPM2, TRPM3, TRPM4, and TRPM5 was reported in mouse islets and TRPM2, TRPV5, TRPC1, and TRPC4 in rat islets or β-cells. Interestingly, no expression could be found in mouse islets for TRPV5, TRPM1, TRPM8, or TRPP3 (Colsoul, Nilius, and Vennekens, unpublished) nor for TRPV1 in rat pancreatic β-cells (Diaz-Garcia et al., 2014). Unfortunately, sparse data are available from human tissue: TRPV5 and TRPV6 are detected in human pancreas, and TRPM2, TRPM4, and TRPM5 transcripts are detected in human islets (Marigo et al., 2009; Qian et al., 2002; Xu et al., 2001).

The pancreatic islet is a complex structure consisting of five different cell types: glucagon-releasing α-cells, insulin-secreting β-cells (which constitute 65%–90% of the islet cell population), somatostatin-producing δ-cells, polypeptide-containing PP-cells, and ghrelin-secreting ε-cells. Insulin is synthesized and secreted into the blood by the β-cells mainly in response to glucose but also in response to other nutrients (such as amino acids and fatty acids), hormones (e.g., the incretin hormones glucagon-like peptide-1 [GLP-1] and gastric inhibitory polypeptide [GIP]) and neurotransmitters (e.g., acetylcholine, ACh) (Winzell and Ahren, 2007; Newsholme, Gaudel, and McClenaghan, 2010). The plasma glucose level is tightly controlled, despite a huge variety of glucose consumption and utilization.

Insulin is the only hormone capable of lowering the plasma glucose level and is secreted upon food ingestion and nutrient absorption. The secretion of insulin by the pancreatic β-cell is a complex process driven by electrical activity and oscillations of the intracellular Ca^{2+} concentration ($[Ca^{2+}]_i$) (MacDonald and Rorsman, 2006). Briefly, glucose enters the β-cell via the high-affinity glucose transporter (GLUT-2), and glucose metabolism increases intracellular ATP levels, which closes ATP-sensitive K^+ channels. This increases the input resistance of the β-cell, allowing a small inward current, whose molecular identity is still elusive, to generate a significant depolarization (Henquin et al., 2009; Henquin, 2009; Rorsman and Trube, 1985). Notably, TRP channels were often hypothesized to be interesting candidates for this background depolarizing current (Jacobson and Philipson, 2007; Drews, Krippeit-Drews, and Düfer, 2010). What follows is a typical pattern of bursts of action potentials from a depolarized plateau, and parallel oscillatory increases of intracellular Ca^{2+} ($[Ca^{2+}]_i$). This pattern results from a complex interplay between ATP-sensitive K^+ channels, voltage-dependent Ca^{2+} and K^+ channels, and the cellular metabolism of the β-cell (Drews, Krippeit-Drews, and Düfer, 2010; Bertram, Sherman, and Satin, 2007; Ashcroft and Rorsman, 1989). Finally, increased $[Ca^{2+}]_i$ triggers exocytosis of insulin-containing vesicles (Lemmens et al., 2001).

12.2.1 TRPCs

TRPC channels are nonselective Ca^{2+}-permeable cation channels, with the selectivity ratio $P_{Ca^{2+}} / P_{Na}$ varying significantly between the different family members (Gees, Colsoul, and Nilius, 2010). TRPC channels are widely expressed, and their characterization is complicated by the possible occurrence of heterotetramers. TRPC channels have been implemented to play a role in store-operated Ca^{2+} entry (Cheng et al., 2013). The calcium homeostasis is important in the regulation of the insulin-secretion from pancreatic β-cells (Sabourin et al., 2015). TRPC1 mRNA could be detected in INS-1 cells and rat β-cells, in mouse islets and MIN6 cells, whereas it could not be detected in another mouse insulinoma cell line βTC3 (Sakura and Ashcroft, 1997; Li and Zhang, 2009). Expression of TRPC1 was found in whole human pancreas (Riccio et al., 2002). Four transcripts of mouse TRPC1 mRNA, representing different splice variants, have been found in mouse islets and the β-cell line MIN6 (Sakura and Ashcroft, 1997). Polymorphisms in the TRPC1 gene have recently been associated with development of type 2 diabetes mellitus (T2DM) (Chen et al., 2013). However, the influence of TRPC1 on insulin release still needs to be determined.

Equally little is known about other TRPC channels in insulin release. TRPC3 is activated downstream of the G protein-coupled receptor (GPCR), GPR40. A nonselective cation current is observed in murine β-cells, which is abolished upon perfusion with the TRPC3 antagonist pyrazole-3. This current contributes to an increased glucose-stimulated insulin secretion (Yamada et al., 2016). TRPC4 could be detected in βTC3, INS-1 cells, rat β-cells, mouse islets, and human pancreas (Li and Zhang, 2009; Riccio et al., 2002; Freichel et al., 2004a; Roe et al., 1998a). However, analysis of blood glucose homeostasis by glucose tolerance tests did not reveal differences between WT and *Trpc4*-deficient mice both regarding basal glucose levels under fasting conditions as well as following intraperitoneal glucose challenge (Freichel et al., 2004b). Since TRPC4 is activated by the phospholipase C pathway, it might be possible that the channel is involved in ACh- or glucagon-induced amplification of insulin release. TRPC6 transcripts have been detected, although to a low level, in βTC-3 insulin-secreting cells. Thus, it is clear that more research is needed in order to clarify the possible function of TRPC channels in the insulin release of the pancreatic β-cell. It has been suggested that TRPC channels mediate the unknown depolarizing current that accounts for the Ca^{2+} release–activated cation current characterized earlier in βTC3 cells (Roe et al., 1998b).

12.2.2   TRPM2

TRPM2 is a nonselective Ca^{2+}-permeable cation channel fused C-terminally to an enzymatic ADP-ribose pyrophosphatase domain (Perraud et al., 2001). The channel is expressed in insulin-secreting cell lines, such as the rat cell lines CRI-G1 and RIN-5F, and in human and mouse pancreatic islets (Qian et al., 2002; Hara et al., 2002; Inamura et al., 2003; Togashi et al., 2006). Moreover, the channel coexpresses with insulin but not with glucagon, indicating expression in β-cells (Togashi et al., 2006). TRPM2 is thought to be activated by various stimuli, including adenine dinucleotides (ADPR, cADPR, NAADP, β-NAD), reactive oxygen species (ROS) such as H_2O_2 and OH^-, and intracellular Ca^{2+} (Perraud et al., 2001; Hara et al., 2002; Du, Xie, and Yue, 2009; Sano et al., 2001). However, it was recently shown that, at least, cADPR, NAADP, and NAD do not directly activate this channel (Tóth, Iordanov, and Csanády, 2015; Rosenbaum, 2015). A current with TRPM2-like properties could be detected in the rat insulinoma cell line CRI-G1, in INS-1 cells, and in primary mouse β-cells. In $Trpm2^{-/-}$ primary β-cells, no ADPR-elicited current could be detected, suggesting that TRPM2 is natively expressed and forms a functional channel in β-cells (Inamura et al., 2003; Lange et al., 2009).

TRPM2 could contribute to insulin release induced by heat, glucose, and incretin hormones (Togashi et al., 2006; Uchida et al., 2011). Indeed, forskolin- (an activator of adenylyl cyclase) and exendin-4- (a GLP-1 receptor agonist) induced insulin release from rat pancreatic islets was significantly reduced in islets after shRNA-mediated knockdown of TRPM2 expression (Togashi et al., 2006). Furthermore, 2-aminoethoxydiphenyl borate (2-APB), a rapid and reversible inhibitor of TRPM2, inhibits both heat- and exendin-4-evoked insulin release from rat pancreatic islets (Togashi, Inada, and Tominaga 2008). These indications are further substantiated in the $Trpm2^{-/-}$ mouse: insulin secretion induced by glucose and GLP-1 is impaired in $Trpm2$-deficient islets, whereas the response to tolbutamide, a K_{ATP} channel inhibitor, is unchanged (Uchida et al., 2011). This results in higher basal glucose levels and an impaired glucose tolerance in $Trpm2$-deficient mice (Uchida et al., 2011). The impairment of insulin secretion is caused by reduced increases in $[Ca^{2+}]_i$, indicating that TRPM2 mediates Ca^{2+} influx upon glucose- and/or GLP-1 stimulation. However, the situation might be more complex, since glucose-stimulated insulin secretion evoked by diazoxide and high K^+ (conditions designed to clamp intracellular Ca^{2+} and inactivate the K_{ATP} channel-mediated pathways) was lost in Trpm2-deficient islets. Since the intracellular Ca^{2+} under these conditions was not altered between wild-type and $Trpm2^{-/-}$ islets, these data suggest that TRPM2 mediates insulin secretion independent of its role as a Ca^{2+} entry channel (Uchida et al., 2011).

TRPM2 is important in β-cell apoptosis, a feature linked to the activation of the channel by H_2O_2 and OH^-. These reactive oxygen species that are produced by oxidative stress are thought to play a central role in β-cell death and the development of type 1 and type 2 diabetes (Mandrup-Poulsen,

2003; Rhodes, 2005). Indeed, activation of TRPM2 by H_2O_2 has been shown to mediate Ca^{2+} influx and β-cell death in a rat β-cell line RIN-5F that natively expresses TRPM2 (Hara et al., 2002; Ishii et al., 2006). Moreover, INS-1 cells with suppressed TRPM2 expression are 72% less affected by H_2O_2-induced cell death (Lange et al., 2009). The H_2O_2-induced Ca^{2+} influx is thought to be mediated by increasing levels of NAD^+ or ADP-ribose (Perraud et al., 2001; Hara et al., 2002).

Finally, TRPM2 has been reported to have an additional role as an intracellular Ca^{2+} release channel in pancreatic β-cells (Lange et al., 2009). Indeed, internally applied ADPR gives rise to a single Ca^{2+} transient both in INS-1 and in primary mouse β-cells, and this effect was completely abolished in $Trpm2^{-/-}$ primary mouse β-cells. Furthermore, TRPM2 colocalizes with lysosome-associated membrane protein-1 (LAMP1), a specific marker for lysosomes. In agreement with this, ADPR-induced intracellular Ca^{2+} release was abolished in INS-1 cells treated with bafilomycin A, a macrolide antibiotic that empties lysosomal calcium stores without affecting ER stores (Bowman, Siebers, and Altendorf, 1988). These data indicate that ADPR-dependent TRPM2-mediated Ca^{2+} release occurs predominantly from a lysosomal store.

12.2.3 TRPM3

The TRPM3 gene encodes for different TRPM3 isoforms due to alternative splicing and exon usage, leading to channels with divergent pore and gating properties (Oberwinkler et al., 2005). We refer here only to TRPM3α2 (1709 amino acids: the pore lacks 12 aa in comparison to the longest form TRPM3α1 of 1721 aa). Interestingly, the TRPM3 gene contains a miRNA sequence (miR-204) in intron 8, which may regulate a variety of target genes at the transcriptional level (Weber, 2005; Oberwinkler and Philipp, 2014). TRPM3 is expressed in a variety of neuronal and nonneuronal tissue (Oberwinkler and Philipp, 2014), including whole pancreas (Grimm et al., 2003; Fonfria et al., 2006), INS-1 cells, and mouse pancreatic islets (Wagner et al., 2008; Klose et al., 2011). TRPM3 channels are directly activated by the neuro-steroid hormone pregnenolone-sulfate (PS). Pancreatic β-cells and INS-1 cells express PS-sensitive channels that share several pharmacological and biophysical properties of recombinant TRPM3 channels (such as sensitivity to nifedipine and block by monovalent cations) (Wagner et al., 2008). Moreover, PS elicits a large Ca^{2+} increase in INS-1 cells and pancreatic islets, an action dependent on TRPM3 expression. This PS-induced Ca^{2+} increase could be blocked by the TRPM3 blocker mefenamic acid in INS-1E cells and mouse pancreatic islets (Klose et al., 2011). Remarkably, mefenamic acid did not block glucose or tolbutamide-induced Ca^{2+} increase, indicating that TRPM3 is not involved in the K_{ATP}-dependent Ca^{2+} signaling of the β-cell. PS did increase, however, glucose-induced insulin secretion from pancreatic islets, an effect abolished by mefenamic acid (Wagner et al., 2008; Klose et al., 2011). A synthetic TRPM3-activating ligand (CIM0216) has recently been described. Upon application of CIM0216, there is a Ca^{2+} increase in the β-cell and a stimulation of the insulin secretion (Held et al., 2015). This proves that TRPM3 is an interesting target for the development of insulin secretagogues. Interestingly, PS activation of β-cells (via TRPM3 and voltage-gated Ca^{2+} channels) induces the biosynthesis of a gene regulatory protein, the zinc finger transcription factor Egr-1, and in this way leads to increased biosynthesis of insulin (Mayer et al., 2011). However, the pharmacological concentrations of PS (50 μM) used to demonstrate enhancement of insulin secretion do not occur in vivo. It is possible that TRPM3 plays a more profound role in conditions where elevated plasma PS levels and changes in glucose homeostasis co-occur, such as pregnancy or 21-hydroxylase-deficiency. TRPM3 channels are also proposed to constitute a regulated Zn^{2+} entry pathway in pancreatic β-cells (Wagner et al., 2010). Zinc is important for insulin release as it is packed into co-crystals with insulin in the exocytotic vesicles. The formation of insulin crystals in β-cells depends, among others, on the ZnT8 transporter, which contributes to the packaging efficiency of stored insulin (Lemaire et al., 2009). Since Zn^{2+} ions are co-released with insulin, pancreatic β-cells need to continuously replenish their Zn^{2+} stores by taking up Zn^{2+} ions from the extracellular space. Insufficient Zn^{2+} uptake leads to impaired insulin synthesis and aggravates diabetic symptoms (Chausmer, 1998). TRPM3 channels in β-cells have been shown to be highly permeable for Zn^{2+} and capable of mediating Zn^{2+} uptake under physiological conditions. The depolarization caused by the activation of TRPM3 channels could lead to the activation of voltage-dependent L-type Ca^{2+} channels and would in this way also lead to Zn^{2+} influx through these channels (Gyulkhandanyan et al., 2006; Wagner et al., 2008). Whether Zn^{2+} influx regulation via TRPM3 channels is functionally relevant in β-cells remains to be elucidated.

12.2.4 TRPM4

TRPM4 is a Ca^{2+}-activated nonselective monovalent cation channel (CAN channel) that is impermeable to divalent cations. TRPM4 is a widely expressed protein present in most adult tissues but abundantly expressed in the heart, prostate, colon, and kidneys. TRPM4 was also found to be endogenously expressed in HEK-293 cells (Launay et al., 2002) and CHO-cells (Yarishkin et al., 2008). Investigating the functional role of TRPM4 has gained momentum in recent years by characterizing TRPM4 knockout mouse models (Mathar et al., 2014). An important role for TRPM4 has been shown in the cardiovascular system, the immune system, and the central nervous system. CAN channels that could be TRPM4 were first identified in cardiac myocytes (Colquhoun et al., 1981) and have been found in a wide range of excitable and nonexcitable cells since. These currents could be mediated by either TRPM4 or TRPM5 or a currently unknown channel. CAN currents have been detected in neuron types (El-Sherif et al., 2001; Liman, 2003; Magistretti and Alonso, 2002; Mironov, 2008; Partridge and Swandulla, 1987; Shalinsky et al., 2002) vascular and smooth muscle cells (Eto et al., 2003; Kim et al., 1998; Miyoshi et al., 2004) and pancreatic tissue (Sturgess et al., 1987), including exocrine cells (Gögelein and Pfannmüller, 1989; Gray and Argent, 1990; Maruyama and Petersen, 1984; Suzuki and Petersen, 1988). Similar Ca^{2+}-activated nonselective cation currents were detected in white and brown adipocytes (Halonen and Nedergaard, 2002; Ringer, Russ, and Siemen, 2000), red blood cells (Kaestner and Bernhardt, 2002; Rodighiero, De Simoni, and Formenti, 2004), cochlear hair cells (Van den Abbeele, Tran Ba Huy, and Teulon, 1996), vascular endothelial cells (Csanády and Adam-Vizi, 2003; Popp and Gogelein, 1992; Suh et al., 2002; Watanabe et al., 2002), and renal tubuli (Hurwitz, Hu, and Segal, 2002). In immune-cells TRPM4-like currents were detected in T cells (Launay et al., 2004) and mast cells (Vennekens et al., 2007).

The role of TRPM4 in insulin release is quite controversial. Protein expression and channel activity are detected in several β-cell lines such as INS-1, HIT-T15, RINm5F, β-TC3, and MIN-6 and also in the α-cell line INR1G9 (Cheng et al., 2007; Marigo et al., 2009). Furthermore, expression of TRPM4 protein was detected in mouse pancreatic islets and human β-cells (Marigo et al., 2009). Ca^{2+} increase and insulin secretion of INS-1 cells after stimulation by glucose, AVP (arginine vasopressin, a Gq-coupled receptor agonist in β-cells), or glyburide (glibenclamide) were decreased when TRPM4 was inhibited by the dominant-negative construct ΔNTRPM4 (Marigo et al., 2009; Cheng et al., 2007), suggesting that depolarizing currents generated by TRPM4 are a component in the control of intracellular Ca^{2+} signals necessary for insulin secretion. Furthermore, it is suggested that TRPM4-containing vesicles are translocated to the plasma membrane via Ca^{2+}-dependent exocytosis, which may represent a regulatory mechanism by which β-cells regulate electrical activity in response to glucose and other nutrients (Marigo et al., 2009; Cheng et al., 2007). However, all these studies have been performed on cell lines. Although TRPM4 protein expression could be found within insulin-producing human β-cells and mouse pancreatic islets (Marigo et al., 2009), studies on $Trpm4^{-/-}$ mice revealed no difference in glucose-induced insulin secretion from freshly isolated pancreatic islets (Vennekens et al., 2007). Moreover, these mice did not suffer from an impaired glucose tolerance after an intraperitoneal injection of glucose. These data suggest that TRPM4 is probably not involved in the signal mechanism following glucose stimulation. Butnthis does not exclude a possible role for TRPM4 in Gq- or Gs-receptor-coupled signaling pathways, for example, during stimulation with glucagon. GLP-1–stimulated insulin secretion is at least partially dependent on protein kinase C (PKC)–dependent TRPM4 activation downstream of the GLP-1R (Shigeto et al., 2015). Additionally, TRPM4 is proposed to be involved in glucagon secretion by the pancreatic α-cell line αTC1-6 (Nelson et al., 2011). TRPM4 inhibition decreased the magnitude of intracellular Ca^{2+} signals and glucagon secretion in response to several agonists such as the G_q-protein coupled receptor agonist AVP and high K^+ (Nelson et al., 2011).

Unfortunately, a major difficulty in the determination of the physiological implications of TRPM4 is the poor specificity of available pharmacological tools. Spermine has TRPM4 inhibiting properties with and $EC_{50} = 35 \pm 11$ µM in inside-out patch clamp experiments (Ullrich et al., 2005). TRPM4 has been shown to be inhibited by flufenamic acid with an $EC_{50} = 2.8 \pm 6.1$ µM; however, in the same concentration range, a potentiation of two-pore potassium channels can be observed (Ullrich et al., 2005; Guinamard, Simard, and Del Negro, 2013). Quinine and quinidine inhibit TRPM4 in the range of 100 µM but are poorly selective (Talavera et al., 2008). Clotrimazole, an antifungal agent, also blocks TRPM4 in the micromolar range, but lacks specificity with respect to TRPM5 and TRPM3 (Ullrich et al., 2005; Vriens et al., 2014). The most commonly used TRPM4 antagonist is the phenanthrene-derivative 9-phenanthrol which shows specificity for TRPM4 over TRPM5 in an overexpression system (Grand et al., 2008). 9-Phenanthrol has been used to show TRPM4-dependent current activity, albeit with difficulties dissecting the TRPM4-dependent part of the current. Glibenclamide, a sulfonylurea, blocks TRPM4-like currents in sinoatrial node cells at 100 µM (Demion et al., 2007). Pretreatment of HEK293 cells with 3.5-bis(trifluoromethyl)pyrazole (BTP2) enhances TRPM4 currents through an unclear mechanism (Takezawa et al., 2006). The main drawback considering pharmacology on TRPM4 is the poor differentiation of the chemicals between TRPM4 and TRPM5 or other off targets effectors. The use of these compounds is valid in cellular overexpression models for TRPM4, but in primary cells or *in vivo* models, one reaches the limits of solubility, bioavailability, and selectivity.

12.2.5   TRPM5

TRPM5 is a monovalent cation-permeable ion channel in the plasma membrane and is activated by a rise of intracellular Ca^{2+} (Prawitt et al., 2003). It is, together with TRPM4, a candidate for Ca^{2+}-activated monovalent cation channels that induce cellular depolarization when increases of $[Ca^{2+}]_i$ occur. Both TRPM5 and TRPM4 are important in Ca^{2+} signaling and membrane excitability. Expression of TRPM5 was initially found in the liver, heart, brain, kidney, spleen, lung, and testes (Enklaar et al., 2000). Further research showed TRPM5 expression in taste tissue, stomach, and small intestine (Pérez et al., 2002). A genetic Cre knock-in TRPM5 mouse was generated and confirmed the expression of TRPM5 in taste bud cells and in the olfactory epithelium (Kusumakshi et al., 2015). A general observation is that TRPM5 is coexpressed with one or more signaling components of the canonical bitter, sweet, and umami taste transduction cascade (PLCβ2, α-gustducin). Type II taste receptor cells in the taste buds express TRPM5 where it is an essential modulator of bitter, sweet, and umami taste (Zhang et al., 2003). It is shown that TRPM5 has a functional role in pancreatic β-cells where it is involved in the regulation of insulin secretion (Colsoul et al., 2010). Surprisingly, when reporting the GFP-TRPM5 knock-in mouse, the authors failed to look for TRPM5-positive cells in the pancreas (Kusumakshi et al., 2015). Additionally, TRPM5 expression has been shown in solitary chemosensory enteroendocrine cells where the functional significance of TRPM5 needs to be further investigated (Kokrashvili, Mosinger, and Margolskee, 2009; Kaske et al., 2007).

Modulation of TRPM5 is possible due to its inherent weak voltage dependency (Liu and Liman, 2003; Nilius et al., 2005). Activation of TRPM5 is temperature dependent as higher temperatures promote the activation of the channel, which is consistent with the observed temperature-dependent taste sensation mediated by TRPM5 (Talavera et al., 2005; Talavera et al., 2007).

The secretion of insulin by the pancreatic β-cell is a complex process driven by electrical activity and oscillations of the intracellular Ca^{2+} concentration, $[Ca^{2+}]_i$ (MacDonald and Rorsman, 2006). ATP produced by the glucose metabolism closes ATP-sensitive K^+ channels, allowing depolarization of the membrane potential and activation of L-type Ca^{2+} channels. This leads to Ca^{2+} increase and secretion of insulin-containing vesicles (Drews, Krippeit-Drews, and Düfer, 2010; Henquin et al., 2009; Henquin, 2009; MacDonald and Rorsman, 2007). As insulin is the only glucose-lowering hormone, its secretion is essential for glucose homeostasis. Although Ca^{2+}-activated nonselective cation channels have been described in β-cells from the pancreas, their contribution to the electrical activity and insulin release was unclear (Roe et al., 1998a; Sturgess, Hales, and Ashford, 1987; Leech and Habener, 1998). A Ca^{2+}-activated nonselective monovalent cation channel could be measured in β-cells and was largely reduced in $Trpm5^{-/-}$ mice, indicating that TRPM5 is an important constituent of the Ca^{2+}-activated cation current in β-cells (Colsoul et al., 2010; Liu and Liman, 2003). TRPM5 essentially functions as a positive regulator of glucose-induced insulin release. The electrical activity in the β-cells is coupled to changes in the intracellular calcium. These changes, in their turn, trigger the insulin secretion. Normal wild-type β-cells respond to glucose stimulation with an oscillatory change in the membrane potential and, in parallel three types of $[Ca^{2+}]_i$ oscillations (slow, mixed, or fast). These oscillations are typically either large-amplitude $[Ca^{2+}]_i$ oscillations with a frequency of 0.2–0.5 min^{-1}(slow) or Ca^{2+} oscillations induced by faster repetitive bursts of action potentials, typically 3–5 min^{-1} (fast) or a complex pattern resulting from the fast $[Ca^{2+}]_i$ oscillations superimposed on the slow oscillations (Liu et al., 1998), $Trpm5^{-/-}$ islets specifically lack fast glucose-induced oscillations in membrane voltage (Vm) and Ca^{2+} (Colsoul et al., 2010). TRPM5 contributes to slow depolarization in the interburst interval of the glucose-induced electrical activity, in this way shortening the interburst interval and leading to faster glucose-induced oscillations in Vm and Ca^{2+}. In line with this, glucose-induced insulin release was reduced in isolated pancreatic islets from $Trpm5^{-/-}$ mice. Moreover, Trpm5-deficient mice display an impaired glucose tolerance during oral and intraperitoneal glucose tolerance tests (Colsoul et al., 2010; Brixel et al., 2010). This is caused by reduced glucose-induced insulin secretion from the β-cells resulting in lower plasma insulin levels. These data suggest that Trpm5-deficient mice display a prediabetic phenotype caused by β-cell dysfunction. The relevance of this prediabetic phenotype during conditions of higher insulin demand (such as pregnancy, obesity, aging, etc.) remains to be shown. In leptin-deficient mice, a model for type II diabetes, a decrease of TRPM5 expression in the pancreatic islets was observed. Due to the reduced TRPM5 expression, the pattern of glucose-induced Ca^{2+} activity in the islets from leptin-deficient mice resembles that of $Trpm5^{-/-}$ mice (Colsoul et al., 2014). The downregulation of $Trpm5$ in the obese (*Ob/Ob*) mutant mice (a model of obesity, insulin resistance, and T2DM) can be reversed by the addition of taurine to the diet. This increased $Trpm5$ expression, compared to Ob/Ob mice without taurine in the diet, ameliorates the systemic glucose homeostasis in these mice (Santos-Silva et al., 2015). It is also interesting to note that $Trpm5$ expression in the small intestine from diabetic patients is negatively correlated with their blood glucose concentrations (Young et al., 2009). Moreover, TRPM5 variants are associated with prediabetic phenotypes in subjects at risk for type 2 diabetes. TRPM5 single nucleotide polymorphisms (SNPs) occurred more often in patients with disturbed insulin secretion, elevated plasma glucose, lower GLP-1 levels, and decreased insulin sensitivity (Ketterer et al., 2011). How these TRPM5 variants might affect insulin sensitivity remains elusive, as $Trpm5^{-/-}$ mice show a normal insulin tolerance (Colsoul et al., 2010). Certain TRPM5 polymorphisms showed increased prevalence in patients with obesity-related metabolic syndrome (Tabur et al., 2015). The functional impact of these mutations on TRPM5 channel activity has not been clarified yet. However, these data indicate a possible link between TRPM5 and type 2 diabetes mellitus.

Screening for new bioactive compounds on TRPM5 opens up a new line of experimental possibilities for research. TRPM5 modulators can be used as lead compounds for insulin secretagogue medication. TRPM5 has recently been implemented as a modulator of GLP-1R signaling in a PKC-dependent way. TRPM5 is shown to be at least partially responsible for the GLP-1 mediated insulinotropic effects (Shigeto et al., 2015). In this way, pharmacological potentiation of TRPM5 should lead to stimulation of the physiological way of insulin secretion. This could be a superior way to modulate insulin secretion compared to the many current antidiabetes drugs that block the K_{ATP} channels independent from the blood glucose levels. Currently, most pharmacological active compounds on TRPM5 show a poor selectivity between TRPM4 and TRPM5 (and other ion channels). Spermine is identified as an intracellular blocking agent of TRPM4 (Nilius et al., 2004) but is equally potent to block TRPM5 currents (Ullrich et al., 2005). Furthermore, flufenamic acid (FFA) blocks both TRPM4 (EC_{50} of 2.5 µM) and TRPM5 (EC_{50} of 24.5 µM) and also clotrimazole and quinine are shown to inhibit both channels (Talavera et al., 2008). Nicotine was identified as an inhibitor of TRPM5 activity but lacks TRPM5 specificity as it has obvious effects on the ACh receptor (Gees et al., 2014). Triphenylphosphine oxide (TPPO) has been identified as an inhibitor of TRPM5 without blocking TRPM4, TRPV1, or TRPA1 and can therefore be considered as a relatively selective blocker for TRPM5 (Palmer et al., 2010).

12.2.6 TRPM8

TRPM8 is a cold-sensing TRP channel that is activated by chemical ligands such as menthol and icilin (McKemy, Neuhausser, and Julius, 2002; Peier et al., 2002). It is highly expressed in prostate tissue, however its functional role there remains to be elucidated (Fonfria et al., 2006). TRPM8 is functionally expressed in dorsal root ganglia (DRG) and trigeminal ganglia (TG). In the sensory nerve system, TRPM8 has a prominent role in the sensing of cold temperatures and maintaining core body temperature (Gavva et al., 2012). There is no expression of TRPM8 in the endocrine pancreas (Fonfria et al., 2006). However, mice lacking *Trpm8* have a normal glucose tolerance but an increased insulin sensitivity. This results from a compensatory mechanism following enhanced insulin clearance in *Trpm8*$^{-/-}$ mice. This compensatory mechanism includes upregulation of the insulin-degrading enzyme (IDE) in the liver, which might result from the loss of TRPM8-mediated neuronal signals in hepatic neural innervation. TRPM8 positive sensory afferents innervate the hepatic vein (McCoy et al., 2013). TRPM8 has recently been indicated in association with pancreatic adenocarcinoma, where it might serve as a potential biomarker (Yee et al., 2012a; Yee et al., 2012b).

12.2.7 TRPV1

TRPV1 is highly expressed in primary sensory neurons where it functions as a polymodal nociceptor detecting noxious heat and pain (Szallasi et al., 2007). It is activated by heat and capsaicin, the pungent compound in chili peppers (Caterina et al., 1997). Some controversy exists about expression of TRPV1 in the pancreas, where it is unclear whether the TRPV1 expression is in the β-cells or merely limited to the innervation of the pancreas (Akiba et al., 2004; Jabin et al., 2012; Gram et al., 2007). Whereas one study reports the expression of TRPV1 in Sprague-Dawley rat islets and rat β-cells lines (RIN-5F and INS-1) and shows capsaicin-induced insulin release from insulinoma cells (Akiba et al., 2004), others failed to show expression in mouse pancreatic islets (Razavi et al., 2006) or in β-cells from Zucker diabetic fatty (ZDF) rat (Gram et al., 2007) and from rat primary cultures (Diaz-Garcia et al., 2014). However, both the exocrine and the endocrine pancreas are innervated by TRPV1-positive neurons (Gram et al., 2007; Winston et al., 2005; Wick et al., 2006). More and more evidence points to a role for these neurons in type 1 and type 2 diabetes mellitus (Szallasi et al., 2007; Suri and Szallasi, 2008).

Treatment of mice with the TRPV1 activator capsaicin increased the postprandial insulin secretion, an effect that was absent in the *Trpv1*$^{-/-}$ mouse. No direct secretagogue effect of TRPV1 activation on the β-cell could be observed. Instead the insulin-promoting effect of activating TRPV1 can be contributed to increased incretin release from GLP-1 secreting intestinal cells (Wang et al., 2012). Antibody reactivity indicates expression of TRPV1 in the afferent neurons innervating the pancreas, but not in the β-cells. In nonobese diabetic (NOD) mice a mutation leading to a hypofunctional TRPV1 channel was identified, which is linked to the onset of T1DM. The onset of islet inflammation and diabetes in these mice might be mediated by the TRPV1+ insulin responsive sensory neurons (Razavi et al., 2006). A similar mutation was found in association with diabetes in a human population (Sadeh et al., 2013). Clearly, TRPV1 could be important to evoke neuronal and hormonal signals to promote β-cell activity. However, current knowledge does not support a direct role of TRPV1 in the β-cell.

There is a local feedback loop between TRPV1-expressing sensory neurons and insulin secreting β-cells. Insulin released by the β-cell will activate insulin receptors on the nerve terminals (Van Buren et al., 2005). This increases TRPV1-mediated membrane currents by enhancing receptor sensitivity and translocation of TRPV1 from cytosol to plasma membrane (Van Buren et al., 2005) and by lowering the thermal activation threshold to room temperature (Sathianathan et al., 2003). The increase in TRPV1 currents leads to local release of neuropeptides such as substance P and CGRP that will sustain β-cell physiology in an optimal range (Razavi et al., 2006): CGRP reduces insulin release from β-cells (Pettersson et al., 1986), whereas substance P promotes neurogenic inflammation in the pancreas (Noble et al., 2006). This local feedback loop has been proposed to be disturbed in type 1 diabetes mellitus.

Ablation of TRPV1-positive neurons by agonist (capsaicin or resiniferatoxin) administration in rat neonates improves glucose tolerance in those made diabetic by streptozotocin treatment (Guillot, Coste, and Angel, 1996). The hypofunctional TRPV1 mutant in NOD mice (Razavi et al., 2006) causes a decreased secretion of substance P by the pancreatic sensory neurons. Interestingly, direct administration of substance P into the pancreas transiently reverses hyperglycemia in NOD mice and clears islet cell inflammation. The congenic NOD.Idd4.1 mouse strain, which expresses the wild-type TRPV1, is protected from diabetes (Razavi et al., 2006; Grattan et al., 2002). However, injecting NOD mice with high doses of capsaicin and thus eliminating TRPV1-positive neurons, also protects mice from diabetes (Razavi et al., 2006). Islet infiltration and the proportions and absolute numbers of effector T-lymphocytes in pancreatic lymph nodes were reduced in capsaicin-treated NOD mice (Razavi et al., 2006). The different effects of the neuropeptides on β-cell function, depending on their concentration, might explain these apparently contradicting results. High concentrations generate a strong trophic signal, promoting β-cell survival and function, whereas low concentrations have the opposite result, with deleterious effects on β-cell function and viability (Hermansen and Ahrén, 1990; Tsui et al., 2007). It has been proposed that a fine balance in the local micro-environment between products such as substance P and islet β-cell stress might determine the ensuing inflammatory reactions (Razavi et al., 2006).

Systemic application of capsaicin ablates TRPV1-expressing nerve fibers and prevents aging-associated obesity and insulin resistance in rats and mice (Melnyk and Himms-Hagen, 1995; Cui and Himms-Hagen, 1992). Furthermore, this treatment prevents the deterioration of glucose homeostasis in Zucker diabetic rats via increased insulin secretion (Gram et al., 2007). The channel might also be linked with appetite regulation and obesity (a well-known risk factor for type-2 diabetes mellitus, T2DM), as obese monkeys show markedly reduced capsaicin-evoked responses in their skin, which may point to downregulated TRPV1 expression and/or loss of TRPV1-expressing nerve fibers (Pare et al., 2007). The anorexic lipid mediator N-oleoylethanolamide is an endogenous TRPV1 ligand that reduces food intake in wild-type (WT) but not in $Trpv1^{-/-}$ mice (Wang, Miyares, and Ahern, 2005). As both obesity and T2DM might in part be an inflammatory disorder (as evidenced by increased levels of inflammatory markers) (Festa et al., 2000; Herder et al., 2007; Wu et al., 2002) and TRPV1 can be activated by inflammatory components (Winston et al., 2005; Pingle, Matta, and Ahern, 2007), it has been suggested that TRPV1-expressing nerves are activated by proinflammatory substances in T2DM patients and that the resulting sustained calcitonin gene-related peptide (CGRP) release and high circulating CGRP levels promote insulin resistance (Suri and Szallasi, 2008). Interestingly, treatment of Ob/Ob mice (a mouse model of obesity, insulin resistance, and T2DM) with a TRPV1 antagonist not only enhances insulin secretion but also decreases insulin resistance (Tanaka et al., 2011).

12.2.8 TRPV2

Expression of TRPV2 could be detected in MIN6 cells and in mouse primary β-cells (Aoyagi et al., 2010; Hisanaga et al., 2009). When MIN6 cells are cultured in a serum-free condition, immunoreactivity of TRPV2 is localized in an intracellular compartment, whereas addition of serum induces translocation of TRPV2 to the plasma membrane (Kanzaki et al., 1999). Both insulin and IGF-1 have been shown to be responsible for the translocation and insertion of TRPV2 from an intracellular compartment into the plasma membrane (Hisanaga et al., 2009; Kanzaki et al., 1999). Translocation of TRPV2 was also induced by insulin secretagogues (including glucose) and knockdown of the insulin receptor attenuated insulin-induced translocation of TRPV2 (Hisanaga et al., 2009). Furthermore, elevation of Ca^{2+} entry caused by pretreatment of either MIN-6 cells or cultured mouse β-cells with insulin, is reduced by inhibition of TRPV2 (Hisanaga et al., 2009). Inhibition of TRPV2 also reduces glucose- or KCl-induced insulin secretion (Hisanaga et al., 2009). These data indicate that insulin released from the β-cells further augments Ca^{2+} entry by recruiting TRPV2 to the plasma membrane and via this feed-forward mechanism accelerates insulin secretion. It has been shown that insulin treatment induces the acceleration of the exocytotic response during the glucose-induced first-phase response by the insertion of TRPV2 into the plasma membrane in a PI_3K-dependent manner (Aoyagi et al., 2010). Although the autocrine effect of insulin on β-cell function has been a matter of debate (Leibiger, Leibiger, and Berggren, 2002), the β-cell specific knockout of the insulin receptor gene shows impaired glucose-induced insulin secretion and reduced β-cell mass, showing an important functional role for the insulin receptor in glucose sensing by the pancreatic β-cell (Kulkarni et al., 1999). In this regard, more knowledge concerning the *in vivo* role of TRPV2 in β-cells would expand the knowledge of insulin signalling in the healthy and the diabetic β-cell.

12.2.9 TRPV4

TRPV4 is a Ca^{2+}-permeable cation channel (Everaerts, Nilius, and Owsianik, 2010). TRPV4 is expressed in MIN6 cells, INS-1E cells, and mouse pancreas (Casas et al., 2007). Activation of TRPV4 enhances glucose-stimulated insulin secretion from INS-1E cells (Skrzypski et al., 2013). Reduction of *Trpv4* expression significantly protected MIN6 cells against human islet amyloid polypeptide (hIAPP)–induced Ca^{2+} elevation, endoplasmic reticulum (ER) stress, and apoptosis (Casas et al., 2008). hIAPP or amylin, the main component of amyloid, is strongly associated with the progressive loss of pancreatic β-cell mass in type 2 diabetes. This may result from disruption of Ca^{2+} homeostasis. hIAPP forms insoluble aggregates and triggers Ca^{2+} changes that are associated with the induction of apoptosis via a pathway involving activation of the ER stress response (Casas et al., 2007). The cytotoxicity of hIAPP is initiated on the cell surface and requires close contact between the aggregation pores and the β-cell plasma membrane (Lorenzo et al., 1994). Since TRPV4 might be mechanosensitive, it has been proposed that TRPV4 senses the physical changes in the plasma membrane induced by hIAPP aggregation and in this way enables Ca^{2+} entry, membrane depolarization, and activation of L-type Ca^{2+} channels, leading to cell death (Casas et al., 2008).

12.2.10 TRPA1

TRPA1 is a Ca^{2+}-permeable nonselective cation channel, which is expressed in TG and DRG neurons. TRPA1 is activated by several reactive electrophilic substances like allyl isothiocyanate (AITC), cinnemaldehyde, allicin, and acrolein, but also by nonreactive compounds like methylsalicylate, menthol, and icilin. TRPA1 is involved in various sensory processes, such as the detection of noxious cold and inflammatory hyperalgesia. There has been evidence that dietary supplementation of the TRPA1 agonist cinnemaldehyde improves the plasma glucose levels in people with T2DM (Solomon and Blannin, 2007; Akilen et al., 2010; Roberts et al., 2016). However, this beneficial effect on the glycemic plasma levels is presumed independent from the activity of TRPA1 in β-cells. Intake of TRPA1 agonists stimulates the incretin secretion from enteroendocrine L-cells. This increased GLP-1 secretion is not observed in primary intestinal cultures of *Trpa1$^{-/-}$* mice (Emery et al., 2014).

TRPA1 is abundantly expressed in a rat pancreatic β-cell line and freshly isolated rat pancreatic β-cells, but not in pancreatic α-cells (Cao et al., 2012; Numazawa et al., 2012). Application of compounds such as mustard oil and 4-hydroxy-2-nonenal, which activate TRPA1, induces an inward membrane current, depolarization, and a rise of the intracellular Ca^{2+} level leading to the release of insulin. This indicates that TRPA1 agonists could function as insulin secretagogues (Cao et al., 2012; Numazawa et al., 2012). Strikingly, it was also reported that glucose-induced insulin release could be inhibited by the TRPA1 antagonist HC030031 (Cao et al., 2012). This might suggest a role for TRPA1 during glucose-induced signaling in the β-cell, though it should be mentioned that data from *Trpa1* knockout mice, which are indispensable controls for the selectivity of the antagonist, are lacking at this point.

12.3   CURRENT THERAPIES FOR DIABETES

The occurrence of obesity and diabetes as a worldwide epidemic accounts for millions of deaths and a huge economic and social cost. There are currently 422 million people with diabetes and their number is still increasing (World Health Organization, 2016; Shi and Hu, 2014). This makes diabetes a major challenge for medical research (Zimmet, Alberti, and Shaw, 2001). The discovery and development of new effective treatments is essential. Diabetes leads to severe reduction of life quality. Microvascular and macrovascular complications lead to increased morbidity and mortality. Diabetes-related retinopathy is the leading cause for new-onset adult blindness, and diabetic nephropathy is responsible for almost half of the cases of end-stage renal disease. Patients with type II diabetes have an increased risk of developing vascular issues as coronary artery disease, hypertension, and hyperlipidemia (Irving et al., 2008; Ohkubo et al., 1995; DCCT, 1993; Stratton et al., 2000). Early diagnosis and proper glycemic control are key elements in delaying disease progress and the associated morbidity. In type I diabetic patients, who lack functional β-cells, the therapy consists of life-long administration of exogenous insulin. Type II diabetic patients are a diverse group all characterized by residual β-cells, incapable of releasing sufficient insulin to reach normoglycemia. Initially, to achieve glycemic control a program of proper diet and exercise is implemented. When this fails to normalize the glucose levels, it is accompanied with oral agents or insulin injections (Wajchenberg, 2010).

Type II diabetes is a complex metabolic disease and pharmacological interventions on different organs can contribute to reach glycemic control. There are several classes of chemicals available for the management of type II diabetes; they exist out of mainly three categories: insulin secretagogues, insulin sensitizers, or insulin receptor agonists. The most commonly used insulin secretagogues are first- and second-generation sulfonylureas (glimepiride, glipizide, and glyburide/glibenclamide). They block K_{ATP} channels in the pancreatic β-cell and induce insulin secretion, independent of the blood glucose level (Proks et al., 2002). First-generation sulfonylurea drugs came to the market more than 50 years ago and are no longer used. The second-generation agents are widely used and are most effective in early stages of type II diabetes where there are still functional β-cells. Second-generation sulfonylureas, which include glimepiride, have a higher selective binding capacity on β-cells that allows lower dosage than first-generation agents (Melander et al., 1989; Rosenstock, 2001). The adverse effects of these agents include nausea, dizziness, and headache, but they are generally mild and reversible. Glimepiride appears to have a distinct binding protein compared to the other sulfonylureas. It binds to a 65 kDa protein-binding site while other sulfonylureas are binding into a 140 kDa protein (Kramer et al., 1994). This, combined with a faster association and dissociation rate from the receptor site, leads to a decreased potential for inducing hypoglycemia

(Holstein, Plaschke, and Egberts, 2001). This makes glimepiride a superior drug for the treatment of elderly patients, where hypoglycaemia is particularly dangerous, or for diabetics who exercise (Davis, 2004). Furthermore, glimepiride was shown to be weight neutral, an advantage considering obesity is a common risk factor for type II diabetes (Bugos et al., 2000).

K_{ATP} inhibition is a common mechanism that the sulfonylurea drugs share with meglitinide analogues as repaglinide and the D-phenylalanine derivatives as nateglinide (Chachin et al., 2003). Glucagon-like peptide-1 (GLP-1) is an incretin hormone that stimulates the GLP-1 receptor on the β-cell to promote insulin secretion. This stimulation pathway is advantageous over sulfonylureas as it has a lower risk of inducing hypoglycemia. GLP-1 receptor agonists (as exenatide and liraglutide) stimulate insulin secretion in the same way as DPP-4 inhibitors which are less effective (Nauck et al., 2009). Biguanides as metformin reduce the glycemia by lowering gluconeogenesis in the liver; it is often used in combination with insulin secretagogues (Lambert, 2013). Thiazolidinediones (pioglitazone and rosiglitazone) are typical insulin sensitizers, making the body's tissue more sensitive to insulin; they are, however, notorious for inducing weight gain and increase the risk of heart failure (Stumvoll, Goldstein, and van Haeften, 2005). Other drugs include the α-glucosidase inhibitors (Acarbose and miglitol) that work by preventing the digestion and uptake of carbohydrates in the intestine (Lebovitz, 2011). A novel approach to reduce the glucose load in the blood is by means of SGLT-2 inhibitors (gliflozin) (Bhartia, Tahrani, and Barnett, 2011). They inhibit the glucose transporter in the kidneys preventing reabsorption of glucose from the pre-urine in the kidneys, thereby eliminating excess glucose out of the body. As functional β-cells are not required for its activity, the drug is convenient for patients with reduced β-cell function.

When all pharmacological interventions fail to gain glycemic control, insulin injections are the final stage for diabetes treatment. There are several types of insulin available and combined with insulin pumps and glycemic control, they provide an "as good as possible" physiological profile of insulin secretion.

12.4 TRP CHANNELS AS DRUG TARGET IN DIABETES

It is generally acknowledged that type II diabetes results from a combination between peripheral insulin resistance and a genetically determined susceptibility to β-cell dysfunction. Emerging evidence indicates that TRP channels regulate insulin release and β-cell function, specifically through their ability to increase intracellular Ca^{2+} levels and membrane depolarization. TRP channels are interesting targets for the development of insulin secretagogues. Although much of the available evidence is generated in cell lines, data are increasingly being confirmed in knockout mice. However, a functional role for any TRP channel in human islets has not been shown to date. This requires selective and potent pharmacology, which is only available for a few TRP channels. Clearly, this is one of the main hurdles that needs to be taken to fully explore the potential of TRP channels in different aspects of non-insulin-dependent diabetes mellitus.

β-cell dysfunction is associated with T2DM so it would be advantageous that antidiabetic drugs also improve β-cell function. Several TRP channels show interesting potential to be used as a target for insulinotropic drugs. TRPM3 activators have been shown to enhance glucose-dependent insulin release (Wagner et al., 2008; Held et al., 2015). TRPV1 might also be an interesting drug target as compounds modulating channel activity influence insulin resistance but also insulin release and GLP-1 secretion (Wang et al., 2012). Pharmacological activation of TRPM5 would enhance insulin release only under conditions of elevated glucose levels with no associated risk to hypoglycemia, in a similar matter as incretin therapy (Shigeto et al., 2015), as TRPM5 is also implicated in transduction of sweet taste, activators of TRPM5 might be expected to increase sweet sensation and in this way reduce intake of sugar (Colsoul et al., 2010; Nakagawa et al., 2009). Considering the side effects of sulfonylureas (risk on hypoglycemia), novel insulin secretagogues are obviously desired, where the glucose dependency of its action would be an essential requirement. Incretin mimetica (GLP-1 receptor agonists that mimic the action of natural GLP-1) and incretin enhancers (inhibitors of the enzyme that degrades the incretin hormones and thus prolong their activity) are potential candidates. However, GLP-1R agonists require subcutaneous administration and both incretin mimetica and enhancers are associated with unwanted effects, such as nausea and diarrhoea for incretin mimetica and higher occurrence of infections for incretin enhancers (Joy, Rodgers, and Scates, 2005; Cernea and Raz, 2011). There clearly is a need for improving diabetes therapies with new molecular targets and suitable pharmaceuticals that are safe to use and have limited side effects.

REFERENCES

Akiba, Y. et al. 2004. Transient receptor potential vanilloid subfamily 1 expressed in pancreatic islet beta cells modulates insulin secretion in rats. *Biochem Biophys Res Commun*, 321(1; August 13): 219–225. doi:10.1016/j.bbrc.2004.06.149.

Akilen, R. et al. 2010. Glycated haemoglobin and blood pressure-lwering effect of cinnamon in multi-ethnic type 2 diabetic patients in the UK: A randomized, placebo-controlled, double-blind clinical trial. *Diabet Med*, 27(10) (October): 1159–1167. doi:10.1111/j.1464-5491.2010.03079.x.

Aoyagi, K. et al. 2010. Insulin/phosphoinositide 3-kinase pathway accelerates the glucose-induced first-phase insulin secretion through TrpV2 recruitment in pancreatic β-cells. *Biochem J*, 432(2; December 1): 375–386. doi:10.1042/BJ20100864.

Ashcroft, F.M. and P. Rorsman. 1989. Electrophysiology of the pancreatic beta-cell. *Progress Biophys Molecul Biol*, 54(2; January): 87–143.

Ashcroft, F.M. and P. Rorsman. 2013. K(ATP) channels and islet hormone secretion: New insights and controversies. *Nat Rev Endocrinol*, 9(11; November): 660–669. doi:10.1038/nrendo.2013.166.

Bertram, R., A. Sherman, and L.S. Satin. 2007. Metabolic and electrical oscillations: Partners in controlling pulsatile insulin secretion. *Am J Physiol Endocrinol Metab*, 293(4; October): E890–E900. doi:10.1152/ajpendo.00359.2007.

Bhartia, M., A.A. Tahrani, and A.H. Barnett. 2011. SGLT-2 inhibitors in development for type 2 diabetes treatment. *Rev Diab Stud*, 8(3): 348–354. doi:10.1900/RDS.2011.8.348.

Bowman, E.J., A. Siebers, and K. Altendorf. 1988. Bafilomycins: A class of inhibitors of membrane ATPases from microorganisms, animal cells, and plant cells. *Proc Natl Acad Sci USA*, 85: 7972–7976. doi:10.1073/pnas.85.21.7972.

Brixel, L.R. et al. 2010. TRPM5 regulates glucose-stimulated insulin secretion. *Pflugers Archiv*, 460: 69–76. doi:10.1007/s00424-010-0835-z.

Bugos, C. et al. 2000. Long-term treatment of type 2 diabetes mellitus with glimepiride Is weight neutral: A meta-analysis. *Diab Res Clin Prac*, 50 (September): 47. doi:10.1016/S0168-8227(00)81616-2.

Cao, D.-S. et al. 2012. Expression of transient receptor potential ankyrin 1 (TRPA1) and Its role in insulin release from rat pancreatic beta cells. *PloS One*, 7(5; January): e38005. doi:10.1371/journal.pone.0038005.

Casas, S. et al. 2007. Impairment of the ubiquitin-proteasome pathway Is a downstream endoplasmic reticulum stress response induced by extracellular human islet amyloid polypeptide and contributes to pancreatic beta-cell apoptosis. *Diabet*, 56(9): 2284–2294. doi:10.2337/db07-0178.

Casas, S. et al. 2008. Calcium elevation in mouse pancreatic beta cells evoked by extracellular human islet amyloid polypeptide involves activation of the mechanosensitive ion channel TRPV4. *Diabetol*, 51(12; December): 2252–2262. doi:10.1007/s00125-008-1111-z.

Caterina, M.J. et al. 1997. The capsaicin receptor: A heat-activated ion channel in the pain pathway. *Nature* 389(6653; October 23): 816–824. doi:10.1038/39807.

Cernea, S. and I. Raz. 2011. Therapy in the early stage: Incretins. *Diabetes Care*, 34(Suppl 2): S264–S271. doi:10.2337/dc11-s223.

Chachin, M. et al. 2003. Nateglinide, a D-phenylalanine derivative lacking either a sulfonylurea or benzamido moiety, specifically inhibits pancreatic beta-cell-type K(ATP) channels. *J Pharmacol Exp Therapeut*, 304(3): 1025–1032. doi:10.1124/jpet.102.044917.1997.

Chausmer, A.B. 1998. Zinc, insulin and diabetes. *J Am Coll Nutrit*, 17: 109–115. doi:10.1080/07315724.1998.10718735.

Chen, K. et al. 2013. Association of TRPC1 gene polymorphisms with type 2 diabetes and diabetic nephropathy in Han Chinese Population. *Endocr Res*, 38(2) (January): 59–68. doi:10.3109/07435800.2012.681824.

Cheng, H. et al. 2007. TRPM4 controls insulin secretion in pancreatic beta-cells. *Cell Calcium*, 41(1) (January): 51–61. doi:10.1016/j.ceca.2006.04.032.

Cheng, K.T. et al. 2013. Contribution and regulation of TRPC channels in store-operated Ca2+ Entry. *Curr Top Membr*, 71: 149–179. doi:10.1016/B978-0-12-407870-3.00007-X.

Colquhoun, D. et al. 1981. Inward current channels activated by intracellular Ca in cultured cardiac cells. *Nature*, 294(5843): 752–754. doi:10.1038/294752a0.

Colsoul, B., R. Vennekens, and B. Nilius. 2011. Transient receptor potential cation channels in pancreatic β cells. *Rev Physiol Biochem Pharmacol*, 161: 87–110. doi:10.1007/112_2011_2.

Colsoul, B. et al. 2010. Loss of high-frequency glucose-induced Ca^{2+} oscillations in pancreatic islets correlates with impaired glucose tolerance in Trpm5$^{-/-}$ mice. *Proc Nat Acad Sci U S A*, 107(11; March 16): 5208–5213. doi:10.1073/pnas.0913107107.

Colsoul, B. et al. 2014. Insulin downregulates the expression of the Ca2+-activated nonselective cation channel TRPM5 in pancreatic islets from leptin-deficient mouse models. *Pflugers Archiv*, 466(November 13): 611–621. doi:10.1007/s00424-013-1389-7.

Csanády, L. and V. Adam-Vizi. 2003. Ca(2+)- and voltage-dependent gating of Ca(2+)- and ATP-sensitive cationic channels in brain capillary endothelium. *Biophys J*, 85(1): 313–327. doi:10.1016/S0006-3495(03)74476-2.

Cui, J. and J. Himms-Hagen. 1992. Long-term decrease in body fat and in brown adipose tissue in capsaicin-desensitized rats. *Am J Phys*, 262(4 Pt 2): R568–R573.

Davis, S.N. 2004. The role of glimepiride in the effective management of type 2 diabetes. *J Diabet Comp*, 18(6): 367–376. doi:10.1016/j.jdiacomp.2004.07.001.

DCCT. 1993. The effect of intensive treatment of diabetes on the development and progression of long-term complications in insulin-dependent diabetes mellitus. The diabetes control and complications trial research group. *N Eng J Med*, 329(14; September 30): 977–986. doi:10.1056/NEJM199309303291401.

Demion, M. et al. 2007. TRPM4, a Ca^{2+}-activated nonselective cation channel in mouse sino-atrial node cells. *Cardiovasc Res*, 73(3): 531–538. doi:10.1016/j.cardiores.2006.11.023.

Diaz-Garcia, C.M. et al. 2014. Role for the TRPV1 channel in insulin secretion from pancreatic beta cells. *J Membr Biol*, 247(6; June): 479–491.

Drews, G., P. Krippeit-Drews, and M. Düfer. 2010. Electrophysiology of islet cells. *Adv Exp Med Biol*, 654: 115–163. doi:10.1007/978-90-481-3271-3_7.

Du, J., J. Xie, and L. Yue. 2009. Intracellular calcium activates TRPM2 and its alternative spliced isoforms. *Proc Nat Acad Sci U S A*, 106: 7239–7244. doi:10.1073/pnas.0811725106.

El-Sherif, Y. et al. 2001. ATP modulates Na+ channel gating and induces a non-selective cation current in a neuronal hippocampal cell line. *Brain Res*, 904(2): 307–317. doi:10.1016/S0006-8993(01)02487-8.

Emery, E.C. et al. 2014. Stimulation of glucagon-like peptide-1 secretion downstream of the ligand-gated ion channel TRPA1. *Diabetes* (October 16). doi:10.2337/db14-0737.

Enklaar, T. et al. 2000. Mtr1, a novel biallelically expressed gene in the center of the mouse distal chromosome 7 imprinting cluster, is a member of the trp gene family. *Genomics* 67(2): 179–187. doi:10.1006/geno.2000.6234.

Eto, W. et al. 2003. Intracellular alkalinization induces Ca^{2+} influx via non-voltage-operated Ca^{2+} channels in rat aortic smooth muscle cells. *Cell Calcium*, 34(6): 477–484. doi:10.1016/S0143-4160(03)00151-9.

Everaerts, W., B. Nilius, and G. Owsianik. 2010. The vanilloid transient receptor potential channel TRPV4: From structure to disease. *Prog Biophys Mol Biol*, 103: 2–17. doi:10.1016/j.pbiomolbio.2009.10.002.

Festa, A. et al. 2000. Chronic subclinical inflammation as part of the insulin resistance syndrome: The Insulin Resistance Atherosclerosis Study (IRAS). *Circulation*, 102(1): 42–47. doi:10.1161/01.CIR.102.1.42.

Fonfria, E. et al. 2006. Tissue distribution profiles of the human TRPM cation channel family. *J Rec Sig Trans Res*, 26(3; January): 159–178. doi:10.1080/10799890600637506.

Freichel, M. et al. 2004a. Functional role of TRPC proteins in vivo: Lessons from TRPC-deficient mouse models. *Biochem Biophys Res Comm*, 322: 1352–1358. doi:10.1016/j.bbrc.2004.08.041.

Freichel, M. et al. 2004b. TRPC4 and TRPC4-deficient mice. *Novartis Foundation Symposium*, 258: 189–199; discussion 199–203, 263–266.

Gavva, N.R. et al. 2012. Transient receptor potential melastatin 8 (TRPM8) channels are involved in body temperature regulation. *Molecular Pain*, 8: 36. doi:10.1186/1744-8069-8-36.

Gees, M., B. Colsoul, and B. Nilius. 2010. The role of transient receptor potential cation channels in Ca^{2+} signaling. *Cold Spring Harb Perspect Biol*, 2: a003962. doi:10.1101/cshperspect.a003962.

Gees, M. et al. 2014. Differential effects of bitter compounds on the taste transduction channels TRPM5 and IP3 receptor type 3. *Chemical Senses* 39(4; May): 295–311. doi:10.1093/chemse/bjt115.

Gögelein, H. and B. Pfannmüller. 1989. The nonselective cation channel in the basolateral membrane of rat exocrine pancreas. Inhibition by 3',5-dichlorodiphenylamine-2-carboxylic acid (DCDPC) and activation by stilbene disulfonates. *Pflugers Archiv*, 413(3): 287–298.

Gram, D.X. et al. 2007. Capsaicin-sensitive sensory fibers in the islets of langerhans contribute to defective insulin secretion in zucker diabetic rat, an animal model for some aspects of human type 2 diabetes. *Eur J Neurosci*, 25(1; January): 213–223. doi:10.1111/j.1460-9568.2006.05261.x.

Grand, T. et al. 2008. 9-phenanthrol inhibits human TRPM4 but not TRPM5 cationic channels. *Br J Pharmacol*, 153(8; April): 1697–1705. doi:10.1038/bjp.2008.38.

Grattan, M. et al. 2002. Congenic mapping of the diabetogenic locus Idd4 to a 5.2-cM region of chromosome 11 in NOD mice: Identification of two potential candidate subloci. *Diabetes* 51(1): 215–223. doi:10.2337/diabetes.51.1.215.

Gray, M.A. and B.E. Argent. 1990. Non-selective cation channel on pancreatic duct cells. *Biochimica et Biophysica Acta*, 1029(1): 33–42. doi:10.1016/0005-2736(90)90433-O.

Grimm, C. et al. 2003. Molecular and functional characterization of the melastatin-related cation channel TRPM3. *J Biol Chem*, 278: 21493–21501. doi:10.1074/jbc.M300945200.

Guillot, E., A. Coste, and I. Angel. 1996. Involvement of capsaicin-sensitive nerves in the regulation of glucose tolerance in diabetic rats. *Life Sciences*, 59(12): 969–977. doi:10.1016/0024-3205(96)00403-1.

Guinamard, R., C. Simard, and C. Del Negro. 2013. Flufenamic acid as an ion channel modulator, 138: 272–284. *Pharmacol Ther*. doi:10.1016/j.pharmthera.2013.01.012.

Gyulkhandanyan, A.V. et al. 2006. The Zn^{2+}-transporting pathways in pancreatic beta-cells: A role for the L-type voltage-gated Ca^{2+} channel. *J Biol Chem*, 281: 9361–9372. doi:10.1074/jbc.M508542200.

Halonen, J. and J. Nedergaard. 2002. Adenosine 5'-monophosphate is a selective inhibitor of the brown adipocyte nonselective cation channel. *J Membr Biol*, 188(3): 183–197. doi:10.1007/s00232-001-0184-0.

Hara, Y. et al. 2002. LTRPC2 Ca^{2+}-permeable channel activated by changes in redox status confers susceptibility to cell death. *Molecular Cell*, 9: 163–173. doi:10.1016/S1097-2765(01)00438-5.

Hariri, N. and L. Thibault. 2010. High-fat diet-induced obesity in animal models. *Nutr Res Rev*, 23(2; December): 270–299. doi:10.1017/S0954422410000168.

Held, K. et al. 2015. Activation of TRPM3 by a potent synthetic ligand reveals a role in peptide release. *Proc Nat Acad Sci U S A*, 112(11): E1363–E1372. doi:10.1073/pnas.1419845112.

Henquin, J.C. 2009. Regulation of insulin secretion: A matter of phase control and amplitude modulation. *Diabetologia* 52: 739–751. doi:10.1007/s00125-009-1314-y.

Henquin, J.C. et al. 2009. Shortcomings of current models of glucose-induced insulin secretion. *Diabetes Obes Metab*, 11(Suppl 4): 168–179. doi:10.1111/j.1463-1326.2009.01109.x.

Herder, C. and M. Roden. 2011. Genetics of type 2 diabetes: Pathophysiologic and clinical relevance. *Eur J Clin Invest*, 41(6) (June): 679–692. doi:10.1111/j.1365-2362.2010.02454.x.

Herder, C. et al. 2007. Low-grade inflammation, obesity, and insulin resistance in adolescents. *J Clin Endocrinol Metab*, 92(12): 4569–4574. doi:10.1210/jc.2007-0955.

Hermansen, K. and B. Ahrén. 1990. Dual effects of calcitonin gene-related peptide on insulin secretion in the perfused dog pancreas. *Regulatory Peptides*, 27(1): 149–157. doi:10.1016/0167-0115(90)90213-G.

Hisanaga, E. et al. 2009. Regulation of calcium-permeable TRPV2 channel by insulin in pancreatic beta-cells. *Diabetes*, 58: 174–184. doi:10.2337/db08-0862.

Holstein, A., A. Plaschke, and E.H. Egberts. 2001. Lower incidence of severe hypoglycaemia in patients with type 2 diabetes treated with glimepiride versus glibenclamide. *Diabetes Metab Res Rev*, 17(6): 467–473. doi:10.1002/dmrr.235.

Hurwitz, C.G., V.Y. Hu, and A.S. Segal. 2002. A mechanogated nonselective cation channel in proximal tubule that is ATP sensitive. *Am J Physiol Renal Physiol*, 283(1): F93–F104. doi:10.1152/ajprenal.00239.2001.

Inamura, K. et al. 2003. Response to ADP-ribose by activation of TRPM2 in the CRI-G1 insulinoma cell line. *J Membr Biol*, 191: 201–207. doi:10.1007/s00232-002-1057-x.

Irving, B.A. et al. 2008. Effect of exercise training intensity on abdominal visceral fat and body composition. *Med Sci Sports Exerc*, 40(11; November): 1863–1872. doi:10.1249/MSS.0b013e3181801d40.

Ishii, M. et al. 2006. Intracellular-produced hydroxyl radical mediates H_2O_2-induced Ca^{2+} influx and cell death in rat beta-cell line RIN-5F. *Cell Calcium*, 39(6; June): 487–494. doi:10.1016/j.ceca.2006.01.013.

Jabin Fågelskiöld, A. et al. 2012. Insulin-secreting INS-1E cells express functional TRPV1 channels. *Islets*, 4(1; January 1): 56–63.

Jacobson, D.A. and L.H. Philipson. 2007. TRP channels of the pancreatic beta cell. *Handb Exp Pharmacol*, (179): 409–424. doi:10.1007/978-3-540-34891-7_24.

Joy, S.V., P.T. Rodgers, and A.C. Scates. 2005. Incretin mimetics as emerging treatments for type 2 diabetes. *Ann Pharmacother*, 39(1; January): 110–118. doi:10.1345/aph.1E245.

Kaestner, L. and I. Bernhardt. 2002. Ion channels in the human red blood cell membrane: Their further investigation and physiological relevance. *Bioelectrochem*, 55: 71–74.

Kanzaki, M. et al. 1999. Translocation of a calcium-permeable cation channel induced by insulin-like growth factor-I. *Nat Cell Biol*, 1(3; July): 165–170. doi:10.1038/11086.

Kaske, S. et al. 2007. TRPM5, a taste-signaling transient receptor potential ion-channel, is a ubiquitous signaling component in chemosensory cells. *BMC Neuroscience*, 8(January): 49. doi:10.1186/1471-2202-8-49.

Ketterer, C. et al. 2011. Genetic variation within the TRPM5 locus associates with prediabetic phenotypes in subjects at increased risk for type 2 diabetes. *Metabolism*, 60: 1325–1333.

Kim, C.J. et al. 1998. Erythrocyte lysate releases Ca^{2+} from IP3-sensitive stores and activates Ca^{2+}-dependent K^+ channels in rat basilar smooth muscle cells. *Neurolog Res*, 20(1): 23–30.

Klose, C. et al. 2011. Fenamates as TRP channel blockers: Mefenamic acid selectively blocks TRPM3. *Br J Pharmacol*, 162(8; April): 1757–1769. doi:10.1111/j.1476-5381.2010.01186.x.

Kokrashvili, Z., B. Mosinger, and R.F. Margolskee. 2009. Taste signaling elements expressed in gut enteroendocrine cells regulate nutrient-responsive secretion of gut hormones. *Am J Clin Nutr*, 90: 822S–825S. doi:10.3945/ajcn.2009.27462T.

Kramer, W. et al. 1994. Differential interaction of glimepiride and glibenclamide with the beta-cell sulfonylurea receptor. II. photoaffinity labeling of a 65 kDa protein by [3H]glimepiride. *Biochimica et Biophysica Acta*, 1191(2): 278–290.

Kulkarni, R.N. et al. 1999. Tissue-specific knockout of the insulin receptor in pancreatic β cells creates an insulin secretory defect similar to that in type 2 diabetes. *Cell*, 96: 329–339. doi:10.1016/S0092-8674(00)80546-2.

Kusumakshi, S. et al. 2015. A binary genetic approach to characterize TRPM5 cells in mice. *Chemical Senses*, 40(6; July 1): 413–425. doi:10.1093/chemse/bjv023.

Lambert, L. 2013. Glucagon-like peptide 1 receptor agonists: A new approach to type 2 diabetes management. *S Afr Fam Pract*, 55(6): 511–513.

Lange, I. et al. 2009. TRPM2 functions as a lysosomal Ca^{2+}-release channel in beta cells. *Sci Signal*, 2: ra23. doi:10.1126/scisignal.2000278.

Launay, P. et al. 2004. TRPM4 regulates calcium oscillations after T cell activation. *Science*, 306(5700): 1374–1377. doi:10.1126/science.1098845.

Launay, P. et al. 2002. TRPM4 Is a Ca^{2+}-activated nonselective cation channel mediating cell membrane depolarization. *Cell*, 109(3; May 3): 397–407.

Lebovitz, H.E. 2011. Type 2 diabetes mellitus—Current therapies and the emergence of surgical options. *Nat Rev Endocr*, 7(7; July): 408–419. doi:10.1038/nrendo.2011.10.

Leech, C.A. and J.F. Habener. 1998. A role for Ca^{2+}-sensitive nonselective cation channels in regulating the membrane potential of pancreatic beta-cells. *Diabetes*, 47(7): 1066–1073. doi:10.2337/diabetes.47.7.1066.

Leibiger, I.B., B. Leibiger, and P.O. Berggren. 2002. Insulin feedback action on pancreatic beta-cell function. *FEBS Lett*, 523: 1–6. doi:10.1016/S0014-5793(02)03627-X.

Lemaire, K. et al. 2009. Insulin crystallization depends on zinc transporter ZnT8 expression, but is not required for normal glucose homeostasis in mice. *Proc Nat Acad Sci U S A*, 106: 14872–14877. doi:10.1073/pnas.0906587106.

Lemmens, R. et al. 2001. Ca^{2+}-induced Ca^{2+} release from the endoplasmic reticulum amplifies the Ca^{2+} signal mediated by activation of voltage-gated L-type Ca^{2+} channels in pancreatic beta-cells. *J Biol Chem*, 276: 9971–9977. doi:10.1074/jbc.M009463200.

Li, F. and Z.-M. Zhang. 2009. Comparative identification of Ca^{2+} channel expression in INS-1 and rat pancreatic beta cells. *World J Gastroenterol*, 15: 3046–3050. doi:10.3748/wjg.15.3046.

Liman, E.R. 2003. Regulation by voltage and adenine nucleotides of a Ca^{2+}-activated cation channel from hamster vomeronasal sensory neurons. *J Physiol*, 548(Pt 3): 777–787. doi:10.1113/jphysiol.2002.037119.

Liu, D. and E.R. Liman. 2003. Intracellular Ca^{2+} and the phospholipid PIP2 regulate the taste transduction ion channel TRPM5. *Proc Nat Acad Sci U S A*, 100(25; December 9): 15160–15165. doi:10.1073/pnas.2334159100.

Liu, Y.-J. et al. 1998. Origin of slow and fast oscillations of Ca^{2+} in mouse pancreatic islets. *J Physiol*, 508(2; April): 471–481. doi:10.1111/j.1469-7793.1998.471bq.x.

Lorenzo, A. et al. 1994. Pancreatic islet cell toxicity of amylin associated with type-2 diabetes mellitus. *Nature*, 368: 756–760. doi:10.1038/368756a0.

MacDonald, P.E. and P. Rorsman. 2007. The ins and outs of secretion from pancreatic beta-cells: Control of single-vesicle exo- and endocytosis. *Physiology*, 22(30): 113–121. doi:10.1152/physiol.00047.2006.

MacDonald, P.E. and P. Rorsman. 2006. Oscillations, intercellular coupling, and insulin secretion in pancreatic beta cells. *PLoS Biol*, 4: e49. doi:10.1371/journal.pbio.0040049.

Magistretti, J. and A. Alonso. 2002. Fine gating properties of channels responsible for persistent sodium current generation in entorhinal cortex neurons. *J Gen Physiol*, 120(6): 855–873. doi:10.1085/jgp.20028676.

Mandrup-Poulsen, T. 2003. Apoptotic signal transduction pathways in diabetes. *Biochem Pharmacol*, 66:1433–1440. doi:10.1016/S0006-2952(03)00494-5.

Marigo, V. et al. 2009. TRPM4 impacts on Ca^{2+} signals during agonist-induced insulin secretion in pancreatic β-cells. *Mol Cell Endocr*, 299: 194–203. doi:10.1016/j.mce.2008.11.011.

Maruyama, Y. and O.H. Petersen. 1984. Single calcium-dependent cation channels in mouse pancreatic acinar cells. *J Membr Biol*, 81(1): 83–87. doi:10.1007/BF01868812.

Mathar, I. et al. 2014. TRPM4. *Hand Exp Pharmacol 223*, 223:461–487. doi:10.1007/978-3-319-05161-1.

Mayer, S.I. et al. 2011. Signal transduction of pregnenolone sulfate in insulinoma cells: Activation of Egr-1 expression involving TRPM3, voltage-gated calcium channels, ERK, and ternary complex factors. *J Biol Chem*, 286: 10084–10096. doi:10.1074/jbc.M110.202697.

McCoy, D.D. et al. 2013. Enhanced insulin clearance in mice lacking TRPM8 channels. *Am J Physiol Endocr Metabol*, 305(1; July 1): E78–E88. doi:10.1152/ajpendo.00542.2012.

McKemy, D.D., W.M. Neuhausser, and D. Julius. 2002. Identification of a cold receptor reveals a general role for TRP channels in thermosensation. *Nature*, 416(6876; March 7): 52–58. doi:10.1038/nature719.

Melander, A. et al. 1989. Sulphonylurea antidiabetic drugs. An update of their clinical pharmacology and rational therapeutic use. *Drugs*, 37(1): 58–72.

Melnyk, A. and J. Himms-Hagen. 1995. Resistance to aging-associated obesity in capsaicin-desensitized rats one year after treatment. *Obes Res*, 3(4): 337–344.

Mironov, S.L. 2008. Metabotropic glutamate receptors activate dendritic calcium waves and TRPM channels which drive rhythmic respiratory patterns in mice. *J Physiol*, 586(9): 2277–2291. doi:10.1113/jphysiol.2007.149021.

Miyoshi, H. et al. 2004. Identification of a non-selective cation channel current in myometrial cells isolated from pregnant rats. *Pflugers Archiv*, 447(4): 457–464. doi:10.1007/s00424-003-1175-z.

Nakagawa, Y. et al. 2009. Sweet taste receptor expressed in pancreatic β-cells activates the calcium and cyclic AMP signaling systems and stimulates insulin secretion. *PLoS ONE*, 4(4): e5106. doi:10.1371/journal.pone.0005106.

Nauck, M.A. et al. 2009. Incretin-based therapies: Viewpoints on the way to consensus. *Diabetes Care*, 32(Suppl 2): S223–S231. doi:10.2337/dc09-S315.

Nelson, P.L. et al. 2011. Regulation of Ca^{2+}-entry in pancreatic α-cell line by transient receptor potential melastatin 4 plays a vital role in glucagon release. *Mole Cell Endocr*, 335(2): 126–134. doi:10.1016/j.mce.2011.01.007.

Newsholme, P., C. Gaudel, and N.H. McClenaghan. 2010. Nutrient regulation of insulin secretion and beta-cell functional integrity. *Adv Exp Med Biol*, 654: 91–114. doi:10.1007/978-90-481-3271-3_6.

Nilius, B. et al. 2004. Decavanadate modulates gating of TRPM4 cation channels. *J Physiol*, 560(Pt 3): 753–765. doi:10.1113/jphysiol.2004.070839.

Nilius, B. et al. 2005. Gating of TRP channels: A voltage connection? *J Physiol*, 567:35–44. doi:10.1113/jphysiol.2005.088377.

Noble, M.D. et al. 2006. Local disruption of the celiac ganglion inhibits substance P release and ameliorates caerulein-induced pancreatitis in rats. *Am J Physiol Gastrointest Liver Physiol*, 291(1): G128–G134. doi:10.1152/ajpgi.00442.2005.

Numazawa, S. et al. 2012. Possible involvement of transient receptor potential channels in electrophile-induced insulin secretion from RINm5F cells. *Biol Pharm Bull*, 35(3; January): 346–354. doi:10.1248/bpb.35.346.

Oberwinkler, J. and S.E. Philipp. 2014. TRPM3. *Hand Exp Pharm*, 222 (January): 427–459. doi:10.1007/978-3-642-54215-2_17.

Oberwinkler, J. et al. 2005. Alternative splicing switches the divalent cation selectivity of TRPM3 channels. *J Biol Chem*, 280: 22540–22548. doi:10.1074/jbc.M503092200.

Ohkubo, Y. et al. 1995. Intensive insulin therapy prevents the progression of diabetic microvascular complications in Japanese patients with non-insulin-dependent diabetes mellitus: A randomized prospective 6-year study. *Diabetes Res Clin Pract*, 28(2): 103–117. doi:10.1016/0168-8227(95)01064-K.

Palmer, R.K. et al. 2010. Triphenylphosphine oxide is a potent and selective inhibitor of the transient receptor potential melastatin-5 ion channel. *Assay Drug Dev Technol*, 8(6; December): 703–713. doi:10.1089/adt.2010.0334.

Pare, M. et al. 2007. Differential hypertrophy and atrophy among all types of cutaneous innervation in the glabrous skin of the monkey hand during aging and naturally occurring type 2 diabetes. *J Comp Neurol*, 501(4; April): 543–567.

Partridge, L.D. and D. Swandulla. 1987. Single Ca-activated cation channels in bursting neurons of helix. *Pflugers Archiv*, 410(6): 627–631. doi:10.1007/BF00581323.

Peier, A.M. et al. 2002. A TRP channel that senses cold stimuli and menthol. *Cell*, 108(5; March 8): 705–715.

Peiris, H. et al. 2012. Increased expression of the glucose-responsive gene, RCAN1, causes hypoinsulinemia, β-cell dysfunction, and diabetes. *Endocr*, 153(11; November): 5212–5221. doi:10.1210/en.2011-2149.

Pende, M. et al. 2000. Hypoinsulinaemia, glucose intolerance and diminished beta-cell size in S6K1-deficient mice. *Nature*, 408(6815): 994–997. doi:10.1038/35050135.

Pérez, C.A. et al. 2002. A transient receptor potential channel expressed in taste receptor cells. *Nat Neurosci*, 5(11; November): 1169–1176. doi:10.1038/nn952.

Perraud, A.L. et al. 2001. ADP-ribose gating of the calcium-permeable LTRPC2 channel revealed by Nudix motif homology. *Nature*, 411: 595–599. doi:10.1038/35079100.

Pettersson, M. et al. 1986. Calcitonin gene-related peptide: Occurrence in pancreatic islets in the mouse and the rat and inhibition of insulin secretion in the mouse. *Endocr*, 119: 865–869.

Phillips, D.I. 1996. Insulin resistance as a programmed response to fetal undernutrition. *Diabetologia*, 39(9; September): 1119–1122.

Pingle, S.C., J.A. Matta, and G.P. Ahern. 2007. Capsaicin receptor: TRPV1 a promiscuous TRP channel. *Hand Exp Pharm*, 179(179): 155–171. doi:10.1007/978-3-540-34891-7_9.

Popp, R. and H. Gogelein. 1992. A calcium and ATP sensitive nonselective cation channel in the antiluminal membrane of rat cerebral capillary endothelial cells. *Biochim Biophys Acta*, 1108(1): 59–66. doi:10.1016/0005-2736(92)90114-2.

Prawitt, D. et al. 2003. TRPM5 is a transient Ca^{2+}-activated cation channel responding to rapid changes in $[Ca^{2+}]i$. *Proc Nat Acad Sci U S A*, 100(25; December 9): 15166–15171. doi:10.1073/pnas.2334624100.

Proks, P. et al. 2002. Sulfonylurea stimulation of insulin secretion. *Diabetes*, 51(Suppl 3): 1–14. doi:10.2337/diabetes.51.2007.S368.

Qian, F. et al. 2002. TRP genes: Candidates for nonselective cation channels and store-operated channels in insulin-secreting cells. *Diabetes*, 51(Suppl 1): S183–S189. doi:10.2337/diabetes.51.2007.S183.

Razavi, R. et al. 2006. TRPV1+ sensory neurons control β cell stress and islet inflammation in autoimmune diabetes. *Cell*, 127(6; December): 1123–1135. doi:10.1016/j.cell.2006.10.038.

Rhodes, C.J. 2005. Type 2 diabetes—A matter of beta-cell life and death? *Science*, 307: 380–384. doi:10.1126/science.1104345.

Riccio, A. et al. 2002. mRNA distribution analysis of human TRPC family in CNS and peripheral tissues. *Mole Brain Res*, 109: 95–104. doi:10.1016/S0169-328X(02)00527-2.

Ringer, E., U. Russ, and D. Siemen. 2000. ??3-Adrenergic stimulation and insulin inhibition of non-selective cation channels in white adipocytes of the rat. *Biochim Biophys Acta*, 1463(2): 241–253. doi:10.1016/S0005-2736(99)00216-3.

Roberts, A. et al. 2016. Chemical-specific adjustment factors (inter-species toxicokinetics) to establish the ADI for steviol glycosides. *Regul Toxicol Pharmacol*, 79(May 13): 91–102. doi:10.1016/j.yrtph.2016.05.017.

Rodighiero, S., A. De Simoni, and A. Formenti. 2004. The voltage-dependent nonselective cation current in human red blood cells studied by means of whole-cell and nystatin-perforated patch-clamp techniques. *Biochim Biophys Acta*, 1660(1): 164–170. doi:10.1016/j.bbamem.2003.11.011.

Roe, M.W. et al. 1998a. Characterization of a Ca^{2+} release-activated nonselective cation current regulating membrane potential and $[Ca^{2+}]i$ oscillations in transgenically derived beta-cells. *J Biol Chem*, 273: 10402–10410. doi:10.1074/jbc.273.17.10402.

Roe, M.W. et al. 1998b. Characterization of a Ca^{2+} release-activated nonselective cation current regulating membrane potential and $[Ca^{2+}]^i$ oscillations in transgenically derived beta-cells. *Biochem*, 273(17): 10402–10410. doi:10.1074/jbc.273.17.10402.

Rorsman, P. and G. Trube. 1985. Glucose dependent K+-channels in pancreatic beta-cells are regulated by intracellular ATP. *Pflugers Archiv*, 405: 305–309. doi:10.1007/BF00595682.

Rosenbaum, T. 2015. Activators of TRPM2: Getting it right. *J Gen Physiol*, 145(6; June): 485–487. doi:10.1085/jgp.201511405.

Rosenstock, J. 2001. Management of type 2 diabetes mellitus in the elderly: Special considerations. *Drugs Aging*, 18(1): 31–44. doi:10.2165/11590570-000000000-00000.

Sabourin, J. et al. 2015. Store-operated Ca^{2+} entry mediated by Orai1 and TRPC1 participates to insulin secretion in rat β-cells. *J Biol Chem*, 290(51; December 18): 30530–30539. doi:10.1074/jbc.M115.682583.

Sadeh, M. et al. 2013. Association of the M3151 variant in the transient receptor potential vanilloid receptor-1 (TRPV1) gene with type 1 diabetes in an ashkenazi jewish population. *Isr Med Assoc J*, 15(9; September): 477–480.

Sakura, H. and F.M. Ashcroft. 1997. Identification of four trp1 gene variants murine pancreatic beta-cells. *Diabetologia*, 40: 528–532. doi:10.1007/s001250050711.

Sano, Y. et al. 2001. Immunocyte Ca^{2+} influx system mediated by LTRPC2. *Science*, 293: 1327–1330. doi:10.1126/science.1062473.

Santos-Silva, J.C. et al. 2015. Taurine supplementation ameliorates glucose homeostasis, prevents insulin and glucagon hypersecretion, and controls B, A, and δ-cell masses in genetic obese mice. *Amino Acids*, 47(8): 1533–1548. doi:10.1007/s00726-015-1988-z.

Sathianathan, V. et al. 2003. Insulin induces cobalt uptake in a subpopulation of rat cultured primary sensory neurons. *Euro J Neurosci*, 18: 2477–2486.

Shalinsky, M.H. et al. 2002. Muscarinic activation of a cation current and associated current noise in entorhinal-cortex layer-II neurons. *J Neurophys*, 88(3): 1197–1211.

Shi, Y. and F.B. Hu. 2014. The global implications of diabetes and cancer. *Lancet*, 383(9933; June 7): 1947–1948. doi:10.1016/S0140-6736(14)60886-2.

Shigeto, M. et al. 2015. GLP-1 stimulates insulin secretion by PKC-dependent TRPM4 and TRPM5 activation. *J Clin Invest*, 125(12; December): 4714–4728. doi:10.1172/JCI81975.beyond.

Skrzypski, M. et al. 2013. Activation of TRPV4 channel in pancreatic INS-1E beta cells enhances glucose-stimulated insulin secretion via calcium-dependent mechanisms. *FEBS Lett*, 587(19; October 1): 3281–3287. doi:10.1016/j.febslet.2013.08.025.

Solomon, T.P.J. and A.K. Blannin. 2007. Effects of short-term cinnamon ingestion on in vivo glucose tolerance. *Diabetes, Obes Metabol*, 9(6; November): 895–901. doi:10.1111/j.1463-1326.2006.00694.x.

Stratton, I.M. et al. 2000. Association of glycaemia with macrovascular and microvascular complications of type 2 diabetes (UKPDS 35): Prospective observational study. *BMJ (Clin Res Ed)*, 321(7258): 405–412. doi:10.1136/bmj.321.7258.405.

Stumvoll, M., B.J. Goldstein, and T.W. van Haeften. 2005. Type 2 diabetes: Principles of pathogenesis and therapy. *Lancet*, 365(9467): 1333–1346. doi:10.1016/S0140-6736(05)61032-X.

Sturgess, N.C., C.N. Hales, and M.L. Ashford. 1987. Calcium and ATP regulate the activity of a non-selective cation channel in a rat insulinoma cell line. *Pflugers Archiv*, 409(6): 607–615. doi:10.1007/BF00584661.

Sturgess, N.C. et al. 1987. Nucleotide-sensitive ion channels in human insulin producing tumour cells. *Pflugers Archiv*, 410(1–2): 169–172. doi:10.1007/BF00581911.

Suh, S.H. et al. 2002. ATP and nitric oxide modulate a Ca^{2+}-activated non-selective cation current in macrovascular endothelial cells. *Pflugers Archiv*, 444(3): 438–445. doi:10.1007/s00424-002-0825-x.

Suri, A. and A. Szallasi. 2008. The emerging role of TRPV1 in diabetes and obesity. *Trends Pharmacol Sci*, 29(1; January): 29–36.

Suzuki, K. and O.H. Petersen. 1988. Patch-clamp study of single-channel and whole-cell K+ currents in guinea pig pancreatic acinar cells. *Am J Phys*, 255(3 Pt 1): G275–G285.

Szallasi, A. et al. 2007. The vanilloid receptor TRPV1: 10 years from channel cloning to antagonist proof-of-concept. *Nat Rev Drug Discov*, 6(5; May): 357–372.

Tabur, S. et al. 2015. Role of the transient receptor potential (TRP) channel gene expressions and TRP melastatin (TRPM) channel gene polymorphisms in obesity-related metabolic syndrome. *Euro Rev Med Pharm Sci*, 19(8): 1388–1397.

Takezawa, R. et al. 2006. A pyrazole derivative potently inhibits lymphocyte Ca^{2+} influx and cytokine production by facilitating transient receptor potential melastatin 4 channel activity. *Mole Pharm*, 69(4): 1413–1420. doi:10.1124/mol.105.021154.

Talavera, K. et al. 2005. Heat activation of TRPM5 underlies thermal sensitivity of sweet taste. *Nature*, 438(7070; December 15): 1022–1025. doi:10.1038/nature04248.

Talavera, K. et al. 2007. Influence of temperature on taste perception. *Cell Mol Life Sci*, 64(4): 377–381. doi:10.1007/s00018-006-6384-0.

Talavera, K. et al. 2008. The taste transduction channel TRPM5 Is a locus for bitter-sweet taste interactions. *FASEB J*, 22(5; May): 1343–1355. doi:10.1096/fj.07-9591com.

Tanaka, H. et al. 2011. Enhanced insulin secretion and sensitization in diabetic mice on chronic treatment with a transient receptor potential vanilloid 1 antagonist. *Life Sci*, 88 (11–12; March): 559–563.

Thompson Coon, J. 2010. Goodman and Gilman's the Pharmacological Basis of Therapeutics. *Focus on Alternative and Complementary Therapies*. Vol. 7. New York, NY: McGraw-Hill. doi:10.1111/j.2042-7166.2002.tb05480.x.

Togashi, K. et al. 2006. TRPM2 activation by cyclic ADP-ribose at body temperature is involved in insulin secretion. *EMBO J*, 25: 1804–1815. doi:10.1038/sj.emboj.7601083.

Togashi, K., H. Inada, and M. Tominaga. 2008. Inhibition of the transient receptor potential cation channel TRPM2 by 2-aminoethoxydiphenyl borate (2-APB). *Br J Pharmacol*, 153(6; March): 1324–1330. doi:10.1038/sj.bjp.0707675.

Tóth, B., I. Iordanov, and L. Csanády. 2015. Ruling out pyridine dinucleotides as true TRPM2 channel activators reveals novel direct agonist ADP-ribose-2'-phosphate. *J Gen Phys*, 145(5; May): 419–430. doi:10.1085/jgp.201511377.

Tsui, H. et al. 2007. 'Sensing' autoimmunity in type 1 diabetes. *Trends Mol Med*, 13(10; October): 405–413.

Uchida, K. et al. 2011. Lack of TRPM2 impaired insulin secretion and glucose metabolisms in mice. *Diabetes*, 60: 119–126. doi:10.2337/db10-0276.

Ullrich, N.D. et al. 2005. Comparison of functional properties of the Ca^{2+}-activated cation channels TRPM4 and TRPM5 from mice. *Cell Calcium*, 37(3; March): 267–278. doi:10.1016/j.ceca.2004.11.001.

Van Buren, J.J. et al. 2005. Sensitization and translocation of TRPV1 by insulin and IGF-I. *Mol Pain*, 1: 17. doi:10.1186/1744-8069-1-17.

Van den Abbeele, T., P. Tran Ba Huy, and J. Teulon. 1996. Modulation by purines of calcium-activated non-selective cation channels in the outer hair cells of the guinea-pig cochlea. *J Phys*, 494(Pt 1): 77–89.

Vennekens, R. et al. 2007. Increased IgE-dependent mast cell activation and anaphylactic responses in mice lacking the calcium-activated nonselective cation channel TRPM4. *Nat Immunol*, 8(3; March): 312–320. doi:10.1038/ni1441.

Vriens, J. et al. 2014. Opening of an alternative ion permeation pathway in a nociceptor TRP channel. *Nat Chem Biol*, 10(January): 1–19. doi:10.1038/nchembio.1428.

Wagner, T.F.J. et al. 2008. Transient receptor potential M3 channels are ionotropic steroid receptors in pancreatic beta cells. *Nat Cell Biol*, 10(12; December): 1421–1430. doi:10.1038/ncb1801.

Wagner, T.F.J.et al. 2010. TRPM3 channels provide a regulated influx pathway for zinc in pancreatic beta cells. *Pflugers Archiv*, 460: 755–765. doi:10.1007/s00424-010-0838-9.

Wajchenberg, B.L. 2010. Clinical approaches to preserve beta-cell function in diabetes. *Adv Exp Med Biol*, 654: 515–535. doi:10.1007/978-90-481-3271-3_23.

Wang, P. et al. 2012. Transient receptor potential vanilloid 1 activation enhances gut glucagon-like peptide-1 secretion and improves glucose homeostasis. *diabetes* 61(8; August): 2155–2165. doi:10.2337/db11-1503.

Wang, X., R.L. Miyares, and G.P. Ahern. 2005. Oleoylethanolamide excites vagal sensory neurones, induces visceral pain and reduces short-term food intake in mice via capsaicin receptor TRPV1. *J Physiol*, 564(Pt 2): 541–547. doi:10.1113/jphysiol.2004.081844.

Watanabe, H. et al. 2002. Heat-evoked activation of TRPV4 channels in a HEK293 cell expression system and in native mouse aorta endothelial cells. *J Biol Chem*, 277(49): 47044–47051. doi:10.1074/jbc.M208277200.

Weber, M.J. 2005. New human and mouse microRNA genes found by homology search. *FEBS J*, 272: 59–73. doi:10.1111/j.1432-1033.2004.04389.x.

Wick, E.C. et al. 2006. Transient receptor potential vanilloid 1, calcitonin gene-related peptide, and substance P mediate nociception in acute pancreatitis. *Am J Phys Gastrointest Liver Phys*, 290(5): G959–G969. doi:10.1152/ajpgi.00154.2005.

Winston, J.H. et al. 2005. Molecular and behavioral changes in nociception in a novel rat model of chronic pancreatitis for the study of pain. *Pain*, 117(1–2): 214–222. doi:10.1016/j.pain.2005.06.013.

Winzell, M.S. and B. Ahrén. 2004. The high-fat diet-fed mouse: A model for studying mechanisms and treatment of impaired glucose tolerance and type 2 diabetes. *Diabetes*, 53(Suppl 3; December): S215–S219.

Winzell, M.S. and B. Ahren. 2007. G-Protein-coupled receptors and islet function-implications for treatment of type 2 diabetes. *Pharmacol Ther*, 116: 437–448. doi:10.1016/j.pharmthera.2007.08.002.

World Health Organization. 2016. Global report on diabetes. *WHO Library*: 6.

Wu, T. et al. 2002. Associations of serum C-reactive protein with fasting insulin, glucose, and glycosylated hemoglobin; The Third National Health and Nutrition Examination Survey, 1988-1994. *Am J Epidemiol*, 155(1): 65–71. doi:10.1093/aje/155.1.65.

Xu, X.Z. et al. 2001. Regulation of melastatin, a TRP-related protein, through interaction with a cytoplasmic isoform. *Proc Nat Acad Sci U S A*, 98(19): 10692–10697. doi:10.1073/pnas.191360198.

Yamada, H. et al. 2016. Potentiation of glucose-stimulated insulin secretion by the GPR40-PLC-TRPC pathway in pancreatic β-Cells. *Scientific Reports*, 6: 25912. doi:10.1038/srep25912.

Yarishkin, O.V. et al. 2008. Endogenous TRPM4-like channel in chinese hamster ovary (CHO) cells. *Biochem Biophys Res Comm*, 369(2): 712–717. doi:10.1016/j.bbrc.2008.02.081.

Yee, N.S. et al. 2012a. TRPM8 ion channel is aberrantly expressed and required for preventing replicative senescence in pancreatic adenocarcinoma: Potential role of TRPM8 as a biomarker and target. *Cancer Biol Therapy* 13 (8; June): 592–599. doi:10.4161/cbt.20079.

Yee, N.S. et al. 2012b. TRPM7 and TRPM8 ion channels in pancreatic adenocarcinoma: Potential roles as cancer biomarkers and targets. *Scientifica*, 2012 (January): 415158. doi:10.6064/2012/415158.

Young, R.L. et al. 2009. Expression of taste molecules in the upper gastrointestinal tract in humans with and without type 2 diabetes. *Gut*, 58(3) (March): 337–346. doi:10.1136/gut.2008.148932.

Zhang, Y. et al. 2003. Coding of sweet, bitter, and umami tastes: Different receptor cells sharing similar signaling pathways. *Cell*, 112(3; February 7): 293–301.

Zimmet, P., K.G. Alberti, and J. Shaw. 2001. Global and societal implications of the diabetes epidemic. *Nature*, 414(6865; December 13): 782–787. doi:10.1038/414782a.

13 Airway Pathogenesis Is Linked to TRP Channels

Helen Wallace

CONTENTS

13.1 INTRODUCTION

Respiratory diseases affect the quality of life of an important population worldwide. In light of this, efforts have been made to determine which ion channels are involved in diseases including asthma, chronic obstructive pulmonary disease (COPD), chronic cough, pulmonary arterial hypertension, cystic fibrosis, allergic and nonallergic rhinitis, acute lung injury, and idiopathic pulmonary fibrosis. In the context of some of these diseases, it has been found that the function and/or the expression of some TRP channels may be altered. Among these the TRPA1, TRPV1, TRPV4, and TRM8 channels have been linked to asthma, COPD, and chronic cough among other maladies. This chapter focuses on the role of these channels in normal and abnormal airway function.

13.2 NORMAL AIRWAY PHYSIOLOGY OF TRP CHANNELS

The sensory nerve innervation of the respiratory system is relatively well understood and includes innervation of the smooth muscle and glands of the larynx, trachea, bronchial tree, and lungs (Widdicombe, 1995; Canning, 2011). When activated, sensory afferent nerve impulses are carried via the vagus nerve to the central nervous system, leading to central reflexes including cough, dyspnea, and changes in breathing pattern. Different subtypes of sensory afferent fibers are known to exist including rapidly adapting receptors (RARs), slowly adapting receptors (SARs), and C-fibers (Canning et al., 2006). Two subpopulations of C-fibers are derived from cell bodies found in the jugular ganglion and nodose ganglion. The C-fibers from nodose neurons primarily innervate intrapulmonary structures, whereas jugular ganglia neurons project to the trachea, extrapulmonary bronchi, and lung paranchymal tissue (Spina et al., 2013).

    Sensory neurons in the upper airway are thought to protect the lower airway by limiting exposure to foreign particles including environmental irritants and noxious chemicals. This is achieved by

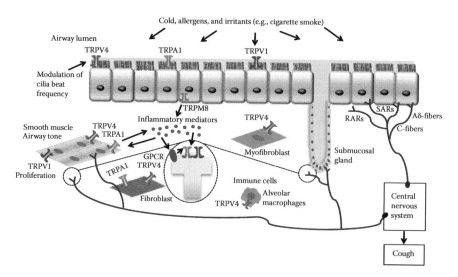

FIGURE 13.1 **(See color insert.)** Localization and function of TRP channels in neuronal and nonneuronal cells of the respiratory tract. (RARs, rapidly adapting receptors; SARs, slowly adapting receptors.)

pathways inducing inflammation, mucus secretion, airway constriction, and reflexes such as cough and sneezing. There is emerging evidence to suggest that TRP channels may play an important role in these airway protective mechanisms. TRP channels have been identified in neuronal and nonneuronal cells of the lung including sensory afferents, epithelia, airway smooth muscle (ASM), immune cells, and inflammatory cells (Nassini et al., 2012) (Figure 13.1). The TRP channels that have been looked at in depth and are discussed in this chapter are TRPA1, TRPV1, TRPV4, and TRPM8.

13.2.1   TRPA1

Transient receptor potential ankyrin 1 (TRPA1) is one of the most extensively studied channels in the airway, and emerging evidence points toward a major role for TRPA1 in the lung's defense system. Under normal conditions, the main functions of TRPA1 are protecting the airway by evoking cough associated with respiratory irritants and minimizing exposure by controlling airway tone causing bronchoconstriction.

Much of the evidence comes from electrophysiology and calcium imaging experiments and utilizing TRPA1 knockout mice and selective TRPA1 antagonists. TRPA1 is referred to as an irritant receptor as it can be activated by exogenous irritants including mustard oil, wasabi, cinnamaldehyde, cigarette smoke, chlorine, aldehydes, and scents (Caceres et al., 2009; Jordt et al., 2004). Activation by endogenous stimuli has also been identified, including reactive oxygen species, hypochlorites, lipid peroxidation products, isoprostanes, and prostaglandins (Bandell et al., 2004; Jha et al., 2015; Bautista, et al., 2005; Andersson et al., 2008).

Quantitative analysis from mouse tissue has shown that TRPA1 channels are mainly expressed in the spinal dorsal root ganglia (DRG) and trigeminal ganglia (TG) (Jang et al., 2012). Expression has also been demonstrated in mouse bronchopulmonary afferent neurons (jugular/nodose) with apparent coexpression of TRPA1 and TRPV1 (Story et al., 2003; Nassenstein et al., 2008). As well as neuronal tissue, evidence also points toward the presence of TRPA1 channels in nonneuronal cells. Using immunohistochemistry, TRPA1 staining has been observed in cultured human and mouse airway epithelial cells and smooth muscle cells (Nassini, 2012). Functional expression has also been demonstrated in human lung fibroblast cells (CCD19-Lu) and in the human pulmonary alveolar epithelial cell line (A549) at both mRNA and protein levels (Mukhopadhyay et al., 2011; Büch et al., 2013).

Activation of TRPA1 leads to neuronal depolarisation and action potential discharge in the afferent fiber. Exogenous compounds such as mustard oil and cinnamaldehyde have been shown to activate TRPA1 channels directly, by covalent modification of reactive cysteine residues in the amino terminus of the protein (Macpherson, 2007; Hinman, 2006). Inflammatory mediators such as bradykinin activate TRPA1 indirectly via G protein-coupled receptors. This leads to a second messenger cascade and increase in intracellular Ca^{2+} (Bandell, 2004). Several studies have demonstrated the importance of Ca^{2+} in the regulation of TRPA1 activity; intracellular Ca^{2+} release may directly activate cation influx through TRPA1 channels (Zurgorg, 2007).

Recent data have demonstrated the importance of the TRPA1 pathway in initiating the normal cough reflex (Birrell et al., 2009; Andrè et al., 2009; Brozmanova et al., 2012; Grace et al., 2012). Isolated vagal nerve preparations have shown that tussive agents activate vagal nerve endings in mouse, guinea pig, and human tissue. A role for TRAP1 was confirmed using TRPA1$^{-/-}$ mice and TRPA1 inhibitors. Furthermore, cough evoked in an *in vivo* guinea pig model and human volunteers was shown to be reduced in the presence of TRPA1 antagonists (Birrell et al., 2009). Studies have identified several triggers of the TRPA1-mediated cough response including exogenous irritants such as those present in cigarette smoke, agents produced endogenously by oxidative stress (Birrell et al., 2009; Andrè et al., 2009; Brozmanova et al., 2012), and the endogenous inflammatory mediators prostaglandin E2 and bradykinin (Grace et al., 2012). As such endogenous mediators are known to be elevated in certain pathological conditions, these findings have implications for the role of TRPA1 channels in disease-associated cough. This is discussed along with the role of pollution in Section 13.3.

A further protective role for TRAP1 has been reported by the modulation of airway tone. Components of cigarette smoke have been shown to cause contraction of isolated guinea pig bronchial rings by activation of TRPA1 channels on sensory neurons. Channel activation triggers tachykinin release, which in turn acts on smooth muscle cells to initiate contraction, a protective mechanism, referred to as the nocifensor system (Andrè, 2008). In addition, the TRPA1 agonist 4-oxo-2-nonenal (4-ONE) and the general anesthetics isoflurane and desflurane have been shown to induce TRPA1-dependent bronchial contraction by a similar mechanism (Satoh et al., 2009). In a more recent study, acrolein a chemical component of smoke, has been shown to relax isolated tracheal smooth muscle (Cheah et al., 2014). Authors suggest relaxation may serve as a defensive mechanism by counteracting the effect of uncontrolled neurogenic excitation and airway narrowing by irritants evoking inflammation and bronchoconstriction. This effect is thought to occur via a pathway involving cross-talk between TRPA1 expressing sensory C-fibers, epithelial cells, and smooth muscle (Cheah et al., 2014). TRPA1 channels have also been implicated in the release of proinflammatory mediators from nonneuronal cells including epithelium, smooth muscle cells, and fibroblasts (Grace et al., 2014).

13.2.2 TRPV1

TRPV1-expressing neurons represent the largest subpopulation of primary sensory neurons. TRPV1-positive neurons belong to cells with C- and Aδ- fibers and are functionally identified as nociceptors. They innervate the entire respiratory tract from the nose to the alveoli including smooth muscle and blood vessels (Watanabe et al., 2006; Grace et al., 2014). TRPA1-positive neurons are most notably stretch insensitive and peptidergic, and are coexpressed with TRPV1 channels. However, TRPV1-postive neurons can be both stretch insensitive and stretch sensitive and either peptidergic or nonpeptidergic. They are also not exclusively expressed with TRPA1 (Bhattacharya et al., 2008).

Observations have reported a positive interaction between TRPA1 and TRPV1 regulating the sensitivity of sensory neurons during an inflammatory reaction (Lee and Hsu, 2015). Like TRPA1, TRPV1 stimulation may also be the consequence of G protein-coupled receptor (GPCR) activation by inflammatory mediators, such as bradykinin (Bautista et al., 2006; Grace et al., 2012). Low levels of TRPV1 in the airway may suggest that TRPV1 has limited physiological function, but its ability to be activated by proinflammatory mediators suggests that this channel may play an important role in inflammatory conditions and be of pathological importance. As is the case with TRPA1, TRPV1 has also been implicated in playing a role in airway defense, and several studies have been carried out investigating the role of TRPV1 sensory nerves in the initiation of the normal cough reflex. Cough induced in human and animal models by substances that activate TRPV1 such as capsaicin and citric acid is inhibited by TRPV1 antagonists (Bhattacharya et al., 2007; Grace et al., 2014). Investigators have also proposed that TRPV1 is involved in modulation of airway tone, and bronchoconstriction occurs by a mechanism involving activation of TRPV1 by inflammatory mediators (Delescluse et al., 2012). In this study, increased airway contraction in a disease model was inhibited by TRPV1 antagonists. Further evidence for a role in defense mechanisms has been shown by a study demonstrating that environmental prototype particulate matter (PM) is sensed by TRPV1 (Deering-Rice et al., 2012). This study showed that TRPV1 mediates the induction of several important proinflammatory cytokine/chemokine genes in lung epithelial cells in response to PM.

It is also important to note the presence and function of TRPV1 in nonneuronal cells in the airway. Functional TRPV1 protein has been demonstrated in cultured primary bronchial epithelial cells by patch-clamp experiments (McGarvey et al., 2014). Stimulation by capsaicin induced a dose-dependent release of interleukin 8 (IL-8), which, being blocked by capsazepine, suggests TRPV1 activation (McGarvey et al., 2014). TRPV1 has also been shown to be expressed in smooth muscle cells and to play a role in airway smooth muscle physiology, including proliferation and apoptosis (Zhao, 2013) and control of airway tone (Grace et al., 2014). Several recent studies have demonstrated expression of TRP channels in inflammatory and immune cells (Parenti et al., 2016). As discussed later in the chapter, emerging evidence suggests that TRP channels may be responsible at least in part for the transition of early defensive immune and inflammatory responses to chronic responses and disease pathology (Parenti et al., 2016). In relation to TRPV1, murine CD4$^+$ T lymphocytes have been shown to express functional TRPV1, which was activated on stimulation of the T cell antigen receptor (TCR), contributed to Ca^{2+} influx and TCR signaling, resulting in T-cell activation. Inhibition of TRPV1 in mouse and human CD4$^+$ T cells, with antagonists or by genetic manipulation, resulted in a Trpv1$^{-/-}$ CD4$^+$ T-cell-like phenotype (Bertin et al., 2014).

13.2.3   TRPV4

Compared to TRPA1 and TRPV1, little is known about TRPV4 expression in peripheral sensory neurons that innervate the lung. However, a recent study identifies the TRPV4-ATP-P2X3 interaction as a key osmosensing pathway involved in airway sensory nerve reflexes (Bonvini, 2016). TRPV4 ligands and hypoosmotic solutions caused depolarization of murine, guinea pig, and human vagus nerves, and firing of Aδ-fibers, which was inhibited by TRPV4 and P2X3 receptor antagonists; both antagonists blocked TRPV4-induced cough. In another study, TRPV4 has also been implicated in the development of neurogenic inflammation (Vergnolle et al., 2010). Hypotonic solutions and 4αPDD have been shown to induce neuropeptide release from afferent nerves in isolated murine airways (Vergnolle et al., 2010). These data suggest that TRPV4 also plays a role in airway defense.

Several studies have demonstrated that TRPV4 is widely expressed in nonneuronal cells in the airways, including mRNA expression in human airway smooth muscle cells (Jia et al., 2004; Dietrich et al., 2006), confirmation of expression by immunohistochemistry in the alveolar septal wall of human, rat and mouse lung (Alvarez et al., 2006), and human nasal, tracheal and bronchial epithelial cells (Alenmyr et al., 2014; Fernández-Fernández et al., 2008). TRPV4 is also present in lung fibroblasts (Rahaman et al., 2014) and inflammatory cells including alveolar macrophages (Hamanaka et al., 2010) and mononuclear cells (Delany et al., 2001).

Compared to other TRP channels, less is also known about external and endogenous ligands that may be responsible for TRPV4 activation and function, and indeed the mechanisms involved. However, known channel activators include osmotic changes, for example, increasing activity in hypotonic solutions, and mechanical stimuli such as shear stress (Garcia-Elias et al., 2014). Both stimuli depend on phospholipase A2 activation and subsequent production of arachidonic acid (AA) metabolites.

TRPV4 channels can also be directly activated by AA (Zheng et al., 2013). Other TRPV4 channel activators include moderate heat (24°C–38°C) and the synthetic phorbol ester 4α-PDD (Watanabe et al., 2002a). Evidence suggests that TRPV4 channel activity is regulated by Ca^{2+}, depending on its concentration. Ca^{2+} can either potentiate or inhibit channel activity (Garcia-Elias et al., 2014).

A role for TRPV4 in airway smooth muscle function has been demonstrated in human bronchial smooth muscle cells, where 4α-PDD, stretch, and hypotonic solutions caused increases in Ca^{2+} and subsequent contraction (Jia et al., 2004). An indirect role for TRPV4 in contraction has been shown in human bronchial smooth muscle segments using the TRPV4 agonist GSK1016790 (McAlexander et al., 2014). This agonist was shown to cause contraction via 5-lipoxygenase activation and production of cysteinyl leukotrienes (McAlexander et al., 2014).

Further evidence for the involvement of TRPV4 in lung defense has been shown by studies demonstrating activation of TRP4 modulating ciliary beat frequency (Lorenzo et al., 2008) and the control of epithelial and endothelial barrier function (Alvarez et al., 2006; Li et al., 2011a). A potential role in macrophage activation by mechanical stress (Hamanaka et al., 2010) has also been suggested.

13.2.4 TRPM8

Evidence to date suggests TRPM8 may play a minimal role in airway physiology with contradictory findings often reported (Grace et al., 2014). One study demonstrated that TRPM8 is not expressed in mouse vagal afferents innervating the airways, and menthol, a TRPM8 agonist, was ineffective at increasing intracellular Ca^{2+} in lung-specific vagal sensory neurons (Nassenstein et al., 2008). Other studies have found evidence of a subpopulation of rat airway vagal afferent nerves expressing TRPM8 receptors and that these receptors are activated by cold temperatures (Xing et al., 2008). It is implied from these data that activation is likely to trigger responses such as airway constriction in response to a reduction in temperature. Molecular analysis has also identified expression of TRPM8 in a subset of nasal trigeminal afferent neurons (Plevkova et al., 2013). However, authors have suggested that activation of TRPM8 has an inhibitory effect; they conclude that menthol suppresses cough evoked in the lower airways primarily through this subset of neurons initiated from the nose (Buday et al., 2012; Plevkova et al., 2013).

TRPM8 is also expressed in airway epithelial cells and has recently been shown to inhibit proliferation and migration in a rat asthma model (Zhang, 2016). TRPM8 has been identified primarily within endoplasmic reticulum membranes of epithelial cells (Sabnis et al., 2008) and may play a role in production of mucus and inflammatory mediators from airway epithelium (Sabnis et al., 2008; Grace et al., 2014).

13.3 TRP CHANNELS AND AIRWAY PATHOPHYSIOLOGY

As already highlighted, TRP channels play a definitive role in protecting the airway from foreign particles. When activated, TRP channels can decrease respiratory drive, trigger cough, induce airway narrowing by modulating airway tone, and induce a coordinated inflammatory response. Evidence is emerging for the involvement of TRP channels in the cross-talk between immunogenic and neurogenic pathways in airway inflammation, as these channels play a key role in the response of sensory neurons to inflammatory mediators. When exaggerated, these same mechanisms may give rise to neurogenic inflammation, airway hyperactivity (AHR), and abnormal cough. Such symptoms are associated with diseases such as asthma, COPD, and chronic cough.

13.3.1 ASTHMA

Asthma is a chronic inflammatory airway disease affecting an estimated 300 million people worldwide. The number of annual deaths attributed to the disease is approximately 250,000. Physiological hallmarks of asthma include AHR and variable airflow obstruction. The disease is defined by symptoms including wheeze, chest tightness, shortness of breath, cough, and dyspnea. The airway inflammatory response can vary between asthmatic patients, and as the disease becomes more persistent, further airway obstruction can occur from edema, mucus hypersecretion, and airway remodeling. Repeated exposure to environmental allergens is thought to be the underlying cause of asthma, the early asthmatic response occurring minutes after initial exposure to allergen. Recognition of the allergen by IgE on the surface of mast cells induces degranulation and release of inflammatory mediators such as histamine and cysteinyl leukotriene (CysLT), causing acute bronchoconstriction (Adelroth et al., 1986). A late asthmatic response (LAR) can follow 3–8 hours after allergen exposure and only occurs in approximately 50% of patients (O'Byrne et al., 2009). The late asthmatic response results in further bronchoconstriction and is also associated with a complex immune and inflammatory response. This response is linked to an increase in inflammatory cells including Th-2 cells, eosinophils, airway, basophils, and less consistently neutrophils and trafficking and activation of myeloid dendritic cells into the airways (Holgate et al., 2008; Gauvreau et al., 2015). Prominent inflammatory mediators include IL-4, IL-5, IL-13, eotaxin (CCL11), and eicosanoids. The LAR phenotype can also occur in response to other triggers such as exercise, cold air, irritants, and stress. The mechanisms regulating the airway response to these factors are less well defined. Isocyanate is another known trigger in patients suffering from occupational asthma (Kenyon et al., 2012). Disease exacerbations in asthmatic patients can also be caused by respiratory infections.

TRP channels are likely candidates in the pathophysiology of asthma, as several exogenous and endogenous TRP channel agonists such as cigarette smoke and reactive oxygen species are also known asthma triggers. Irritants can trigger asthma-type symptoms and subsequent hypersensitivity to chemical and physical stimuli (Preti et al., 2012). Endogenous mediators produced by infiltrating immune cells or inflamed airway tissue can reach elevated levels high enough to chronically activate TRPA1 in airway sensory neurons (Trevisani et al., 2007). A critical role for TRPA1 in the pathogenesis of asthma has been shown in several studies using the TRPA1 knockout mouse. These mice have been shown to be deficient in the neuronal detection of proinflammatory asthma agents (Bessac et al., 2008; Andrè et al., 2008). Lack of TRPA1 may prevent neuronal excitation and Ca^{2+} influx normally activated by these mediators, which is likely to occur during inflammatory progression in asthma (Caceres et al., 2009). Further evidence of a role for TRPA1 in asthmatic airway inflammation has been shown in the mouse ovalbumin model of asthma (Caceres et al., 2009). In this study, pharmacological inhibition and genetic deletion of TRPA1 diminish allergen-induced inflammatory leukocyte infiltration, mucus production, cytokine and chemokine levels, and AHR. TRPAP1 knockout mice also show impaired acute and inflammatory neuropeptide release in the airways required for leukocyte infiltration. A role for TRPV1 is less clear as data from asthma models have reported conflicting results since lack of TRPV1 did not demonstrate any change in the mouse ovalbumin model (Caceres et al., 2009). However, other asthma models have shown that modulation of TRPV1 attenuated the asthma model inflammatory phenotype (Mori et al., 2011; Rehman et al., 2013).

Evidence of a role for TRP channels in control of airway tone was discussed earlier in the chapter. It is therefore not surprising to find studies linking TRP channel activation and AHR in asthma models. The results of one such study using ovalbumin-sensitized animals, show abolished LAR response in the presence of TRP channel inhibitors (Raemdonck et al., 2012). The authors conclude that allergen challenge activates TRPA1 channels on sensory neurons, which in turn triggers a central reflex leading to parasympathetic cholinergic activation of airway smooth muscle. Although this study ruled out a role for TRPV1 in AHR, data from unanesthetized, ovalbumin-sensitized guinea pig showed inhibition of AHR by TRPV1 antagonists (Delescluse et al., 2012). A recent study has demonstrated a link between TRPA1 and TRPV1 in the development of AHR in response to the chemical sensitizer toluene-2,4-diisocyanate (TDI) (Devos et al., 2016). Authors used a mouse model of TDI sensitization to induce AHR, and blocked TRPA1- and TRPV1-channel activity using pharmacological methods and knockout mice. They put forward a neuroimmune pathway as a possible mechanism after demonstrating that TRPA1 and TRPV1 channels and mast cell knockout mice did not exhibit AHR despite being sensitized by TDI.

The discovery of genetic variants of certain TRP channels has led to studies demonstrating a link between the presence of certain polymorphisms and altered disease presentation. The outcome of one such study showed an association between a TRPV1 single nucleotide poly-morphism (SNP), and a protective effect against the presence of wheezing in a group of asthma patients (Cantero-Recasens et al., 2010). Fluorescent microscopy revealed a lower increase in Ca^{2+} concentration upon stimulation indicating a decrease in channel activity. These results provide evidence for a role of TRPV1 channels in altered Ca^{2+} signaling underlying asthma pathophysiol-ogy. Another study has implicated TRPA1 in the modulation of asthma in children exposed to high levels of pollution (Deering-Rice et al., 2015). Expression of variant forms of TRPA1 may increase the sensitivity of TRPA1 to insoluble particulate matter including 3,5-ditert butylphe-nol, a soluble nonelectrophilic agonist and component of diesel exhaust particles. Such polymor-phisms of the TRPA1 channel may lead to an increase in activation and a correlation with reduced asthma control.

As the role of TRP channels in asthma pathophysiology becomes more apparent, changes in expression levels or patterns of TRP channel expression become increasingly likely. A search of the literature revealed a study demonstrating altered TRPV1 expression in a rat model of chronic inflam-mation induced by an ovalbumin sensitization (Zhang et al., 2008). Authors report an increase in the proportion of TRPV1 expressing neurons in pulmonary myelinated afferents in the nodose gan-glia. Data also showed an increase in sensitivity to capsaicin of vagal bronchopulmonary myelin-ated afferents. Further studies using a similar model in guinea pigs have shown that inflammation causes a "phenotypic switch in vagal tracheal cough-causing, low-threshold mechanosensitive Aδ neurons, such that they begin expressing TRPV1 channels" (Lieu et al., 2012). Authors also provide evidence to suggest a role for neurotrophic factors in the modulation of gene expression of TRPV1 and nerve phenotype, which could contribute to excessive coughing caused by allergens. Other results have shown an increase in sensitivity to continuous capsaicin inhalation in cough-variant asthma (Nakajima et al., 2006). Other evidence has highlighted the role of TRPV1 and TRPA1 in sensitization resulting in exaggerated sensory responses to ROS causing airway hypersensitivity—a characteristic feature of asthma (Ruan et al., 2014).

13.3.2 Chronic Obstructive Pulmonary Disease

Chronic obstructive pulmonary disease (COPD) is a term used to describe a number of condi-tions including emphysema that affects the alveoli, and chronic bronchitis affecting the bronchi. According to the Global Initiative for COPD guidelines, it is characterized by persistent airflow limitation that is usually progressive and associated with an enhanced chronic inflammatory response to noxious particles or gases (Pauwels et al., 2001). It is well established that tobacco smoke increases susceptibility to COPD, but other factors including genetic factors and prolonged exposure to occupational dusts and chemicals can also increase the risk of developing the disease (Salvi and Barnes, 2009). Symptoms include chronic cough, dyspnea, and excess sputum produc-tion; exacerbations also often occur. Examination of smokers with chronic bronchitis has shown an increased number of CD8+ T lymphocytes, neutrophils, and macrophages in surgical pulmonary specimens (Saetta et al., 1997; Saetta et al., 1998). Increased levels of proinflammatory mediators have also been shown in patients with COPD, which are significantly associated with disease severity (Barnes, 2016; Selvarajah et al., 2016).

Recent evidence suggests that TRPA1 is the primary TRP channel implicated in the patho-genesis of COPD. As described earlier, it is activated by several components of cigarette smoke, including acrolein and crotonaldehyde and nicotine (Lin et al., 2010; Andrè et al., 2008; Talavera et al., 2009), so a role of TRPA1 is highly likely. Cigarette smoke aqueous extract (CSE) has been shown to mobilize Ca^{2+} in cultured guinea pig jugular ganglia neurons and promote contraction of isolated guinea pig bronchi (Andrè et al., 2008). CSE and unsaturated aldehydes contract the guinea pig bronchus by a neurogenic mechanism mediated by TRPA1 stimulation, Ca^{2+} mobilization, and release of neuropeptides. Another study by Shapiro and colleagues has investigated the effects of wood smoke particle material (WSPM), which has been shown to have a clear impact on human health causing a progressive decline in lung function and development of chronic COPD (Naeher et al., 2007; Laumbach et al., 2012; Shapiro et al., 2013). This study demonstrated that TRPA1 in trigeminal sensory neurons was activated by WSPM resulting in Ca^{2+} flux shown by imaging experiments. Continued exposure to WSPM therefore has implications for the development of neurogenic inflammation, decreased respiratory drive, chronic cough, and bronchoconstriction by activating this pathway (Shapiro et al., 2013).

There is emerging evidence to suggest the involvement of other TRP channels in respiratory symptoms associated with COPD. The suggestion of a possible role for TRPV1 comes from the effects of tiotropium, a widely prescribed drug for its bronchodilator effects in COPD and asthma. It has also been shown to attenuate cough in preclinical and tussive challenge studies (Bateman et al., 2009). Recent data have shown that tiotropium inhibited capsaicin-induced Ca^{2+} responses and currents in isolated guinea pig vagal tissue, implying a TRPV1 neuronal mediated effect (Birrell et al., 2014). Increased TRPV1 and TRPV4 mRNA expression has also been demonstrated in patients with COPD (Baxter et al., 2014). In the same study, TRPV1 and TRPV4 were shown to play a role in cigarette smoke (CS)–induced release of extracellular ATP, which is elevated in the COPD airway and is thought to be intrinsic to CS-induced inflammation. Moreover, SNPs in TRVP4 have been associated with COPD (Zhu et al., 2009).

13.3.3 CHRONIC COUGH

Chronic cough is typically defined as a cough that lasts longer than 8 weeks and occurs in approximately 40% of the population. As discussed above, chronic cough is present in diseases such as COPD and asthma, but can also occur as a persistent cough after a viral or bacterial infection or in isolation with an unknown cause. The underlying mechanisms of chronic cough are complex, and causes of exaggerated cough in disease pathology are not completely understood. As already described, TRP channels play an important role in the cough reflex as a protective function in airway defense so are likely to be involved in abnormal cough presentation. A role for TRPV1 in chronic cough is a likely scenario due to the protussive effects of the TRPV1 agonist capsaicin. Increased expression of TRPV1 in sensory nerves has been demonstrated in patients with chronic cough (Mitchell et al., 2005; Groneberg et al., 2004). Moreover, in a recent study an association has been made between the presence of TRPV1 SNPs and a higher risk of developing chronic cough in patients who smoke, or with occupational exposure (Smit et al., 2012). Studies have also shown an increase in sensitivity to capsaicin in patients with chronic cough (Doherty et al., 2000; Pecova et al., 2008). The TRPA1 channel is also involved in cough associated with asthma and COPD as described above. TRPA1 is activated by reactive electrophilic molecules including acrolein that is a by-product of oxidative stress that can lead to activation of the cough reflex (Bautista et al., 2006; Trevisani et al., 2007). As oxidative stress can be initiated by environmental irritants and as a by-product of inflammation, this is a possible pathway whereby exaggerated cough is initiated. This is made more likely during inflammation by the increase of other endogenous ligands that directly activate TPA1, such as prostaglandins (Grace and Belvisi, 2011).

Patients suffering from idiopathic cough have been reported to have suffered from a viral infection preceding the onset of their cough (Haque et al., 2005). Human rhinovirus is the major cause of the common cold and has also been shown to be responsible for asthma exacerbations (Arden et al., 2006; Nicholson et al., 1993). A recent study has demonstrated the ability of rhinovirus to infect neuronal cells, and that infection causes an increase in expression of TRPV1, TRPA1, and TRPM8 mRNA and protein (Abdullah et al., 2014). Virus-induced soluble factors, for example, IL-8, IL-16, and nerve growth factor, were shown to be sufficient for TRPV1 and TRPA1 upregulation in culture,

Interestingly, mechanisms of upregulation differed for TRPM8 which required replicating virus. This suggests that direct interaction with virus particles and neuronal cells is required for TRPM8 upregulation (Abdullah et al., 2014). Taken together these data suggest a role for certain TRP channels on sensory afferents in postviral cough.

13.3.4 OTHER RESPIRATORY DISEASES

Nonallergic rhinitis (NAR) is inflammation of the nose that affects approximately 10% of people worldwide. Symptoms include sneezing, irritation, and nasal congestion that are induced by hypersensitivity to stimuli including smoke, temperature changes, and irritants. TRPV1 channels on C-fiber afferents innervating the nasal mucosa are thought to play a role in NAR, since the activation of this ion channel causes the release of neuropeptides substance P and calcitonin G-related peptides from nerve terminals resulting in a local inflammatory response (Van Gerven et al., 2012). Nasal application of capsaicin has been shown to alleviate the symptoms of TRPV1. This is thought to occur by desensitization after strong excitation of the TRPV1 neurons, or massive influx of Ca^{2+} resulting in nonfunctional afferents or degeneration of nerve terminals (Anand and Bley, 2011; Van Gerven et al., 2014). The role of neuronal TRP channels in allergic rhinitis is less clear. It has been hypothesized that allergic rhinitis induced by exposure to allergens may have increased sensitivity to stimulation of TRP channels (Alenmyr, 2009). This study went on to demonstrate an increase in itch response to stimulation of TRPV1 at seasonal allergen exposure. However, the TRPV1 blocker SB-705498 had no effect on nasal symptoms of patients with seasonal allergic rhinitis including itch (Alenmyr et al., 2012). More recent studies have concentrated on alternative target cells, demonstrating localization of TRPV1 expression to CD4[+] T cells (Samivel et al., 2016). TRPV1 has been shown to be involved in T-cell receptor signaling that is altered in TRPV1 knockout allergic rhinitis mice (Samivel et al., 2016).

Apneic responses (lethal ventilator arrest) have recently been shown to be associated with upregulation of TRPV1 channels (Zhuang et al., 2015). Prenatal exposure to nicotine has been shown to trigger apneic responses during severe hypoxia (Zhuang et al., 2014). This is caused by lack of inspiratory drive from the CNS but the underlying mechanisms are poorly understood. The study by Zhuang et al. (2015) has demonstrated an association between apneic responses and sensitization of bronchopulmonary C-fibers shown by increased firing rate. Upregulation of TRPV1 was associated with increased gene expression of the tropomyosin receptor kinase A (TrkA) in the nodose/jugular ganglia.

13.4 NONNEURONAL TRP CHANNELS CAUSING DISEASE

As described in Section 13.2, TRP channels are expressed in several nonneuronal cells in the airway including epithelial cells, smooth muscle cells, fibroblasts and inflammatory cells. It is important to note that increasing evidence highlights the role of these TRP channels in airway disease. Increased airway smooth muscle contractility and subsequent airway narrowing is associated with obstructive disease states including asthma and COPD and has been shown to be due in part to increased activation of sensory afferents. Airway smooth muscle has also been shown to undergo structural changes contributing to airway modeling, and to orchestrate the inflammatory process in chronic airway disease. TRP channels expressed by ASM cells including members of the TRPC, TRPM, TRPV, and TRPA superfamilies are thought to be important for regulating abnormal SM structural functional changes during disease (Ong et al., 2003; Jha et al., 2015). Recent studies in the cystic fibrosis airway have shown that TRPC6-mediated Ca^{2+} influx was abnormally increased in cystic fibrosis (CF) epithelial cells (Antigny et al., 2011). Authors conclude that abnormal Ca_{2+} signaling contributing to CF airway pathophysiology is due to a loss of functional coupling between TRPC6 and CFTR, which is dysfunctional in CF. A recent study has identified a role for TRPA1 in the inflammatory response to infection in the CF airway. Results showed that inhibition of TRPA1 expression resulted in a reduction of release of IL-8, IL-1β, and TNF-α, from CF primary bronchial epithelial cells exposed to *Pseudomonas aeruginosa* (Prandini et al., 2016) TRPV4 channels have recently been implicated in airway pathologies including acute lung injury (ALI), idiopathic pulmonary fibrosis (IPF), and pulmonary hypertention (De Logu, 2016). TRPV4 expressed by neutrophils has been identified as a novel regulator of neutrophil activation and response to proinflammatory stimuli in ALI pathophysiology (Yin et al., 2016). The TRPV4 channel has also been linked to pulmonary fibrogenesis in IPF as the mechanical sensor that controls myofibroblast differentiation (Rahaman et al., 2014). TRPV4 is also thought to be involved in the development of pulmonary hypertension due to the role of this channel in vasoconstriction and regulation of arterial tone (Xia et al., 2013).

13.5   CONCLUSION

In summary, TRP channels play an integral part in protecting the airway from environmental challenges. It is unclear what causes the transition from early defensive immune and inflammatory responses to chronic responses and disease pathology, but it is thought that TRP channels may be at least in part responsible. Enhanced neuronal sensory responses in the diseased airway may be caused by increased expression of TRP channels or altered expression patterns resulting in a change in phenotype. Increased channel activity may be a result of sensitization, increasing excitability of afferent nerves reducing the threshold for activation or alternatively an increase in inflammatory mediators that can activate TRP channels. Further work will help to understand the communication between neurogenic and immune responses and the development of new therapeutic strategies.

REFERENCES

Abdullah, H. et al. 2014. Rhinovirus upregulates transient receptor potential channels in a human neuronal cell line: Implications for respiratory virus-induced cough reflex sensitivity. *Thorax*, 69(1): 46–54.

Adelroth, E. et al. 1986. Airway responsiveness to leukotrienes C4 and D4 and to methacholine in patients with asthma and normal controls. *N Engl J Med*, 315: 480–484.

Alenmyr, L. et al. 2009. TRPV1-mediated itch in seasonal allergic rhinitis. *Allergy*, 64(5): 807–810.

Alenmyr, L. et al. 2012. Effect of mucosal TRPV1 inhibition in allergic rhinitis. *Basic Clin Pharmacol Toxicol*, 110(3): 264–268.

Alenmyr, L. et al. 2014. TRPV4-mediated calcium influx and ciliary activity in human native airway epithelial cells. *Basic Clin Pharmacol Toxicol*, 114: 210–216.

Alvarez, D.F. et al. 2006. Transient receptor potential vanilloid 4-mediated disruption of the alveolar septal barrier: A novel mechanism of acute lung injury. *Circ Res*, 99: 988–995.

Anand, P. and K. Bley. 2011. Topical capsaicin for pain management: Therapeutic potential and mechanisms of action of the new high-concentration capsaicin 8% patch. *Br J Anaesth*, 107: 490–502.

Andersson, D.A. et al. 2008. Transient receptor potential A1 is a sensory receptor for multiple products of oxidative stress. *J Neurosci*, 28(10): 2485–2494.

Andrè, E., B. Campi, and S. Materazzi. 2008. Cigarette smoke–induced neurogenic inflammation is mediated by α, β-unsaturated aldehydes and the TRPA1 receptor in rodents. *J Clin Invest*, 118: 2574–2582.

Andrè, E. et al. 2009. Transient receptor potential ankyrin receptor 1 is a novel target for pro-tussive agents. *Br J Pharmacol*, 158: 1621–1628.

Antigny, F. et al. 2011. Transient receptor potential canonical channel 6 links Ca^{2+} mishandling to cystic fibrosis transmembrane conductance regulator channel dysfunction in cystic fibrosis. *Am J Respir Cell Mol Biol*, 44(1): 83–90.

Arden, K. E. et al. 2006. Frequent detection of human rhinoviruses, paramyxoviruses, coronaviruses, and bocavirus during acute respiratory tract infections. *J Med Virol*, 78: 1232–1240.

Bandell, M. et al. 2004. Noxious cold ion channel TRPA1 is activated by pungent compounds and bradykinin. *Neuron*, 41: 849–857.

Barnes, P.J. 2016. Inflammatory mechanisms in patients with chronic obstructive pulmonary disease. *J Allergy Clin Immunol*, 138(1):16-27.

Bateman, E.D. et al. 2009. Alternative mechanisms for tiotropium. *Pulm Pharmacol Ther*, 22(6): 533–542.

Bautista, D.M. et al. 2005. Pungent products from garlic activate the sensory ion channel TRPA1. *Proc Natl Acad Sci USA*, 102: 12248–12252.

Bautista, D.M. et al. 2006. TRPA1 mediates the inflammatory actions of environmental irritants and proalgesic agents. *Cell*, 124(6): 1269–1282.

Baxter, M. et al. 2014. Role of transient receptor potential and pannexin channels in cigarette smoke-triggered ATP release in the lung. *Thorax*, 69: 1080–1089.

Bertin, S. et al. 2014. The ion channel TRPV1 regulates the activation and proinflammatory properties of CD4+ T cells. *Nat Immunol*, 15: 1055–1063.

Bessac, B.F. et al. 2008. TRPA1 is a major oxidant sensor in murine airway sensory neurons. *J Clin Invest*, 118(5): 1899–1910.

Bhattacharya, A. et al. 2007. Pharmacology and antitussive efficacy of 4-(3-trifluoromethyl-pyridin-2-yl)-peperazine-1-carboxylic acid (5-trifluoromethyl-pyridin-2-yl)-amide (JNJ17203212), a transient receptor potential vanilloid 1 antagonist in guinea pigs. *J Pharmacol Exp Ther*, 323: 665–674.

Bhattacharya, M.R. et al. 2008. Radial stretch reveals distinct populations of mechanosensitive mammalian somatosensory neurons. *Proc Natl Acad Sci USA*, 105: 20015–20020.

Birrell, M.A. et al. 2009. TRPA1 agonists evoke coughing in guinea-pig and human volunteers. *Am J Respir Crit Care Med*, 180: 1042–1047.

Birrell, M.A. et al. 2014. Tiotropium modulates transient receptor potential V1 (TRPV1) in airway sensory nerves: A beneficial off-target effect? *J Allergy Clin Immunol*, 133(3): 679–687.

BonvinI, S.J. et al 2016. Transient receptor potential cation channel, subfamily V, member 4 and airway sensory afferent activation: Role of adenosine triphosphate. *J Allergy Clin Immunol*, 138(1): 249–261.

Brozmanova, M. et al. 2012. Comparison of TRPA1-versus TRPV1-mediated cough in guinea pigs. *Eur J Pharmacol*, 689: 211–218.

Büch, T.R.H. et al. 2013. Functional expression of the transient receptor potential channel TRPA1, a sensor for toxic lung inhalants, in pulmonary epithelial cells. *Chem Biol Interact*, 206(3): 462–471.

Buday, T. et al. 2012. Modulation of cough response by sensory inputs from the nose—Role of trigeminal TRPA1 versus TRPM8 channels. *Cough*, 8: 11.

Caceres, A.I. et al. 2009. A sensory neuronal ion channel essential for airway inflammation and hyperreactivity in asthma. *Proc Natl Acad Sci U S A*, 106: 9099–9104.

Canning, B.J. 2011. Functional implications of the multiple afferent pathways regulating cough. *Pulm Pharmacol Ther*, 24: 295–299.

Canning, B.J., N. Mori, and S.B. Mazzone. 2006. Vagal afferent nerves regulating the cough reflex. *Respir Physiol Neurobiol*, 152(3): 223–242.

Cantero-Recasens, G. et al. 2010. Loss of function of transient receptor potential vanilloid 1 (TRPV1) genetic variant is associated with lower risk of active childhood asthma. *J Biol Chem*, 285: 27532–27535.

Cheah, E.Y. et al. 2014. Acrolein relaxes mouse isolated tracheal smooth muscle via a TRPA1-dependent mechanism. *Biochem Pharmacol*, 89(1): 148–156.

Deering-Rice, C.E. et al. 2012. Transient receptor potential vanilloid-1 (TRPV1) is a mediator of lung toxicity for coal fly ash particulate material. *Mol Pharmacol*, 81: 411–419.

Deering-Rice, C.E. et al. 2015. Activation of transient receptor potential ankyrin-1 by insoluble particulate material and association with asthma. *Am J Respir Cell Mol Biol*, 53(6): 893–901.

Delany, N.S. et al. 2001. Identification and characterization of a novel human vanilloid receptor-like protein, VRL-2. *Physiol Genomics*, 4(3): 165–174.

De Logu, F. et al. 2016 TRP functions in the broncho-pulmonary system. *Semin Immunopathol*, 38(3): 321–329.

Delescluse, I., H. Mace, and J.J. Adcock. 2012. Inhibition of airway hyper-responsiveness by TRPV1 antagonists (SB-705498 andPF-04065463) in the unanaesthetized, ovalbumin-sensitized guinea pig. *Br J Pharmacol*, 166: 1822–1832.

Devos, F.C. et al. 2016. Neuro-immune interactions in chemical-induced airway hyperreactivity TRPA1 channels–bronchoconstriction and AHR *Eur Respir J*, 48(2): 380–392.

Dietrich, A. et al. 2006. Cation channels of the transient receptor potential superfamily: Their role in physiological and pathophysiological processes of smooth muscle cells. *Pharmacol Ther*, 112: 744–760.

Doherty, M.J. et al. 2000.Capsaicin responsiveness and cough in asthma and chronic obstructive pulmonary disease. *Thorax*, 55(8): 643–649.

Fernández-Fernández, J.M. et al. 2008. Functional coupling of TRPV4 cationic channel and large conductance, calcium-dependent potassium channel in human bronchial epithelial cell lines. *Pflugers Arch*, 457: 149–159.

Garcia-Elias, A. et al. 2014. The TRPV4 channel. *Handb Exp Pharmacol*, 222: 293–319.

Gauvreau, G.M., A.I. El-Gammal, and P.M. O'Byrne. 2015. Allergen-induced airway responses. *Eur Respir J*, 46(3): 819–831.

Grace, M.S. and M.G. Belvisi. 2011. TRPA1 receptors in cough. *Pulm Pharmacol Ther*, 24(3): 286–288.

Grace, M. et al. 2012. Transient receptor potential channels mediate the tussive response to prostaglandin E2 and bradykinin. *Thorax*, 67: 891–900.

Grace, M.S. et al. 2014. Transient receptor potential (TRP) channels in the airway: Role in airway disease. *Br J Pharmacol*, 171(10): 2593–2607.

Groneberg, D.A. et al. 2004. Increased expression of transient receptor potential vanilloid-1 in airway nerves of chronic cough. *Am J Respir Crit Care Med*, 170: 1276–1280.

Hamanaka, K., M.Y. Jian, and M.I. Townsley. 2010. TRPV4 channels augment macrophage activation and ventilator-induced lung injury. *Am J Physiol Lung Cell Mol Physiol*, 299(3): L353–L362.

Haque, R.A., O.S. Usmani, and P.J. Barnes. 2005. Chronic idiopathic cough: A discrete clinical entity? *Chest*, 127: 1710–1713.

Hinman, A. et al. 2006. TRP channel activation by reversible covalent modification. *Proc Natl Acad Sci U S A*, 103: 19564–19568.

Holgate, S.T. 2008. Pathogenesis of asthma. *Clin Exp Allergy*, 38(6): 872–897.

Jang, Y. et al. 2012. Quantitative analysis of TRP channel genes in mouse organs. *Arch Pharm Res*, 35(10): 1823–1830.

Jha, A. et al. 2015. A role for transient receptor potential ankyrin 1 cation channel (TRPA1) in airway hyper-responsiveness? *Can J Physiol Pharmacol*, 93(3): 171–176.

Jia, Y. et al. 2004. Functional TRPV4 channels are expressed in human airway smooth muscle cells. *Am J Physiol Lung Cell Mol Physiol*, 287(2): L272–L278.

Jordt, S.E. et al. 2004. Mustard oils and cannabinoids excite sensory nerve fibres through the TRP channel ANKTM1. *Nature*, 427: 260–265.

Kenyon, N.J. et al. 2012. Occupational asthma. *Clin Rev Allergy Immunol*, 43(1–2): 3–13.

Laumbach, R.J. and H.M. Kipen. 2012. Respiratory health effects of air pollution: Update on biomass smoke and traffic pollution. *J Allergy Clin Immunol*, 129: 3–11.

Lee, L.Y. et al. 2015. Interaction between TRPA1 and TRPV1: Synergy on pulmonary sensory nerves. *Pulm Pharmacol Ther*, 35: 87–93.

Lieu, T.M. et al. 2012. TRPV1 induction in airway vagal low-threshold mechanosensory neurons by allergen challenge and neurotrophic factors. *Am J Physiol Lung Cell Mol Physiol*, 302: L941–L948.

Li, J. et al. 2011. TRPV4-mediated calcium influx into human bronchial epithelia upon exposure to diesel exhaust particles. *Environ Health Perspect*, 119(6): 784–793.

Lin, Y.S. et al. 2010. Activations of TRPA1 and P2X receptors are important in ROS-mediated stimulation of capsaicin-sensitive lung vagal afferents by cigarette smoke in rats. *J Appl Physiol*, 108: 1293–1303.

Lorenzo, I.M. et al. 2008. TRPV4 channel participates in receptor-operated calcium entry and ciliary beat frequency regulation in mouse airway epithelial cells. *Proc Natl Acad Sci U S A*, 105(34): 12611–12616.

Macpherson, L.J. et al. 2007. Noxious compounds activate TRPA1 ion channels through covalent modification of cysteines. *Nature*, 445: 541–545.

McAlexander, M.A. et al. 2014. Transient receptor potential vanilloid 4 activation constricts the human bronchus via the release of cysteinyl leukotrienes. *J Pharmacol Exp Ther*, 349(1): 118–125.

McGarvey, L.P. et al. 2014. Increased expression of bronchial epithelial transient receptor potential vanilloid 1 channels in patients with severe asthma. *J Allergy Clin Immunol*, 133(3): 704–712.

Mitchell, J.E. et al. 2005. Expression and characterization of the intracellular vanilloid receptor (TRPV1) in bronchi from patients with chronic cough. *Exp Lung Res*, 31: 295–306.

Mori, T. 2011. Lack of transient receptor potential vanilloid-1 enhances Th2-biased immune response of the airways in mice receiving intranasal, but not intraperitoneal, sensitization. *Int Arch Allergy Immunol*, 156: 305–312.

Mukhopadhyay, P. et al. 2011. Expression of functional TRPA1 receptor on human lung fibroblast and epithelial cells. *J Recept Signal Transduct Res*, 31(5): 350–358.

Naeher, L.P. et al. 2007. Woodsmoke health effects. *Inhal Toxicol*, 19: 67–106.

Nakajima, T. et al. 2006. Cough sensitivity in pure cough variant asthma elicited using continuous capsaicin inhalation. *Allergol Int*, 55: 149–155.

Nassenstein, C. et al. 2008. Expression and function of the ion channel TRPA1 in vagal afferent nerves innervating mouse lungs. *J Physiol*, 586: 1595–1604.

Nassini, R. et al. 2012. Transient receptor potential ankyrin 1 channel localized to non-neuronal airway cells promotes non-neurogenic inflammation. *PLOS ONE*, 7(8): e42454.

Nicholson, K.G., J. Kent, and D.C. Ireland. 1993. Respiratory viruses and exacerbations of asthma in adults. *BMJ*, 307: 982–986.

O'Byrne, P.M., G.M. Gauvreau, and J.D. Brannan. 2009. Provoked models of asthma: What have we learnt? *Clin Exp Allergy*, 39: 181e92.

Ong, H.L. et al. 2003. Evidence for the expression of transient receptor potential proteins in guinea pig airway smooth muscle cells. *Respirology*, 8(1): 23–32.

Parenti, A. et al. 2016. What is the evidence for the role of TRP channels in inflammatory and immune cells? *Br J Pharmacol*, 173(6): 953–969.

Pauwels, R.A. et al. 2001. Global strategy for the diagnosis, management, and prevention of chronic obstructive pulmonary disease. NHLBI/WHO global initiative for chronic obstructive lung disease (GOLD) workshop summary. *Am J Respir Crit Care Med*, 163: 1256–1276.

Pecova, R. et al. 2008. Cough reflex sensitivity testing in seasonal allergic rhinitis patients and healthy volunteers. *J Physiol Pharmacol*, 59(6): 557–564.

Plevkova, J. et al. 2013. The role of trigeminal nasal TRPM8-expressing afferent neurons in the antitussive effects of menthol. *J Appl Physiol*, 115: 268–274.

Prandini, P. et al. 2016. TRPA1 channels modulate inflammatory response in respiratory cells from cystic fibrosis patients. *Am J Respir Cell Mol Biol*, 55(5): 645–656.

Preti, D., A. Szallasi, and R. Patacchini. 2012. TRP channels as therapeutic targets in airway disorders: A patent review. *Expert Opin Ther Pat*, 22: 663–695.

Raemdonck, K. et al. 2012. A role for sensory nerves in the late asthmatic response. *Thorax*, 67: 19–25.

Rahaman, S.O. et al. 2014. TRPV4 mediates myofibroblast differentiation and pulmonary fibrosis in mice. *J Clin Invest*, 124: 5225–5238.

Rehman, R. et al. 2013 TRPV1 inhibition attenuates IL-13 mediated asthma features in mice by reducing airway epithelial injury. *Int Immunopharmacol*, 15: 597–605.

Ruan, T. et al. 2014. Sensitization by pulmonary reactive oxygen species of rat vagal lung C-fibers: The roles of the TRPV1, TRPA1, and P2X receptors. *PLOS ONE*, 9: e91763.

Saetta, M. et al. 1997. Inflammatory cells in the bronchial glands of smokers with chronic bronchitis. *Am J Respir Crit Care Med*, 156: 1633–1639.

Saetta, M. et al. 1998. CD8+ T-lymphocytes in peripheral airways of smokers with chronic obstructive pulmonary disease. *Am J Respir Crit Care Med*, 157(3): 822–826.

Sabnis, A.S. et al. 2008. Human lung epithelial cells express a functional cold-sensing TRPM8 variant. *Am J Respir Cell Mol Biol*, 39(4): 466–474.

Salvi, S.S. and P.J. Barnes. 2009. Chronic obstructive pulmonary disease in non-smokers. *Lancet*, 374: 733–743.

Samivel, R. et al. 2016. The role of TRPV1 in the CD4+ T cell mediated inflammatory response of allergic rhinitis. *Oncotarget*, 7(1): 14860.

Satoh, J. and M. Yamakage. 2009. Desflurane induces airway contraction mainly by activating transient receptor potential A1 of sensory C-fibers. *J Anesth*, 23(4): 620–623.

Selvarajah, S. et al. 2016. Multiple circulating cytokines are coelevated in chronic obstructive pulmonary disease. *Mediators Inflamm*, 2016:3604842.

Shapiro, D. et al. 2013. Activation of transient receptor potential ankyrin-1 (TRPA1) in lung cells by wood smoke particulate material. *Chem Res Toxicol*, 26: 750–758.

Smit L.A. et al. 2012. Transient receptor potential genes, smoking, occupational exposures and cough in adults. *Respir Res*, 13: 26.

Spina, D. and C.P. Page. 2013. Regulating cough through modulation of sensory nerve function in the airways. *Pulm Pharmacol Ther*, 26(5): 486–490.

Story, G.M. et al. 2003. ANKTM1, a TRP-like channel expressed in nociceptive neurons, is activated by cold temperatures. *Cell*, 112(6): 819–829.

Talavera, K. et al. 2009. Nicotine activates the chemosensory cation channel TRPA1. *Nat Neurosci*, 12: 1293–1300.

Trevisani, M. et al. 2007. 4-Hydroxynonenal, an endogenous aldehyde, causes pain and neurogenic inflammation through activation of the irritant receptor TRPA1. *Proc Natl Acad Sci U S A*, 104: 13519–13524.

Van Gerven, L., G. Boeckxstaens, and P. Hellings. 2012. Up-date on neuro-immune mechanisms involved in allergic and non-allergic rhinitis. *Rhinology*, 50: 227–235.

Van Gerven, L. et al. 2014. Capsaicin treatment reduces nasal hyperreactivity and transient receptor potential cation channel subfamily V, receptor 1 (TRPV1) overexpression in patients with idiopathic rhinitis. *J Allergy Clin Immunol*, 133(5): 1332–1339.

Vergnolle, N. et al. 2010. A role for transient receptor potential vanilloid 4 in tonicity-induced neurogenic inflammation. *Br J Pharmacol*, 159(5): 1161–1173.

Watanabe, H., J.B. Davis, and D. Smart. 2002a. Activation of TRPV4 channels (hVRL-2/mTRP12) by phorbol derivatives. *J Biol Chem*, 277: 13569–13577.

Watanabe, N. et al. 2006. Immunohistochemical co-localization of transient receptor potential vanilloid (TRPV) 1 and sensory neuropeptides in the guinea-pig respiratory system. *Neuroscience*, 141: 1533–1543.

Widdicombe, J.G. 1995. Neurophysiology of the cough reflex. *Eur Respir J*, 8: 1193–1202.

Xia, Y. et al. 2013. TRPV4 channel contributes to serotonin-induced pulmonary vasoconstriction and the enhanced vascular reactivity in chronic hypoxic pulmonary hypertension. *Am J Physiol Cell Physiol*, 305: C704–C715.

Xing, H. et al. 2008. TRPM8 mechanism of autonomic nerve response to cold in respiratory airway. *Mol Pain*, 5(4): 22.

Yin, J. et al. 2016. Role of transient receptor potential vanilloid 4 in neutrophil activation and acute lung injury. *Am J Respir Cell Mol Biol*, 54(3): 370–383.

Zhang, G. et al. 2008. Altered expression of TRPV1 and sensitivity to capsaicin in pulmonary myelinated afferents following chronic airway inflammation in the rat. *J Physiol*, 586: 5771–5786.

Zhang, L. et al. 2016. Activation of cold-sensitive channels TRPM8 and TRPA1 inhibits the proliferative airway smooth muscle cell phenotype. *Lung*, 194(4): 595–603.

Zhao, L. et al. 2013. Effect of TRPV1 channel on the proliferation and apoptosis in asthmatic rat airway smooth muscle cells. *Exp Lung Res*, 39(7): 283–294.

Zheng, X. et al 2013. Arachidonic acid-induced dilation in human coronary arterioles: Convergence of signaling mechanisms on endothelial TRPV4- mediated Ca^{2+} entry. *J Am Heart Assoc*, 2: e000080.

Zhuang, J., L. Zhao, and F. Xu. 2014. Maternal nicotinic exposure produces a depressed hypoxic ventilatory response and subsequent death in postnatal rats. *Physiol Rep*, 2: 1–12.

Zhuang, J. et al. 2015. Prenatal nicotinic exposure augments cardiorespiratory responses to activation of bronchopulmonary C-fibers. *Am J Physiol Lung Cell Mol Physiol*, 308: L922–L930.

Zhu, G. et al. 2009. Association of TRPV4 gene polymorphisms with chronic obstructive pulmonary disease. *Hum Mol Genet*, 18(11): 2053–2062.

Zurgorg, S. et al. 2007. Direct activation of the ion channel TRPA1 by Ca2þ. *Nat Neurosci*, 10: 277e9.

14 ThermoTRPs
Role in Aging

Celine Emmanuelle Riera

CONTENTS

14.1 INTRODUCTION

An organism's health depends on the integrity of molecular and biochemical networks responsible for ensuring homeostasis within its cells and tissues. However, upon aging, a progressive failure in the maintenance of this homeostatic balance occurs in response to various insults, allowing the accumulation of damage, the physiological decline of individual tissues, and susceptibility to diseases. Despite the complex nature of the aging process, simple genetic and environmental alterations can cause an increase in healthy lifespan or "healthspan" in laboratory model organisms. Genetic manipulations of model organisms including yeast, worms, flies, and mice have revealed signaling elements involved in DNA damage, stem cells maintenance, proteostasis, energy, and oxidative metabolism (Riera et al., 2016).

However, one of the most intriguing discoveries made in these models resides in the ability of environmental factors to profoundly alter the aging process by remodeling some of the genetic programs mentioned above (Riera and Dillin, 2016). The first line of evidence that an external cue could powerfully regulate longevity was obtained by performing dietary restriction in rodents, a reduction in food intake without malnutrition. Dietary restriction is the most robust intervention to increase lifespan in model organisms including rodents and primates, and delays the emergence of age-related diseases (Mair and Dillin, 2008). How dietary restriction extends lifespan remains an open question, but decades of research are evidencing molecular pathways embedded in the response to reduce energy availability, resulting in the emergence of an altered metabolic state that promotes health and longevity. Nonetheless, the discovery of dietary restriction opened a new avenue of research in the aging field, and in particular in the understanding of how animals deal with fluctuating energy levels in their natural environment, and how their longevity is affected by such factors. This is particularly relevant for the nematode *Caenorhabditis elegans,* which survives in a changing environment and must be able to coordinate energy-demanding processes including basal cellular functions, growth, reproduction, and physical activity with available external resources. In order to sense their environment, *C. elegans* possess ciliated sensory neurons located primarily in sensory organs in the head and tail regions. Cilia function as sensory receptors, expressing many G protein-coupled receptors (GPCRs) and transient receptor potential (TRP) channels, and mutants with defective sensory cilia have impaired sensory perception (Bargmann, 2006). Cilia are membrane-bound microtubule-based structures and in *C. elegans* are only found at the dendritic endings of sensory neurons.

Sensory neurons provide nematodes with a remarkable form of developmental plasticity, allowing them to assess food availability, temperature, and crowding information (worm density) in order to arrest their development if required, thus forming long-lived and stress-resistant dauer larvae (Bargmann, 2006; Golden and Riddle, 1982). When favorable times return, worms assess the same cues to recover and resume normal development. As the entry and exit of the dauer larval stage suggest, worm sensory neurons truly function as neuroendocrine organs, being implicated in many physiological functions in addition to their behavioral role (Bargmann, 2006). Much information on these neurons has been gathered from laser ablation experiments and analysis of mutants presenting defects in sensory cilia. A seminal discovery in the aging field was achieved when the laboratory of Cynthia Kenyon showed in 1999 that mutations that cause various defects in cilia formation, including the absence of cilia, deletion of middle and distal segments, or impair chemosensory signal transduction increase longevity profoundly (Apfeld and Kenyon, 1999). Later, this group also demonstrated that laser ablation of specific pairs of gustatory and olfactory chemosensory neurons was sufficient to extend lifespan (Alcedo and Kenyon, 2004). What is the role of TRP channels in modulating these neuroendocrine processes, and what kind of stimuli are these receptors detecting to control aging? This chapter summarizes relevant discoveries that clarify some of the roles of TRP channels in the aging process.

14.2   *C. ELEGANS* TRPA1 IN THE REGULATION OF LONGEVITY AT LOW TEMPERATURES

14.2.1   CORE BODY TEMPERATURE AND AGING

In 1916, Loeb and Northrop asked whether the duration of life depends on a definite temperature coefficient for each species. Their work demonstrated that lower temperatures could dramatically extend the lifespan of the fruit fly, *Drosophila* (Loeb and Northrop, 1916). Other poikilothermic animals, whose internal temperature varies considerably, including *C. elegans* and the fish *Cynolebias adloffi*, also present increased lifespan upon modest temperature reduction (Conti, 2008). Additionally, lowering the core body temperature of homeothermic animals, such as mice, also increases lifespan (Conti et al., 2006), highlighting a general role of temperature reduction in lifespan extension in both poikilotherms and homeotherms. Reduction in core body temperature has been proposed to mediate the longevity benefits of dietary restriction (Lane et al., 1996). Conversely, raising the culturing temperature (e.g., to 25°C) greatly shortens nematode lifespan (Lee and Kenyon, 2009). This phenomenon is mediated by a pair of amphid thermosensory neurons with finger-like ciliated endings termed *AFD neurons*, which allow the animals to migrate toward temperatures previously associated with food or thermotaxis (Hedgecock and Russell, 1975; Mori and Ohshima, 1995).

14.2.2   MOLECULAR BASIS OF LIFESPAN EXTENSION UPON REDUCED CORE BODY TEMPERATURE

How is the cold-dependent lifespan extension mediated? One prominent model assumes that lowering the body temperature would reduce the rate of chemical reactions, thereby leading to a slower pace of living. This model suggests that the extended lifespan observed at low temperatures is simply a passive thermodynamic process. It takes a longer time for worms to develop from embryos to adults at lower temperatures, a phenomenon seemingly consistent with this model. However, a more attractive hypothesis suggests that specific genetic programs might be engaged to actively promote longevity at cold temperatures, as observed upon dietary restriction or other paradigms. Xiao et al. reasoned that a cold sensor of the TRP channel family might be recruited in this process (Xiao et al., 2013). The best-known mammalian cold sensors are TRPA1 and TRPM8; however, TRPM8 does not have a *C. elegans* homolog (Peier et al., 2002; Story et al., 2003; McKemy et al., 2002), thus ruling this receptor out of the candidate-based approach. But, TRPA1 has one ortholog in *C. elegans* referred to as TRPA-1, which becomes active under 20°C (Chatzigeorgiou et al., 2010) and therefore constitutes an attractive candidate to mediate the longevity extension observed under cold temperature.

Three temperatures (15°C, 20°C, and 25°C) are common laboratory conditions for culturing worms. If TRPA-1 is involved in promoting longevity at low temperatures, one would expect that mutant worms lacking TRPA-1 should have a shorter lifespan at 15°C and 20°C than wild-type worms, but not at 25°C. This is because this cold-sensitive channel is expected to be functional at 15°C and 20°C but remains closed at 25°C. Consistent with this prediction, *trpa-1* null mutant worms showed a significantly shorter lifespan than wild-type worms at 15°C and 20°C but not 25°C (Xiao et al., 2013). Similarly, transgenic expression of TRPA-1 under its own promoter increased lifespan at 15°C and 20°C but not at 25°C (Xiao et al., 2013).

Lifespan extension at cold temperatures depends on the Ca^{2+} permeability of TRPA-1, as point mutants E1018A, which are Ca^{2+} impermeable but retain Na^+ or K^+ permeability, fail to extend lifespan at low temperature (Xiao et al., 2013). Calcium signaling is therefore critical to mediate the effects of TRPA-1, and suggest that canonical signaling cascades function downstream of the channel to regulate lifespan. Mutation of the Ca^{2+}-sensitive kinase protein kinase C-2 (PKC-2), which is the sole classical PKC in *C. elegans*, fully suppressed the long-lived phenotype of TRPA-1 transgenic animals, indicating that PKC-2 is required for the function of TRPA-1 in the pathway (Xiao et al., 2013). Using genetic epistasis, Xiao et al. showed that TRPA-1 acts specifically upstream on the transcription factor daf-16, a FOXO longevity master regulator (Xiao et al., 2013). How are Ca^{2+} signals transmitted to DAF-16? Mutation of the Ca^{2+}-sensitive kinase PKC-2, which is the sole classical PKC in *C. elegans*, fully suppressed the long-lived phenotype of TRPA-1 transgenic animals, indicating that PKC-2 is required for the function of TRPA-1 in the pathway. More specifically, these authors were able to show using genetic evidence that PKC-2 acts upstream of SGK-1, a serine/threonine kinase that directly phosphorylates DAF-16, and is linked to increased DAF-16 nuclear activity in these conditions (Figure 14.1, Xiao et al., 2013). Analysis of tissue specificity revealed that both the nervous and intestine systems were required for the low temperature–dependent longevity

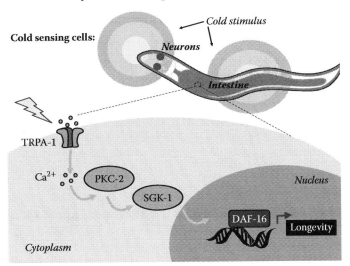

FIGURE 14.1 (See color insert.) A genetic pathway that promotes longevity at cold temperatures in *C. elegans* upon TRPA-1 activation in cold sensing tissues (neurons and intestine). Calcium signaling triggers canonical Ca^{2+}-signaling cascade leading to the FOXO transcription factor DAF-16 to promote transcriptional programs that repress aging.

increase observed in TRPA-1–overexpressing worms, but the specific function of each of these tissues remains to be determined. Both tissues are critical for lifespan extension in insulin/IGF-1-pathway mutants, with the classical view being that the intestine integrates anti-aging cues provided by the neurons and also signals to other tissues to propagate a body-wise response (Kenyon, 2010; Libina et al., 2003).

14.3   ROLE OF THE TRPVS OCR-2 AND OSM-9 IN AGING

14.3.1   Sensory Function of the TRPV OCR-2 and OSM-9

In nematodes, many amphid sensory neurons signal through channels encoded by the TRPV *osm-9* and *ocr-2* genes (Colbert et al., 1997; Tobin et al., 2002). OCR-2 and OSM-9 are coexpressed in the sensory cilia and plasma membrane of four pairs of chemosensory neurons: ADF, AWA, ASH, and ADL (Colbert et al., 1997; Tobin et al., 2002).

Osm-9 and *ocr-2* mutants are defective in all forms of AWA olfaction and ASH nociception, and may play additional roles in other amphid sensory neurons that do not contain the cyclic nucleotide–gated (CNG) channels TAX-4/TAX-2 (Colbert et al., 1997; Tobin et al., 2002). OSM-9 and OCR-2 proteins are localized to the AWA and ASH cilia and are mutually required for each other's cilia localization, suggesting that the two proteins assemble into a single channel complex (Tobin et al., 2002). These channels are also coexpressed in the ADF and ADL amphid neurons, where less is known about their sensory functions. Loss of function of this channel complex results in downregulation of the gene encoding the serotonin (5HT) synthesis enzyme tryptophan hydroxylase (tph-1) in serotonergic ADF neurons through cell autonomous regulation of *tph-1* transcription (Zhang, 2004). The nature of the sensory cues and activation mechanisms of OCR-2/OSM-9 in ADF neurons is not yet determined.

14.3.2   Molecular Basis of Lifespan Extension Downstream of OCR-2/OSM-9

Loss of OCR-2/OSM-9 in the worm results in increased longevity (Riera et al., 2014). Null mutants of either *osm-9(ky4)* or *ocr-2(ak47)* yield to a modest increase in longevity, consistent with the functional redundancy of this receptor pair (Colbert et al., 1997; Tobin et al., 2002). Lifespan extension by *ocr-2(ak47)* mutation has previously been shown to depend on daf-16, and to extend larval starvation survival (Lee and Ashrafi, 2008). However, loss of both *osm-9* and *ocr-2* resulted in a robust longevity extension up to 32% compared to control animals. The lifespan extension observed in worms lacking OCR-2/OSM-9 channels relies on reduced Ca^{2+} signaling within affected cells, and utilize one of the major transponders of Ca^{2+} flux in the cell, the phosphatase calcineurin (Mellstrom et al., 2008). The worm calcineurin ortholog, the Ca^{2+}-activated calcineurin catalytic A subunit, *tax-6*, plays an intricate role in the aging process (Dong et al., 2007; Mair et al., 2011). Loss of *tax-6* results in long-lived animals, and hyperactivation results in short lifespan (Dong et al., 2007). One essential target of *tax-6* to regulate the aging process in worms is the highly conserved CRTC1 (CREB-regulated transcriptional coactivator 1). Dephosphorylation of CRTC1 on serines 76 and 179 by *tax-6* results in nuclear localization, modulation of CREB transcriptional targets, and increased longevity (Mair et al., 2011). Opposing *tax-6*, AMP-activated protein kinase (AMPK) monitors energy sources and phosphorylates CRTC1, retaining CRTC1 in the cytoplasm (Mair et al., 2011). Consistent with loss of *tax-6* resulting in increased longevity, increased activity of AMPK results in increased longevity through phosphorylation of CRTC1 at serines 76 and 179, sites counteracted by *tax-6* (Mair et al., 2011).

Upon tricaine treatment, a drug that increases intracellular Ca^{2+} in cells, CRTC1 shuttles to the nucleus in wild-type animals but remains strictly cytoplasmic in *tax-6(ok2065)* mutants (Mair et al., 2011). Similarly to *tax-6* mutant worms, *trpv* mutants (*osm-9; ocr-2* double mutant animals) retained cytoplasmic localization of CRTC1 upon tricaine treatment, suggesting that OCR-2/OSM-9 function

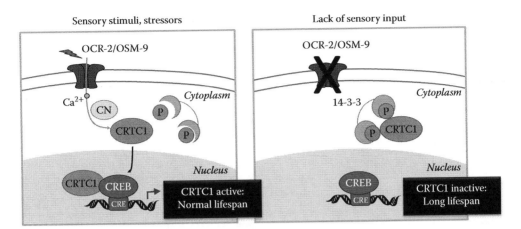

FIGURE 14.2   Model for the sensory regulation of aging by OCR-2/OSM-9–expressing neurons. Stimulation of OCR-2/OSM-9 by external stimuli results in Ca^{2+} influx and activation of the calcineurin TAX-6 (CN), allowing dephosphorylation of CRTC1 and release from 14-3-3 proteins, resulting in nuclear internalization of CRTC1 and transcription of its targets, resulting in normal lifespan. In contrast, loss of OCR-2/OSM-9 promotes lifespan extension through inactivating of the CRTC1/CREB signaling cascade.

within the tax-6/CRTC pathway (Riera et al., 2014). The increased longevity caused by loss of OCR-2/OSM-9 in the worm is completely dependent on the CRTC1 longevity pathway. Inactivating *tax-6*, which extends lifespan in wild-type animals, did not further increase the lifespan of the *trpv* mutants, suggesting that *tax-6* and *osm-9/ocr-2* function in the same pathway. Concordant with *tax-6* modulating longevity through post-translational modifications of CTRC1, the increased longevity of the *trpv* mutants was abrogated when CRTC1 is mutated at the calcineurin dephosphorylation sites S76A, S179A, making it constitutively nuclear (Riera et al., 2014). Therefore, the lifespan extension caused by loss of *trpv* signaling depends on nuclear exclusion of the CREB-regulated transcriptional coactivator CRTC1 at the same phosphorylation sites used for regulation by AMPK and calcineurin (Riera et al., 2014). Taken together, these results indicate that a subset of chemosensory neurons utilizes a TRPV Ca^{2+} signaling cascade to adjust the worm metabolism with environmental conditions by modulating CREB activity that ultimately dictates longevity of the animal (Figure 14.2).

14.4   ROLE OF TRPV1 IN MAMMALIAN AGING

The ability to affect aging by manipulation of TRP channels in invertebrate models such as *C. elegans* provides evidence for evolutionary conservation and argues for the investigation of homologous and analogous circuits in mammalian models. Recently, evidence of the conserved function of chemosensory neurons in the regulation of longevity has been provided through the study of the capsaicin receptor TRPV1 (Riera et al., 2014).

14.4.1   TRPV1 Mutation Increases Mouse Lifespan

Impairment of TRPV1 sensory receptors is sufficient to extend mouse lifespan and improve many aspects of health in aging mice such as metabolic decline, cognitive impairment, and cancer incidence (Riera et al., 2014). Under normal fed ad libitum conditions, the TRPV1 mutation is not sex specific in its effects: longevity in both genders was extended to a similar extent, with 11.9% increase in male TRPV1 mutants and 15.9% increase in median female lifespan compared to wild-type, isogenic C57BL/6 controls (WT). The longevity increase observed in these animals is not due to previously established mouse longevity paradigms such as reduced growth hormone (GH) and/or insulin growth factor (IGF-1) signaling, often resulting in delayed growth and small adult animals (Bluher, 2003; Ortega-Molina et al., 2012; Selman et al., 2008). TRPV1 mutants show no growth delay and do not differ in body composition compared to control animals. TRPV1 mutant mice also do not present core body temperature differences with controls, arguing that their long lifespan is not due to a dietary restriction mimetic mechanism.

14.4.2 Visceral Role of TRPV1 in Lifespan

How can a mutation in a sensory TRPV result in increased lifespan? TRPV1 is highly expressed in sensory nerves innervating the abdominal viscera (such as stomach, pancreas, small intestine) arising from the vagus and spinal nerves with cell bodies (NG) and dorsal root ganglia (DRG) (Christianson and Davis, 2010). In particular, DRG afferents innervating the pancreas, stomach, duodenum, and jejunum are largely peptidergic, expressing calcitonin gene-related peptide (CGRP) and substance P (Christianson and Davis, 2010). A fundamental output of activating TRPV1 receptors in spinal nerves from the DRG is the secretion of multiple neuropeptides from the terminals of primary sensory neurons including the tachynins, CGRP, neurokinin A (NKA), and substance P (SP), involved in neurogenic inflammation (Benemei et al., 2009). Among these substances, CGRP is the main neurotransmitter in the nociceptive C sensory nerves and a potent vasodilator and hypotensive agent implicated in chronic pain and migraines (Springer et al., 2003). Unmyelinated C-fibers of spinal afferents form a dense meshwork innervating the pancreas, as observed in retrograde labeling studies from the pancreas 75% of these DRG afferents are positive for TRPV1, among them 65% reacting for CGRP (Fasanella et al., 2008). In contrast, very few NG afferent innervating the same viscera are peptidergic and the TRPV1/CGRP-positive neurons represent only 35% of the NG population (Fasanella et al., 2008). The secretion of CGRP and substance P occurs in a TRPV1-dependent manner and has been associated with neurogenic inflammation (Noble et al., 2006) and insulin release inhibition in animal models, respectively (Ahrén et al., 1987; Akiba et al., 2004; Asahina et al., 1995; Gram, 2005; Gram et al., 2007; Kogire et al., 1991; Lewis et al., 1988; Pettersson et al., 1986; Tanaka et al., 2011; Melnyk and Himms-Hagen, 1995).

Consistent with a role of TRPV1 and CGRP in antagonizing insulin secretion, mice presenting TRPV1 mutation display a greater ability to secrete insulin upon glucose challenge coupled to enhanced beta cell mass at an advanced age (Riera et al., 2014). Very strikingly, TRPV1 mutant mice present improved glucose tolerance throughout life, as well as increased oxygen consumption as measured in metabolic cages. The respiratory exchange ratio (RER), obtained by indirect calorimetry, compares the volume of carbon dioxide an organism produces to the volume of oxygen consumed over a given time and varies inversely with lipid oxidation. In young and healthy wild-type mice, the RER displays a youthful circadian shift from night to day reflecting the daily transition between carbohydrates to lipid metabolism. Old mice, however, develop a substrate preference toward lipids, losing the capacity to switch between fuel sources also known as metabolic inflexibility (Riera and Dillin, 2015). Old TRPV1 mutants maintain a youthful RER with age, and are protected from age-associated disease, presenting both reduced cancer incidence and delayed onset of cognitive decline with age.

The insulin antagonizing capacity of TRPV1 fibers appears to rely on the neuropeptide CGRP, which locally inhibits insulin secretion from the pancreatic β-cells microenvironment as presented in many *in vitro* and *in vivo* assays (Riera et al., 2014; Ahrén et al., 1987; Akiba et al., 2004; Asahina et al., 1995; Gram, 2005; Gram et al., 2007; Kogire et al., 1991; Lewis et al., 1988; Pettersson et al., 1986; Tanaka et al., 2011; Melnyk and Himms-Hagen, 1995), whereas *in vitro* assays show that substance P does not affect glucose-dependent insulin secretion (Riera et al., 2014). Additionally, CGRP levels appear to fluctuate with age and become elevated in aging animals (Riera et al., 2014; Melnyk and Himms-Hagen, 1995), whereas they remain youthful in old TRPV1 mutant animals (Riera et al., 2014). Similarly, obese and diabetic rodent models show sustained CGRP levels associated with impaired insulin secretion, and reduction of CGRP through TRPV1 inhibition or sensory denervation improved metabolic function in these animals (Gram, 2005; Tanaka et al., 2011). Taken together, these findings suggest that sustained TRPV1 activation and corresponding high CGRP levels are detrimental to metabolic health in aged animals (Figure 14.3). To test this directly, 22-month-old mice were implanted with osmotic pumps diffusing the CGRP receptor antagonist $CGRP_{8-37}$ (Poyner et al., 1998). After 6 weeks of treatment, pharmacologic inhibition of CGRP receptors restores the RER in old mice as observed upon genetic deletion of TRPV1 (Riera et al., 2014), thus improving these animals' age-induced metabolic inflexibility.

14.4.3 TRPV1 AND CREB TRANSCRIPTIONAL ACTIVITY WITH AGE

The lifespan extension of mice lacking TRPV1 appears to be regulated by inactivation of the CRTC1/CREB pathway in DRG sensory neurons, conserved with results in the worm (Riera et al., 2014). Application of capsaicin to cultured DRG neurons provoked accumulation of CRTC1 in the nuclei of CGRP-positive cells of WT DRGs cultures. Capsaicin-induced CRTC1 shuttling is abolished in TRPV1 mutant DRG neurons or in the presence of SB-366791, a selective TRPV1 antagonist (Riera et al., 2014). The ability of CRTC1 to shuttle to the nucleus under TRPV1 activity demonstrates the existence of a plastic transcriptional mechanism adapting rapidly to external outputs. The nuclear exclusion of CRTC1 in TRPV1 mutant DRG neurons suggests that CREB transcriptional activity

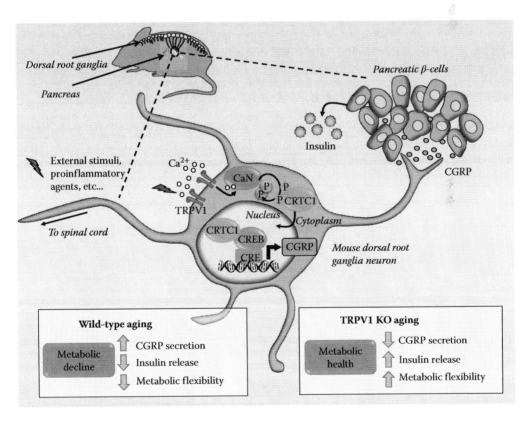

FIGURE 14.3 (See color insert.) Model for the neuroendocrine regulation of metabolism by TRPV1-expressing neurons. Stimulation of TRPV1 by external stimuli promotes CGRP secretion from DRG neurons onto the pancreatic β-cells and inhibition of insulin release. TRPV1 activation results in Ca^{2+} influx and activation of calcineurin, allowing dephosphorylation of CRTC1 and release from 14-3-3 proteins, resulting in nuclear internalization of CRTC1 and transcription of its targets, such as CGRP. CGRP accumulation has detrimental effects on energy expenditure, glucose tolerance, and aging. In contrast, loss of TRPV1 promotes lifespan extension through increased insulin secretion, metabolic health by inactivation of the CRTC1/CREB signaling cascade.

is likely to be altered in the DRG neurons of TRPV1 mutant mice. Under inflammatory conditions, TRPV1 expressing DRG neurons utilize a CREB signaling cascade to induce neurogenic inflammation through the release of CGRP, by the binding of CREB onto the CGRP promoter (Nakanishi et al., 2010). CREB transcriptional activity is downregulated due to the nuclear exclusion of CRTC1 in the TRPV1 mutant mice, and results in downregulation of many CREB target genes including calcitonin-related polypeptide α (calca) and tachykinin 1 (tac1) transcripts, precursors of two TRPV1 secreted neuropeptides, CGRP and substance P.

14.4.4 TRPV1 AND METAINFLAMMATION WITH AGE

These findings raise the question as to which potential age-dependent factors may cause increased TRPV1 activation and lead to sustained CGRP secretion during aging. Accumulation of systemic low-grade inflammation is a hallmark of aging, and increased levels of multiple inflammatory cytokines including tumor necrosis factor-α (tnf-α), interleukin-6 (IL-6), IL-1β, cytokine antagonists, and acute phase proteins such as C-reactive protein (CRP), may underlie the activation of pathological senescence processes (Bruunsgaard et al., 2000). The accumulation of these proinflammatory agents or "inflammaging" characterizes multiple age-induced pathologies, such as sarcopenia, neurodegeneration, arthritis, atherosclerosis, and insulin resistance (Salvioli et al., 2013). Both age-derived adipose tissue expansion and macrophage recruitment in inflamed tissues ramp up the levels of proinflammatory cytokines, which contribute to chronic insulin resistance and metabolic inflexibility (Riera and Dillin, 2015). The presence of low-grade chronic inflammation, common of obesity-associated diseases, has been termed "metainflammation" (Lumeng and Saltiel, 2011). Because TRPV1 is a polymodal receptor activated by many reagents in the inflammatory milieu (Suri and Szallasi, 2008), it is plausible that the low-grade inflammation observed during obesity, diabetes, and aging sustains TRPV1 activation and exacerbates CGRP release, thus impacting negatively on metabolic health. Mutation of α-CGRP protects against diet-induced obesity by increasing energy expenditure, as observed in the TRPV1 mutant mice (Walker et al., 2010). Similarly, TRPV1 mutant animals present reduced metainflammation in the brain and skeletal muscle tissues (Riera et al., 2014), both shown to be critically involved in aging and insulin resistance upon inflammatory activation (Zhang et al., 2008, 2013). In addition to the regulation of insulin secretion from β-cells, CGRP mediates distinct pro- and anti-inflammatory immune activities that implicate this peptide in neuroimmunological communication (Assas et al., 2014; Harzenetter et al., 2007). The broad distribution of CGRP fibers and their association with immune cells including dendritic cells, mast cells, and T cells places CGRP as a key mediator of neuroimmune communication with the sensory fibers participating in both the mediation of sensory signals as well as a controller of immune function (Assas et al., 2014). Future studies investigating the neural-immune interaction involving TRPV1 fibers and CGRP secretion will uncover key mechanisms to understand age-dependent metainflammation.

14.5 CONCLUSION

In light of the evidence reviewed here, multiple members of the TRP channel superfamily have already been implicated in processes that drive the aging process. TRPA-1 functions as a cold sensor in nematodes in which activation drives daf-16 transcriptional activity, activating a genetic program associated with increased lifespan. Additionally, TRPV channels that are recruited for sensory perception of the environment appear to be tightly connected with regulation of neuroendocrine processes that affect aging in both nematodes and mice. TRPV1 afferent fibers secrete the neuropeptide CGRP, a natural inhibitor of insulin secretion with age. However, TRPV1 mutation results in enhanced insulin secretion with age and a youthful metabolic profile that leads to increased lifespan in mice. In accordance with these findings, the insulin secretion capacity of the beta cell, but not insulin resistance, has been shown to be the limiting factor that predicts the onset of diabetes (Goldfine et al., 2003) and appears to be a major gatekeeper of metabolic health in humans (Ahrén and Larsson, 2002). Remarkably, if the causal role of CGRP in regulating longevity remains unknown, considerable lifespan extension is observed in a rodent naturally lacking CGRP, the naked mole rat, an exceptionally long-lived rodent, with a lifespan that can reach 30 years. In comparison, mice that are of a similar size have a maximum lifespan of 4 years. Naked mole rats are fully resistant to cancer, which is reduced in TRPV1 knockout mice (Riera et al., 2014). However, whether CGRP plays a role in the extreme longevity of the naked mole-rat is unknown, and other mediators of this exceptional lifespan have been suggested. For example, naked mole-rat fibroblasts secrete extremely high-molecular-mass hyaluronan, which is over five times larger than the human or mouse homologs, and prevents tumorigenesis in this species (Seluanov et al., 2009; Tian et al., 2013). Nonetheless, these preliminary discoveries established a strong role for TRP channels in the regulation of aging, leading to the mobilization of intracellular Ca^{2+} within target cells to affect different transcriptional profiles associated with aging. Whether other TRP channels also play a role in the control of age-dependent health in sensory or nonsensory tissues remains unknown and will provide an exciting avenue of research for future studies.

REFERENCES

Ahrén, B. and H. Larsson. 2002. Quantification of insulin secretion in relation to insulin sensitivity in nondiabetic postmenopausal women. *Diabetes*, 51: S202–S211.

Ahrén, B., H. Mårtensson, and A. Nobin. 1987. Effects of calcitonin gene-related peptide (CGRP) on islet hormone secretion in the pig. *Diabetologia*, 30: 354–359.

Akiba, Y. et al. 2004. Transient receptor potential vanilloid subfamily 1 expressed in pancreatic islet β cells modulates insulin secretion in rats. *Biochem Biophys Res Commun*, 321: 219–225.

Alcedo, J. and C. Kenyon. 2004. Regulation of *C. elegans* longevity by specific gustatory and olfactory neurons. *Neuron*, 41: 45–55.

Apfeld, J. and C. Kenyon. 1999. Regulation of lifespan by sensory perception in *Caenorhabditis elegans*. *Nature*, 402: 804–809.

Asahina, A. et al. 1995. Specific induction of cAMP in Langerhans cells by calcitonin gene-related peptide: Relevance to functional effects. *Proc Natl Acad Sci U S A*, 92: 8323–8327.

Assas, B.M., J.I. Pennock, and J.A. Miyan. 2014. Calcitonin gene-related peptide is a key neurotransmitter in the neuro-immune axis. *Front Neurosci*, 8: 23.

Bargmann, C. I. 2006. Chemosensation in *C. elegans.WormBook*, 1-29.

Benemei, S. et al. 2009. CGRP receptors in the control of pain and inflammation. *Curr Opin Pharmacol*, 9: 9–14.

Bluher, M. 2003. Extended longevity in mice lacking the insulin receptor in adipose tissue. *Science*, 299: 572–574.

Bruunsgaard, H. et al. 2000 Ageing, tumour necrosis factor-alpha (TNF-alpha) and atherosclerosis. *Clin Exp Immunol*, 121: 255–260.

Chatzigeorgiou, M. et al. 2010. Specific roles for DEG/ENaC and TRP channels in touch and thermosensation in *C. elegans* nociceptors. *Nat Neurosci*, 13: 861–868.

Christianson, J.A. and B.M. Davis. 2010. In *Translational Pain Research: From Mouse to Man*, edited by L. Kruger and A.R. Light. Boca Raton, FL: CRC Press.

Colbert, H.A., T.L. Smith, and C.I. Bargmann. 1997. OSM-9, a novel protein with structural similarity to channels, is required for olfaction, mechanosensation, and olfactory adaptation in *Caenorhabditis elegans*. *J Neurosci*, 17: 8259–8269.

Conti, B. 2008. Considerations on temperature, longevity and aging. *Cell Mol Life Sci*, 65: 1626–1630.

Conti, B. et al. 2006. Transgenic mice with a reduced core body temperature have an increased life span. *Science*, 314: 825–828.

Dong, M.Q. et al. 2007. Quantitative mass spectrometry identifies insulin signaling targets in *C. elegans*. *Science*, 317: 660–663.

Fasanella, K.E. et al. 2008. Distribution and neurochemical identification of pancreatic afferents in the mouse. *J Comp Neurol*, 509: 42–52.

Golden, J.W. and D.L. Riddle. 1982. A pheromone influences larval development in the nematode *Caenorhabditis elegans*. *Science*, 218: 578–580.

Goldfine, A.B. et al. 2003. Insulin resistance is a poor predictor of type 2 diabetes in individuals with no family history of disease. *Proc Natl Acad Sci U S A*, 100: 2724–2729.

Gram, D.X. 2005. Plasma calcitonin gene-related peptide is increased prior to obesity, and sensory nerve desensitization by capsaicin improves oral glucose tolerance in obese Zucker rats. *Eur J Endocrinol*, 153: 963–969.

Gram, D.X. et al. 2007. Capsaicin-sensitive sensory fibers in the islets of Langerhans contribute to defective insulin secretion in Zucker diabetic rat, an animal model for some aspects of human type 2 diabetes. *Eur J Neurosci*, 25: 213–223.

Harzenetter, M.D. et al. 2007. Negative regulation of TLR responses by the neuropeptide CGRP is mediated by the transcriptional repressor ICER. *J Immunol*, 179: 607–615.

Hedgecock, E.M. and R.L. Russell. 1975. Normal and mutant thermotaxis in the nematode *Caenorhabditis elegans*. *Proc Natl Acad Sci U S A*, 72: 4061–4065.

Kenyon, C.J. 2010. The genetics of ageing. *Nature*, 464: 504–512.

Kogire, M. et al. 1991. Inhibitory action of islet amyloid polypeptide and calcitonin gene-related peptide on release of insulin from the isolated perfused rat pancreas. *Pancreas*, 6: 459–463.

Lane, M.A. et al. 1996. Calorie restriction lowers body temperature in rhesus monkeys, consistent with a postulated anti-aging mechanism in rodents. *Proc Natl Acad Sci U S A*, 93: 4159–4164.

Lee, B.H. and K. Ashrafi. 2008. A TRPV channel modulates *C. elegans* neurosecretion, larval starvation survival, and adult lifespan. *PLOS GENET*, 4: e1000213.

Lee, S.J. and C. Kenyon. 2009. Regulation of the longevity response to temperature by thermosensory neurons in *Caenorhabditis elegans*. *Curr Biol*, 19: 715–722.

Lewis, C.E. et al. 1988. Calcitonin gene-related peptide and somatostatin inhibit insulin release from individual rat B cells. *Mol Cell Endocrinol*, 57: 41–49.

Libina, N., J.R. Berman, and C. Kenyon. 2003. Tissue-specific activities of *C. elegans* DAF-16 in the regulation of lifespan. *Cell*, 115: 489–502.

Loeb, J. and J.H. Northrop. 1916. Is there a temperature coefficient for the duration of life? *Proc Natl Acad Sci U S A*, 2: 456–457.

Lumeng, C.N. and A.R. Saltiel. 2011 Inflammatory links between obesity and metabolic disease. *J Clin Invest*, 121: 2111–2117.

Mair, W. and A. Dillin. 2008. Aging and survival: The genetics of life span extension by dietary restriction. *Annu Rev Biochem*, 77: 727–754.

Mair, W. et al. 2011. Lifespan extension induced by AMPK and calcineurin is mediated by CRTC-1 and CREB. *Nature*, 470: 404–408.

McKemy, D.D., W.M. Neuhausser, and D. Julius. 2002. Identification of a cold receptor reveals a general role for TRP channels in thermosensation. *Nature*, 416: 52–58.

Mellstrom, B. et al. 2008. Ca^{2+}-Operated transcriptional networks: Molecular mechanisms and in vivo models. *Physiol Rev*, 88: 421–449.

Melnyk, A. and J. Himms-Hagen. 1995. Resistance to aging-associated obesity in capsaicin-desensitized rats one year after treatment. *Obes Res*, 3: 337–344.

Mori, I. and Y. Ohshima. 1995. Neural regulation of thermotaxis in *Caenorhabditis elegans*. *Nature*, 376: 344–348.

Nakanishi, M. et al. 2010. Acid activation of trpv1 leads to an up-regulation of calcitonin gene-related peptide expression in dorsal root ganglion neurons via the CaMK-CREB cascade: A potential mechanism of inflammatory pain. *Mol Biol Cell*, 21: 2568–2577.

Noble, M.D. et al. 2006. Local disruption of the celiac ganglion inhibits substance P release and ameliorates caerulein-induced pancreatitis in rats. *Am J Physiol Gastrointest Liver Physiol*, 291: G128–G134.

Ortega-Molina, A. et al. 2012. Pten positively regulates brown adipose function, energy expenditure, and longevity. *Cell Metab*, 15: 382–394.

Peier, A.M. et al. 2002. A TRP channel that senses cold stimuli and menthol. *Cell*, 108: 705–715.

Pettersson, M. et al. 1986. Calcitonin gene-related peptide: Occurrence in pancreatic islets in the mouse and the rat and inhibition of insulin secretion in the mouse. *Endocrinology*, 119: 865–869.

Poyner, D.R. et al. 1998. Structural determinants for binding to CGRP receptors expressed by human SK-N-MC and Col 29 cells: Studies with chimeric and other peptides. *Br J Pharmacol*, 124: 1659–1666.

Riera, C.E. and A. Dillin. 2015. Tipping the metabolic scales towards increased longevity in mammals. *Nat Cell Biol*, 17(3): 196–203.

Riera, C.E. and A. Dillin. 2016. Emerging role of sensory perception in aging and metabolism. *Trends Endocrinol Metab TEM*, 27(5): 294–303. doi:10.1016/j.tem.2016.03.007.

Riera, C.E. et al. 2014. TRPV1 pain receptors regulate longevity and metabolism by neuropeptide signaling. *Cell*, 157: 1023–1036.

Riera, C.E. et al. 2016. Signaling networks determining life span. *Annu Rev Biochem*, 85: 35–64.

Salvioli, S. et al. 2013. Immune system, cell senescence, aging and longevity—Inflamm-aging reappraised. *Curr Pharm Des*, 19: 1675–1679.

Selman, C. et al. 2008. Evidence for lifespan extension and delayed age-related biomarkers in insulin receptor substrate 1 null mice. *FASEB J*, 22: 807–818.

Seluanov, A. et al. 2009. Hypersensitivity to contact inhibition provides a clue to cancer resistance of naked mole-rat. *Proc Natl Acad Sci*, 106: 19352–19357.

Springer, J. et al. 2003. Calcitonin gene-related peptide as inflammatory mediator. *Pulm Pharmacol Ther*, 16: 121–130.

Story, G.M. et al. 2003. ANKTM1, a TRP-like channel expressed in nociceptive neurons, is activated by cold temperatures. *Cell*, 112: 819–829.

Suri, A. and A. Szallasi. 2008 The emerging role of TRPV1 in diabetes and obesity. *Trends Pharmacol Sci*, 29: 29–36.

Tanaka, H. et al. 2011. Enhanced insulin secretion and sensitization in diabetic mice on chronic treatment with a transient receptor potential vanilloid 1 antagonist. *Life Sci*, 88: 559–563.

Tian, X. et al. 2013. High-molecular-mass hyaluronan mediates the cancer resistance of the naked mole rat. *Nature*, 499(7458): 346–349.

Tobin, D.M. et al. 2002. Combinatorial expression of TRPV channel proteins defines their sensory functions and subcellular localization in *C. elegans* neurons. *Neuron*, 35: 307–318.

Walker, C.S. et al. 2010. Mice lacking the neuropeptide alpha-calcitonin gene-related peptide are protected against diet-induced obesity. *Endocrinology*, 151: 4257–4269.

Xiao, R. et al. 2013. A genetic program promotes *C. elegans* longevity at cold temperatures via a thermosensitive TRP channel. *Cell*, 152: 806–817.

Zhang, G. et al. 2013. Hypothalamic programming of systemic ageing involving IKK-β, NF-κB and GnRH. *Nature*, 497: 211–216.

Zhang, S. 2004. *Caenorhabditis elegans* TRPV ion channel regulates 5HT biosynthesis in chemosensory neurons. *Development*, 131: 1629–1638.

Zhang, X. et al. 2008. Hypothalamic IKKβ/NF-κB and ER stress link overnutrition to energy imbalance and obesity. *Cell*, 135: 61–73.

15 Roles of Neuronal TRP Channels in Neuroimmune Interactions

Alejandro López-Requena, Brett Boonen, Laura Van Gerven,
Peter W. Hellings, Yeranddy A. Alpizar, and Karel Talavera

CONTENTS

15.1 INTRODUCTION

In order to survive, organisms must adapt to environmental conditions that may subject them to changes in physical variables such as temperature, pressure, light, concentration of chemicals, and so on. In addition, they must be able to react appropriately to their biological milieu and especially to the challenge presented by pathogens that colonize tissues, producing injury, depletion of resources, and interference with vital metabolic functions. The adaptation processes typically entail the initial detection of external stimuli and the posterior implementation of reflex reactions. While in the classical view the sensing of physical stimuli is carried out mainly by sensory neurons and the detection of pathogen-derived cues by the immune cells, in recent years it has become clear that there is a functional interplay between the nervous and immune systems. These neuroimmune interactions arise not only from an intense biochemical cross-talk between neurons and immune cells, but also from the overlap in the sensory functions of these cells.

As exciting and important as it is, the study of neuroimmune interactions seems to be in its infancy, mainly due to insufficient interaction between immunologists and neuroscientists. There is, however, one concept—inflammation—that serves as a meeting point for these research communities. But again it is interesting to see that for immunologists, inflammation is mainly related to the action and regulation of immune cells, without much consideration given to neural functions. For neuroscientists, and especially for sensory neuroscientists, inflammation is mostly associated with "neurogenic inflammation." This concept refers to the consequences of the local release of inflammatory mediators upon activation of sensory nerve endings, which include vasodilation, plasma extravasation, but also recruitment of leukocytes and degranulation of mast cells (Chiu et al., 2012; Engel et al., 2011a; Fernandes et al., 2009; Geppetti et al., 2008; Russell et al., 2014).

Neuroimmune interactions can take place at systemic levels, as occurs when acetylcholine is released to the circulation by the parasympathetic innervation thus influencing the function of distal immune cells, but also at the level of close-range cross-talks between nerve and immune cells (Andersson and Tracey, 2012; Chiu et al., 2012; Ordovas-Montanes et al., 2015). Neuronal signals such as neurotransmitters and neuropeptides influence hematopoiesis, priming and migration of immune cells, while cytokines and histamine produced by the latter can lead to neuronal activation and sensitization. In the central nervous system, the two-sided interplay between neurons and immune cells is particularly evident for microglia and mast cells, which have important roles in neuroinflammatory conditions (Skaper et al., 2014).

Several advances in the field of neuroimmunology have been driven by the need of considering, and in some cases reconsidering, the function of the nervous system to understand pathologies that were mostly defined as immune disorders (see, e.g., Bautista et al., 2014; Belvisi et al., 2016; Halliez and Buret, 2015; Hyland et al., 2014; Ji, 2015; Mayer et al., 2015; Mazzone and Undem, 2016; O'Malley, 2015; Undem and Taylor-Clark, 2014). Increasing evidence indicates that neuroimmune interactions are implicated in hypersensitivity and hyperreactivity conditions such as irritable bowel syndrome, rhinitis, and asthma. Additional momentum has been gained through the recent identification of some of the molecules and receptors implicated in the mechanisms of cellular neuroimmune cross-talk.

In this chapter, we focus on transient receptor potential (TRP) proteins (Flockerzi and Nilius, 2014; Nilius and Flockerzi, 2014), which compose a superfamily of cation channels that play increasingly acknowledged roles in the pathophysiology of all of the vertebrate systems, including the nervous (Gerhold and Bautista, 2009; Julius, 2013; Mickle et al., 2015; Talavera et al., 2008; Vennekens et al., 2012) and immune (Parenti et al., 2016) systems. They are implicated in inflammatory responses, not only through the phenomenon of neurogenic inflammation (Geppetti et al., 2008; Xanthos and Sandkuhler, 2014), but also via their function in proinflammatory immune cells either resident in the nervous tissues or infiltrating, such as microglia (Echeverry et al., 2016; Eder, 2010; Sharma and Ping, 2014), macrophages (Isami et al., 2013), neutrophils (Gelderblom et al., 2014), and mast cells (Freichel et al., 2012). The two main protagonists are the capsaicin receptor TRPV1 and the broadly tuned noxious chemosensor TRPA1. The functional expression of these channels in nociceptive neurons and their roles in neurogenic inflammation are clearly established (Bevan et al., 2014; Nilius et al., 2012; Zygmunt and Hogestatt, 2014). In the following, we discuss how several neuronal TRP channels may serve to sense exogenous pathogen-derived cues, as well as endogenous effector molecules released by immune cells, and review the implication of these channels in neuroimmune interactions during inflammatory processes.

15.2   TRP CHANNELS AS EFFECTORS OF PATHOGEN-DERIVED CUES

A primary function of immune cells is to detect infection, and they do this via specialized pattern recognition receptors that recognize damage- and pathogen-associated molecular patterns (DAMPs and PAMPs). The DAMPs are factors released upon tissue injury (e.g., ATP), whereas the PAMPs are usually components of the pathogens that have essential functions for their survival. For instance, lipopolysaccharides (LPS) and lipoteichoic acids (LTA) are key structural elements of the wall of Gram-negative and Gram-positive bacteria, respectively. LPS and LTA released during division and lyses are detected by mammalian cells via toll-like receptors (TLR) 4 and 2, leading to immune inflammatory responses that eventually kill and clear bacteria (Abbas et al., 2015).

Notably, the ability of detecting pathogens does not rely entirely on the signaling mechanisms of pattern recognition receptors. It has been recently shown that some bacterial components can activate sensory neurons directly, leading to protective responses. For instance, N-formylated peptides and the pore-forming toxin α-hemolysin produced by Gram-positive bacteria elicit Ca^{2+} influx and action potential firing in mouse nociceptive neurons. These effects lead to pain sensation and were proposed to limit innate immune inflammatory responses (Chiu et al., 2013).

Particular attention has been given to LPS, which are the most important molecules for the sensing of Gram-negative bacteria (Neyen and Lemaitre, 2016). The TLR4 complex is described as the key cellular component for the recognition of LPS (Park and Lee, 2013). In addition to inflammation, bacterial infections are accompanied by somatic or visceral pain. These symptoms were generally attributed to activation of nociceptors secondary to immune activation (Ren and Dubner, 2010). However, neuronal activity in the vagal ganglia was shown to occur before the intercellular signaling cascades of the immune system had a chance to mature (Goehler et al., 2005). Immunohistochemical analyses presented a capsaicin-sensitive subclass of trigeminal nociceptors that express TLR4, making neurons a probable direct target of bacterial components (Lin et al., 2015; Wadachi and Hargreaves, 2006). The interaction of LPS with TLR4 expressed in rat trigeminal neurons was found to sensitize coexpressed TRPV1 to capsaicin, resulting in release of calcitonin gene-related peptide (CGRP) (Diogenes et al., 2011; Ferraz et al., 2011). However, dorsal root ganglion (DRG) neurons could still be excited in the absence of a functional TLR4 (Ochoa-Cortes, 2010). Although TLR4 is expressed in nociceptive sensory neurons, its role in excitability remains unclear, as often its enhancing nociceptive action depends on an indirect mechanism based on sensitization of TRPV1, instead of direct triggering of action potentials. In that sense, responses to capsaicin in sensory neurons from *Tlr4* knockout mice were found to be reduced, while the effect of the TRPA1 agonist allyl isothiocyanate (AITC) remained unaltered (Min et al., 2014). Additionally, increases in intracellular Ca^{2+} concentration in rat DRG neurons, shortly after LPS challenge, were proven necessary for subsequent CGRP release (Hou and Wang, 2001). Furthermore, hyperalgesia induced by intraplantar injection of LPS was reduced by a TRPA1 inhibitor and in *Trpa1* knockout mice. This effect was proposed to be related to hydrogen sulfide, which activates TRPA1, as it was prevented by blocking the production of this compound (Andersson et al., 2012).

Ultimately, the TRPA1 ion channel was revealed to act as an acute LPS effector in mouse nociceptive neurons (Meseguer et al., 2014). It was found that *Escherichia coli* LPS activates both mouse and human TRPA1 isoforms in heterologous expression systems and that the acute responses of mouse sensory neurons to this LPS are significantly reduced by genetic ablation of *Trpa1* but not of *Tlr4*. Furthermore, it was found that TRPA1 mediates several acute responses to LPS in mice, including pain, hindpaw inflammation, CGRP release from the airways, and dilation of mesenteric arteries (Figure 15.1a). From the mechanistic point of view, it was proposed that TRPA1 is a direct sensor of LPS by detecting the mechanical perturbations induced by the insertion of the lipid A moiety of this molecule in the plasma membrane. The degree of TRPA1 activation and the acute inflammation induced by LPS derived from different bacterial species correlated with the level of symmetry of these molecules, the conical ones being more effective than the semiconical and the cylindrical ones (Figure 15.1b and c; Meseguer et al., 2014). Although these data are in line with the proposed mechanosensory properties of TRPA1 (Brierley et al., 2009; Kwan et al., 2006; Kwan et al., 2009), further research is required to elucidate the molecular mechanisms underlying the interactions between LPS, the lipid bilayer, and TRPA1.

All these findings have established that nociceptive neurons function as detectors of pathogenic cues and as modulators of innate immune responses, a novel notion that has raised significant interest in the field of neuroimmune interactions (Gonzalez-Navajas et al., 2014; Grace et al., 2014; Nizet and Yaksh, 2013; Steinberg et al., 2014). Such early neural activity may provide the host a critical time period for the protective reaction before the adaptive immune response, which requires additional signals such as cytokines. The discovery of the involvement of ion channels and more specifically TRPs on LPS sensing offers new insights into the modulatory role of sensory neurons in bacterial inflammation and paves the way for potential pharmacological intervention.

It is important to notice that the experimental paradigms used to establish the implication of sensory neurons in the detection of bacterial components *in vivo* consisted of infection models (Chiu et al., 2013) or entailed the injection of the bacterial components (Chiu et al., 2013; Meseguer et al., 2014). Thus, they are relevant only for scenarios in which pathogens have already

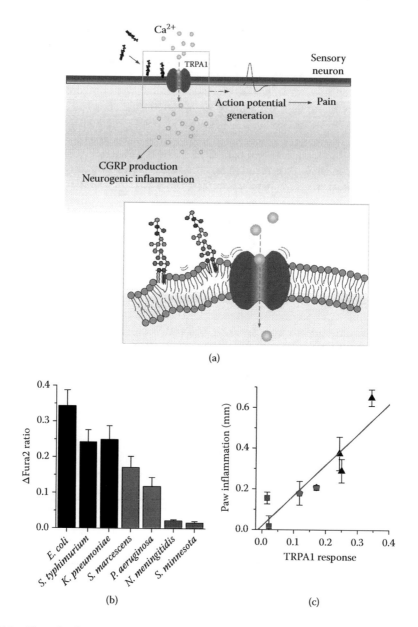

FIGURE 15.1 **(See color insert.)** TRPA1 as a neuronal sensor of bacterial lipopolysaccharides. (a) Interactions between LPS and TRPA1 in the membrane of mammalian nociceptive neurons. Insertion of the lipid A moiety of LPS may induce mechanical perturbations in the plasma membrane, which are sensed by TRPA1. Activation of this channel leads to membrane depolarization and action potential firing, which results in pain, and to the release of CGRP and neurogenic inflammation. (b) Distinct effects of different LPS molecules on TRPA1. The blue bars represent cylindrical or lamellar LPS, the red bars semiconical LPS and the black bars conical LPS. (c) Correlation between the ability of different LPS to activate TRPA1 and to induce paw inflammation in the mouse. The blue squares represent cylindrical or lamellar LPS, the red pentagons semi-conical LPS, and the black triangles conical LPS. (Reproduced from Meseguer, V. et al., *Nat Commun.*, 5, 3125, 2014. With permission.)

infected the tissue. These studies therefore left open the question of whether animals are endowed with mechanisms allowing pathogen detection before infection actually takes place. Recently, *Drosophila* fruit flies have been shown to possess an olfactory circuit that senses and confers avoidance to geosmin, an earthy-smelling volatile compound produced by *Penicillium* fungi, actinomyces Gram-positive bacteria, and cyanobacteria (Stensmyr et al., 2012). However, fruit flies, like a plethora of other animals, can also be infected by ubiquitous bacteria, and it remains unknown whether they are able to detect and avoid such pathogens prior to infection. Considering that TRPA1 is a conserved sensor of some chemical irritants (i.e., electrophilic compounds) (Kang et al., 2010), a subsequent study hypothesized that this channel endows *Drosophila* fruit flies with the ability to detect and avoid LPS and Gram-negative bacteria. It was found that LPS also activates *Drosophila* TRPA1 and that expression of this channel in bitter-sensing gustatory neurons is both necessary and sufficient for eliciting avoidance behavior toward LPS-contaminated surfaces during feeding and egg laying (Soldano et al., 2016). The TRPA1-mediated avoidance of LPS and *E. coli* during egg laying may be extremely important, because it was found that food contamination with *E. coli* strongly affects survival in early developmental stages, when animals are not able to move (in the embryonic stage) or to avoid LPS embedded in their food (early larval stages). These findings are in line with the idea that the selection of the oviposition site may be the only manner in which female flies take care of their progeny (Laturney and Billeter, 2014). In the adults the avoidance of LPS during ingestion may serve to prevent infection with Gram-negative bacteria, lowering the burden of the much simpler immune system of these animals, which are devoid of mechanisms of adaptive immunity.

Taken together these data indicate that the detection of bacterial endotoxins is an evolutionary conserved property of sensory neurons that express TRPA1, irrespective of the chemosensory modality (general chemosensation in mammals and taste in flies). The new evidence of sensory bases for avoidance of contamination prior to infection in the flies underscores the need to consider the function of the sensory nervous system in the mechanistic understanding of pathogen-host interactions.

15.3 NEURONAL TRP CHANNELS AS SENSORS OF IMMUNE SYSTEM EFFECTORS

Macrophages and neutrophils recruited at infection and injury sites produce reactive nitrogen and oxygen species (RNS and ROS) such as nitric oxide (NO) and H_2O_2. These chemicals have direct antimicrobial actions (killing and repellence) and regulate wound healing processes and leukocyte attraction (De Deken et al., 2014). However, secondary injury may occur when the concentrations of these reactive species are such that their actions exceed the capacity of the detoxification mechanisms of the host (Droge, 2002; Uehara et al., 2015). Crucial to the control of the actions of ROS and RNS within the physiological limits is the proper detection of these species by cellular systems with the capability of triggering regulatory loops for the production and/or the neutralization of these compounds (Droge, 2002). Several of these systems, featuring scavenging and/or sensing properties, have been identified, for instance, those based on the functions of superoxide dismutases, catalase, peroxiredoxin enzymes, glutathione peroxidase and reductase, and thioredoxin reductases for the control of ROS (Bhattacharyya et al., 2014; Holmstrom and Finkel, 2014; Zuo et al., 2015), as well as NO-dependent self-inactivation of the neuronal NO synthase for RNS (Droge, 2002). Notably, several TRP channels (TRPM2, TRPA1, TRPV1, TRPC5, and TRPM7) have been found to be activated by RNS and ROS, and to contribute to a multiplicity of cellular responses to endogenous and exogenous stimuli (Kozai et al., 2014). All these TRPs are expressed in sensory neurons, and with the exception of TRPM7, have been ascribed sensory functions.

TRPM2, initially named TRPC7 or LTRPC2, was the first TRP channel to be identified as sensor of H_2O_2 (Hara et al., 2002). The mechanism of activation of TRPM2 by this compound is yet to be clarified, as it has been proposed to occur via direct interactions or through an indirect mechanism, mediated by ADP-ribose (Kozai et al., 2014; Shimizu et al., 2014). This channel has been detected in multiple cell types, including pancreatic β-cells (Togashi et al., 2006), immune cells (Knowles et al., 2013; Yamamoto and Shimizu, 2016), and rat DRG neurons (Naziroglu et al., 2011). Activation of TRPM2 by H_2O_2 is thought to be implicated in mechanisms of cell death, and immune cell-mediated inflammation (Kozai et al., 2014). Interestingly, recent studies have shown that TRPM2 functions as a heat sensor in the hypothalamus (Song et al., 2016) and in sensory neurons and sympathetic neurons from the superior cervical ganglion of the mouse (Tan and McNaughton, 2016), thus confirming the sensory functionality of this channel.

TRPA1 is the most promiscuous chemosensor known to date. This channel is activated by a myriad of exogenous and endogenous chemical species that can be grouped in reactive electrophiles, ROS, RNS, reactive carbonyl species, Ca^{2+}, heavy metals (Ni^{2+}, Zn^{2+}, Cd^{2+}, Cu^{2+}), and a wide variety of membrane-permeable nonelectrophilic compounds (see, for review, Boonen et al., 2016; Nilius et al., 2012; Zygmunt and Hogestatt, 2014). H_2O_2, NO, nitrooleic acid, and hypochlorite are among the ROS and RNS that activate TRPA1 (see, for review, Kozai et al., 2014). Other TRPA1 agonists relevant for neuroimmune interactions are 15d-PGJ$_2$, a metabolite of prostaglandin D2 produced during inflammation (Andersson et al., 2008; Cruz-Orengo et al., 2008; Maher et al., 2008; Taylor-Clark et al., 2008b; Uchida and Shibata, 2008), and 4-hydroxynonenal and 4-oxononenal, which are endogenous aldehydes released upon tissue damage (Andersson et al., 2008; Taylor-Clark et al., 2008a; Trevisani et al., 2007).

The chemical promiscuity of TRPA1 implies that modulation of this channel during chemical neuroimmune cross-talk might be very complex. Many of the agonists referred above may be produced simultaneously and may therefore compete for common interaction sites in the channel. Indeed, RNS, ROS, and reactive electrophiles induce channel activation by covalent modification of intracellular amino acid residues, some of which are common for these compounds (Hinman et al., 2006; Kozai et al., 2014; Macpherson et al., 2007). Further complexity is expected from the effects of channel sensitization and desensitization, both of which are dependent on intracellular Ca^{2+} concentration and therefore on channel activation (Doerner et al., 2007; Karashima et al., 2008; Nagata et al., 2005; Nilius et al., 2011; Wang et al., 2008; Zurborg et al., 2007). Furthermore, many TRPA1 agonists have actually a bimodal effect, inducing channel inhibition at high concentrations when applied alone, or inhibition at low concentrations when applied after TRPA1 is prestimulated (see, e.g., Alpizar et al., 2013b; Everaerts et al., 2011; Karashima et al., 2007; Talavera et al., 2009). How the complexity of the chemical modulation of TRPA1 may play out in real contexts of infection and injury remains to be investigated.

TRPV1 has a long list of endogenous modulators (Morales-Lazaro et al., 2013) and is characterized by its capacity to integrate several noxious stimuli into channel activation, including capsaicin, heat, acidosis, and ethanol (Aneiros et al., 2011; Tominaga et al., 1998), as well as H_2O_2 (Chuang and Lin, 2009; Ruan et al., 2014) and electrophilic compounds acting (allicin; Salazar et al., 2008) or not (Alpizar et al., 2013a; Everaerts et al., 2011) via covalent modification. TRPV1 was shown to be sensitized by oxidizing agents (Susankova et al., 2006), and activated by NO (Miyamoto et al., 2009; Yoshida et al., 2006), nitrooleic acid (Artim et al., 2011; Sculptoreanu et al., 2010), and H_2O_2 (DelloStritto et al., 2016a). Activation of TRPV1 in sensory neurons by NO leads to peripheral sensitization and nociception (Miyamoto et al., 2009) and might be involved in cutaneous active vasodilation (Wong and Fieger, 2012) and thermal hyperemia (Wong and Fieger, 2010), whereas activation by H_2O_2 has been associated with neurogenic vasodilation (Starr et al., 2008), inflammatory pain (Ibi et al., 2008), and hyperalgesia (Keeble et al., 2009). Interestingly, it has been recently reported that TRPV1 activation is inhibited by 4-hydroxynonenal (DelloStritto et al., 2016b) and that prolonged exposure to H_2O_2 actually reduces TRPV1-mediated vascular signaling (DelloStritto et al., 2016a). This indicates that TRPV1 may be bidirectionally modulated in the context of tissue damage.

The reported activation of TRPC5 by NO (Yoshida et al., 2006) and its cold- and mechano-sensory functions in neurons (Gomis et al., 2008; Jemal et al., 2014; Zimmermann et al., 2011) make this channel another candidate for NO sensing in neuroimmune interactions. However, it should be noticed that TRPC5 activation by NO has been contested (Wong et al., 2010; Xu et al., 2008), and that the reason for these divergences has been ascribed to heterogeneous experimental conditions (Takahashi et al., 2012). Of note, as for TRPA1 and TRPV1, activation of TRPC5 has been linked to NO production (Aubdool et al., 2014; Aubdool et al., 2016; Birder et al., 2001; Foster et al., 2006; Leonelli et al., 2013; Poblete et al., 2005; Takahashi et al., 2012), suggesting that these channels may play central roles in positive feedback mechanisms in NO signaling.

Considering the above, we argue that the chemosensory functions of neuronal TRP channels may serve in physiological conditions as triggers of regulatory loops that, through protective reflexes, result in behavioral and humoral responses that limit the secondary injury induced by overactivation of the immune system. It is also possible that aberrant activities of TRP channels in pathological conditions potentiate the inflammatory response, thereby worsening tissue damage (Figure 15.2). However, much research needs to be done before the boundaries between beneficial and detrimental consequences of TRP activation are clearly defined in each pathological condition. Neurogenic inflammation seems to be one of the most obvious scenarios in which to pursue the study of the role of neuronal TRPs in neuroimmune interactions, because of its obvious importance and for being a phenomenon in which the role of these channels is well characterized.

In addition to the reactive species, other factors released by immune cells have been shown to modulate the function of neuronal TRP channels (e.g., prostaglandins, bradykinin, serotonin, cytokines, and histamine). In the following, we summarize several studies in which their effects have been described in the context of neuroimmune interactions. In some cases it is exemplified how the pharmacological targeting of elements of TRP-immune axes has exhibited positive results in therapies against hypersensitivity and hyperreactivity conditions.

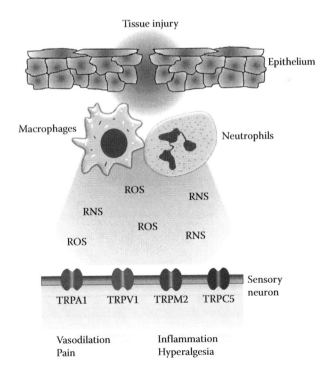

FIGURE 15.2 (See color insert.) Neuronal TRP channels as sensors of immune effectors. Activation of macrophages and neutrophils by tissue injury results in the local release of reactive oxygen and nitrogen species that activate multiple TRP channels in sensory neurons. This leads to vasodilation, local inflammation, pain, and hyperalgesia.

15.4 NEURONAL TRP CHANNELS IN IMMUNE-RELATED DISORDERS IN THE GUT AND AIRWAYS

The gastrointestinal system provides for good examples of the implication of neuronal TRP channels in neuroimmune interactions. The pathogenesis of inflammatory bowel disease (IBD) and irritable bowel syndrome (IBS) features the interaction of several factors, including immune activation (de Souza and Fiocchi, 2016; Mayer et al., 2015). TRPs are important contributors to gastrointestinal sensory transduction and inflammatory nociception (Yu et al., 2016).

In a model of colitis induced by azoxymethane/dextran sodium sulfate (DSS), transcription of the proinflammatory enzyme cyclooxygenase-2 (COX-2) gene, as well as expression of the proinflammatory cytokines interleukin (IL)-6 and IL-11, was increased in *Trpv1* knockout mice. These animals exhibited higher numbers of infiltrating lymphocytes and myeloid cells. The enhanced inflammation was likely due to the absence of TRPV1 in the sensory neurons, as dendritic cells from both wild-type and TRPV1-lacking animals equally responded to LPS stimulation in terms of IL-6 mRNA production, while the expression of the anti-inflammatory neuropeptides vasoactive intestinal peptide (VIP) and pituitary adenylate cyclase–activating peptide (PACAP) was reduced in the knockout mice (Vinuesa et al., 2012). In contrast, colitis induced by transfer of T-helper (Th) cells to immunodeficient animals was prevented by previous ablation of TRPV1-expressing nociceptive enteric neurons with toxic doses of the TRPV1 agonist capsaicin (Gad et al., 2009). There are in fact conflicting results on the effect of TRPV1 ligands in the development of colitis (Engel et al., 2011a). Besides, activation of TRPA1 has also been shown to play a role in the 2,4,6-trinitrobenzene-sulfonic-acid (TNBS) and DSS models of the disease (Engel et al., 2011b). In the same line, blocking of TRPA1 but not of TRPV1 with specific inhibitors hampered the increase in mechanical hypersensitivity caused in mouse colonic sensory nerves by the inflammatory cytokine TNF, overexpressed in the supernatants of peripheral blood mononuclear cells from diarrheic IBS patients (Hughes et al., 2013).

The gut microbiota has a role in gastrointestinal pathophysiology (de Souza and Fiocchi, 2016; Kamada et al., 2013; Mayer et al., 2015) but also influences the nervous system (Bauer et al., 2016; Mu et al., 2016). It is known that the equilibrium of the gut microbiota can be altered by antibiotic treatment (Kaiko and Stappenbeck, 2014; Pamer, 2016). Pain-related behaviors induced by intracolonic administration of the TRPV1 agonist capsaicin were found to be attenuated in antibiotic-treated mice (Aguilera et al., 2015).

Several examples of the role of neuronal TRPs in neuroimmune interactions have been found in contexts featuring cross-talk between visceral innervation and mast cells. For instance, stress-induced colonic hypersensitivity could be prevented in maternally separated rats by a mast cell stabilizer or a neutralizing antiserum against nerve growth factor (NGF), a mast cell mediator. This effect was reduced or reversed by TRPV1 inhibitors (van den Wijngaard et al., 2009). TRPV1-expressing colon submucosal neurons from IBS patients showed increased responsiveness to capsaicin when compared with healthy volunteers, due to sensitization of histamine receptor H1-expressing neurons with a histamine metabolite (Wouters et al., 2016). Moreover, antagonists of this receptor ameliorated disease symptoms (Klooker et al., 2010; Wouters et al., 2016). Also, the expression of both the protease-activated receptor-2 (PAR2), receptor of the mast cell product tryptase, as well as of TRPA1, increased in the bladder of mice subjected to water avoidance stress (Pierce et al., 2016).

It is known that TRPV1 plays an important role in histamine-induced itch (Shim et al., 2007). Interestingly, scratching responses to histamine were affected in *Tlr4* knockout mice, although LPS was found not to be pruritogenic. The expression of histamine receptors and TRPV1 in sensory neurons was not altered, indicating a regulatory activity of TLR4 over TRPV1 in histamine-mediated pruritus (Min et al., 2014).

The role of sensory TRP channels has been also probed in classical and chemically induced models of asthma. *Trpa1* knockout mice were reported to have impaired leukocyte infiltration and airway hyperreactivity in a model of ovalbumin (OVA)-induced asthma. These parameters were also reduced by treating challenged mice with an inhibitor of TRPA1 (Caceres et al., 2009). Alternatively, ablation of TRPV1-expressing neurons in the vagal ganglia abolished airway hyperreactivity in the same model without affecting the immune reaction, although as proven with *Trpv1* knockout animals, the channel itself was not required for the increase in airway resistance. Instead, the expression of a receptor for sphingosine-1-phosphate was implicated in this response (Trankner et al., 2014).

Nonallergic airway hyperreactivity could be induced in mice lacking lymphocytes by nasal instillation of hypochlorite and OVA. This effect, mediated by the neuropeptide substance P(SP) was dependent on mast cells and TRPA1, as it was absent in both mast cell–deficient and *Trpa1* knockout animals. In contrast, *Trpv1* knockout mice developed the hiperreactive response. Hypochlorite, but not OVA, was shown to activate recombinant TRPA1. No functional expression of TRPA1 could be found in mast cells, indicating a link between TRPA1-expressing sensory neurons and activation of these immune cells (Hox et al., 2013).

The occupational asthmogen toluene-2,4-diisocyanate (TDI) was found to activate TRPA1 but not TRPV1 (Devos et al., 2016; Taylor-Clark et al., 2009). However, airway hyperreactivity was absent in both *Trpa1* and *Trpv1* knockout animals. This response to TDI sensitization and challenge was independent of cellular inflammation in the airways but dependent on SP, as it could be inhibited by a blocker of its neurokinin 1 receptor. Airway hyperreactivity did not develop by TDI sensitization followed by TRPA1 or TRPV1 agonist challenge or in mast cell-deficient animals, or in animals treated with a mast cell stabilizer (Devos et al., 2016). These data indicate that hyperreactivity to TDI involves a complex interplay between TRPA1, TRPV1, and mast cells, via mechanisms that are yet to be fully defined.

In the upper airways, capsaicin is used as a therapeutic agent since 1991 in patients with nonallergic noninfectious rhinitis (NANIR) (Gevorgyan et al., 2015; Lacroix et al., 1991; Mounsey and Feller, 2016). Capsaicin was shown to reduce nasal hyperreactivity (Gerth van Wijk et al., 1999), which is a key feature in NANIR patients, and the associated nasal symptoms (Van Rijswijk et al., 2003). Later it was shown that patients with idiopathic rhinitis, a subgroup of NANIR patients, had increased expression levels of TRPV1 in the nasal mucosa in association with increased SP levels in their nasal secretions. Symptom scores as well as nasal hyperreactivity triggered by cold-dry air challenge were reduced after capsaicin treatment, indicating that capsaicin exerts its therapeutic action by ablating the TRPV1-SP nociceptive signaling pathway in the nasal mucosa (Van Gerven et al., 2014). Furthermore, intranasal administration of a selective TRPV1 antagonist (SB-705498) in patients with idiopathic rhinitis induced a marked reduction in total symptom scores triggered by nasal capsaicin challenge (Holland et al., 2014). However, the beneficial effects of capsaicin treatment in certain forms of chronic rhinitis are not applicable in allergic rhinitis. In the latter group the capsaicin nasal spray turned out not to be effective (Gerth Van Wijk et al., 2000).

15.5 NEURONAL TRPS IN OTHER INFLAMMATORY PROCESSES

The complement system component C5a was described to be involved in the model of inflammatory pain by complete Freund's adjuvant (CFA) injection in the mouse paw, through NGF-mediated sensitization of TRPV1. Animals devoid of the C5a receptor C5a1R did not exhibit thermal hyperalgesia, an effect that could be achieved by direct injection of C5a. This response was dependent on TRPV1 and macrophages, as it was not verified in *Trpv1* knockout mice or by treatment with a TRPV1 inhibitor, nor in macrophage-depleted animals. These C5aR1-positive cells were seen in the vicinity of TRPV1-expressing skin nerves. C5a administration led to upregulation of NGF, which, when intraplantarly injected, also led to thermal hyperalgesia even in macrophage-deficient animals. Primary macrophages were shown to transcribe the NGF gene (Shutov et al., 2016). In the same CFA model but in rats, *Trpv1* gene transcription increased in DRG, a phenomenon that was mimicked in cultured neurons by addition of TNF (Lin et al., 2011). TRPV2 expression also increased in this model, while intraplantar injection of NGF led to higher expression of TRPV1 but not TRPV2 (Shimosato et al., 2005). In LPS-treated or aged rats, neuronal overexpression of TRPV2 was found to be associated with increased activation of neurons producing arginine vasopressin (Sauvant et al., 2014), a neuropeptide that regulates water absorption (Treschan and Peters, 2006). Both LPS treatment and aging caused an increase in *Tnf* and *Il-1β* gene transcription in the supraoptic nuclei of the hypothalamo-neurohypophysial system, likely due to activated microglia (Sauvant et al., 2014). TRPC6-targeting hyperforin (Friedland and Harteneck, 2015) was found to inhibit rat microglia activation, and thus NF-κB-driven TNF production, as well as monocyte infiltration following pilocarpine-induced status epilepticus (Lee et al., 2014). Also, TNF production by satellite glial cells in rat DRG was found to be stimulated by the anticancer agent paclitaxel, in a TLR4-dependent manner. TNF increased the expression of TRPA1 and TRPV4 in DRG neurons. A recombinant truncated soluble TNF receptor blocked both this effect and paclitaxel-induced pain (Wu et al., 2015).

The participation of TRPM2 in pathological pain has been evidenced in several models, including paclitaxel-induced neuropathic pain. Mechanical allodynia in *Trpm2* knockout animals was reduced in comparison with wild-type mice. The same result was observed in monosodium iodoacetate-induced osteoarthritis pain, mechanical allodynia associated to experimental autoimmune encephalomyelitis, and streptozotocin-induced diabetic neuropathic pain (So et al., 2015).

There is also evidence for the implication of TRPV1 in neuroimmune interactions at the level of the skin. Th2-derived IL-31 and its receptor were found to be upregulated in affected skin samples from pruritic atopic dermatitis patients. The expression of the receptor was demonstrated in human and mouse DRG neurons coexpressing TRPV1. Mast cell–independent induction of itch by IL-31 was reduced by administration of capsaicin and in *Trpv1* knockout mice, as well as in *Trpa1* knockout mice. In addition, the proportion of IL-31-responsive DRG neurons was reduced in these animals (Cevikbas et al., 2014). In another model of itch, induced by intradermal injection of leukotriene B$_4$ (LTB$_4$), it was found that TRPV1 activation and production of O$_2^-$ mediated the accumulation of polymorphonuclear leukocytes, which resulted in further release of O$_2^-$ leading to activation of TRPA1 (Fernandes et al., 2013). Furthermore, it has been recently reported that oleic acid reduces histamine-induced itch by inhibiting TRPV1 (Morales-Lazaro et al., 2016).

A role of TRPV1 was also found in interactions between nociceptive neurons and dendritic cells. Ablation of TRPV1- and Na$_V$1.8-expressing sensory nerves reduced the psoriasis-like imiquimod-induced inflammation in mouse skin. Nociceptor denervation impaired IL-23 production by dermal dendritic cells, which initiates the inflammatory cascade leading to skin lesions. Although dermal dendritic cells were found in close contact to cutaneous nerves, the communication pathway between these cells remains to be determined (Riol-Blanco et al., 2014).

TRPs are also involved in the activation of the inflammasome (Santoni et al., 2015). The inflammatory response induced in rat paws injected with monosodium urate (MSU) crystals was reduced by coadministration of a TRPA1 inhibitor. Specifically, levels of IL-1β and infiltrating neutrophils decreased in response to this treatment (Trevisan et al., 2013). Similarly, a TRPV1 inhibitor ameliorated pain induced by ankle injection of MSU or capsaicin. Again, IL-1β production, leukocyte infiltration, and neutrophil activity were reduced. Moreover, systemic or local defunctionalization of TRPV1-expressing neurons had an effect similar to that of the TRPV1 inhibitor on the nociceptive responses induced by MSU or capsaicin (Hoffmeister et al., 2014).

TRPV1 function has been found to have an impact on the systemic manifestations of infection. In a mouse model of sepsis by cecal ligation and puncture administration of a TRPV1 inhibitor reduced the levels and activity of COX-2 in the lungs (Ang et al., 2011), and genetic ablation of this channel resulted in decreased defensive mechanisms dependent on macrophage function and enhanced organ failure (Fernandes et al., 2012b). In fact, TRPV1 was postulated to have a key role at the site of infection (Bodkin and Fernandes, 2013), but it remains unclear to what extent the expression in immune cells or in sensory nerves contributed to these functions.

15.6 CONCLUDING REMARKS

For the sake of focus, the scope of this chapter was restricted to neuronal TRPs, but increasing evidence suggests that functional expression of these channels in other cell types, such as epithelial, glial, and immune cells themselves, may be also implicated in the mechanisms of neuroimmune interactions. For instance, TRPV2 is expressed in immune cells (Kojima and Nagasawa, 2014), TRPV3 is expressed in skin keratinocytes (Yang and Zhu, 2014), and TRPV4 is ubiquitously expressed in epithelial cells and endothelium (Garcia-Elias et al., 2014). The function of TRPV4 in particular has been shown to regulate the permeability of cellular barriers in several tissues (Akazawa et al., 2013; Janssen et al., 2016; Sokabe et al., 2010; Sokabe and Tominaga, 2010; Villalta and Townsley, 2013). Moreover, there are reports on the function of TRPV1 and TRPA1 in epithelial cells, vascular smooth muscle, and endothelium, although with activities that are clearly less prominent than in sensory neurons (Fernandes et al., 2012a; Nassini et al., 2012). Hence, given the clear importance of epithelial barriers in immunity (Georas and Rezaee, 2014; Hammad and Lambrecht, 2015; Hiemstra et al., 2015; O'Brien et al., 2015), it is possible that the function of sensory TRP channels in epithelial cells has an indirect but determinant contribution to neuroimmune interactions.

The main idea that we would like to emphasize here is that sensory TRP channels are probably the most relevant targets of multiple external stimuli, including changes in temperature and exogenous chemicals such as spices and environmental contaminants, and of endogenous compounds generated in the context of immune responses to infection and tissue damage. Furthermore, the notion that these channels may serve as sensors of pathogen-derived cues, as in the case of LPS-induced activation of TRPA1 in sensory neurons, suggests that TRPs themselves can drive protective innate responses, independently from the activation of classical pattern recognition receptors. There is, however, a vast amount of basic research to be done before the roles of TRP channels in neuroimmunity are definitely established. Most critically, we need to further clarify the patterns of expression, the functional properties, and the mechanisms of regulation of these channels in physiological and pathological conditions. Nonetheless, several recent studies have yielded clinical evidence that sensory TRPs can be used as direct or indirect targets for the treatment of immune-related disorders, especially those featuring hypersensitivity or hyperreactivity symptoms. These studies illustrate that neuroimmune research is picking up significant momentum, and above all, that the TRP channel field serves as a very good context for successful interactions between immunologists and neuroscientists.

ACKNOWLEDGMENTS

This work was supported by grants from the Belgian Federal Government (Belspo; IUAP P7/13), the Research Council of the KU Leuven (GOA/14/011, OT/12/091 and PF-TRPLe), and the Research Foundation-Flanders (FWO) (G.0C77.15, G.0C68.15 and G.0702.12). Y.A.A. held a Postdoctoral Mandate from KU Leuven and is currently a Postdoctoral Fellow of the FWO.

REFERENCES

Abbas, A.K., A.H. Lichtman, and S. Pillai. 2015. Innate immunity. In *Cellular and Molecular Immunology*, *51-86*. Philadelphia, PA:Elsevier Saunders.

Aguilera, M., M. Cerda-Cuellar, and V. Martinez. 2015. Antibiotic-induced dysbiosis alters host-bacterial interactions and leads to colonic sensory and motor changes in mice. *Gut Microbes*, 6: 10–23.

Akazawa, Y. et al. 2013. Activation of TRPV4 strengthens the tight-junction barrier in human epidermal keratinocytes. *Skin Pharmacol Physiol*, 26: 15–21.

Alpizar, Y.A. et al. 2013a. Allyl isothiocyanate sensitizes TRPV1 to heat stimulation. *Pflugers Arch*, 466: 507–515.

Alpizar, Y.A. et al. 2013b. Bimodal effects of cinnamaldehyde and camphor on mouse TRPA1. *Pflugers Arch*, 465: 853–864.

Andersson, D.A., C. Gentry, and S. Bevan. 2012. TRPA1 has a key role in the somatic pro-nociceptive actions of hydrogen sulfide. *PLOS ONE*, 7: e46917.

Andersson, D.A. et al. 2008. Transient receptor potential A1 is a sensory receptor for multiple products of oxidative stress. *J Neurosci*, 28: 2485–2494.

Andersson, U. and K.J. Tracey. 2012. Reflex principles of immunological homeostasis. *Annu Rev Immunol*, 30: 313–335.

Aneiros, E. et al. 2011. The biophysical and molecular basis of TRPV1 proton gating. *EMBO J*, 30: 994–1002.

Ang, S.F. et al. 2011. Hydrogen sulfide upregulates cyclooxygenase-2 and prostaglandin E metabolite in sepsis-evoked acute lung injury via transient receptor potential vanilloid type 1 channel activation. *J Immunol*, 187: 4778–4787.

Artim, D.E. et al. 2011. Nitro-oleic acid targets transient receptor potential (TRP) channels in capsaicin sensitive afferent nerves of rat urinary bladder. *Experimental Neurol*, 232: 90–99.

Aubdool, A.A. et al. 2014. TRPA1 is essential for the vascular response to environmental cold exposure. *Nat Commun*, 5: 5732.

Aubdool, A.A. et al. 2016. TRPA1 activation leads to neurogenic vasodilatation: Involvement of reactive oxygen nitrogen species in addition to CGRP and NO. *Br J Pharmacol*, 173: 2419–2433.

Bauer, K.C., K.E. Huus, and B.B. Finlay. 2016. Microbes and the mind: Emerging hallmarks of the gut microbiota-brain axis. *Cell Microbiol*, 18: 632–644.

Bautista, D.M., S.R. Wilson, and M.A. Hoon. 2014. Why we scratch an itch: The molecules, cells and circuits of itch. *Nat Neurosci*, 17: 175–182.

Belvisi, M.G. et al. 2016. Neurophenotypes in airway diseases: Insights from translational cough studies. *Am J Respir Crit Care Med*, 193: 1364–1372.

Bevan, S., T. Quallo, and D.A. Andersson. 2014. TRPV1. *Handb Exp Pharmacol*, 222: 207–245.

Bhattacharyya, A. et al. 2014. Oxidative stress: An essential factor in the pathogenesis of gastrointestinal mucosal diseases. *Physiol Rev*, 94: 329–354.

Birder, L.A. et al. 2001. Vanilloid receptor expression suggests a sensory role for urinary bladder epithelial cells. *Proc Natl Acad Sci U S A*, 98: 13396–13401.

Bodkin, J.V. and E.S. Fernandes. 2013. TRPV1 and SP: Key elements for sepsis outcome? *Br J Pharmacol*, 170: 1279–1292.

Boonen, B., J.B. Startek, and K. Talavera. 2016. Chemical activation of sensory TRP channels. In *Topics in Medicinal Chemistry,* Dietmar Krautwurst (editor), 1–41. Switzerland: Springer International Publishing.

Brierley, S.M. et al. 2009. The ion channel TRPA1 is required for normal mechanosensation and is modulated by algesic stimuli. *Gastroenterology*, 137: 2084–2095, e2083.

Caceres, A.I. et al. 2009. A sensory neuronal ion channel essential for airway inflammation and hyperreactivity in asthma. *Proc Natl Acad Sci U S A*, 106: 9099–9104.

Cevikbas, F. et al. 2014. A sensory neuron-expressed IL-31 receptor mediates T helper cell-dependent itch: Involvement of TRPV1 and TRPA1. *J Allergy Clin Immunol*, 133: 448–460.

Chiu, I.M. et al. 2013. Bacteria activate sensory neurons that modulate pain and inflammation. *Nature*, 501: 52–57.

Chiu, I.M., C.A. von Hehn, and C.J. Woolf. 2012. Neurogenic inflammation and the peripheral nervous system in host defense and immunopathology. *Nat Neurosci*, 15: 1063–1067.

Chuang, H.H. and S. Lin. 2009. Oxidative challenges sensitize the capsaicin receptor by covalent cysteine modification. *Proc Natl Acad Sci U S A*, 106: 20097–20102.

Cruz-Orengo, L. et al. 2008. Cutaneous nociception evoked by 15-delta PGJ$_2$ via activation of ion channel TRPA1. *Mol Pain*, 4: 30.

De Deken, X. et al. 2014. Roles of DUOX-mediated hydrogen peroxide in metabolism, host defense, and signaling. *Antioxid Redox Signal*, 20: 2776–2793.

de Souza, H.S. and C. Fiocchi. 2016. Immunopathogenesis of IBD: Current state of the art. Nature reviews. *Gastroenterol Hepatol*, 13: 13–27.

DelloStritto, D.J. et al. 2016a. Differential regulation of TRPV1 channels by H$_2$O$_2$: Implications for diabetic microvascular dysfunction. *Basic Res Cardiol*, 111: 21.

DelloStritto, D.J. et al. 2016b. 4-Hydroxynonenal dependent alteration of TRPV1-mediated coronary microvascular signaling. *Free Radic Biol Med*, 101: 10–19.

Devos, F.C. et al. 2016. Neuro-immune interactions in chemical-induced airway hyperreactivity. *Eur Respir J*, 48: 380–392.

Diogenes, A. et al. 2011. LPS sensitizes TRPV1 via activation of TLR4 in trigeminal sensory neurons. *J Dent Res*, 90: 759–764.

Doerner, J.F. et al. 2007. Transient receptor potential channel A1 is directly gated by calcium ions. *J Biol Chem*, 282: 13180–13189.

Droge, W. 2002. Free radicals in the physiological control of cell function. *Physiol Rev*, 82: 47–95.

Echeverry, S., M.J. Rodriguez, and Y.P. Torres. 2016. Transient receptor potential channels in microglia: Roles in physiology and disease. *Neurotox Res*, 30: 467–478.

Eder, C. 2010. Ion channels in monocytes and microglia/brain macrophages: Promising therapeutic targets for neurological diseases. *J Neuroimmunol*, 224: 51–55.

Engel, M.A. et al. 2011a. Role of sensory neurons in colitis: Increasing evidence for a neuroimmune link in the gut. *Inflamm Bowel Dis*, 17: 1030–1033.

Engel, M.A. et al. 2011b. TRPA1 and substance P mediate colitis in mice. *Gastroenterology*, 141: 1346–1358.

Everaerts, W. et al. 2011. The capsaicin receptor TRPV1 is a crucial mediator of the noxious effects of mustard oil. *Curr Biol*, 21: 316–321.

Fernandes, E.S., M.A. Fernandes, and J.E. Keeble. 2012a. The functions of TRPA1 and TRPV1: Moving away from sensory nerves. *Br J Pharmacol*, 166: 510–521.

Fernandes, E.S., S.M. Schmidhuber, and S.D. Brain. 2009. Sensory-nerve-derived neuropeptides: Possible therapeutic targets. *Handb Exp Pharmacol*, 194: 393–416.

Fernandes, E.S. et al. 2012b. TRPV1 deletion enhances local inflammation and accelerates the onset of systemic inflammatory response syndrome. *J Immunol*, 188: 5741–5751.

Fernandes, E.S. et al. 2013. Superoxide generation and leukocyte accumulation: Key elements in the mediation of leukotriene B(4)-induced itch by transient receptor potential ankyrin 1 and transient receptor potential vanilloid 1. *FASEB J*, 27: 1664–1673.

Ferraz, C.C. et al. 2011. Lipopolysaccharide from *Porphyromonas gingivalis* sensitizes capsaicin-sensitive nociceptors. *J Endod*, 37: 45–48.

Flockerzi, V. and B. Nilius. 2014. TRPs: Truly remarkable proteins. *Handb Exp Pharmacol*, 222: 1–12.

Foster, M.W., D.T. Hess, and J.S. Stamler. 2006. S-nitrosylation TRiPs a calcium switch. *Nat Chem Biol*, 2: 570–571.

Freichel, M., J. Almering, and V. Tsvilovskyy. 2012. The role of TRP proteins in mast cells. *Front Immunol*, 3: 150.

Friedland, K. and C. Harteneck. 2015. Hyperforin: To be or not to be an activator of TRPC6. *Rev Physiol Biochem Pharmacol*, 169: 1–24.

Gad, M. et al. 2009. Blockage of the neurokinin 1 receptor and capsaicin-induced ablation of the enteric afferent nerves protect SCID mice against T-cell-induced chronic colitis. *Inflamm Bowel Dis*, 15: 1174–1182.

Garcia-Elias, A. et al. 2014. The TRPV4 channel. *Handb Exp Pharmacol*, 222: 293–319.

Gelderblom, M. et al. 2014. Transient receptor potential melastatin subfamily member 2 cation channel regulates detrimental immune cell invasion in ischemic stroke. *Stroke*, 45: 3395–3402.

Georas, S.N. and F. Rezaee. 2014. Epithelial barrier function: At the front line of asthma immunology and allergic airway inflammation. *J Allergy Clin Immunol*, 134: 509–520.

Geppetti, P. et al. 2008. The concept of neurogenic inflammation. *BJU Int*, 101(Suppl 3): 2–6.

Gerhold, K.A. and D.M. Bautista. 2009. Molecular and cellular mechanisms of trigeminal chemosensation. *Ann N Y Acad Sci*, 1170: 184–189.

Gerth van Wijk, R. et al. 2000. Intranasal capsaicin is lacking therapeutic effect in perennial allergic rhinitis to house dust mite. A placebo-controlled study. *Clin Exp Allergy*, 30: 1792–1798.

Gerth van Wijk, R.G., C. de Graaf-in 't Veld, and I.M. Garrelds. 1999. Nasal hyperreactivity. *Rhinology*, 37: 50–55.

Gevorgyan, A. et al. 2015. Capsaicin for non-allergic rhinitis. *Cochrane Database Syst Rev*, CD010591.

Goehler, L.E. et al. 2005. Activation in vagal afferents and central autonomic pathways: Early responses to intestinal infection with *Campylobacter jejuni. Brain Behav Immun*, 19: 334–344.

Gomis, A. et al. 2008. Hypoosmotic- and pressure-induced membrane stretch activate TRPC5 channels. *J Physiol*, 586: 5633–5649.

Gonzalez-Navajas, J.M., M.P. Corr, and E. Raz. 2014. The immediate protective response to microbial challenge. *Eur J Immunol*, 44: 2536–2549.

Grace, M.S. et al. 2014. Transient receptor potential (TRP) channels in the airway: Role in airway disease. *Br J Pharmacol*, 171: 2593–2607.

Halliez, M.C. and A.G. Buret. 2015. Gastrointestinal parasites and the neural control of gut functions. *Front Cell Neurosci*, 9: 452.

Hammad, H. and B.N. Lambrecht. 2015. Barrier epithelial cells and the control of type 2 immunity. *Immunity*, 43: 29–40.

Hara, Y. et al. 2002. LTRPC2 Ca^{2+}-permeable channel activated by changes in redox status confers susceptibility to cell death. *Mol Cell*, 9: 163–173.

Hiemstra, P.S., P.B. McCray, Jr., and R. Bals. 2015. The innate immune function of airway epithelial cells in inflammatory lung disease. *Eur Respir J*, 45: 1150–1162.

Hinman, A. et al. 2006. TRP channel activation by reversible covalent modification. *Proc Natl Acad Sci U S A*, 103: 19564–19568.

Hoffmeister, C. et al. 2014. Participation of the TRPV1 receptor in the development of acute gout attacks. *Rheumatology (Oxford)*, 53: 240–249.

Holland, C. et al. 2014. Inhibition of capsaicin-driven nasal hyper-reactivity by SB-705498, a TRPV1 antagonist. *Br J Clin Pharmacol*, 77: 777–788.

Holmstrom, K.M. and T. Finkel. 2014. Cellular mechanisms and physiological consequences of redox-dependent signalling. *Nat Rev Mol Cell Biol*, 15: 411–421.

Hou, L. and X. Wang. 2001. PKC and PKA, but not PKG mediate LPS-induced CGRP release and $[Ca^{2+}](i)$ elevation in DRG neurons of neonatal rats. *J Neurosci Res*, 66: 592–600.

Hox, V. et al. 2013. Crucial role of transient receptor potential ankyrin 1 and mast cells in induction of nonallergic airway hyperreactivity in mice. *Am J Respir Crit Care Med*, 187: 486–493.

Hughes, P.A. et al. 2013. Sensory neuro-immune interactions differ between irritable bowel syndrome subtypes. *Gut*, 62: 1456–1465.

Hyland, N.P., E.M. Quigley, and E. Brint. 2014. Microbiota-host interactions in irritable bowel syndrome: Epithelial barrier, immune regulation and brain-gut interactions. *World J Gastroenterol*, 20: 8859–8866.

Ibi, M. et al. 2008. Reactive oxygen species derived from NOX1/NADPH oxidase enhance inflammatory pain. *J Neurosci*, 28: 9486–9494.

Isami, K. et al. 2013. Involvement of TRPM2 in peripheral nerve injury-induced infiltration of peripheral immune cells into the spinal cord in mouse neuropathic pain model. *PLOS ONE*, 8: e66410.

Janssen, D.A. et al. 2016. TRPV4 channels in the human urogenital tract play a role in cell junction formation and epithelial barrier. *Acta Physiol*, 218: 38–48.

Jemal, I. et al. 2014. G protein-coupled receptor signalling potentiates the osmo-mechanical activation of TRPC5 channels. *Pflugers Arch*, 466: 1635–1646.

Ji, R.R. 2015. Neuroimmune interactions in itch: Do chronic itch, chronic pain, and chronic cough share similar mechanisms? *Pulm Pharmacol Ther*, 35: 81–86.

Julius, D. 2013. TRP channels and pain. *Annu Rev Cell Dev Biol*, 29: 355–384.

Kaiko, G.E. and T.S. Stappenbeck. 2014. Host-microbe interactions shaping the gastrointestinal environment. *Trends Immunol*, 35: 538–548.

Kamada, N. et al. 2013. Role of the gut microbiota in immunity and inflammatory disease. *Nat Rev Immunol*, 13: 321–335.

Kang, K. et al. 2010. Analysis of *Drosophila* TRPA1 reveals an ancient origin for human chemical nociception. *Nature*, 464: 597–600.

Karashima, Y. et al. 2007. Bimodal action of menthol on the transient receptor potential channel TRPA1. *J Neurosci*, 27: 9874–9884.

Karashima, Y. et al. 2008. Modulation of the transient receptor potential channel TRPA1 by phosphatidylinositol 4,5-biphosphate manipulators. *Pflugers Arch*, 457: 77–89.

Keeble, J.E. et al. 2009. Hydrogen peroxide is a novel mediator of inflammatory hyperalgesia, acting via transient receptor potential vanilloid 1-dependent and independent mechanisms. *Pain*, 141: 135–142.

Klooker, T.K. et al. 2010. The mast cell stabiliser ketotifen decreases visceral hypersensitivity and improves intestinal symptoms in patients with irritable bowel syndrome. *Gut*, 59: 1213–1221.

Knowles, H., Y. Li, and A.L. Perraud. 2013. The TRPM2 ion channel, an oxidative stress and metabolic sensor regulating innate immunity and inflammation. *Immunol Res*, 55: 241–248.

Kojima, I. and M. Nagasawa. 2014. TRPV2. *Handb Exp Pharmacol*, 222: 247–272.

Kozai, D., N. Ogawa, and Y. Mori. 2014. Redox regulation of transient receptor potential channels. *Antioxid Redox Signal*, 21: 971–986.

Kwan, K.Y. et al. 2006. TRPA1 contributes to cold, mechanical, and chemical nociception but is not essential for hair-cell transduction. *Neuron*, 50: 277–289.

Kwan, K.Y. et al. 2009. TRPA1 modulates mechanotransduction in cutaneous sensory neurons. *J Neurosci*, 29: 4808–4819.

Lacroix, J.S. et al. 1991. Improvement of symptoms of non-allergic chronic rhinitis by local treatment with capsaicin. *Clin Exp Allergy*, 21: 595–600.

Laturney, M. and J.C. Billeter 2014. Neurogenetics of female reproductive behaviors in *Drosophila melanogaster*. *Adv Genet*, 85: 1–108.

Lee, S.K. et al. 2014. Hyperforin attenuates microglia activation and inhibits p65-Ser276 NFκB phosphorylation in the rat piriform cortex following status epilepticus. *Neurosci Res*, 85: 39–50.

Leonelli, M., D.O. Martins, and L.R. Britto. 2013. Retinal cell death induced by TRPV1 activation involves NMDA signaling and upregulation of nitric oxide synthases. *Cell Mol Neurobiol*, 33: 379–392.

Lin, J.J. et al. 2015. Toll-like receptor 4 signaling in neurons of trigeminal ganglion contributes to nociception induced by acute pulpitis in rats. *Sci Rep*, 5: 12549.

Lin, Y.T. et al. 2011. Up-regulation of dorsal root ganglia BDNF and TrkB receptor in inflammatory pain: An in vivo and in vitro study. *J Neuroinflammation*, 8: 126.

Macpherson, L.J. et al. 2007. Noxious compounds activate TRPA1 ion channels through covalent modification of cysteines. *Nature*, 445: 541–545.

Maher, M. et al. 2008. Activation of TRPA1 by farnesyl thiosalicylic acid. *Mol Pharmacol*, 73: 1225–1234.

Mayer, E.A. et al. 2015. Towards a systems view of IBS. Nature reviews. *Gastroenterol Hepatol*, 12: 592–605.

Mazzone, S.B. and B.J. Undem. 2016. Vagal afferent innervation of the airways in health and disease. *Physiol Rev*, 96: 975–1024.

Meseguer, V. et al. 2014. TRPA1 channels mediate acute neurogenic inflammation and pain produced by bacterial endotoxins. *Nat Commun*, 5: 3125.

Mickle, A.D., A.J. Shepherd, and D.P. Mohapatra. 2015. Sensory TRP channels: The key transducers of nociception and pain. *Prog Mol Biol Transl Sci*, 131: 73–118.

Min, H. et al. 2014. TLR4 enhances histamine-mediated pruritus by potentiating TRPV1 activity. *Mol Brain*, 7: 59.

Miyamoto, T. et al. 2009. TRPV1 and TRPA1 mediate peripheral nitric oxide-induced nociception in mice. *PLOS ONE*, 4: e7596.

Morales-Lazaro, S.L., S.A. Simon, and T. Rosenbaum. 2013. The role of endogenous molecules in modulating pain through transient receptor potential vanilloid 1 (TRPV1). *J Physiol*, 591: 3109–3121.

Morales-Lazaro, S.L. et al. 2016. Inhibition of TRPV1 channels by a naturally occurring omega-9 fatty acid reduces pain and itch. *Nat Commun*, 7: 13092.

Mounsey, A.L. and C.M. Feller. 2016. Capsaicin for nonallergic rhinitis. *Am Fam Physician*, 94: 217–218.

Mu, C., Y. Yang, and W. Zhu. 2016. Gut microbiota: The brain peacekeeper. *Front Microbiol*, 7: 345.

Nagata, K. et al. 2005. Nociceptor and hair cell transducer properties of TRPA1, a channel for pain and hearing. *J Neurosci*, 25: 4052–4061.

Nassini, R. et al. 2012. Transient receptor potential ankyrin 1 channel localized to non-neuronal airway cells promotes non-neurogenic inflammation. *PLOS ONE*, 7: e42454.

Naziroglu, M. et al. 2011. Aminoethoxydiphenyl borate and flufenamic acid inhibit Ca^{2+} influx through TRPM2 channels in rat dorsal root ganglion neurons activated by ADP-ribose and rotenone. *J Membr Biol*, 241: 69–75.

Neyen, C. and B. Lemaitre. 2016. Sensing Gram-negative bacteria: A phylogenetic perspective. *Curr Opin Immunol*, 38: 8–17.

Nilius, B., G. Appendino, and G. Owsianik. 2012. The transient receptor potential channel TRPA1: From gene to pathophysiology. *Pflugers Arch*, 464: 425–458.

Nilius, B. and V. Flockerzi. 2014. What do we really know and what do we need to know: Some controversies, perspectives, and surprises. *Handb Exp Pharmacol*, 223: 1239–1280.

Nilius, B., J. Prenen, and G. Owsianik. 2011. Irritating channels: The case of TRPA1. *J Physiol*, 589: 1543–1549.

Nizet, V. and T. Yaksh. 2013. Neuroscience: Bacteria get on your nerves. *Nature*, 501: 43–44.

O'Brien, V.P. et al. 2015. Are you experienced? Understanding bladder innate immunity in the context of recurrent urinary tract infection. *Curr Opin Infect Dis*, 28: 97–105.

O'Malley, D. 2015. Immunomodulation of enteric neural function in irritable bowel syndrome. *World J Gastroenterol*, 21: 7362–7366.

Ochoa-Cortes, F. et al. 2010. Bacterial cell products signal to mouse colonic nociceptive dorsal root ganglia neurons. *Am J Physiol Gastrointest Liver Physiol*, 299: G723–G732.

Ordovas-Montanes, J. et al. 2015. The regulation of immunological processes by peripheral neurons in homeostasis and disease. *Trends Immunol*, 36: 578–604.

Pamer, E.G. 2016. Resurrecting the intestinal microbiota to combat antibiotic-resistant pathogens. *Science*, 352: 535–538.

Parenti, A. et al. 2016. What is the evidence for the role of TRP channels in inflammatory and immune cells? *Br J Pharmacol*, 173: 953–969.

Park, B.S. and J.O. Lee. 2013. Recognition of lipopolysaccharide pattern by TLR4 complexes. *Exp Mol Med*, 45: e66.

Pierce, A.N. et al. 2016. Urinary bladder hypersensitivity and dysfunction in female mice following early life and adult stress. *Brain Res*, 1639: 58–73.

Poblete, I.M. et al. 2005. Anandamide elicits an acute release of nitric oxide through endothelial TRPV1 receptor activation in the rat arterial mesenteric bed. *J Physiol*, 568: 539–551.

Ren, K. and R. Dubner. 2010. Interactions between the immune and nervous systems in pain. *Nat Med*, 16: 1267–1276.

Riol-Blanco, L. et al. 2014. Nociceptive sensory neurons drive interleukin-23-mediated psoriasiform skin inflammation. *Nature*, 510: 157–161.

Ruan, T. et al. 2014. Sensitization by pulmonary reactive oxygen species of rat vagal lung C-fibers: The roles of the TRPV1, TRPA1, and P2X receptors. *PLOS ONE*, 9: e91763.

Russell, F.A. et al. 2014. Calcitonin gene-related peptide: Physiology and pathophysiology. *Physiol Rev*, 94: 1099–1142.

Salazar, H. et al. 2008. A single N-terminal cysteine in TRPV1 determines activation by pungent compounds from onion and garlic. *Nat Neurosci*, 11: 255–261.

Santoni, G. et al. 2015. Danger- and pathogen-associated molecular patterns recognition by pattern-recognition receptors and ion channels of the transient receptor potential family triggers the inflammasome activation in immune cells and sensory neurons. *J Neuroinflammation*, 12: 21.

Sauvant, J. et al. 2014. Mechanisms involved in dual vasopressin/apelin neuron dysfunction during aging. *PLOS ONE*, 9: e87421.

Sculptoreanu, A. et al. 2010. Nitro-oleic acid inhibits firing and activates TRPV1- and TRPA1-mediated inward currents in dorsal root ganglion neurons from adult male rats. *J Pharmacol Exp Ther*, 333: 883–895.

Sharma, P. and L. Ping. 2014. Calcium ion influx in microglial cells: Physiological and therapeutic significance. *J Neurosci Res*, 92: 409–423.

Shim, W.S. et al. 2007. TRPV1 mediates histamine-induced itching via the activation of phospholipase A2 and 12-lipoxygenase. *J Neurosci*, 27: 2331–2337.

Shimizu, S., N. Takahashi, and Y. Mori. 2014. TRPs as chemosensors (ROS, RNS, RCS, gasotransmitters). *Handb Exp Pharmacol*, 223: 767–794.

Shimosato, G. et al. 2005. Peripheral inflammation induces up-regulation of TRPV2 expression in rat DRG. *Pain*, 119: 225–232.

Shutov, L.P. et al. 2016. The complement system component C5a produces thermal hyperalgesia via macrophage-to-nociceptor signaling that requires NGF and TRPV1. *J Neurosci*, 36: 5055–5070.

Skaper, S.D., L. Facc, and P. Giusti. 2014. Mast cells, glia and neuroinflammation: Partners in crime? *Immunology*, 141: 314–327.

So, K. et al. 2015. Involvement of TRPM2 in a wide range of inflammatory and neuropathic pain mouse models. *J Pharmacol Sci*, 127: 237–243.

Sokabe, T. and M. Tominaga. 2010. The TRPV4 cation channel: A molecule linking skin temperature and barrier function. *Commun Integr Biol*, 3: 619–621.

Sokabe, T. et al. 2010. The TRPV4 channel contributes to intercellular junction formation in keratinocytes. *J Biol Chem*, 285: 18749–18758.

Soldano, A. et al. 2016. Gustatory-mediated avoidance of bacterial lipopolysaccharides via TRPA1 activation in *Drosophila*. eLife, 5: e13133.

Song, K. et al. 2016. The TRPM2 channel is a hypothalamic heat sensor that limits fever and can drive hypothermia. *Science*, 353: 1393–1398.

Starr, A. et al. 2008. A reactive oxygen species-mediated component in neurogenic vasodilatation. *Cardiovasc Res*, 78: 139–147.

Steinberg, B.E., K.J. Tracey, and A.S. Slutsky. 2014. Bacteria and the neural code. *N Engl J Med*, 371: 2131–2133.

Stensmyr, M.C. et al. 2012. A conserved dedicated olfactory circuit for detecting harmful microbes in *Drosophila*. *Cell*, 151: 1345–1357.

Susankova, K. et al. 2006. Reducing and oxidizing agents sensitize heat-activated vanilloid receptor (TRPV1) current. *Mol Pharmacol*, 70: 383–394.

Takahashi, N., D. Kozai, and Y. Mori. 2012. TRP channels: Sensors and transducers of gasotransmitter signals. *Front Physiol*, 3: 324.

Talavera, K. et al. 2009. Nicotine activates the chemosensory cation channel TRPA1. *Nat Neurosci*, 12: 1293–1299.

Talavera, K., B. Nilius, and T. Voets. 2008. Neuronal TRP channels: Thermometers, pathfinders and life-savers. *Trends Neurosci*, 31: 287–295.

Tan, C.H. and P.A. McNaughton. 2016. The TRPM2 ion channel is required for sensitivity to warmth. *Nature*, 536: 460–463.

Taylor-Clark, T.E. et al. 2008a. Relative contributions of TRPA1 and TRPV1 channels in the activation of vagal bronchopulmonary C-fibres by the endogenous autacoid 4-oxononenal. *J Physiol*, 586: 3447–3459.

Taylor-Clark, T.E. et al. 2008b. Prostaglandin-induced activation of nociceptive neurons via direct interaction with transient receptor potential A1 (TRPA1). *Mol Pharmacol*, 73: 274–281.

Taylor-Clark, T.E. et al. 2009. Transient receptor potential ankyrin 1 mediates toluene diisocyanate-evoked respiratory irritation. *Am J Respir Cell Mol Biol*, 40: 756–762.

Togashi, K. et al. 2006. TRPM2 activation by cyclic ADP-ribose at body temperature is involved in insulin secretion. *EMBO J*, 25: 1804–1815.

Tominaga, M. et al. 1998. The cloned capsaicin receptor integrates multiple pain-producing stimuli. *Neuron*, 21: 531–543.

Trankner, D. et al. 2014. Population of sensory neurons essential for asthmatic hyperreactivity of inflamed airways. *Proc Natl Acad Sci U S A*, 111: 11515–11520.

Treschan, T.A. and J. Peters. 2006. The vasopressin system: Physiology and clinical strategies. *Anesthesiology*, 105: 599–612; quiz 639–540.

Trevisan, G. et al. 2013. Transient receptor potential ankyrin 1 receptor stimulation by hydrogen peroxide is critical to trigger pain during monosodium urate-induced inflammation in rodents. *Arthritis Rheum*, 65: 2984–2995.

Trevisani, M. et al. 2007. 4-Hydroxynonenal, an endogenous aldehyde, causes pain and neurogenic inflammation through activation of the irritant receptor TRPA1. *Proc Natl Acad Sci U S A*, 104: 13519–13524.

Uchida, K. and T. Shibata. 2008. 15-Deoxy-Delta(12,14)-prostaglandin J2: An electrophilic trigger of cellular responses. *Chem Res Toxicol*, 21: 138–144.

Uehara, E.U., S. Shida Bde, and C.A. de Brito. 2015. Role of nitric oxide in immune responses against viruses: Beyond microbicidal activity. *Inflamm Res*, 64: 845–852.

Undem, B.J. and T. Taylor-Clark. 2014. Mechanisms underlying the neuronal-based symptoms of allergy. *J Allergy Clin Immunol*, 133: 1521–1534.

van den Wijngaard, R.M. et al. 2009. Essential role for TRPV1 in stress-induced (mast cell-dependent) colonic hypersensitivity in maternally separated rats. *Neurogastroenterol Motil*, 21: e1107–e1194.

Van Gerven, L. et al. 2014. Capsaicin treatment reduces nasal hyperreactivity and transient receptor potential cation channel subfamily V, receptor 1 (TRPV1) overexpression in patients with idiopathic rhinitis. *J Allergy Clin Immunol*, 133: 1332–1339. e1331–1333.

Van Rijswijk, J.B. et al. 2003. Intranasal capsaicin reduces nasal hyperreactivity in idiopathic rhinitis: A double-blind randomized application regimen study. *Allergy*, 58: 754–761.

Vennekens, R., A. Menigoz, and B. Nilius. 2012. TRPs in the brain. *Rev Physiol Biochem and Pharmacol*, 163: 27–64.

Villalta, P.C. and M.I. Townsley. 2013. Transient receptor potential channels and regulation of lung endothelial permeability. *Pulm Circ*, 3: 802–815.

Vinuesa, A.G. et al. 2012. Vanilloid receptor-1 regulates neurogenic inflammation in colon and protects mice from colon cancer. *Cancer Res*, 72: 1705–1716.

Wadachi, R. and K.M. Hargreaves. 2006. Trigeminal nociceptors express TLR-4 and CD14: A mechanism for pain due to infection. *J Dent Res*, 85: 49–53.

Wang, Y.Y. et al. 2008. The nociceptor ion channel TRPA1 is potentiated and inactivated by permeating calcium ions. *J Biol Chem*, 283: 32691–32703.

Wong, B.J. and S.M. Fieger. 2010. Transient receptor potential vanilloid type-1 (TRPV-1) channels contribute to cutaneous thermal hyperaemia in humans. *J Physiol*, 588: 4317–4326.

Wong, B.J. and S.M. Fieger. 2012. Transient receptor potential vanilloid type 1 channels contribute to reflex cutaneous vasodilation in humans. *J Appl Physiol*, 112: 2037–2042.

Wong, C.O. et al. 2010. Nitric oxide lacks direct effect on TRPC5 channels but suppresses endogenous TRPC5-containing channels in endothelial cells. *Pflugers Arch*, 460: 121–130.

Wouters, M.M. et al. 2016. Histamine receptor H1-mediated sensitization of TRPV1 mediates visceral hypersensitivity and symptoms in patients with irritable bowel syndrome. *Gastroenterology*, 150: 875–887 e879.

Wu, Z. et al. 2015. Activation of TLR-4 to produce tumour necrosis factor-alpha in neuropathic pain caused by paclitaxel. *Eur J Pain*, 19: 889–898.

Xanthos, D.N. and J. Sandkuhler. 2014. Neurogenic neuroinflammation: Inflammatory CNS reactions in response to neuronal activity. *Nat Rev Neurosci*, 15: 43–53.

Xu, S.Z. et al. 2008. TRPC channel activation by extracellular thioredoxin. *Nature*, 451: 69–72.

Yamamoto, S. and S. Shimizu. 2016. Targeting TRPM2 in ROS-coupled diseases. *Pharmaceuticals (Basel)*, 9: E57.

Yang, P. and M.X. Zhu. 2014. TRPV3. *Handb Exp Pharmacol*, 222: 273–291.

Yoshida, T. et al. 2006. Nitric oxide activates TRP channels by cysteine S-nitrosylation. *Nat Chem Biol*, 2: 596–607.

Yu, X. et al. 2016. TRP channel functions in the gastrointestinal tract. *Semin Immunopathol*, 38: 385–396.

Zimmermann, K. et al. 2011. Transient receptor potential cation channel, subfamily C, member 5 (TRPC5) is a cold-transducer in the peripheral nervous system. *Proc Natl Acad Sci U S A*, 108: 18114–18119.

Zuo, L. et al. 2015. Biological and physiological role of reactive oxygen species—The good, the bad and the ugly. *Acta Physiol (Oxf)*, 214: 329–348.

Zurborg, S. et al. 2007. Direct activation of the ion channel TRPA1 by Ca^{2+}. *Nat Neurosci*, 10: 277–279.

Zygmunt, P.M. and E.D. Hogestatt. 2014. TRPA1. *Handb Exp Pharmacol*, 222: 583–630.

16 TRP Channels in the Brain
What Are They There For?

Seishiro Sawamura, Hisashi Shirakawa, Takayuki Nakagawa, Yasuo Mori, and Shuji Kaneko

CONTENTS

16.1 TRANSIENT RECEPTOR POTENTIAL CHANNELS: A SHORT INTRODUCTION

16.1.1 BRIEF INTRODUCTION

Transient receptor potential (TRP) family proteins form tetrameric nonselective cation channels. Upon activation, TRP channels depolarize the membrane potential, which can lead to activation or inactivation of voltage-gated ion channels, and regulate Ca^{2+} signaling, which controls diverse cellular functions (Wu et al., 2010; Nilius and Szallasi, 2014). It is well known that some members of the TRP canonical (TRPC), TRP melastatin (TRPM), and TRP vanilloid (TRPV) subfamilies of TRP channels are highly expressed and play important roles in the brain (Vennekens et al., 2012; Nilius and Szallasi, 2014). They regulate diverse neuronal and glial functions including developmental and homeostatic functions of the brain. Recent studies show that dysregulation of the TRP channel functions is involved in various pathological events of neurological and psychiatric disorders. Here, we review the current insights of the physiological roles of the TRPC, TRPM, and TRPV channels, mainly TRPC3/TRPC6/TRPC7, TRPM2, and TRPV1 in neurons and glia, and their pathophysiological roles in neurological and psychiatric disorders.

16.1.2 TRPC CHANNELS IN THE BRAIN

The TRPC subfamily contains seven members: TRPC1–TRPC7. All TRPC channels form Ca^{2+}-permeable nonselective cation channels and are highly expressed in diverse regions of the brain (Vennekens et al., 2012). They generally work as receptor-operated cation channels and have been implicated in various cellular functions including neuronal firing, synapse transmission, gene expression, migration, neurite elongation, and growth cone guidance. Based on sequence homology and functional similarity, TRPC members are classified into three groups: the TRPC1/TRPC4/TRPC5, TRPC2, and TRPC3/TRPC6/TRPC7 groups (Vazquez et al., 2004; Wu et al., 2010). TRPC3, TRPC6, and TRPC7 form the diacylglycerol (DAG)-sensitive group, which is activated by direct action of DAG upon phospholipase C (PLC)-coupled receptor stimulation such as $G_{q/11}$ type G protein-coupled receptors (GPCRs) or receptor tyrosine kinases (RTKs). TRPC3, TRPC6, and TRPC7 are highly expressed in various regions of the brain (Tables 16.1 through 16.3) and play

TABLE 16.1
Expression of TRPC3 in Brain Tissues or Cells and Methods of Detection

Expressed Tissues or Cells	Detection Methods	References (species)
Cerebrum cortex	RT-PCR, qRT-PCR, WB, IHC	Mouse (Kunert-Keil et al., 2006), rat (Li et al., 1999; Mizuno et al., 1999; Roedding et al., 2013)
Hippocampus	RT-PCR, qRT-PCR, ISH, WB, IHC, ICC	Mouse (Kunert-Keil et al., 2006; Li et al., 2012a; Zeng et al., 2015a), rat (Li et al., 1999; Singh et al., 2004; Chung et al., 2006)
Cerebellum	RT-PCR, WB, IHC	Human (Roedding et al., 2009), mouse (Kunert-Keil et al., 2006), rat (Li et al., 1999; Huang et al., 2007)
Purkinje cell	ISH, WB, IHC	Mouse (Becker et al., 2009; Kato et al., 2012; Mitsumura et al., 2011; Kim et al., 2012), rat (Huang et al., 2007)
Granule neuron	RT-PCR, WB	Rat (Li et al., 2005; Jia et al., 2007; Sawamura et al., 2016)

(Continued)

TABLE 16.1 (*Continued*)

Expression of TRPC3 in Brain Tissues or Cells and Methods of Detection

Expressed Tissues or Cells	Detection Methods	References (species)
Striatum	RT-PCR, single cell RT-PCR, ISH, WB, IHC	Mouse (Kim et al., 2012), rat (Kim et al., 2012; Xie and Zhou, 2014), guinea pig (Kim et al., 2012)
Substantia nigra	RT-PCR, single-cell RT-PCR, IHC, EM	Mouse (Zhou et al., 2008a), rat (Sylvester et al., 2001; Fusco et al., 2004)
Amygdala	WB, IHC	Rat (Li et al., 1999)
Pontine neuron	WB, IHC	Rat (Li et al., 1999)
Astrocyte	qRT-PCR, WB, IHC, EM	Rat (Fusco et al., 2004; Shirakawa et al., 2010)
Microglia	IHC, ICC, EM	Rat (Fusco et al., 2004; Mizoguchi et al., 2014)
Oligodendrocyte	IHC, EM	Rat (Fusco et al., 2004)
Basilar artery	qRT-PCR, IHC	Rat (Song et al., 2013)
Pituitary cell	qRT-PCR	Rat (Tomic et al., 2011)

Note: Abbreviations: EM, electron microscopy; ICC, immunocytochemistry; IHC, immunohistochemistry; ISH, in situ hybridization; RT-PCR, reverse transcriptase polymerase chain reaction; qRT-PCR, quantitative RT-PCR; WB, western blotting.

TABLE 16.2

Expression of TRPC6 in Brain Tissues or Cells and Methods of Detection

Expressed Tissues or Cells	Detection Methods	References (species)
Cerebrum cortex	RT-PCR, ISH, WB, IHC	Mouse (Kunert-Keil et al., 2006; Boisseau et al., 2009; Gibon et al., 2013), rat (Mizuno et al., 1999; Du et al., 2010; Lin et al. 2013a; Yao et al., 2013)
Hippocampus	RT-PCR, qRT-PCR, ISH, WB, IHC, ICC, EM	Mouse (Kunert-Keil et al., 2006; Xu et al., 2012; Gibon et al., 2013; Nagy et al., 2013; Zeng et al., 2015a), rat (Chung et al., 2006; Tai et al., 2008; Zhou et al., 2008b; He et al. 2012; Liu et al., 2015)
Cerebellum	RT-PCR, ISH, WB, IHC	Mouse (Kunert-Keil et al., 2006; Jia et al., 2007), rat (Huang et al., 2007)
Purkinje cell	IHC	Rat (Huang et al., 2007)
Granule neuron	RT-PCR, WB, IHC, ICC	Mouse (Jia et al., 2007), rat (Li et al. 2005; Huang et al., 2007; Sawamura et al., 2016)
Substantia nigra	IHC	Rat (Giampa et al., 2007)
Astrocyte	WB	Mouse (Beskina et al., 2007)
Microglia	IHC	Rat (Lee et al., 2014)
Cerebral artery	qRT-PCR, WB, IHC	Mouse (Toth et al., 2013), rat (Kim et al., 2013; Johansson et al., 2015)
Pituitary cell	qRT-PCR	Rat (Tomic et al., 2011)

TABLE 16.3

Expression of TRPC7 in Brain Tissues or Cells and Methods of Detection

Expressed Tissues or Cells	Detection Methods	References (species)
Cerebral cortex	qRT-PCR, ISH	Mouse (Okada et al., 1999; Boisseau et al., 2009)
Hippocampus	ISH	Mouse (Okada et al., 1999)
Cerebellum	ISH	Mouse (Okada et al., 1999)
Granule neuron	RT-PCR	Rat (Sawamura et al., 2016)
Striatum	ISH, IHC	Rat (Berg et al., 2007)
Median preoptic nucleus	RT-PCR	Mouse (Tabarean, 2012)
Hypothalamus	RT-PCR	Rat (Cvetkovic-Lopes et al., 2010)
Pons	ISH	Mouse (Okada et al., 1999)

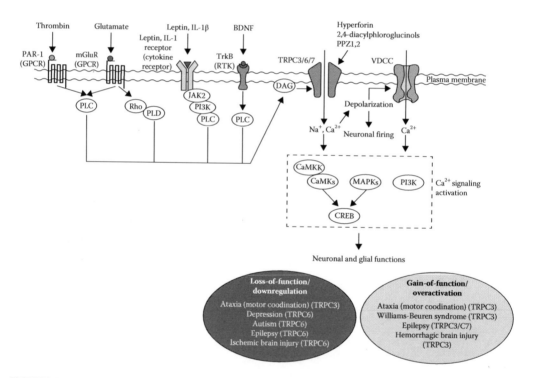

FIGURE 16.1 Activation mechanisms and functions of TRPC3, TRPC6, and TRPC7 in the brain. Stimulation of GPCR, RTK, and cytokine receptor activate these TRPC channels probably via production of DAG. The activity of TRPC3, TRPC6, and TRPC7 regulates neuronal and glial functions through membrane depolarization and Ca^{2+} signaling activation. Loss-of-function mutations/downregulation and gain-of-function mutations/overactivation of these TRPC channels may contribute to the pathology of neurological and psychiatric disorders.

important roles in many neuronal functions in response to neuronal receptor stimulation; they are also involved in various neuronal diseases and psychiatric disorders (Figure 16.1) (Abramowitz and Birnbaumer, 2009; Chahl, 2011; Takada et al., 2013). In contrast, the TRPC1/TRPC4/TRPC5 group of channels, which is also activated by receptor stimulation, is unresponsive to DAG (Venkatachalam et al., 2003). TRPC1 is thought to be involved in store-operated Ca^{2+} (SOC) entry (Wes et al., 1995; Mori et al., 2002), while TRPC5 is regulated by Ca^{2+} and oxidative stress (Yoshida et al., 2006; Blair et al., 2009). These channels are also involved in certain physiological and pathological events in the brain. TRPC2 is primarily expressed in the vomeronasal organ and is a pseudogene in humans (Lucas et al., 2003).

16.1.3 TRPM CHANNELS IN THE BRAIN

The TRPM subfamily contains eight members: TRPM1–TRPM8. The activity of some TRPM proteins is modulated by oxidative stress (Simon et al., 2013). Specifically, TRPM2 is highly expressed in the brain (Table 16.4), and its role in the central nervous system (CNS) has been studied. It is the first identified redox-sensitive TRP channel (Hara et al., 2002) and forms a Ca^{2+}-permeable cation channel, which is activated by oxidative stress mediated by reactive oxygen species (ROS), such as hydrogen peroxide (H_2O_2) through the production of nicotinamide adenine dinucleotide (NAD^+), and its metabolites such as ADP-ribose (ADPR) and cyclic ADPR (Takahashi et al., 2011), although some controversy exists on whether NAD^+ and cADPR are direct activators of TRPM2 (Toth et al., 2015; Rosenbaum, 2015). Studies indicate that TRPM2 mediates H_2O_2-induced Ca^{2+} influx, which modulates physiological and pathological cellular functions. TRPM2 is expressed in both neurons and glia, and oxidative stress-induced TRPM2 activation is implicated in neuronal diseases (Figure 16.2).

16.1.4 TRPV CHANNELS IN THE BRAIN

The TRPV subfamily contains six members: TRPV1–TRPV6. All TRPV channels form Ca^{2+}-permeable cation channels. The function of TRPV1 has been investigated using highly selective exogenous agonists such as capsaicin and resiniferatoxin (RTX). Physiologically, TRPV1 is thought to be activated by physical or chemical stimuli such as noxious heat (>43°C), mechanical force, pH change, ROS, and endogenous compounds such as arachidonic acid metabolites and endocannabinoids (Rosenbaum and Simon, 2007; Takahashi and Mori, 2011; Nilius and Szallasi; 2014). TRPV1 was initially thought to be expressed mainly in sensory neurons; however, research has shown broad expression in the brain (Table 16.5) and physiological roles of TRPV1 in central neurons and glia. It has also been suggested that TRPV1 plays a role in neuronal and psychiatric disorders. TRPV4 is also activated by wide variety of physical and chemical stimuli including hypotonicity,

TABLE 16.4
Expression of TRPM2 in Brain Tissues or Cells and Methods of Detection

Expressed Tissues or Cells	Detection Methods	References (species)
Cerebrum cortex	RT-PCR, qRT-PCR, northern blot, WB, IHC, ICC	Human (Uemura et al., 2005), mouse (Uemura et al., 2005; Jia et al., 2011), rat (Kaneko et al., 2006; Cook et al. 2010; Roedding et al., 2013)
Hippocampus	RT-PCR, qRT-PCR, northernblot, ISH, WB, IHC	Human (Uemura et al., 2005), mouse (Uemura et al., 2005; Belrose et al. 2012; Katano et al., 2012), rat (Lipski et al., 2006; Bai and Lipski, 2010; Cook et al., 2010)
Cerebellum	Northern blot, ISH	Human, mouse (Uemura et al. 2005)
Striatum	Northern blot, ISH	Human, mouse (Uemura et al., 2005)
Substantia nigra	RT-PCR ISH, WB, IHC	Mouse (Mrejeru et al 2011), guinea pig (Lee et al., 2013)
Locus coeruleus	qRT-PCR	Mouse (Cui et al., 2011)
Microglia	RT-PCR	Rat (Kraft et al., 2004)
Endothelial cell	WB, ICC	Mouse (Park et al., 2014)

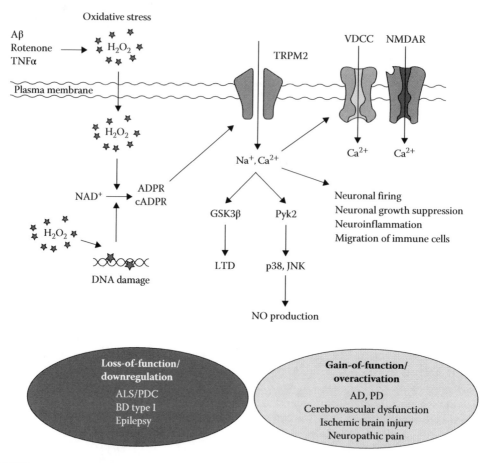

FIGURE 16.2 Activation mechanisms and functions of TRPM2 in the brain. The activity of TRPM2 is modulated by oxidative stress via production of ADPR. TRPM2 regulates synaptic plasticity, VDCC and NMDAR functions, neuronal firing, neurite outgrowth, NO production, neuroinflammation and migration of immune cells via Ca^{2+} signaling and depolarizing effects. Dysregulation of TRPM2 activity may contribute to the pathology of neurological and psychiatric disorders.

heat (>25°C), and endocannabinoids (Everaerts et al., 2010; Wu et al., 2010). TRPV4 expression has been detected in several regions of the brain, and its contribution to neuronal and glial functions has been investigated.

16.2 THE TRANSIENT RECEPTOR POTENTIAL CANONICAL SUBFAMILY

16.2.1 PHYSIOLOGICAL ROLES OF TRPC CHANNELS IN NEURONS

TRPC channels are involved in diverse neuronal functions via neuronal receptor stimulation by neurotrophic factors or neuropeptides. Activation of TRPC channels induces cation influx, which controls cellular functions and supports depolarization to regulate neuronal activity in the brain.

TRPC1 is involved in brain-derived neurotrophic factor (BDNF)- and netrin-1-induced axon guidance (Li et al., 2005; Shim et al., 2009). TRPC1 is activated by peptidic hormone leptins through a Janus kinase-2 (JAK2)/phosphoinositide 3-kinase (PI3K)/PLCγ pathway, and its activation mediates the depolarizing effects of leptins in hypothalamic proopiomelanocortin neurons (Qiu et al., 2010). In hippocampal neurons, TRPC1 and TRPC3 are required for leptin-induced current and spine formation (Dhar et al., 2014). Furthermore, knockdown of TRPC1 reduced the degree of neuronal progenitor cell proliferation by cell cycle arrest (Li et al., 2012b).

TABLE 16.5
Expression of TRPV1 in Brain Tissues or Cells and Methods of Detection

Expressed Tissues or Cells	Detection Methods	References (species)
Cerebrum cortex	RT-PCR, ISH, WB, IHC, RTX autoradiography	Human (Mezey et al., 2000), mouse (Roberts et al., 2004), rat (Mezey et al., 2000; Toth et al., 2005),
Limbic system	ISH	Rat (Mezey et al., 2000)
Hippocampus	RT-PCR, ISH, WB, IHC, genetic reporter, RTX autoradiography	Mouse (Roberts et al., 2004; Cristino et al., 2006; Cavanaugh et al., 2011), rat (Mezey et al., 2000; Toth et al., 2005)
Cerebellum	ISH, WB, RTX autoradiography	Mouse (Roberts et al., 2004; Cristino et al., 2006), rat (Mezey et al., 2000; Toth et al., 2005)
Striatum	WB	Mouse (Maccarrone et al., 2008)
Substantia nigra	RT-PCR, ISH, WB, IHC	Human (Nam et al., 2015), mouse (Cristino et al., 2006), rat (Mezey et al., 2000; Kim et al., 2005; Kim et al., 2006; Nam et al., 2015)
Basal ganglia	IHC	Mouse (Cristino et al., 2006)
Thalamus	IHC, RTX autoradiography	Mouse (Roberts et al., 2004; Cristino et al., 2006)
Hypothalamus	RT-PCR, ISH, WB, IHC, genetic reporter, RTX autoradiography	Mouse (Roberts et al., 2004; Cristino et al., 2006; Cavanaugh et al., 2011; Prager-Khoutorsky et al., 2014), rat (Cavanaugh et al., 2011)
Midbrain	RT-PCR, WB, RTX autoradiography	Mouse (Roberts et al., 2004), rat (Marinelli et al., 2003; Toth et al., 2005)
Periaqueductal gray	IHC, genetic reporter, RTX autoradiography	Mouse (Roberts et al., 2004; Cristino et al., 2006; Cavanaugh et al. 2011)
Hindbrain	WB	Rat (Toth et al., 2005)
Pontine nuclei	IHC	Mouse (Cristino et al., 2006)
Olfactory bulb	WB, genetic reporter, RTX autoradiography	Mouse (Roberts et al., 2004; Cavanaugh et al. 2011), rat (Toth et al., 2005)
Inferior olive	ISH, IHC	Rat (Mezey et al., 2000)
Circumventricular organs	RT-PCR, WB, IHC	Mouse (Mannari et al., 2013)
Astrocyte	qRT-PCR, WB, IHC, ICC	Human (Nam et al., 2015), mouse (Mannari et al., 2013), rat (Huang et al., 2010; Nam et al., 2015)
Microglia	RT-PCR, WB, IHC, ICC	Mouse (Miyake et al., 2015), rat (Kim et al., 2006; Park et al., 2012)

TRPC4 expression inhibits neurite growth and branching in cultured hippocampal neurons via a Ca^{2+}/calmodulin-dependent kinase (CaMK)II-dependent pathway (Jeon et al., 2013). TRPC5 is also known to have an inhibitory role in neurite growth and growth cone morphology (Greka, 2003). TRPC5 activation by neurotrophin-3 decreases dendritic growth and arborization in hippocampal and cerebellar granule neurons via a CaMKII-dependent pathway (Puram et al., 2011; He et al., 2012). In *Trpc5* null mice, semaphorin 3A-induced growth cone collapse was reduced; in this event, TRPC5 is cleaved by calpain to its functionally activated state (Kaczmarek et al., 2012). These findings suggest that Ca^{2+} signaling from TRPC4/C5 channels is important for negative regulation of neurite growth necessary for proper development of the brain. TRPC5-mediated localized Ca^{2+} influx may play different roles in regulating the signaling cascades on the neurite outgrowth of the neurons in different stages; other studies have found that TRPC5 can promote neurite outgrowth of hippocampal neurons via CaMK kinase/CaMKIγ pathway in early phases (Davare et al., 2009).

TRPC5 is suggested to mediate muscarinic receptor–induced slow afterdepolarization in pyramidal cells of the cerebral cortex (Yan et al., 2009). Enhancement of neuronal excitability induced by neuropeptide cholecystokinin (CCK) in the entorhinal cortex was significantly suppressed by inhibition of TRPC5, suggesting its involvement (Wang et al., 2011). Knockout (KO) studies in mice showed that both TRPC4 and TRPC5 are involved in innate fear response. Lack of *Trpc4* or *Trpc5* resulted in a reduction in response mediated by the synaptic activation of the group I muscarinic glutamate receptor (mGluR) and CCK-2 receptors in neurons of the amygdala, which is responsible for fear conditioning in the brain. The observed reductions led to a decrease of innate fear levels (Riccio et al., 2009, 2014).

TRPC3 is the most abundantly expressed TRPC subunit in Purkinje cells, where it is activated by mGluR1 stimulation via the subsequent Rho/Phospholipase D (PLD) pathway but not via the PLC pathway (Glitsch, 2010). TRPC3 plays an essential role in slow excitatory postsynaptic potential and induction of long-term depression (LTD) (Hartmann et al., 2008; Nelson and Glitsch, 2012; Kim, 2013). Although TRPC3 KO mice did not show the expected alterations in brain development, they exhibited loss of mGluR-dependent inward current and impaired walking behavior (Hartmann et al., 2008, 2011). These results suggest that TRPC3 is a postsynaptic channel that mediates mGluR-dependent synaptic transmission in cerebellar Purkinje cells and is important for motor coordination.

In several types of neurons, TRPC3 is activated by BDNF and possibly interacts with the BDNF receptor tropomyosin related kinase B (TrkB) (Li et al., 1999, 2010). Activation of TRPC3 by BDNF-TrkB signaling appears to be essential for the guidance of nerve growth cones and dendritic spine formation (Li et al., 2005; Amaral and Pozzo-Miller, 2007). It was also shown that TRPC3 and TRPC6 play an essential role in BDNF-induced neuronal protection in cerebellar granule neurons via the mitogen-associated protein kinase (MAPK), CaMK/cAMP-response-element binding protein (CREB)–dependent pathway (Jia et al., 2007). A recent study suggests that Ca^{2+} signaling via TRPC3 and TRPC6 is required for BDNF-induced mitochondrial accumulation in presynaptic sites, which leads to enhancement of neurotransmitter release (Su et al., 2014).

TRPC6 is localized at excitatory postsynapses in the rat hippocampus and promotes their formation via a CaMKIV/CREB-dependent pathway. Overexpression of TRPC6 increases the number of spines in hippocampal neurons and enhances spatial learning and memory in the Morris water maze test, suggesting a functional role for TRPC6 in synaptic and behavioral plasticity (Tai et al., 2008; Zhou et al, 2008b). TRPC6 KO mice showed no significant differences in anxiety behavior in a marble burying test, but demonstrated reduced exploration in the square open field and the elevated star maze tests (Beis et al., 2011). This suggests that TRPC6 may have an important role in exploration behavior. The physiological importance of TRPC6 has also been investigated through studies of hyperforin, an active ingredient of the widely used and well-studied medicinal plant St. John's wort (Wolfle et al., 2014) and its analogues. Hyperforin is a multitarget drug that modulates various molecular functions and has antimicrobial, anti-inflammatory, anticancer, and antidepressant activity (Russo et al., 2014). It has been shown that hyperforin activates TRPC6 without affecting other TRPC channels, resulting in its antidepressant activity (Leuner et al., 2007). It also increases neurite outgrowth through MAPK, PI3K, and CaMK pathways in neuronal PC12 cells (Leuner et al., 2007; Heiser et al., 2013). Synthetic analogues of hyperforin, 2,4-diacylphloroglucinols, activated TRPC6 without affecting the closely related TRPC3 and TRPC7, and induced neurite outgrowth of PC12 cells (Leuner et al., 2010). In hippocampal neurons, hyperforin modulated dendritic spine morphology but had no effect on dendritic length and arborization (Leuner et al., 2013). These results suggest that TRPC6 but not TRPC3/TRPC7 induces neurotrophic effects triggered by hyperforin in neuronal cells. However, the activity of hyperforin via TRPC6 remains controversial; recent studies attribute the protonophore activity of hyperforin to its pharmacological effects (Tu et al., 2010; Sell et al., 2014).

We recently reported piperazine-derived small activators for the DAG-sensitive TRPC channels TRPC3, TRPC6, and TRPC7, derived using chemical library screening. We found that two activators [4-(5-chloro-2-methylphenyl)piperazin-1-yl](3-fluorophenyl)methanone (PPZ1) and 2-[4-(2,3-dimethylphenyl)-piperazin-1-yl]-*N*-(2-ethoxyphenyl)acetamide (PPZ2) selectively activated the TRPC3/TRPC6/TRPC7 group in a dose-dependent manner. PPZ2 stimulated cation currents and Ca^{2+} influx in cultured rat cerebellar granule neurons, and both PPZ1 and PPZ2 induced BDNF-like promotion of neurite outgrowth and neuronal survival, which was abolished by a knockdown or inhibition of TRPC3/TRPC6/TRPC7 in cultured neurons. We also found contribution of several Ca^{2+} signaling pathways, same with BDNF except for calcineurin, in PPZ-induced neurite outgrowth. These results also support neurotrophic functions of TRPC3, TRPC6, and TRPC7 via Ca^{2+} signaling activation. TRPC7 is the most elusive TRPC channel; although it is highly expressed in the brain, its physiological function is still not well understood. It has been suggested that TRPC3 and TRPC7 act as ion channel targets for mGluR1/5 signaling in rat striatal cholinergic interneurons (Berg et al., 2007).

16.2.2 Physiological Roles of TRPC Channels in Glia

It has been reported that a subtype of TRPC is expressed in glial cells, including astrocytes, microglia, and oligodendrocytes. In astrocytes, which are the most abundant cells in the CNS, various patterns of fluctuations in intracellular Ca^{2+} concentration ($[Ca^{2+}]_i$) are also observed with respect to the activation of nonelectrically excitable astrocytes (Bazargani and Attwell, 2016). Several isoforms of TRPC channels are expressed in astrocytes (Pizzo et al., 2001) and several studies have revealed the functional properties of TRPC channels in astrocytes. Song et al. (2005) have reported that TRPC4 interacts with the PDZ1 domain of ZO-1 and can regulate Ca^{2+} homeostasis in astrocytes, particularly as part of a signaling complex that forms at junctional sites between astrocytes. Golovina (2005) has reported that selective inhibition of TRPC1 protein expression using an antisense oligonucleotide targeted to the TRPC1 gene attenuated the SOC channel-mediated rise of $[Ca^{2+}]_i$ and the consequent astrocyte proliferation. Malarkey et al. (2008) have reported that TRPC1-mediated Ca^{2+} entry induced by mechanical stimulation or ATP contributes to intracellular Ca^{2+} dynamics and the consequent release of glutamate from rat astrocytes. Moreover, Akita and Okada (2011) have reported that TRPC1 and TRPC3 channels are involved in Ca^{2+} entry. Importantly, entry through TRPC1 channels is strongly involved in the bradykinin-induced activation of volume-sensitive outwardly rectifying (VSOR) anion channels, which involves local cell volume regulation in mouse astrocytes. Additionally, Liang et al. (2014) have reported that ammonium-induced increases in SOC entry are the result of an upregulation in mRNA and protein expression of TRPC1 in astrocytes *in vitro* and *in vivo*, implying a role for ammonia neurotoxicity. Grimaldi et al. (2003) have shown that TRPC3 mediates complex and heterogeneous $[Ca^{2+}]_i$ responses to thrombin in rat cultured astrocytes and C6 glioma cells. In addition, Nakao et al. (2008) have reported that thrombin-induced Ca^{2+} mobilization and actin rearrangement are both mediated by TRPC3 in 1321N1 human astrocytoma cells. Moreover, Shirakawa et al. (2010) have reported that the expression of TRPC3 protein increases in thrombin-activated rat primary cortical astrocytes. They also reported that multiple TRPC3-mediated thrombin-induced cellular responses, namely morphological changes, S100B expression, and cell proliferation, which are associated with astrogliosis, indicate that TRPC3-mediated Ca^{2+} signaling may contribute to cellular responses in reactive astrocytes. Additionally, TRPC3 mediates ATP-induced Ca^{2+} waves and transients in primary striatal astrocytes, which are inhibited by low micromolar levels of Mn^{2+} (Streifel et al., 2013). This implies that the loss of normal astrocytic responses to purinergic signals due to the accumulation of Mn^{2+} may be a critical homeostatic function necessary for metabolic and trophic support of neurons. The proinflammatory cytokine interleukin-1β (IL-1β) can evoke TRPC6-dependent changes in $[Ca^{2+}]_i$ in mouse astrocytes (Beskina et al., 2007).

There are only a few studies regarding the functional expression of TRPC channels in microglia, which are the resident immune cells of the CNS. Quantitative comparisons of mRNA expression have shown that TRPC1, TRPC3, and TRPC6 are highly expressed in cultured microglia (Ohana et al., 2009). In line with this finding, double-label immunofluorescence studies have shown that TRPC3 and TRPC6 are weakly localized in microglia *in vivo* (Fusco et al., 2004; Lee et al., 2014). Recently, Mizoguchi et al. (2014) have reported that BDNF induces sustained intracellular Ca^{2+} elevation through the upregulation of surface TRPC3 channels. They also found that TRPC3 channels may be important for the BDNF-induced suppression of nitric oxide (NO) production in activated microglia.

TRPC3 is mainly localized in oligodendrocytes (Fusco et al. 2004). However, the physiological and functional significance of this protein remains unknown. Paez et al. (2011) have reported that in oligodendrocyte progenitor cells (OPCs), TRPC1 is involved in SOC influx, which is modulated by Golli proteins, which are products of the myelin basic protein gene. In addition, the authors indicate that Golli is probably associated with TRPC1 in processes of OPCs and that TRPC1 expression is essential for the effects of Golli on OPC proliferation.

16.2.3 PATHOPHYSIOLOGICAL ROLES OF TRPC CHANNELS IN THE BRAIN

Considering their function as Ca^{2+} influx channels, and the critical role of intracellular Ca^{2+} dynamics in neuronal development, functional signaling, and neuronal death/survival, it is clear that dysfunctional TRP channels can be expected to have a profound effect on neuronal health.

16.2.3.1 Neurodegenerative Disorders

Neurodegenerative disorders are an important area of research in both science and medicine; although much research has been done on the causes of these disorders, only marginal clinical progress has been made, and an effective treatment is still elusive. These disorders cause progressive neuronal loss, which can be associated with oxidative stress, neuroinflammation, and perturbed neuronal Ca^{2+} homeostasis (Barnham et al., 2004; Bezprozvanny, 2009). Some members of the TRP channel family have been implicated in the pathophysiology of various neurodegenerative disorders (Takada et al., 2013).

As mentioned previously, TRPC3 plays an essential role in cerebellar Purkinje cells. TRPC3 KO mice showed absence of mGluR-dependent cation current in Purkinje cells and impaired walking behavior, suggesting that TRPC3 is involved in motor control and coordination (Hartmann et al., 2008, 2011). A gain-of-function mutant in TRPC3 (T635A) results in a cerebellar ataxia-like phenotype in mice, the so-called Moonwalker (Mwk) mice (Becker et al., 2009). This mutation causes an increase of mGluR-dependent current in the cerebellum slice and degeneration of Purkinje cells as well as severe motor coordination defects, including impaired gait and balance, similar to those seen in patients with ataxia. KO and a gain-of-function of the same TRPC3 channel lead to similar defects in mice. KO mice showed loss of a depolarizing current, leading to defects in mGluR signaling, whereas the gain-of-function of TRPC3 in Mwk mice disturbs the normal Ca^{2+} and Na^+ homeostasis, causing altered dendritic development and Purkinje cell loss (Trebak, 2010). These findings implicate the importance of TRPC3 channel in the regulation of motor behavior mediated by the cerebellum, although a genetic screen for TRPC3 mutations in patients with late-onset cerebellar ataxia failed to reveal the contribution of TRPC3 mutants to this disease (Becker et al., 2011).

TRPC3 is thought to be involved in autosomal dominant spinocerebellar ataxia type 14, caused by mutations in protein kinase C (PKC)γ, which negatively regulates TRPC3 in the recombinant system. The mutant PKCγ showed reduced membrane residence time and activity at the plasma membrane, which may lead to reduction of TRPC3 phosphorylation, causing abnormal activity of the protein; this is in agreement with the results of an increase in slow excitatory postsynaptic current (EPSC) in mutant PKCγ-expressed Purkinje cells (Shuvaev et al., 2011). Studies have also shown that activation and inhibition of conventional PKC did not affect mGluR-dependent slow

EPSC in rat Purkinje cells (Nelson and Glitsch, 2012). Further studies are needed to clarify the involvement of TRPC3 in this disease.

TRPC3 may also be indirectly involved in Williams-Beuren syndrome, a neurodevelopmental disorder associated with distinctive physical characteristics, moderate mental retardation, strong emotional activity, heart or blood vessel problems, and hypercalcemia. The main genetic defect underlies deletion of the transcription factor IIi gene encoding TFII-I, which normally suppresses cell surface expression of TRPC3. TFII-I mutations increase expression of TRPC3, which results in TRPC3 gain-of-function phenotype. Although functional expression of TRPC3 in the brain of the patients has not been established, these findings may suggest its involvement in the disease (Letavernier et al., 2012).

Parkinsonian movement disorders are often associated with abnormalities in the intensity and pattern of γ-aminobutyric acid (GABA) neuron firing in the substantia nigra. TRPC3 is expressed in substantia nigra GABA projection neurons and controls the firing of the neurons which is crucial to movement control, suggesting involvement of TRPC3 in the pathology of Parkinson's disease (PD) (Zhou et al, 2008a). Hyperforin, a TRPC6 activator, has been shown to have cognitive and memory enhancing properties. It has been suggested to have neuroprotective effects against Alzheimer's disease (AD) neuropathology, including the ability to disassemble amyloid-β (Aβ) aggregates *in vitro*, decrease astrogliosis and microglia activation, as well as improve spatial memory *in vivo* (Griffith et al., 2010).

16.2.3.2 Mental Disorders

As mentioned in Section 2.1, hyperforin has been proposed to activate TRPC6 in its antidepressant and neurotrophic effects. In the stress-induced depression model along with downregulation of TRPC6, hyperforin ameliorated impaired dendritic morphology, synapse formation, impaired synaptic transmission, and spatial cognition (Liu et al., 2015). Recently, balanced translocation disruption of the TRPC6 gene has been identified in the individual with nonsyndromic autism. In induced pluripotent stem cell–derived neurons of the individual, TRPC6 haploinsufficiency causes dysregulation of Ca^{2+} signaling, which leads to altered neuronal development, morphology, and function. These changes were reversed by application of hyperforin. Using genetic screening, loss-of-function mutations of TRPC6 with incomplete penetrance was found in two families with autism spectral disorder. These findings suggest that TRPC6 may contribute to the onset of autism and become a potential therapeutic target (Griesi-Oliveira et al., 2015).

16.2.3.3 Epilepsy

Epilepsy is a result of disturbed electrical activity in the brain, characterized by recurrent synchronized electrical discharges that prevent normal neuronal functions, which may be caused by a variety of factors. Although multiple factors contribute to the onset of epilepsy, ion channels play a pivotal role as they regulate many neuronal properties, including membrane excitability, and dysregulation of these channels leads to generalized cortical spike and wave seizures, a central feature of epilepsy (Sander, 2003; Morelli et al., 2013). Recent studies suggest involvement of TRPC channels in the pathology of epilepsy.

TRPC3 is sensitive to extracellular divalent cations including Ca^{2+} and Mg^{2+}. Decreasing concentrations of divalent cations activate, while increasing concentrations inhibit TRPC3 channels (Zhou et al, 2008a). TRPC3 is highly expressed in immature and dysplastic cortices. Combinations of low Ca^{2+} and low Mg^{2+} induced larger depolarization in pyramidal neurons and greater susceptibility to epileptiform activity in immature and dysplastic cortices than in mature and control cortices. The blockade of TRPC3 significantly diminished these effects, suggesting that enhanced TRPC3 activity may contribute to susceptibility to epilepsy (Zhou and Roper, 2014). TRPC3 and TRPC6 are also involved in epilepsy-induced neuronal damages. Following status epilepticus (SE), TRPC3 expression is significantly elevated in CA1 and CA3 pyramidal cells and dentate granule cells, while TRPC6 expression is reduced in these regions. The selective TRPC3 inhibitor pyrazole-3 (Pyr3) prevented upregulation of TRPC3 induced by SE, and the TRPC6 activator hyperforin prevented downregulation of TRPC6 induced by SE. Both Pyr3 and hyperforin effectively blocked neuronal damage resulting from SE (Kim et al., 2013).

Temporal lobe epilepsy (TLE) is the most common form of intractable epilepsy and is accompanied by hippocampal sclerosis. Cortical TRPC3 and TRPC6 protein expression was increased in TLE patients and in the CA3 region of the hippocampus of TLE model mice. Inhibition of TRPC3 prevented aberrant mossy fiber sprouting, a possible pathological basis of TLE in the CA3 region, while inhibition of TRPC6 reduced the dendritic complexity of CA3 pyramidal neurons (Zeng et al., 2015a). TRPC6 expression is downregulated in chronic epileptic rats. TRPC6 knockdown increases seizure susceptibility, the excitability ratio, and paired-pulse inhibition in the dentate gyrus. Furthermore, TRPC6 knockdown enhanced programmed neuronal necrosis in dentate granule cells following SE, but suppressed it in CA1 and CA3 neurons. Thus, TRPC6 may inhibit seizure susceptibility and neuronal vulnerability in the rat dentate gyrus (Kim and Kang, 2015). These findings suggest that enhanced TRPC3 activity or reduced TRPC6 activity may contribute to the pathogenesis of epilepsy. TRPC7 is also involved in the initiation of seizures. Genetic ablation of TRPC7 reduces pilocarpine-induced gamma wave activity and occurrence of SE. TRPC7 plays a critical role in long-term potentiation (LTP) at the CA3 recurrent collateral synapses and Schaffer collateral-CA1 synapses as well as in the generation of spontaneous epileptiform burst firing in CA3 pyramidal neurons in brain slices (Phelan et al., 2014).

16.2.3.4 Stroke

Stroke, which is also called a brain attack, is a cerebrovascular accident caused by a sudden interruption in blood flow within the brain. There are two main types of stroke: ischemic (lack of blood flow) and hemorrhagic (bleeding).

In ischemic stroke, N-methyl-D-aspartate (NMDA)–type glutamate receptor-mediated excitotoxicity has been studied extensively, because the Ca^{2+} overload contributes to neuronal death in cerebral ischemia. However, TRPC3 and TRPC6 expressed in neurons are strongly involved in neuroprotection. As mentioned in Section 2.1, Ca^{2+} entry through TRPC3 and TRPC6 can be beneficial to neurons. This notion is supported by the finding that TRPC3 and TRPC6 are important in promoting neuronal survival. For example, BDNF can stimulate TRPC3 and TRPC6 opening and allow the elevation of $[Ca^{2+}]_i$ to activate the CREB pathway (Jia et al., 2007). In line with this idea, Du et al. (2010) have found that in a rat transient middle cerebral artery occlusion (tMCAO) model, TRPC6 protein levels in neurons are greatly reduced via NMDA receptor (NMDAR)-dependent calpain proteolysis of the N-terminal domain of TRPC6. In addition, a fusion peptide derived from the calpain cleavage site in TRPC6 can inhibit the degradation of TRPC6, which was shown to reduce infarct size and improve behavioral performance measures via the CREB signaling pathway. Moreover, Li et al. (2012a) have demonstrated that TRPC6 inhibits NMDA-induced $[Ca^{2+}]_i$ elevation and that increasing TRPC6 levels protects neurons against excitotoxicity and ischemic brain damage in TRPC6 transgenic mice. Additionally, hyperforin, which is a key constituent of St. John's wort and specifically activates TRPC6, has been shown to increase TRPC6 and CREB expression and attenuate the brain damage induced by tMCAO in rats via the inhibition of TRPC6 channel degradation (Lin et al., 2013b). A recent report indicates that IL-17A, which is a proinflammatory cytokine involved in cerebral ischemic injury, contributes to brain ischemia reperfusion injury through the calpain-TRPC6 degradation pathway in mice (Zhang et al., 2014). Taken together, the above observations indicate that TRPC6 in neurons may be a new therapeutic target for the prevention of neuronal injury after cerebral ischemia.

Intracerebral hemorrhage (ICH) is a subtype of bleeding stroke with high morbidity and mortality. Basic research has demonstrated that various drugs with antioxidative, anti-inflammatory, or neurotrophic/neuroprotective properties exert therapeutic effects in ICH animal models. The modulation of inflammatory processes mediated by astrocytes, microglia/macrophages, neutrophils, and so on, may provide an opportunity for restricting the expansion of ICH-induced tissue damage.

Astrocytes accumulate in the perihematomal region and lead to toxic edema, provoke inflammation, release cytotoxins, and form scars following ICH. Moreover, neutrophils, macrophages, and microglia are major CNS sources of cytokines, chemokines, and other immunomolecules, and are thought to provoke secondary brain damage after ICH (Zhou et al., 2014). In this context, Munakata et al. (2013) have demonstrated that in the collagenase/autologous blood infusion mouse models of ICH, the numbers of astrocytes and microglia increase in the perihematomal region and that Pyr3, which is a selective TRPC3 inhibitor, can reduce the perihematomal accumulation of astrocytes and attenuate neurological deficits, neuronal injury, and brain edema. This suggests that TRPC3 contributes to astrocyte activation and the resulting outcomes after ICH (Munakata et al., 2013). Since thrombin has been shown to activate TRPC3 in rat cultured astrocytes *in vitro* (Shirakawa et al., 2010), thrombin in the hematoma may induce TRPC3-mediated astrocyte activation and result in neuroinflammation and neurodegeneration *in vivo*. This indicates that inhibition of astrocyte activation by the selective TRPC3 inhibitor Pyr3 may lead to improvements in neurological function (Munakata et al., 2013). These data suggest that TRPC3 is involved in the development of brain injury after ICH and implicate TRPC3 in astrocytes as a new therapeutic target for the prevention of secondary brain injury and neurological deficits after ICH.

16.3 THE TRANSIENT RECEPTOR POTENTIAL MELASTATIN SUBFAMILY

16.3.1 PHYSIOLOGICAL ROLES OF TRPM CHANNELS IN NEURONS

TRPM2 plays a pivotal role in H_2O_2-induced neuronal death as redox-sensitive Ca^{2+}-permeable channels expressed in primary cultured neurons (Kaneko et al., 2006). TRPM2 is linked to neuronal cell death after oxidative stress induced by glutathione (GSH). GSH inhibits TRPM2 channels through a thiol-independent mechanism, which plays an important role in aging and neurological diseases associated with depletion of GSH (Belrose et al., 2012). TRPM2 may be a modulator of hippocampal synaptic plasticity. TRPM2 KO hippocampus slices showed that the LTD is selectively impaired because of inhibition of the glycogen synthase kinase 3β (GSK3β), confirming the role of TRPM2 in hippocampal synaptic plasticity (Xie et al., 2011). TRPM2 is also involved in neuronal development and has an inhibitory role in neurite outgrowth; TRPM2 inhibition increased, while TRPM2 overexpression decreased axonal growth. The neurons from TRPM2 KO mice had longer neurites and more spines than those of wild-type mice. TRPM2 is suggested to mediate lysophosphatidic acid-induced suppression of axonal growth (Jang et al., 2014). Spontaneous firing rate and burst activity of substantia nigra pars reticulata GABAergic neurons are modulated by H_2O_2 acting via TRPM2 channels (Lee et al., 2013). TRPM2 has been suggested as a regulator of voltage-dependent Ca^{2+} channels (VDCC) and the NMDAR, as a result of increase in $[Ca^{2+}]_i$ and membrane potential depolarization in hippocampal pyramidal neurons (Olah et al., 2009).

16.3.2 PHYSIOLOGICAL ROLES OF TRPM CHANNELS IN GLIA

Several studies have targeted glial cells as a potential therapeutic avenue by focusing on intracellular Ca^{2+} signaling, which is essential for glial cell function. These studies have revealed that a subtype of TRPM channels is expressed in glial cells, including astrocytes, microglia, and oligodendrocytes.

The presence of an oxidative stressor, *tert*-butyl hydroperoxide, leads to the upregulation of TRPM2 mRNA in rat cultured astrocytes (Bond and Greenfield, 2007). In addition, inhibiting intracellular GSH biosynthesis with D,L-buthionine-(*S*,*R*)-sulfoximine (BSO) causes oxidative stress and results in Ca^{2+} influx through TRPM2 channels, leading to a neuroinflammatory response in human astrocytes (Lee et al., 2010). Moreover, TRPM7 is important for cell proliferation and migration in mouse cortical astrocytes (Zeng et al., 2015b) and glioblastomas (Chen et al., 2015).

Several studies have focused on the physiological and pathophysiological roles of TRPM2 in microglia, as microglia are of myeloid-monocytic lineage. Many studies have clearly demonstrated that TRPM2 functions as a Ca^{2+}-permeable channel in mouse monocytes (Yamamoto et al., 2008; Knowles et al., 2011) and human monocytes (Wehrhahn et al., 2010). Kraft et al. (2004) reported that TRPM2 is functionally expressed at high levels in rat cultured microglia. Using cultured microglia derived from wild-type and TRPM2-KO mice, Haraguchi et al. (2012) have shown that TRPM2 is involved in NO production. Subsequently, Miyake et al. (2014) examined the intracellular signaling mechanisms underlying these phenomena and demonstrated that combined application of lipopolysaccharide (LPS) and interferon-γ can stimulate TRPM2-mediated extracellular Ca^{2+} influx in cultured microglia. They also showed that activation of TRPM2 results in protein tyrosine kinase 2 (Pyk2)–mediated activation of p38 MAPK and Jun amino-terminal kinase (JNK) signaling, leading to increased NO production in microglia.

The sulfonylurea receptor 1-TRPM channel regulates NO synthase 2 transcription in toll-like receptor 4–activated microglia (Kurland et al., 2016). In addition, TRPM7 is functionally expressed in rat cultured microglia (Jiang et al., 2003) and is essential for the enhanced ability of microglia to migrate and invade during anti-inflammatory states (Siddiqui et al., 2014).

TRPM3 is expressed in myelinating oligodendrocytes *in vitro* and *in vivo*. In fact, sphingosine-sensitive Ca^{2+}-permeable channels, presumably TRPM3, have been detected in rat cultured oligodendrocytes (Hoffmann et al., 2010). However, the physiological significance of these TRPM channels in glia remains unknown.

16.3.3 PATHOPHYSIOLOGICAL ROLES OF TRPM CHANNELS IN THE BRAIN

16.3.3.1 Neurodegenerative Disorders

It has been suggested that TRPM2 may be involved in AD (Yamamoto et al., 2007). Neuronal death induced by Aβ accumulation in the brain is recognized as a crucial factor underlying AD (Bertram and Tanzi, 2008). Also, Aβ accumulation is known for its association with ROS production and dysregulation of Ca^{2+} homeostasis (Barnham et al., 2004; Bezprozvanny, 2009). In AD models, striatal TRPM2 is upregulated, and knockdown of this channel ameliorated Aβ-induced neurotoxicity (Fonfria et al., 2005). It has been shown that Aβ oligomer augmented TRPM2 activity in cultured hippocampal neurons. Moreover, genetic elimination of TRPM2 reduced endoplasmic reticulum stress responses, synaptic loss, and abnormal microglial activation in AD model mice. Critically, lack of TRPM2 rescues spatial memory deficits in the model mice. These results suggest the contribution of abnormal TRPM2 activation to the pathological events of AD model mice and the potential of TRPM2 to be a therapeutic target for AD (Ostapchenko et al., 2015). Aβ-induced cerebrovascular dysfunction is also mediated by DNA damage caused by oxidative stress in cerebral endothelial cells. This DNA damage activates the DNA repair enzyme poly(ADP)-ribose polymerase, resulting in TRPM2 activation by production of ADPR, which leads to intracellular Ca^{2+} overload and endothelial dysfunction (Park et al., 2014). TRPM2, expressed in the rat striatal neurons, is activated by ADPR (Hill et al., 2006). TRPM2 is also expressed in dopaminergic neurons of the rat substantia nigra, which is considered as a major lesion of PD (Schapira and Jenner, 2011), and is thought to be responsible for H_2O_2-induced Ca^{2+} influx (Chung et al., 2011). TRPM2 activity has also been recorded in dopaminergic neurons by injection of rotenone, used as a model of PD. The rotenone-induced ROS production may be the key factor in the activation of TRPM2 in this model (Freestone et al., 2009).

Considering that oxidative stress–induced TRPM2 activation can lead to cell death in neurons, TRPM2 may be involved in the pathology of PD. Linkage analyses revealed that a subset of patients with Western Pacific Guamanian amyotrophic lateral sclerosis-parkinsonism dementia complex (ALS/PDC) are heterozygotes for a missense mutation in TRPM2 (P1018L) and TRPM7 (T1482I). Although the TRPM2 (P1018L) mutant forms functional channels, which are activated by H_2O_2 and ADPR, mutant channels inactivate quickly. In contrast, the heterologously expressed TRPM7 (T1482I) mutant produces functional TRPM7 channels that show an increased sensitivity to inhibitory intracellular Mg^{2+} (Hermosura et al., 2005, 2008; Hermosura and Garruto, 2007). These results suggest that ion influx through both channels is physiologically important, and its disruption may contribute to the onset of ALS/PDC (Hermosura et al., 2008). Clinical studies also suggested high incidence and high familial occurrence of ALS/PDC in the Kii Peninsula, Japan. However, linkage analysis in a large extended family with ALS/PDC did not reveal any evidence supporting the linkage of the TRPM7 locus, indicating that TRPM7 may not be associated with ALS/PDC in the Kii Peninsula (Hara et al., 2010).

16.3.3.2 Mental Disorders

TRPM2 may be involved in bipolar disorder (BD) type I, characterized by mood swings between manic and depressed states. Linkage analyses of bipolar families have confirmed that there is a susceptibility locus near the telomere on chromosome 21q (Straub et al., 1994; McQuillin et al., 2006). Several studies revealed conserved single nucleotide polymorphisms (SNPs) of TRPM2 in BD type I patients including rs1556314 (D543E mutation) (Liu et al., 2001; McQuillin et al., 2006; Xu et al., 2006, 2009). The TRPM2 (D543E) mutant showed decreased channel activity in response to ADPR, and TRPM2-deficient mice exhibited BD-related behavior such as increased anxiety and decreased social responses, along with disrupted electroencephalogram functional connectivity (Jang et al., 2015). Loss of function of TRPM2 leads to dysregulation of phosphorylation of GSK3β, which is a key element in BD-like behavior. These findings suggest a link between TRPM2 function and the pathology of BD.

16.3.3.3 Epilepsy

Juvenile myoclonic epilepsy (JME) is the most common form of idiopathic generalized epilepsy, accounting for 10%–30% of all epilepsies. The gene encoding EFHC1 has been identified in the region associated with JME. We found that EFHC1 is coexpressed with TRPM2 in hippocampal neurons and ventricle cells and interacts with TRPM2. Coexpression of EFHC1 significantly potentiates H_2O_2- and ADPR-induced Ca^{2+} responses and cationic currents via recombinant TRPM2. Furthermore, EFHC1 enhances TRPM2-conferred susceptibility to H_2O_2-induced cell death, which is reversed by JME mutations. These results suggest that TRPM2 may contribute to the expression of JME phenotypes by mediating the disruptive effects of the JME mutations of EFHC1 to biological processes, including cell death (Katano et al., 2012).

16.3.3.4 Stroke

TRPM2 may be involved in stroke, especially in ischemic cerebral infarction, which is the leading cause of death and permanent disability in adults worldwide. This protein may thus be the target of new and better medications. The injury mechanisms following ischemic stroke are multifaceted and complicated. Oxidative stress induced by cerebral ischemia-reperfusion injury is considered to be the main event leading to neuronal death. It has been reported that TRPM2 is activated by intracellular ADPR that is overproduced in response to oxidative stress and ROS, such as H_2O_2 (Hara et al., 2002). Therefore, several lines of evidence indicated that TRPM2 mediates ROS-induced neuronal death. For example, TRPM2 acts as a redox-sensitive Ca^{2+}-permeable channel and therefore has a pivotal role in H_2O_2-induced neuronal death in primary cultured neurons (Kaneko et al., 2006) and organotypic hippocampal culture (Bai and Lipski, 2010). Additionally, TRPM2 channel activation following *in vitro* ischemia contributes to male hippocampal cell death (Verma et al., 2012). In line with these findings, several studies indicate that TRPM2 mediates ischemic brain damage. For example, Shimizu et al. (2013) have reported that the genetic ablation of TRPM2 or the administration of the TRPM2 inhibitor clotrimazole 2 hours after the onset of ischemia reduces infarct volume in male mice. Alim et al. (2013) have used wild-type and TRPM2-KO mice to show that the genetic ablation of TRPM2 causes a shift in the expression ratio of NMDAR GluN2A/GluN2B subunits, which may then selectively upregulate survival pathways and result in neuroprotection from cerebral ischemia-induced neuronal cell death *in vivo*. More recently, Shimizu et al. (2016) have reported the presence of an extended therapeutic window for a novel peptide inhibitor of TRPM2 channels following focal cerebral ischemia, suggesting that TRPM2 is a promising candidate for therapeutics for acute cerebral ischemic infarction.

Recent studies have focused extensively on the pathophysiological role of TRPM2 expressed in immune cells. Treatment of monocytes with LPS increases TRPM2 expression and ADPR-induced currents (Wehrhahn et al., 2010). In addition, TRPM2-mediated Ca^{2+} influx induces the production of proinflammatory cytokines/chemokines in monocytes (Yamamoto et al., 2008). Moreover, TRPM2 is involved in the adhesion of neutrophils to endothelial cells (Hiroi et al., 2013) and in CXCL2 and NO production in cultured macrophages and microglia (Haraguchi et al., 2012; Miyake et al., 2014). In line with these findings, TRPM2 regulates the migratory capacities of neutrophils and macrophages in response to ischemic brain injury, thereby secondarily perpetuating brain injury within 3 days (Gelderblom et al., 2014). TRPM2 mRNA is shown to increase in the chronic stage from 1 to 4 weeks postoperation in a tMCAO model (Fonfria et al., 2006). In addition, there are sex differences in neuroprotection due to the inhibition of TRPM2 channels following experimental stroke (Jia et al., 2011; Shimizu et al., 2016). Taken together, the above results indicate that further studies are needed to identify the precise role of TRPM2 in the development of cerebral ischemic injury.

16.3.3.5 Neuropathic Pain

TRPM2 may be involved in a range of pathological pain, including neuropathic pain (Haraguchi et al., 2012; Isami et al., 2013; So et al., 2016). Accumulating evidence suggests that peripheral and spinal neuroinflammation mediated by the interaction between nociceptive neurons and immune/glial cells plays a pivotal role in neuropathic pain (Ren and Dubner, 2010). Following peripheral nerve injury, pro-nociceptive inflammatory mediators, such as proinflammatory cytokines, chemokines, and excess ROS produced by peripheral tissues and the spinal cord can lead to peripheral and central sensitization of nociceptive neurons. It is reported that functional TRPM2 is expressed in the dorsal root ganglion (DRG) and trigeminal ganglion neurons, and contributes to the oxidative stress-induced cell death of sensory neurons (Nazıroğlu et al., 2011; Chung et al., 2015). This may be involved in diabetic neuropathy (Sözbir and Nazıroğlu, 2016). However, TRPM2 deficiency fails to change the basal sensitivity to mechanical and thermal stimuli and the nociceptive response evoked by exogenous H_2O_2 injections (Haraguchi et al., 2012; So et al., 2015). It is noted that the Ca^{2+} response of cultured DRG neurons to low concentrations of H_2O_2 is fully mediated through TRPA1 (So et al., 2016), which has the highest sensitivity to oxidation among the redox-sensitive TRP channels (Takahashi et al., 2011). These findings suggest that TRPM2 expressed in sensory neurons may be responsible for the oxidative damage of sensory neurons in response to excess ROS production, while its physiological roles are limited. By contrast, TRPM2 expressed in macrophages and microglia plays a critical role in neuropathic pain. TRPM2 is responsible for CXCL2 and NO production in macrophages and microglia, which aggravate peripheral and spinal pronociceptive inflammatory responses, respectively, in inflammatory and neuropathic pain models (Haraguchi et al., 2012; Miyake et al., 2014). Furthermore, TRPM2 may play a role in the peripheral nerve injury–induced infiltration of peripheral immune cells into the spinal cord (Isami et al., 2013). Consequently, TRPM2 is involved in a wide range of pathological pain induced by peripheral and spinal neuroinflammation, such as inflammatory pain, osteoarthritic pain, peripheral nerve injury–induced neuropathic pain, chemotherapy-induced peripheral neuropathy, and diabetic painful neuropathy, rather than physiological nociceptive pain (So et al., 2015).

16.4 THE TRANSIENT RECEPTOR POTENTIAL VANILLOID SUBFAMILY

16.4.1 Physiological Roles of TRPV Channels in Neurons

TRPV1 plays a role in the regulation of neuronal activity and synaptic plasticity. TRPV1 activation triggers LTD at excitatory synapses on hippocampal interneurons. Brain slices of TRPV1 KO mice did not show LTD in the synapse even when treated with TRPV1 activators (Gibson et al., 2008).

Synaptic modifications in the nucleus accumbens (NAc) are important for adaptive and pathological reward-dependent learning. In medium spiny neurons, a major cell type of NAc, mGluR-induced endocannabinoid production led postsynaptic TRPV1 to trigger LTD through promotion of AMPA receptor endocytosis (Grueter et al., 2010). In the dentate gyrus, TRPV1 activity modulates somatic inhibitory synaptic transmission. TRPV1-mediated depression of transmission is dependent on postsynaptic Ca^{2+} influx, calcineurin activation, and clathrin-dependent internalization of GABA receptors (Chavez et al., 2014).

It has also been reported that TRPV1 facilitates excitatory transmission and promotes LTP in the brain. TRPV1 activity facilitated glutamate transmission in dopaminergic neurons of the substantia nigra via administration of TRPV1 agonists such as capsaicin and anandamide (Marinelli et al., 2003). In the nucleus of the solitary tract, TRPV1 activity triggered asynchronous glutamate release from solitary tract afferents, leading to potentiation of the duration of postsynaptic excitatory periods. This may be a central mechanism to mediate powerful C-fiber reflex responses (Peters et al., 2010; Shoudai et al., 2010). Acute stress facilitated LTD and suppressed LTP in CA1 slices of the hippocampus and impaired spatial memory retrieval in juvenile rats. TRPV1 agonists diminished the effects of stress on synaptic plasticity and spatial learning. The TRPV1 channel is a potential target to facilitate LTP and suppress LTD, in turn protecting spatial learning from the effects of acute stress (Li et al., 2008).

In the human brain, TRPV1 SNP variants (rs222747), which were found to enhance channel activity, affected cortical excitability studied by paired-pulse transcranial magnetic stimulation of the primary motor cortex (Mori et al., 2012). TRPV4 also plays a key role in neuronal excitability. TRPV4 is constitutively active at physiological temperatures and depolarizes the resting membrane potential in hippocampal neurons. These characteristics were not observed in the neurons from TRPV4 KO mice, which also required a larger injection current to be depolarized. Consistent with these results, TRPV4 KO mice showed abnormal cortical activity in wake periods. Moreover, TRPV4 KO mice exhibited reduced depression-like and social behaviors. TRPV4 may be a key factor for homeothermic animals to control various complicated behaviors (Shibasaki et al., 2007, 2015).

Body temperature increase stimulates sweat production and renal water reabsorption through the release of vasopressin. TRPV1 contributes to the thermosensitivity of hypothalamic vasopressin neurons. TRPV1 KO mice showed an impaired response in heat-activated cation current of vasopressin neurons and in vasopressin release in response to hyperthermia (Sharif-Naeini et al., 2008). Both TRPV1 and TRPV4 play a role in the basal excitability of magnocellular neurosecretory cells (MNCs) isolated from the rat supraoptic nucleus at physiological temperature. Genetic deletion of TRPV1, but not TRPV4, interfered with thermally induced firing of MNCs, suggesting a dynamic role of TRPV1 in thermosensation mediated by MNCs (Sudbury and Bourque, 2013).

TRPV1 may promote neuronal death by Ca^{2+} influx. Capsaicin-induced TRPV1 activation triggered apoptotic cell death in TRPV1-expressing cortical neurons via L-type Ca^{2+} channel activation, extracellular signal-regulated kinase (ERK) phosphorylation, and ROS production (Shirakawa et al., 2008). TRPV1 and cannabinoid receptor type 1 (CB$_1$) agonist-induced cell death of dopaminergic neurons was observed *in vitro* using mesencephalic culture and *in vivo* by injecting these agonists intranigrally. These cell deaths were induced by the Ca^{2+}-dependent apoptosis pathway and prevented by TRPV1 or CB$_1$ antagonists (Kim et al., 2005, Kim et al., 2008). The role of TRPV1 in enhancing cell death can be universal among different cell types (Badr et al., 2016).

The crucial role of TRPV1 in peripheral nociception has been well established (Julius, 2013). TRPV1 also plays an important role in complete Freund's adjuvant- and capsaicin-induced mechanical allodynia, which are pain models presumably mediated by central sensitization (Cui et al., 2006). TRPV1 may regulate pain sensation by both peripheral and central mechanisms.

In osmosensory neurons of the hypothalamic supraoptic nucleus, induced TRPV1 activation by cell volume decrease regulates the electrical activity. TRPV1 is activated in response to cell volume change via mechanical force transduced by physically interacting microtubule scaffolds (Prager-Khoutorsky et al., 2014).

16.4.2   Physiological Roles of TRPV Channels in Glia

It has been reported that a subtype of TRPV is expressed in glial cells, especially astrocytes and microglia.

In astrocytes, several isoforms of TRPV channels, especially TRPV1, TRPV2, and TRPV4 are expressed, and several studies have revealed the functional properties of the TRPV channels. TRPV1 is localized in the plasma membrane of astrocyte in the brain tissues revealed by a double-immunofluorescence-labeling study (Huang et al., 2010). Mannari et al. (2013) have reported that TRPV1 is expressed preferentially at thick cellular processes of astrocytes rather than fine cellular processes and cell bodies, and astrocytic TRPV1 is an important sensing and signaling pathway to detect bloodborne molecules in the sensory circumventricular organs of adult mouse brains. TRPV1 also plays a role in acid-evoked cation current in cortical astrocytes (Huang et al., 2010). Ho et al. (2014) have reported retinal astrocytes could sense mechanical stimuli in part mediated by TRPV1 activation so that antagonism of TRPV1 reduces the rate of astrocyte migration through a reduction in Ca^{2+} influx and in cytoskeletal remodeling. In the pathophysiological condition, Nam et al. (2015) have clearly demonstrated that activation of astrocytic TRPV1 produces endogenous ciliary neurotrophic factor *in vivo*, which prevents degeneration of nigral dopamine neurons in animal models of PD.

TRPV2 is functionally expressed in astrocytes *in vitro* and *in vivo* (Shibasaki et al., 2013). Zhang et al. (2016) have reported that oxygen-glucose deprivation and reoxygenation treatment enhances the expression level of TRPV2 in cortical astrocytes and blocking the TRPV2 promotes proliferation and increases nerve growth factor secretion via MAPK/JNK signaling pathway in the astrocytes. In addition, Benfenati et al. (2011) reported that an aquaporin-4 (AQP4) and TRPV4 complex is essential for cell-volume control in astrocytes. Moreover, activation of TRPV4 in astrocytes contributes to amplify flow/pressure-evoked parenchymal arteriole vasoconstriction (Dunn et al., 2013), regulates neuronal excitability via release of gliotransmitters (Shibasaki et al., 2014), and modulates volume regulation, swelling, and AQP4 gene expression (Jo et al., 2015).

Microglia also express several isoforms of TRPV channels, especially TRPV1, TRPV2, and TRPV4. Activation of TRPV1 triggered Ca^{2+} signaling-dependent cell death in microglia (Kim et al., 2006). Schilling and Eder (2010) have demonstrated a contribution of TRPV1 to nicotinamide adenine dinucleotide phosphate (NADPH) oxidase-mediated ROS production in microglia. Miyake et al. (2015) have demonstrated that TRPV1 activation enhances chemotactic activity of microglia via mitochondrial ROS production. In an experimental model of PD, TRPV1 activator capsaicin inhibited microglia-mediated ROS production and blocked death of dopaminergic neurons. Sappington and Calkins (2008) have reported that TRPV1 partially regulates cytokine production induced by elevated hydrostatic pressure in retinal microglia.

With respect to TRPV2 and TRPV4, activation of TRPV1 and TRPV2 by cannabidiol results in enhanced microglial phagocytosis (Hassan et al., 2014), and stimulation of TRPV4 suppresses LPS-induced tumor necrosis factor-α release, galectin-3 upregulation, and augmented amplitude of voltage-dependent K^+ current in rat cultured microglia (Konno et al., 2014).

Finally, TRPV channels are described in chemotherapeutics sensitivity. For example, TRPV1 stimulation triggers tumor cell death through the branch of the endoplasmic reticulum stress pathway (Stock et al., 2012). TRPV2 channel negatively regulates glioma cell survival, proliferation, and resistance to Fas-induced apoptosis in ERK-dependent manner (Nabissi et al., 2010).

16.4.3   Pathophysiological Roles of TRPV Channels in the Brain

16.4.3.1   Mental Disorders

TRPV1 activation may produce anxiogenic effects. Administration of a TRPV1 agonist had anxiogenic effects, while a TRPV1 antagonist had anxiolytic effects in behavioral tests (Kasckow et al., 2004; Santos et al., 2008; Aguiar et al., 2009; Terzian et al., 2009). TRPV1 KO mice showed less anxiety-related behavior in the light-dark test and elevated plus maze. The mice also showed reduced innate and conditioned fear, which was mirrored by a decrease in LTP in the Schaffer collateral-commissural pathway to CA1 hippocampal neurons (Marsch et al., 2007). These insights suggest that TRPV1 is a potential target for the development of anxiolytic drugs.

16.4.3.2 Epilepsy

TRPV1 plays a regulatory role in neuronal activity and synaptic plasticity and is a potential target for antiepileptic drugs (Fu et al., 2009). Several studies have demonstrated the proepileptic activity of TRPV1 and antiepileptic potential of TRPV1 inhibition. Capsaicin enhanced but the TRPV1 antagonist suppressed 4-aminopyridine induced seizure-like activity in the hippocampus (Gonzalez-Reyes et al., 2013). Anandamide, which is an endocannabinoid, showed anticonvulsant effects by activating the CB_1 receptor and showed proconvulsant effects by activating TRPV1 (Manna and Umathe, 2012). Capsaicin significantly enhanced spontaneous and miniature EPSC in mice with TLE. Anandamide also increased glutamate release in the presence of a CB_1 receptor antagonist, while it reduced EPSC in the presence of a TRPV1 antagonist. These results suggest that anandamide exerts antiepileptic effects via CB_1 receptor activation and proepileptic effects via TRPV1 activation (Bhaskaran and Smith, 2010).

16.5 CONCLUSION

Recent studies emphasize the significance of TRP channels in brain functions and as well in the pathologies of numerous neurological and psychiatric disorders. TRPC3 and TRPC6 channels play an essential role in the neurotrophic actions of BDNF, a promising therapeutic agent for these diseases. Notably, hyperforin, which activates TRPC6, and PPZ1 and PPZ2, activators for DAG-sensitive TRPC channels, exert neurotrophic effects on neurons. TRPM2 also plays important roles in brain. Not only loss of TRPM2 function but also oxidative stress–induced overactivation of TRPM2 is involved in various diseases of the brain, suggesting that modulators of the TRPM2 channel may be effective therapeutic agents for these diseases. TRPV1 has a key role in the regulation of neuronal excitability and synaptic plasticity. Agonists and antagonists of TRPV1 also exhibit therapeutic potential in the broad range of neuronal diseases. Although the properties of TRP modulators should be improved and further *in vivo* studies are needed, we have reasonable grounds to believe that activators and/or inhibitors of the above TRP channels will be developed as therapeutic tools for neurological and psychiatric disorders.

ACKNOWLEDGMENTS

This work was supported by a Grant-in-Aid for Scientific Research on Innovative Areas "Oxygen Biology: A New Criterion for Integrated Understanding of Life" [Grant 26111004] of The Ministry of Education, Culture, Sports, Science and Technology, Japan.

REFERENCES

Abramowitz, J. and L. Birnbaumer. 2009. Physiology and pathophysiology of canonical transient receptor potential channels. *FASEB J*, 23(2): 297–328.

Aguiar, D.C. et al. 2009. Anxiolytic-like effects induced by blockade of transient receptor potential vanilloid type 1 (TRPV1) channels in the medial prefrontal cortex of rats. *Psychopharmacology*, 205(2): 217–225.

Akita, T. and Y. Okada. 2011. Regulation of bradykinin-induced activation of volume-sensitive outwardly rectifying anion channels by Ca^{2+} nanodomains in mouse astrocytes. *J Physiol*, 589(Pt 16): 3909–3927.

Alim, I. et al. 2013. Modulation of NMDAR subunit expression by TRPM2 channels regulates neuronal vulnerability to ischemic cell death. *J Neurosci*, 33(44): 17264–17277.

Amaral, M.D. and L. Pozzo-Miller. 2007. TRPC3 channels are necessary for brain-derived neurotrophic factor to activate a nonselective cationic current and to induce dendritic spine formation. *J Neurosci*, 27(19): 5179–5189.

Badr, H. et al. 2016. Different contribution of redox-sensitive transient receptor potential channels to acetaminophen-induced death of human hepatoma cell line. *Front Pharmacol*, 7: 19.

Bai, J.Z. and J. Lipski. 2010. Differential expression of TRPM2 and TRPV4 channels and their potential role in oxidative stress-induced cell death in organotypic hippocampal culture. *Neurotoxicology*, 31(2): 204–214.

Barnham, K.J. et al. 2004. Neurodegenerative diseases and oxidative stress. *Nat Rev Drug Discov*, 3(3): 205–214.

Bazargani, N. and D. Attwell. 2016. Astrocyte calcium signaling: The third wave. *Nat Neurosci*, 19(2): 182–189.

Becker, E.B. et al. 2009. A point mutation in TRPC3 causes abnormal Purkinje cell development and cerebellar ataxia in moonwalker mice. *Proc Natl Acad Sci U S A*, 106(16): 6706–6711.

Becker, E.B. et al. 2011. Candidate screening of the TRPC3 gene in cerebellar ataxia. *Cerebellum*, 10(2): 296–299.

Beis, D. et al. 2011. Evidence for a supportive role of classical transient receptor potential 6 (TRPC6) in the exploration behavior of mice. *Physiol Behav*, 102(2): 245–250.

Belrose, J.C. et al. 2012. Loss of glutathione homeostasis associated with neuronal senescence facilitates TRPM2 channel activation in cultured hippocampal pyramidal neurons. *Mol Brain*, 5: 11.

Benfenati, V. et al. 2011. An aquaporin-4/transient receptor potential vanilloid 4 (AQP4/TRPV4) complex is essential for cell-volume control in astrocytes. Proc Natl Acad Sci U S A, 108(6): 2563-2568.

Berg, A.P., N. Sen, and D.A. Bayliss. 2007. TrpC3/C7 and Slo2.1 are molecular targets for metabotropic glutamate receptor signaling in rat striatal cholinergic interneurons. *J Neurosci*, 27(33): 8845–8856.

Bertram, L. and R.E. Tanzi. 2008. Thirty years of Alzheimer's disease genetics: The implications of systematic meta-analyses. *Nat Rev Neurosci*, 9(10): 768–778.

Beskina, O. et al. 2007. Mechanisms of interleukin-1β-induced Ca^{2+} signals in mouse cortical astrocytes: Roles of store- and receptor-operated Ca^{2+} entry. *Am J Physiol Cell Physiol*, 293(3): C1103–C1111.

Bezprozvanny, I. 2009. Calcium signaling and neurodegenerative diseases. *Trends Mol Med*, 15(3): 89–100.

Bhaskaran, M.D. and B.N. Smith. 2010. Effects of TRPV1 activation on synaptic excitation in the dentate gyrus of a mouse model of temporal lobe epilepsy. *Exp Neurol*, 223(2): 529–536.

Blair, N.T. et al. 2009. Intracellular calcium strongly potentiates agonist-activated TRPC5 channels. *J Gen Physiol*, 133(5): 525–546.

Boisseau, S. et al. 2009. Heterogeneous distribution of TRPC proteins in the embryonic cortex. *Histochem Cell Biol*, 131(3): 355–363.

Bond, C.E. and S.A. Greenfield. 2007. Multiple cascade effects of oxidative stress on astroglia. *Glia*, 55(13): 1348–1361.

Cavanaugh, D.J. et al. 2011. *Trpv1* reporter mice reveal highly restricted brain distribution and functional expression in arteriolar smooth muscle cells. *J Neurosci*, 31(13): 5067–5077.

Chahl, L A. 2011. TRP channels and psychiatric disorders. *Adv Exp Med Biol*, 704: 987–1009.

Chavez, A.E. et al. 2014. Compartment-specific modulation of GABAergic synaptic transmission by TRPV1 channels in the dentate gyrus. *J Neurosci*, 34(50): 16621–16629.

Chen, W.L. et al. 2015. Inhibition of TRPM7 by carvacrol suppresses glioblastoma cell proliferation, migration and invasion. *Oncotarget*, 6(18): 16321–16340.

Chung, K.K., P.S. Freestone, and J. Lipski. 2011. Expression and functional properties of TRPM2 channels in dopaminergic neurons of the substantia nigra of the rat. *J Neurophysiol*, 106(6): 2865–2875.

Chung, M.K. et al. 2015. The role of TRPM2 in hydrogen peroxide-induced expression of inflammatory cytokine and chemokine in rat trigeminal ganglia. *Neuroscience*, 297: 160–169.

Chung, Y.H. et al. 2006. Immunohistochemical study on the distribution of TRPC channels in the rat hippocampus. *Brain Res*, 1085(1): 132–137.

Cook, N.L. et al. 2010. Transient receptor potential melastatin 2 expression is increased following experimental traumatic brain injury in rats. *J Mol Neurosci*, 42(2): 192–199.

Cristino, L. et al. 2006. Immunohistochemical localization of cannabinoid type 1 and vanilloid transient receptor potential vanilloid type 1 receptors in the mouse brain. *Neuroscience*, 139(4): 1405–1415.

Cui, M. et al. 2006. TRPV1 receptors in the CNS play a key role in broad-spectrum analgesia of TRPV1 antagonists. *J Neurosci*, 26(37): 9385–9393.

Cui, N. et al. 2011. Involvement of TRP channels in the CO_2 chemosensitivity of locus coeruleus neurons. *J Neurophysiol*, 105(6): 2791–2801.

Cvetkovic-Lopes, V. et al. 2010. Rat hypocretin/orexin neurons are maintained in a depolarized state by TRPC channels. *PLOS ONE*, 5(12): e15673.

Davare, M.A. et al. 2009. Transient receptor potential canonical 5 channels activate Ca^{2+}/calmodulin kinase Iγ to promote axon formation in hippocampal neurons. *J Neurosci*, 29(31): 9794–9808.

Dhar, M. et al. 2014. Leptin-induced spine formation requires TrpC channels and the CaM kinase cascade in the hippocampus. *J Neurosci*, 34(30): 10022–10033.

Du, W. et al. 2010. Inhibition of TRPC6 degradation suppresses ischemic brain damage in rats. *J Clin Invest*, 120(10): 3480–3492.

Dunn, K.M. et al. 2013. TRPV4 channels stimulate Ca^{2+}-induced Ca^{2+} release in astrocytic endfeet and amplify neurovascular coupling responses. *Proc Natl Acad Sci U S A*, 110(15): 6157–6162.

Everaerts, W., B. Nilius, and G. Owsianik. 2010. The vanilloid transient receptor potential channel TRPV4: From structure to disease. *Prog Biophys Mol Biol*, 103(1): 2–17.

Fonfria, E. et al. 2005. Amyloid beta-peptide (1–42) and hydrogen peroxide-induced toxicity are mediated by TRPM2 in rat primary striatal cultures. *J Neurochem*, 95(3): 715–723.

Fonfria, E. et al. 2006. TRPM2 is elevated in the tMCAO stroke model, transcriptionally regulated, and functionally expressed in C13 microglia. *J Recept Signal Transduct Res*, 26(3): 179–198.

Freestone, P.S. et al. 2009. Acute action of rotenone on nigral dopaminergic neurons—Involvement of reactive oxygen species and disruption of Ca^{2+} homeostasis. *Eur J Neurosci*, 30(10): 1849–1859.

Fu, M., Z. Xie, and H. Zuo. 2009. TRPV1: A potential target for antiepileptogenesis. *Med Hypotheses*, 73(1): 100–102.

Fusco, F. R. et al. 2004. Cellular localization of TRPC3 channel in rat brain: Preferential distribution to oligodendrocytes. *Neurosci Lett*, 365(2): 137–142.

Gelderblom, M. et al. 2014. Transient receptor potential melastatin subfamily member 2 cation channel regulates detrimental immune cell invasion in ischemic stroke. *Stroke*, 45(11): 3395–3402.

Giampa, C. et al. 2007. Immunohistochemical localization of TRPC6 in the rat substantia nigra. *Neurosci Lett*, 424(3): 170–174.

Gibon, J. et al. 2013. The antidepressant hyperforin increases the phosphorylation of CREB and the expression of TrkB in a tissue-specific manner. *Int J Neuropsychopharmacol*, 16(1): 189–198.

Gibson, H.E. et al. 2008. TRPV1 channels mediate long-term depression at synapses on hippocampal interneurons. *Neuron*, 57(5): 746–759.

Glitsch, M.D. 2010. Activation of native TRPC3 cation channels by phospholipase D. *FASEB J*, 24(1): 318–325.

Golovina, V.A. 2005. Visualization of localized store-operated calcium entry in mouse astrocytes. Close proximity to the endoplasmic reticulum. *J Physiol*, 564(Pt 3): 737–749.

Gonzalez-Reyes, L.E. et al. 2013. TRPV1 antagonist capsazepine suppresses 4-AP-induced epileptiform activity *in vitro* and electrographic seizures *in vivo*. *Exp Neurol*, 250: 321–332.

Greka, A. et al. 2003. TRPC5 is a regulator of hippocampal neurite length and growth cone morphology. *Nat Neurosci*, 6(8): 837–845.

Griesi-Oliveira, K. et al. 2015. Modeling non-syndromic autism and the impact of TRPC6 disruption in human neurons. *Mol Psychiatry*, 20(11): 1350–1365.

Griffith, T.N. et al. 2010. Neurobiological effects of hyperforin and its potential in Alzheimer's disease therapy. *Curr Med Chem*, 17(5): 391–406.

Grimaldi, M., M. Maratos, and A. Verma. 2003. Transient receptor potential channel activation causes a novel form of $[Ca^{2+}]_i$ oscillations and is not involved in capacitative Ca^{2+} entry in glial cells. *J Neurosci*, 23(11): 4737–4745.

Grueter, B.A., G. Brasnjo, and R. C. Malenka. 2010. Postsynaptic TRPV1 triggers cell type-specific long-term depression in the nucleus accumbens. *Nat Neurosci*, 13(12): 1519–1525.

Hara, K. et al. 2010. TRPM7 is not associated with amyotrophic lateral sclerosis-parkinsonism dementia complex in the Kii peninsula of Japan. *Am J Med Genet B Neuropsychiatr Genet*, 153b (1): 310–313.

Hara, Y. et al. 2002. LTRPC2 Ca^{2+}-permeable channel activated by changes in redox status confers susceptibility to cell death. *Mol Cell*, 9(1): 163–173.

Haraguchi, K. et al. 2012. TRPM2 contributes to inflammatory and neuropathic pain through the aggravation of pronociceptive inflammatory responses in mice. *J Neurosci*, 32(11): 3931–3941.

Hartmann, J. et al. 2008. TRPC3 channels are required for synaptic transmission and motor coordination. *Neuron*, 59(3): 392–398.

Hartmann, J., H.A. Henning, and A. Konnerth. 2011. mGluR1/TRPC3-mediated synaptic transmission and calcium signaling in mammalian central neurons. *Cold Spring Harb Perspect Biol*, 3(4): 1–16.

Hassan, S. et al. 2014. Cannabidiol enhances microglial phagocytosis via transient receptor potential (TRP) channel activation. *Br J Pharmacol*, 171(9): 2426–2439.

He, Z. et al. 2012. TRPC5 channel is the mediator of neurotrophin-3 in regulating dendritic growth via CaMKIIα in rat hippocampal neurons. *J Neurosci*, 32(27): 9383–9395.

Heiser, J.H. et al. 2013. TRPC6 channel-mediated neurite outgrowth in PC12 cells and hippocampal neurons involves activation of RAS/MEK/ERK, PI3K, and CAMKIV signaling. *J Neurochem*, 127(3): 303–313.

Hermosura, M.C. et al. 2008. Altered functional properties of a TRPM2 variant in Guamanian ALS and PD. *Proc Natl Acad Sci U S A*, 105(46): 18029–18034.

Hermosura, M.C. and R.M. Garruto. 2007. TRPM7 and TRPM2-candidate susceptibility genes for Western Pacific ALS and PD? *Biochim Biophys Acta*, 1772(8): 822–835.

Hermosura, M.C. et al. 2005. A TRPM7 variant shows altered sensitivity to magnesium that may contribute to the pathogenesis of two Guamanian neurodegenerative disorders. *Proc Natl Acad Sci U S A*, 102(32): 11510–11515.

Hill, K. et al. 2006. Characterisation of recombinant rat TRPM2 and a TRPM2-like conductance in cultured rat striatal neurones. *Neuropharmacology*, 50(1): 89–97.

Hiroi, T. et al. 2013. Neutrophil TRPM2 channels are implicated in the exacerbation of myocardial ischaemia/reperfusion injury. *Cardiovasc Res*, 97(2): 271–281.

Ho, K.W., W.S. Lambert, and D.J. Calkins. 2014. Activation of the TRPV1 cation channel contributes to stress-induced astrocyte migration. *Glia*, 62(9): 1435–1451.

Hoffmann, A. et al. 2010. TRPM3 is expressed in sphingosine-responsive myelinating oligodendrocytes. *J Neurochem*, 114(3): 654–665.

Huang, C. et al. 2010. Existence and distinction of acid-evoked currents in rat astrocytes. *Glia*, 58(12): 1415–1424.

Huang, W.C., J.S. Young, and M.D. Glitsch. 2007. Changes in TRPC channel expression during postnatal development of cerebellar neurons. *Cell Calcium*, 42(1): 1–10.

Isami, K. et al. 2013. Involvement of TRPM2 in peripheral nerve injury-induced infiltration of peripheral immune cells into the spinal cord in mouse neuropathic pain model. *PLOS ONE*, 8(7):e66410.

Jang, Y. et al. 2014. TRPM2 mediates the lysophosphatidic acid-induced neurite retraction in the developing brain. *Pflugers Arch*, 466(10): 1987–1998.

Jang, Y. et al. 2015. TRPM2, a susceptibility gene for bipolar disorder, regulates glycogen synthase kinase-3 activity in the brain. *J Neurosci*, 35(34): 11811–11823.

Jeon, J.P. et al. 2013. Activation of TRPC4 β by G α_i subunit increases Ca^{2+} selectivity and controls neurite morphogenesis in cultured hippocampal neuron. *Cell Calcium*, 54(4): 307–319.

Jia, J. et al. 2011. Sex differences in neuroprotection provided by inhibition of TRPM2 channels following experimental stroke. *J Cereb Blood Flow Metab*, 31(11): 2160–2168.

Jia, Y. et al. 2007. TRPC channels promote cerebellar granule neuron survival. *Nat Neurosci*, 10(5): 559–567.

Jiang, X., E.W. Newell, and L.C. Schlichter. 2003. Regulation of a TRPM7-like current in rat brain microglia. *J Biol Chem*, 278(44): 42867–42876.

Jo, A.O. et al. 2015. TRPV4 and AQP4 channels synergistically regulate cell volume and calcium homeostasis in retinal Müller glia. *J Neurosci*, 35(39): 13525–13537.

Johansson, S.E. et al. 2015. Cerebrovascular endothelin-1 hyper-reactivity is associated with transient receptor potential canonical channels 1 and 6 activation and delayed cerebral hypoperfusion after forebrain ischaemia in rats. *Acta Physiol*, 214(3): 376–389.

Julius, D. 2013. TRP channels and pain. *Annu Rev Cell Dev Biol*, 29: 355–384.

Kaczmarek, J.S., A. Riccio, and D.E. Clapham. 2012. Calpain cleaves and activates the TRPC5 channel to participate in semaphorin 3A-induced neuronal growth cone collapse. *Proc Natl Acad Sci U S A*, 109(20): 7888–7892.

Kaneko, S. et al. 2006. A critical role of TRPM2 in neuronal cell death by hydrogen peroxide. *J Pharmacol Sci*, 101(1): 66–76.

Kasckow, J.W., J.J. Mulchahey, and T.D. Geracioti, Jr. 2004. Effects of the vanilloid agonist olvanil and antagonist capsazepine on rat behaviors. *Prog Neuropsychopharmacol Biol Psychiatry*, 28(2): 291–295.

Katano, M. et al. 2012. The juvenile myoclonic epilepsy-related protein EFHC1 interacts with the redox-sensitive TRPM2 channel linked to cell death. *Cell Calcium*, 51(2): 179–185.

Kato, A.S. et al. 2012. Glutamate receptor δ2 associates with metabotropic glutamate receptor 1 (mGluR1), protein kinase Cδ, and canonical transient receptor potential 3 and regulates mGluR1-mediated synaptic transmission in cerebellar Purkinje neurons. *J Neurosci*, 32(44): 15296–15308.

Kim, D.S. et al. 2013. The reverse roles of transient receptor potential canonical channel-3 and -6 in neuronal death following pilocarpine-induced status epilepticus. *Cell Mol Neurobiol*, 33(1): 99–109.

Kim, S.J. 2013. TRPC3 channel underlies cerebellar long-term depression. *Cerebellum*, 12(3): 334–337.

Kim, S.R. et al. 2005. Transient receptor potential vanilloid subtype 1 mediates cell death of mesencephalic dopaminergic neurons *in vivo* and *in vitro*. *J Neurosci*, 25(3): 662–671.

Kim, S.R. et al. 2006. Transient receptor potential vanilloid subtype 1 mediates microglial cell death in vivo and in vitro via Ca^{2+}-mediated mitochondrial damage and cytochrome C release. *J Immunol*, 177(7): 4322–4329.

Kim, S.R. et al. 2008. Interactions between CB_1 receptors and TRPV1 channels mediated by 12-HPETE are cytotoxic to mesencephalic dopaminergic neurons. *Br J Pharmacol*, 155(2): 253–264.

Kim, Y.J. and T.C. Kang. 2015. The role of TRPC6 in seizure susceptibility and seizure-related neuronal damage in the rat dentate gyrus. *Neuroscience*, 307: 215–230.

Kim, Y. et al. 2012. Alternative splicing of the TRPC3 ion channel calmodulin/IP_3 receptor-binding domain in the hindbrain enhances cation flux. *J Neurosci*, 32(33): 11414–11423.

Knowles, H. et al. 2011. Transient receptor potential melastatin 2 (TRPM2) ion channel is required for innate immunity against *Listeria* monocytogenes. *Proc Natl Acad Sci U S A*, 108(28): 11578–11583.

Konno, M. et al. 2012. Stimulation of transient receptor potential vanilloid 4 channel suppresses abnormal activation of microglia induced by lipopolysaccharide. *Glia*, 60(5): 761–770.

Kraft, R. et al. 2004. Hydrogen peroxide and ADP-ribose induce TRPM2-mediated calcium influx and cation currents in microglia. *Am J Physiol Cell Physiol*, 286(1): C129–C137.

Kunert-Keil, C. et al. 2006. Tissue-specific expression of TRP channel genes in the mouse and its variation in three different mouse strains. *BMC Genomics*, 7: 159.

Kurland, D.B. et al. 2016. The Sur1-Trpm4 channel regulates NOS2 transcription in TLR4-activated microglia. *J Neuroinflammation*, 13(1): 130.

Lee, C.R. et al. 2013. TRPM2 channels are required for NMDA-induced burst firing and contribute to H_2O_2-dependent modulation in substantia nigra pars reticulata GABAergic neurons. *J Neurosci*, 33(3): 1157–1168.

Lee, M. et al. 2010. Depletion of GSH in glial cells induces neurotoxicity: Relevance to aging and degenerative neurological diseases. *FASEB J*, 24(7): 2533–2545.

Lee, S.K. et al. 2014. Hyperforin attenuates microglia activation and inhibits p65-Ser276 NF γ B phosphorylation in the rat piriform cortex following status epilepticus. *Neurosci Res*, 85: 39–50.

Letavernier, E. et al. 2012. Williams-Beuren syndrome hypercalcemia: Is TRPC3 a novel mediator in calcium homeostasis? *Pediatrics*, 129(6): e1626–e1630.

Leuner, K. et al. 2007. Hyperforin—A key constituent of St. John's wort specifically activates TRPC6 channels. *FASEB J*, 21(14): 4101–4111.

Leuner, K. et al. 2010. Simple 2,4-diacylphloroglucinols as classic transient receptor potential-6 activators—Identification of a novel pharmacophore. *Mol Pharmacol*, 77(3): 368–377.

Leuner, K. et al. 2013. Hyperforin modulates dendritic spine morphology in hippocampal pyramidal neurons by activating Ca^{2+}-permeable TRPC6 channels. *Hippocampus*, 23(1): 40–52.

Li, H.B. et al. 2008. Antistress effect of TRPV1 channel on synaptic plasticity and spatial memory. *Biol Psychiatry*, 64(4): 286–292.

Li, H.S., X.Z. Xu, and C. Montell. 1999. Activation of a TRPC3-dependent cation current through the neurotrophin BDNF. *Neuron*, 24(1): 261–273.

Li, H. et al. 2012a. TRPC6 inhibited NMDA receptor activities and protected neurons from ischemic excitotoxicity. *J Neurochem*, 123(6): 1010–1018.

Li, M. et al. 2012b. A TRPC1-mediated increase in store-operated Ca^{2+} entry is required for the proliferation of adult hippocampal neural progenitor cells. *Cell Calcium*, 51(6): 486–496.

Li, Y. et al. 2005. Essential role of TRPC channels in the guidance of nerve growth cones by brain-derived neurotrophic factor. *Nature*, 434(7035): 894–898.

Li, Y. et al. 2010. Activity-dependent release of endogenous BDNF from mossy fibers evokes a TRPC3 current and Ca^{2+} elevations in CA3 pyramidal neurons. *J Neurophysiol*, 103(5): 2846–2856.

Liang, C. et al. 2014. Ammonium increases Ca^{2+} signalling and up-regulates expression of TRPC1 gene in astrocytes in primary cultures and in the in vivo brain. *Neurochem Res*, 39(11): 2127–2135.

Lin Y. et al. 2013b. Hyperforin attenuates brain damage induced by transient middle cerebral artery occlusion (MCAO) in rats via inhibition of TRPC6 channels degradation. *J Cereb Blood Flow Metab*, 33(2): 253–262.

Lin, Y. et al. 2013a. Neuroprotective effect of resveratrol on ischemia/reperfusion injury in rats through TRPC6/CREB pathways. *J Mol Neurosci*, 50(3): 504–513.

Lipski, J. et al. 2006. Involvement of TRP-like channels in the acute ischemic response of hippocampal CA1 neurons in brain slices. *Brain Res*, 1077(1): 187–199.

Liu, B. et al. 2001. Molecular consequences of activated microglia in the brain: Overactivation induces apoptosis. *J Neurochem*, 77(1): 182–189.

Liu, Y. et al. 2015. The change of spatial cognition ability in depression rat model and the possible association with down-regulated protein expression of TRPC6. *Behav Brain Res*, 294: 186–193.

Lucas, P. et al. 2003. A diacylglycerol-gated cation channel in vomeronasal neuron dendrites is impaired in TRPC2 mutant mice: Mechanism of pheromone transduction. *Neuron*, 40(3): 551–561.

Maccarrone, M. et al. 2008. Anandamide inhibits metabolism and physiological actions of 2-arachidonoylglycerol in the striatum. *Nat Neurosci*, 11(2): 152–159.

Malarkey, E., B.Y. Ni, and V. Parpura. 2008. Ca^{2+} entry through TRPC1 channels contributes to intracellular Ca^{2+} dynamics and consequent glutamate release from rat astrocytes. *Glia*, 56(8): 821–835.

Manna, S.S. and S.N. Umathe. 2012. Involvement of transient receptor potential vanilloid type 1 channels in the pro-convulsant effect of anandamide in pentylenetetrazole-induced seizures. *Epilepsy Res*, 100(1–2): 113–124.

Mannari, T. et al. 2013. Astrocytic TRPV1 ion channels detect blood-borne signals in the sensory circumventricular organs of adult mouse brains. *Glia*, 61(6): 957–971.

Marinelli, S. et al. 2003. Presynaptic facilitation of glutamatergic synapses to dopaminergic neurons of the rat substantia nigra by endogenous stimulation of vanilloid receptors. *J Neurosci*, 23(8): 3136–3144.

Marsch, R. et al. 2007. Reduced anxiety, conditioned fear, and hippocampal long-term potentiation in transient receptor potential vanilloid type 1 receptor-deficient mice. *J Neurosci*, 27(4): 832–839.

McQuillin, A. et al. 2006. Fine mapping of a susceptibility locus for bipolar and genetically related unipolar affective disorders, to a region containing the C21ORF29 and TRPM2 genes on chromosome 21q22.3. *Mol Psychiatry*, 11(2): 134–142.

Mezey, E. et al. 2000. Distribution of mRNA for vanilloid receptor subtype 1 (VR1), and VR1-like immunoreactivity, in the central nervous system of the rat and human. *Proc Natl Acad Sci U S A*, 97(7): 3655–3660.

Mitsumura, K. et al. 2011. Disruption of metabotropic glutamate receptor signalling is a major defect at cerebellar parallel fibre-Purkinje cell synapses in staggerer mutant mice. *J Physiol*, 589(Pt 13): 3191–3209.

Miyake, T. et al. 2014. TRPM2 contributes to LPS/IFN γ-induced production of nitric oxide via the p38/JNK pathway in microglia. *Biochem Biophys Res Commun*, 444(2): 212–217.

Miyake, T. et al. 2015. Activation of mitochondrial transient receptor potential vanilloid 1 channel contributes to microglial migration. *Glia*, 63(10): 1870–1882.

Mizoguchi, Y. et al. 2014. Brain-derived neurotrophic factor (BDNF) induces sustained intracellular Ca^{2+} elevation through the up-regulation of surface transient receptor potential 3 (TRPC3) channels in rodent microglia. *J Biol Chem*, 289(26): 18549–18555.

Mizuno, N. et al. 1999. Molecular cloning and characterization of rat *trp*, homologues from brain. *Brain Res Mol Brain Res*, 64(1): 41–51.

Morelli, M.B. et al. 2013. TRP channels: New potential therapeutic approaches in CNS neuropathies. *CNS Neurol Disord Drug Targets*, 12(2): 274–293.

Mori, F. et al. 2012. TRPV1 channels regulate cortical excitability in humans. *J Neurosci*, 32(3): 873–879.

Mori, Y. et al. 2002. Transient receptor potential 1 regulates capacitative Ca^{2+} entry and Ca^{2+} release from endoplasmic reticulum in B lymphocytes. *J Exp Med*, 195(6): 673–681.

Mrejeru, A., A. Wei, and J.M. Ramirez. 2011. Calcium-activated non-selective cation currents are involved in generation of tonic and bursting activity in dopamine neurons of the substantia nigra pars compacta. *J Physiol*, 589(Pt 10): 2497–2514.

Munakata, M. et al. 2013. Transient receptor potential canonical 3 inhibitor Pyr3 improves outcomes and attenuates astrogliosis after intracerebral hemorrhage in mice. *J Neurosci*, 44(7): 1981–1987.

Nabissi, M. et al. 2010. TRPV2 channel negatively controls glioma cell proliferation and resistance to Fas-induced apoptosis in ERK-dependent manner. *Carcinogenesis*, 31(5): 794–803.

Nagy, G.A. et al. 2013. DAG-sensitive and Ca^{2+} permeable TRPC6 channels are expressed in dentate granule cells and interneurons in the hippocampal formation. *Hippocampus*, 23(3): 221–232.

Nakao, K. et al. 2008. Ca^{2+} mobilization mediated by transient receptor potential canonical 3 is associated with thrombin-induced morphological changes in 1321N1 human astrocytoma cells. *J Neurosci Res*, 86(12): 2722–2732.

Nam, J.H. et al. 2015. TRPV1 on astrocytes rescues nigral dopamine neurons in Parkinson's disease via CNTF. *Brain*, 138(Pt 12): 3610–3622.

Nazıroğlu, M. et al. 2011. Aminoethoxydiphenyl borate and flufenamic acid inhibit Ca^{2+} influx through TRPM2 channels in rat dorsal root ganglion neurons activated by ADP-ribose and rotenone. *J Membr Biol*, 241(2): 69–75.

Nelson, C. and M. D. Glitsch. 2012. Lack of kinase regulation of canonical transient receptor potential 3 (TRPC3) channel-dependent currents in cerebellar Purkinje cells. *J Biol Chem*, 287(9): 6326–6335.

Nilius, B. and A. Szallasi. 2014. Transient receptor potential channels as drug targets: From the science of basic research to the art of medicine. *Pharmacol Rev*, 66(3): 676–814.

Ohana, L. et al. 2009. The Ca^{2+} release-activated Ca^{2+} current (I_{CRAC}) mediates store-operated Ca^{2+} entry in rat microglia. *Channels*, 3(2): 129–139.

Okada, T. et al. 1999. Molecular and functional characterization of a novel mouse transient receptor potential protein homologue TRP7: Ca^{2+}-permeable cation channel that is constitutively activated and enhanced by stimulation of g protein-coupled receptor. *J Biol Chem*, 274(39): 27359–27370.

Olah, M.E. et al. 2009. Ca^{2+}-dependent induction of TRPM2 currents in hippocampal neurons. *J Physiol*, 587(Pt 5): 965–979.

Ostapchenko, V.G. et al. 2015. The transient receptor potential melastatin 2 (TRPM2) channel contributes to β-amyloid oligomer-related neurotoxicity and memory impairment. *J Neurosci*, 35(45): 15157–15169.

Paez, P.M. et al. 2011. Modulation of canonical transient receptor potential channel 1 in the proliferation of oligodendrocyte precursor cells by the golli products of the myelin basic protein gene. *J Neurosci*, 31(10): 3625–3637.

Park, E.S., S.R. Kim, and B.K. Jin. 2012. Transient receptor potential vanilloid subtype 1 contributes to mesencephalic dopaminergic neuronal survival by inhibiting microglia-originated oxidative stress. *Brain Res Bull*, 89(3–4): 92–96.

Park, L. et al. 2014. The key role of transient receptor potential melastatin-2 channels in amyloid-beta-induced neurovascular dysfunction. *Nat Commun*, 5: 5318.

Peters, J.H. et al. 2010. Primary afferent activation of thermosensitive TRPV1 triggers asynchronous glutamate release at central neurons. *Neuron*, 65(5): 657–669.

Phelan, K.D. et al. 2014. Critical role of canonical transient receptor potential channel 7 in initiation of seizures. *Proc Natl Acad Sci U S A*, 111(31): 11533–11538.

Pizzo, P. et al. 2001. Role of capacitative calcium entry on glutamate-induced calcium influx in type-I rat cortical astrocytes. *J Neurochem*, 79(1): 98–109.

Prager-Khoutorsky, M., A. Khoutorsky, and C.W. Bourque. 2014. Unique interweaved microtubule scaffold mediates osmosensory transduction via physical interaction with TRPV1. *Neuron*, 83(4): 866–878.

Puram, S.V. et al. 2011. A TRPC5-regulated calcium signaling pathway controls dendrite patterning in the mammalian brain. *Genes Dev*, 25(24): 2659–2673.

Qiu, J. et al. 2010. Leptin excites proopiomelanocortin neurons via activation of TRPC channels. *J Neurosci*, 30(4): 1560–1565.

Ren, K. and R. Dubner. 2010. Interactions between the immune and nervous systems in pain. *Nat Med*, 16: 1267–1276.

Riccio, A. et al. 2009. Essential role for TRPC5 in amygdala function and fear-related behavior. *Cell*, 137(4): 761–772.

Riccio, A. et al. 2014. Decreased anxiety-like behavior and $G\alpha_{q/11}$-dependent responses in the amygdala of mice lacking TRPC4 channels. *J Neurosci*, 34(10): 3653–3667.

Roberts, J.C., J.B. Davis, and C.D. Benham. 2004. [^{3}H]Resiniferatoxin autoradiography in the CNS of wild-type and TRPV1 null mice defines TRPV1 (VR-1) protein distribution. *Brain Res*, 995(2): 176–183.

Roedding, A.S. et al. 2009. TRPC3 protein is expressed across the lifespan in human prefrontal cortex and cerebellum. *Brain Res*, 1260: 1–6.

Roedding, A.S. et al. 2013. Chronic oxidative stress modulates TRPC3 and TRPM2 channel expression and function in rat primary cortical neurons: Relevance to the pathophysiology of bipolar disorder. *Brain Res*, 1517: 16–27.

Rosenbaum, T. 2015. Activators of TRPM2: Getting it right. *J Gen Physiol*, 145(6): 485–487.

Rosenbaum, T. and S.A. Simon. 2007. TRPV1 receptors and signal transduction. In *TRP Ion Channel Function in Sensory Transduction and Cellular Signaling Cascades*, edited by W. B. Liedtke and S. Heller, 69–84. Boca Raton, FL: CRC Press/Taylor & Francis.

Russo, E. et al. 2014. *Hypericum perforatum*: Pharmacokinetic, mechanism of action, tolerability, and clinical drug-drug interactions. *Phytother Res*, 28(5): 643–655.

Sander, J.W. 2003. The epidemiology of epilepsy revisited. *Curr Opin Neurol*, 16(2): 165–170.

Santos, C.J., C.A. Stern, and L.J. Bertoglio. 2008. Attenuation of anxiety-related behaviour after the antagonism of transient receptor potential vanilloid type 1 channels in the rat ventral hippocampus. *Behav Pharmacol*, 19(4): 357–360.

Sappington, R.M. and D.J. Calkins. 2008. Contribution of TRPV1 to microglia-derived IL-6 and NFκB translocation with elevated hydrostatic pressure. *Invest Ophthalmol Vis Sci*, 49(7): 3004–3017.

Sawamura, S. et al. 2016. Screening of transient receptor potential canonical channel activators identifies novel neurotrophic piperazine compounds. *Mol Pharmacol*, 89(3): 348–363.

Schapira, A.H. and P. Jenner. 2011. Etiology and pathogenesis of Parkinson's disease. *Mov Disord*, 26(6): 1049–1055.

Schilling, T. and C. Eder. 2010. Stimulus-dependent requirement of ion channels for microglial NADPH oxidase-mediated production of reactive oxygen species. *J Neuroimmunol*, 225(1–2): 190–194.

Sell, T.S. et al. 2014. Protonophore properties of hyperforin are essential for its pharmacological activity. *Sci Rep*, 4: 7500.

Sharif-Naeini, R., S. Ciura, and C.W. Bourque. 2008. TRPV1 gene required for thermosensory transduction and anticipatory secretion from vasopressin neurons during hyperthermia. *Neuron*, 58(2): 179–185.

Shibasaki, K. et al. 2007. Effects of body temperature on neural activity in the hippocampus: Regulation of resting membrane potentials by transient receptor potential vanilloid 4. *J Neurosci*, 27(7): 1566–1575.

Shibasaki, K. et al. 2014. A novel subtype of astrocytes expressing TRPV4 (transient receptor potential vanilloid 4) regulates neuronal excitability via release of gliotransmitters. *J Biol Chem*, 289(21): 14470–14480.

Shibasaki, K. et al. 2015. TRPV4 activation at the physiological temperature is a critical determinant of neuronal excitability and behavior. *Pflugers Arch*, 467(12): 2495–2507.

Shibasaki, K., Y. Ishizaki, and S. Mandadi. 2013. Astrocytes express functional TRPV2 ion channels. *Biochem Biophys Res Commun*, 441(2): 327–332.

Shim, S. et al. 2009. Peptidyl-prolyl isomerase FKBP52 controls chemotropic guidance of neuronal growth cones via regulation of TRPC1 channel opening. *Neuron*, 64(4): 471–483.

Shimizu, T. et al. 2013. Androgen and PARP-1 regulation of TRPM2 channels after ischemic injury. *J Cereb Blood Flow Metab* ,33(10): 1549–1555.

Shimizu, T. et al. 2016. Extended therapeutic window of a novel peptide inhibitor of TRPM2 channels following focal cerebral ischemia. *Exp Neurol*, 283(Pt A):151–156.

Shirakawa, H. et al. 2008. TRPV1 stimulation triggers apoptotic cell death of rat cortical neurons. *Biochem Biophys Res Commun*, 377(4): 1211–1215.

Shirakawa, H. et al. 2010. Transient receptor potential canonical 3 (TRPC3) mediates thrombin-induced astrocyte activation and upregulates its own expression in cortical astrocytes. *J Neurosci*, 30(39): 13116–13129.

Shoudai, K. et al. 2010. Thermally active TRPV1 tonically drives central spontaneous glutamate release. *J Neurosci*, 30(43): 14470–14475.

Shuvaev, A.N. et al. 2011. Mutant PKCγ in spinocerebellar ataxia type 14 disrupts synapse elimination and long-term depression in Purkinje cells *in vivo*. *J Neurosci*, 31(40): 14324–14334.

Siddiqui, T. et al. 2014. Expression and contributions of TRPM7 and KCa2.3/SK3 channels to the increased migration and invasion of microglia in anti-inflammatory activation states. *PLOS ONE*, 9(8): e106087.

Simon, F., D. Varela, and C. Cabello-Verrugio. 2013. Oxidative stress-modulated TRPM ion channels in cell dysfunction and pathological conditions in humans. *Cell Signal*, 25(7): 1614–1624.

Singh, B.B. et al. 2004. VAMP2-dependent exocytosis regulates plasma membrane insertion of TRPC3 channels and contributes to agonist-stimulated Ca^{2+} influx. *Mol Cell*, 15(4): 635–646.

So, K. et al. 2015. Involvement of TRPM2 in a wide range of inflammatory and neuropathic pain mouse models. *J Pharmacol Sci*, 127(3): 237–243.

So, K. et al. 2016. Hypoxia-induced sensitisation of TRPA1 in painful dysesthesia evoked by transient hindlimb ischemia/reperfusion in mice. *Sci Rep*, 6: 23261.

Song, J.N. et al. 2013. Potential contribution of SOCC to cerebral vasospasm after experimental subarachnoid hemorrhage in rats. *Brain Res*, 1517: 93–103.

Song, X. et al. 2005. Canonical transient receptor potential channel 4 (TRPC4) co-localizes with the scaffolding protein ZO-1 in human fetal astrocytes in culture. *Glia*, 49(3): 418–429.

Sözbir, E. and M. Nazıroğlu. 2016. Diabetes enhances oxidative stress-induced TRPM2 channel activity and its control by N-acetylcysteine in rat dorsal root ganglion and brain. *Metab Brain Dis*, 31(2): 385–393.

Stock, K. et al. 2012. Neural precursor cells induce cell death of high-grade astrocytomas through stimulation of TRPV1. *Nat Med*, 18(8): 1232–1238.

Straub, R.E. et al. 1994. A possible vulnerability locus for bipolar affective disorder on chromosome 21q22.3. *Nat Genet*, 8(3): 291–296.

Streifel, K.M. et al. 2013. Manganese inhibits ATP-induced calcium entry through the transient receptor potential channel TRPC3 in astrocytes. *Neurotoxicology*, 34: 160–166.

Su, B. et al. 2014. Brain-derived neurotrophic factor (BDNF)-induced mitochondrial motility arrest and presynaptic docking contribute to BDNF-enhanced synaptic transmission. *J Biol Chem*, 289(3): 1213–1226.

Sudbury, J.R. and C.W. Bourque. 2013. Dynamic and permissive roles of TRPV1 and TRPV4 channels for thermosensation in mouse supraoptic magnocellular neurosecretory neurons. *J Neurosci*, 33(43): 17160–17165.

Sylvester, J.B., J. Mwanjewe, and A.K. Grover. 2001. Transient receptor potential protein mRNA expression in rat substantia nigra. *Neurosci Lett*, 300(2): 83–86.

Tabarean, I.V. 2012. Persistent histamine excitation of glutamatergic preoptic neurons. *PLoS One*, 7(10): e47700.

Tai, Y. et al. 2008. TRPC6 channels promote dendritic growth via the CaMKIV-CREB pathway. *J Cell Sci*, 121(Pt 14): 2301–2307.

Takada, Y., T. Numata, and Y. Mori. 2013. Targeting TRPs in neurodegenerative disorders. *Curr Top Med Chem*, 13(3): 322–334.

Takahashi, N. and Y. Mori. 2011. TRP channels as sensors and signal integrators of redox status changes. *Front Pharmacol*, 2: 58.

Takahashi, N. et al. 2011. TRPA1 underlies a sensing mechanism for O_2. *Nat Chem Biol*, 7(10): 701–711.

Terzian, A.L. et al. 2009. Modulation of anxiety-like behaviour by transient receptor potential vanilloid type 1 (TRPV1) channels located in the dorsolateral periaqueductal gray. *Eur Neuropsychopharmacol*, 19(3): 188–195.

Tomic, M. et al. 2011. Role of nonselective cation channels in spontaneous and protein kinase A-stimulated calcium signaling in pituitary cells. *Am J Physiol Endocrinol Metab*, 301(2): E370–E379.

Toth, A. et al. 2005. Expression and distribution of vanilloid receptor 1 (TRPV1) in the adult rat brain. *Brain Res Mol Brain Res*, 135(1–2): 162–168.

Toth, B., I. Iordanov, and L. Csanady. 2015. Ruling out pyridine dinucleotides as true TRPM2 channel activators reveals novel direct agonist ADP-ribose-2'-phosphate. *J Gen Physiol*, 145(5): 419–430.

Toth, P. et al. 2013. Age-related autoregulatory dysfunction and cerebromicrovascular injury in mice with angiotensin II-induced hypertension. *J Cereb Blood Flow Metab*, 33(11): 1732–1742.

Trebak, M. 2010. The puzzling role of TRPC3 channels in motor coordination. *Pflugers Arch*, 459(3): 369–375.

Tu, P., J. Gibon, and A. Bouron. 2010. The TRPC6 channel activator hyperforin induces the release of zinc and calcium from mitochondria. *J Neurochem*, 112(1): 204–213.

Uemura, T. et al. 2005. Characterization of human and mouse TRPM2 genes: Identification of a novel N-terminal truncated protein specifically expressed in human striatum. *Biochem Biophys Res Commun*, 328(4): 1232–1243.

Vazquez, G. et al. 2004. The mammalian TRPC cation channels. *Biochim Biophys Acta*, 1742(1–3): 21–36.

Venkatachalam, K., F. Zheng, and D.L. Gill. 2003. Regulation of canonical transient receptor potential (TRPC) channel function by diacylglycerol and protein kinase C. *J Biol Chem*, 278(31): 29031–29040.

Vennekens, R., A. Menigoz, and B. Nilius. 2012. TRPs in the brain. *Rev Physiol Biochem Pharmacol*, 163: 27–64.

Verma, S. et al. 2012. TRPM2 channel activation following in vitro ischemia contributes to male hippocampal cell death. *Neurosci Lett*, 530(1): 41–46.

Wang, S. et al. 2011. Cholecystokinin facilitates neuronal excitability in the entorhinal cortex via activation of TRPC-like channels. *J Neurophysiol*, 106(3): 1515–1524.

Wehrhahn, J. et al. 2010. Transient receptor potential melastatin 2 is required for lipopolysaccharide-induced cytokine production in human monocytes. *J Immunol*, 184(5): 2386–2393.

Wes, P.D. et al. 1995. TRPC1, a human homolog of a *Drosophila* store-operated channel. *Proc Natl Acad Sci U S A*, 92(21): 9652–9656.

Wolfle, U., G. Seelinger, and C.M. Schempp. 2014. Topical application of St. John's wort (*Hypericum perforatum*). *Planta Med*, 80(2–3): 109–120.

Wu, L.J., T.B. Sweet, and D.E. Clapham. 2010. International Union of Basic and Clinical Pharmacology. LXXVI. Current progress in the mammalian TRP ion channel family. *Pharmacol Rev*, 62(3): 381–404.

Xie, Y.F. and F. Zhou. 2014. TRPC3 channel mediates excitation of striatal cholinergic interneurons. *Neurol Sci*, 35(11): 1757–1761.

Xie, Y.F. et al. 2011. Dependence of NMDA/GSK-3β mediated metaplasticity on TRPM2 channels at hippocampal CA3-CA1 synapses. *Mol Brain*, 4: 44.

Xu, C. et al. 2006. Association of the putative susceptibility gene, transient receptor potential protein melastatin type 2, with bipolar disorder. *Am J Med Genet B Neuropsychiatr Genet*, 141b(1): 36–43.

Xu, C. et al. 2009. TRPM2 variants and bipolar disorder risk: Confirmation in a family-based association study. *Bipolar Disord*, 11(1): 1–10.

Xu, P. et al. 2012. Expression of TRPC6 in renal cortex and hippocampus of mouse during postnatal development. *PLOS ONE*, 7(6): e38503.

Yamamoto S. et al. 2008. TRPM2-mediated Ca^{2+} influx induces chemokine production in monocytes that aggravates inflammatory neutrophil infiltration. *Nat Med*, 14(7): 738–747.

Yamamoto, S. et al. 2007. Transient receptor potential channels in Alzheimer's disease. *Biochim Biophys Acta* ,1772(8): 958–967.

Yan, H.D., C. Villalobos, and R. Andrade. 2009. TRPC channels mediate a muscarinic receptor-induced after-depolarization in cerebral cortex. *J Neurosci*, 29(32): 10038–10046.

Yao, C. et al. 2013. Neuroprotectin D1 attenuates brain damage induced by transient middle cerebral artery occlusion in rats through TRPC6/CREB pathways. *Mol Med Rep*, 8(2): 543–550.

Yoshida, T. et al. 2006. Nitric oxide activates TRP channels by cysteine S-nitrosylation. *Nat Chem Biol*, 2(11): 596–607.

Zeng, C. et al. 2015a. Upregulation and diverse roles of TRPC3 and TRPC6 in synaptic reorganization of the mossy fiber pathway in temporal lobe epilepsy. *Mol Neurobiol*, 52(1): 562–572.

Zeng, Z. et al. 2015b. Silencing TRPM7 in mouse cortical astrocytes impairs cell proliferation and migration via ERK and JNK signaling pathways. *PLOS ONE*, 10(3): e0119912.

Zhang, H. et al. 2016. Blocking transient receptor potential vanilloid 2 channel in astrocytes enhances astrocyte-mediated neuroprotection after oxygen-glucose deprivation and reoxygenation. *Eur J Neurosci*, 44(7): 2493–2503.

Zhang, J. et al. 2014. IL-17A contributes to brain ischemia reperfusion injury through calpain-TRPC6 pathway in mice. *Neuroscience*, 274: 419–428.

Zhou, F.W. and S.N. Roper. 2014. TRPC3 mediates hyperexcitability and epileptiform activity in immature cortex and experimental cortical dysplasia. *J Neurophysiol*, 111(6): 1227–1237.

Zhou, F.W., S.G. Matta, and F.M. Zhou. 2008a. Constitutively active TRPC3 channels regulate basal ganglia output neurons. *J Neurosci*, 28(2): 473–482.

Zhou, J. et al. 2008b. Critical role of TRPC6 channels in the formation of excitatory synapses. *Nat Neurosci* 11(7): 741–743.

Zhou, Y. et al. 2014. Inflammation in intracerebral hemorrhage: From mechanisms to clinical translation. *Prog Neurobiol*, 115: 25–44.

Index